Handbook of Materials Science and Engineering

Handbook of Materials Science and Engineering

Editor: Heather Dale

NY RESEARCH PRESS

New York

Published by NY Research Press
118-35 Queens Blvd., Suite 400,
Forest Hills, NY 11375, USA
www.nyresearchpress.com

Handbook of Materials Science and Engineering
Edited by Heather Dale

© 2019 NY Research Press

International Standard Book Number: 978-1-63238-633-5 (Hardback)

Cataloging-in-Publication Data

Handbook of materials science and engineering / edited by Heather Dale.
p. cm.
Includes bibliographical references and index.
ISBN 978-1-63238-633-5
1. Materials science. 2. Materials. I. Dale, Heather.
TA403 .H36 2019
620.11--dc23

Contents

Preface...IX

Chapter 1   **Experimental Investigation of the Influence of Burnishing Parameters on Surface Roughness and Hardness of Brass Alloy**...1
             Frihat MH, Al Quran FMF and Al-Odat MQ

Chapter 2   **Poly(Itaconate) Derivatives, at the Air-Water Interface: Case of Poly(Monobenzyl) and Poly(Dibenzyl) Itaconate**...5
             Radic D, Gargallo L and Leiva A

Chapter 3   **Low-Cycle Fatigue Behaviors of Pre-Corroded Q345R Steel under Wet H_2S Environments**...12
             Luo Yun-Rong, Fu Lei, Zeng Tao, Lin Haibo and Chen Yanqiang

Chapter 4   **Observation of Optical Properties of Gold Thin Films Using Spectroscopic Ellipsometry**...16
             Pradhan SK

Chapter 5   **Optimization of Reaction Parameters for Silver Nanoparticles Synthesis from _Fusarium Oxysporum_ and Determination of Silver Nanoparticles Concentration**...21
             Khan NT and Jameel J

Chapter 6   **Influence of Dopants on Mechanical Properties of Steel: A Spin-Polarized Pseudopotential Study**...25
             Zavodinsky V and Kabaldin Y

Chapter 7   **Polymer Electrode Material for Microbial Bioelectrochemical Systems**...28
             Moutcine A, Akhramez S, Maallah R, Hafid A and Chtaini A

Chapter 8   **Obtainment of Sorbitol on Ferroalloy Promoted Nickel Catalysts**...31
             Kedelbaev BS, Korazbekova KU and Kudasova DE

Chapter 9   **Polysulfone/Cellulose Acetate Butyrate Environmentally Friendly Blend to minimize the Impact of UV Radiation**...35
             Raouf RM, Wahab ZA, Ibrahim NA and Talib ZA

Chapter 10   **Mechanical Behavior of Long Carbon Fiber Reinforced Polyarylamide at Elevated Temperature**...42
             Wang Q, Ning H, Vaidya U and Pillay S

Chapter 11   **Fourier Transform Infrared Spectroscopy and Liquid Chromatography– Mass Spectrometry Study of Extracellular Polymer Substances Produced on Secondary Sludge Fortified with Crude Glycerol**...48
             Nouha K, Hoang NV and Tyagi RD

Chapter 12   **Preparation of a Photoactive 3D Polymer Pillared with Metalloporphyrin** 58
Zargari S, Rahimi R and Rahimi A

Chapter 13   **Microstructure Evolution and Mangnetique Proprieties of Nanocrystalline
Fe$_{60}$Cu$_{30}$Al$_{10}$ Prepared by Combustion Processes** 63
Hafs A, Benaldijia A and Aitbara A

Chapter 14   **Investigation of the Flexoelectric Coupling Effect on the 180° Domain Wall
Structure and Interaction with Defects** 69
Mbarki R, Borvayeh L and Sabati M

Chapter 15   **Polarization Dependent Reflectivity and Transmission for Cd$_{1-x}$zn$_x$te/GaAs (001)
Epifilms in the Far-Infrared and Near-Infrared to Ultraviolet Region** 75
Talwar DN and Becla P

Chapter 16   **Iriartea Deltoidea and Socratea Exhorriza: Sustainable Production Alternatives
for Integrated Biosystems** 84
Sánchez LM and Quiñonez MF

Chapter 17   **Raman Spectroscopy of Iron Oxide of Nanoparticles (Fe$_3$O$_4$)** 87
Panta PC and Bergmann CP

Chapter 18   **Technology Geometric Decoration with Silver and/or Bronze Wires on Magnetite
Patinas in the Armament of the Preroman Peoples of the Iberian Peninsula** 90
Sánchez LG, Portal AJC, Valenzuela FP,
Salazar Y Caso De Los Cobos JMG and Martínez García JA

Chapter 19   **Simulation to Study the Effect of Carrier Concentration on I-V Characteristics
of Schottky Diode** 95
Sharma R

Chapter 20   **Structural, Electronic and Magnetic Properties of Geometrically Frustrated
Spinel CdCr$_2$O$_4$ from First-principles Based on Density Functional Theory** 99
Bolandhemat N, Rahman M and Shuaibu A

Chapter 21   **Studying of Species Composition of Activators of Skin and Visceral
Leishmaniasis in the South of Kazakhstan** 103
Kozhabayevich KM, Dossanuly SR and Zhanabayevich OA

Chapter 22   **Rheology Properties of Castor Oil: Temperature and Shear Rate-dependence
of Castor Oil Shear Stress** 106
Abdelraziq IR and Nierat TH

Chapter 23   **The Free Energy of Solvation for N-Decane in Ethanol-Water Solutions** 112
Mongelli GF

Chapter 24   **Study of the Adsorption of Bright Green by a Natural Clay and Modified** 115
Ltifi I, Ayari F, Hassen Chehimi DB and Ayadi MT

Chapter 25   **The Benefit of 3D Printing in Medical Field: Example Frontal Defect
Reconstruction** 122
Singare S, Shenggui C and Nan Li

Chapter 26   **Size-Exclusion Chromatography and Its Optimization for Material Science** 125
Netopilík M and Trhlíková O

Chapter 27   **Removal of Chromium from Industrial Wastewater by Adsorption Using Coffee Husk** 130
Dessalew Berihun

Chapter 28   **The Effect on Extracting Solvents using Natural Dye Extracts from *Hyphaene thebaica* for Dye-sensitized Solar Cells** 136
Mohammed IK, Kasim Uthman ISAH, Yabagi JA and Taufiq S

Chapter 29   **Stress-Strain Characterization in Fiber-Reinforced Composites by Digital Image Correlation** 139
Heredia AS, Aguilar PAM and Ocampo AM

Chapter 30   **Scratch Wear Resistance of TiALN And AlCrN Coated EN-353 Steel** 145
Chandrashekhar A, Kabadi VR and Bhide R

Chapter 31   **Synthesis of Pyramid-Shaped NiO Nanostructures using Low-Temperature Composite- Hydroxide- Mediated Approach** 154
Shahid T, Khan TM, Zakria M, Shakoor RI, Arfan M and Khursheed S

Chapter 32   **Self-Recovering Section of RPV Steel Radiation Embrittlement Kinetics as Indication of Material Smart Behavior** 160
Evgenii K

Chapter 33   **Translational Development of Biocompatible X-Ray Visible Microspheres for Use in Transcatheter Embolization Procedures** 164
Benzina A, Aldenhoff YB, Heijboer R and Koole LH

Chapter 34   **Using Flixweed Seed as a Pore-former to Prepare Porous Ceramics** 169
Hedayat N and Du Y

Chapter 35   **Total Excess Lifetime Cancer Risk Estimation from Enhanced Heavy Metals Concentrations Resulting from Tailings in Katsina Steel Rolling Mill, Nigeria** 173
Bello S, Muhammad BG and Bature B

Chapter 36   **Two Binary Liquid Critical Mixtures Belong to Class of Universality** 178
Ata BN and Abderaziq IR

Chapter 37   **Use of the Additive Based on Amorphous Silica-Alumina in the Adhesive Dry Mixes** 184
Loganina VI and Zhegera CV

Chapter 38   **Wetting of Olivine Sand against Steel Alloys** 187
Rastgoo Oskoui P and Payam RO

Chapter 39   **Visible Light assisted Photocatalytic Acitivity of Zinc Titanate in Presence of Metallic Sodium** 191
Sirajudheen P

Chapter 40   **Universality of Graphene as 2-D Material**...**194**
Solanki Bhaumikkumar Ketansinh

Permissions

List of Contributors

Index

Preface

The purpose of the book is to provide a glimpse into the dynamics and to present opinions and studies of some of the scientists engaged in the development of new ideas in the field from very different standpoints. This book will prove useful to students and researchers owing to its high content quality.

This book elucidates the concepts and innovative models around prospective developments with respect to materials science and engineering. It is an interdisciplinary field of science which involves the study of structures of materials and their mechanisms. Its applications vary across a number of scientific fields and industries such as studying the formation of new metals, construction of alloys, study of nanostructures, etc. This book is compiled to provide a detailed overview of some upcoming concepts related to the field of materials science. It also elaborates the modern advancements that are taking place within scientific research. With its comprehensive language and extensive use of examples, this book will serve as a valuable source of knowledge to students and experts alike.

At the end, I would like to appreciate all the efforts made by the authors in completing their chapters professionally. I express my deepest gratitude to all of them for contributing to this book by sharing their valuable works. A special thanks to my family and friends for their constant support in this journey.

Editor

Experimental Investigation of the Influence of Burnishing Parameters on Surface Roughness and Hardness of Brass Alloy

Frihat MH*, Al Quran FMF and Al-Odat MQ

Department of Mechanical Engineering, Al-Huson University College, Al-Balqa Applied University, Al-Huson-Irbid, Jordan

Abstract

Burnishing is a cold working surface finishing process in which peaks and valleys on machined surfaces deformed plastically by the application of hard and finished ball or roller against to it. Recently burnishing is becoming popular among post finishing processes because of its many advantages along its primary role i.e., increasing surface finish. In this paper the effect of burnishing process parameters on surface roughness and hardness of brass alloy is experimentally investigated. A simple roller burnishing tool was used for the experiential work of the present study, surface test and micro hardness test were used to demonstrate the effects of the burnishing force and feed rate on surface roughness and surface hardness of a brass alloy. The smoothing process under consideration can be performed on standard machine tools without additional reconfiguration tasks. Process is very useful for any workshop and can be carried out without coolant. The results proved that all the parameters have significant effect on the above said two surface characteristics. The results revealed that improvements in the surface finish and increase in the surface hardness are obtained by reduce the burnishing feed and increase the burnishing force.

Keywords: Burnishing; Hardness; Surface roughness; Burnishing feed; Burnishing force

Introduction

Roller burnishing is an economical and feasible mechanical treatment for the quality improvement of rotating components, not only in sense of surface roughness but in compressive residual stresses as well. Burnishing is a plastic deformation process in which the force is applied to a workpiece surface by a hard smooth ball or roller. The mechanism of a burnishing operation is that asperities peaks of the workpiece are compressed to fill in the valleys and thus characteristics of the surface under consideration are changed [1-4]. The characteristics of a burnished surface depend upon controlling burnishing parameters such as applied burnishing force (burnishing depth), burnishing speed, burnishing feed rate, and number of passes, geometry and material of burnishing tool, as well as the material of burnished surface.

Burnishing is considered as a cold-working finishing process, differs from other cold-working, surface treatment processes such sand blasting, in that it produces a good surface finish and also induces residual compressive stresses at the metallic surface layers [5]. Also, Burnishing is one of the important finishing operations carried out generally to enhance the fatigue resistance characteristics of components. Additionally, burnishing is economically beneficial, because it is a simple and less costly process, requiring less time. High quality surface finish can be obtained with semi skilled operators [6].

Roller burnishing is a fine machining process that is used to improve certain physical and mechanical properties, such as surface roughness, corrosion resistance, friction coefficient, wear, and fatigue resistance. Surface roughness and hardness plays an important role in many areas and is factor of great importance for the functioning of machine parts such as valve seats, internal surface of hydraulic cylinders, pistons, bearings, etc.,.

Many researchers have undertaken extensive work on burnishing processes. Hassan [7] explained the effects of ball and roller burnishing on the surface roughness and hardness of some non ferrous metals. It was suggested by many investigators that an improvement in wear resistance can be achieved by burnishing process. Siva Prasad [8] investigated the roller burnishing process on aluminum components

and concluded that surface finish improvement is better with the burnishing process. Thamizhmnaii [6] presented the improvement of surface hardness and roughness of titanium alloy by burnishing process.

Hassan [9] investigated the effect of burnishing parameters on surface roughness and hardness of non-ferrous metals like Al, Brass etc. Further, he also conducted experiments to find out the effect of burnishing process on wear resistance property of non-ferrous metals [10]. El-Axir [11] detailed description about roller burnishing process and effect of its parameter on surface finish and roughness using design of experiments. The principle of the burnishing process, shown in Figure 1, is based on the rolling movement of a tool (a ball or a roller) over the work piece's surface. With application of roller burnishing process, plastic deformation of machining surface and allocation of material starts from peaks to valleys. Roller burnishing is a material micro-displacement process which is shown in Figure 1 in comparison with other finishing processes, like grinding process, also lowers the surface roughness height but the burnishing process can be achieved by applying a highly polished and hard roll on to a metallic surface under pressure. Microscopic "peaks" on the machined surface are during roller burnishing process exposed to cold flow into the "valleys," creating a plateau-like profile in which sharpness is reduced or eliminated in the contact plane. The presence of compressive stresses and improved hardness in chip less finishing process i.e., burnishing further enhances properties like fatigue, wear resistance and corrosion resistance on surface. Reducing surface roughness reduces the friction

*Corresponding author: Frihat MH, Department of Mechanical Engineering, Al-Huson University College, Al-Balqa Applied University, Al-Huson-Irbid, Jordan, E-mail: Mohamed_frihat@yahoo.com

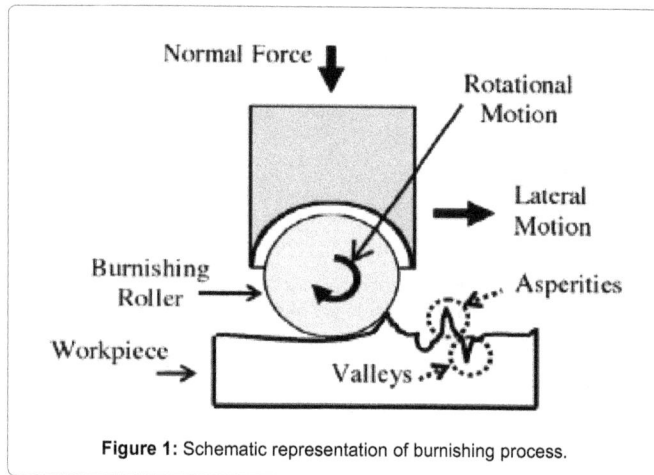

Figure 1: Schematic representation of burnishing process.

Cu %	Pb %	Fe %	Zn %
68.5	0.07	0.05	remaining

Table 1: Chemical composition in weight percent for (C2600) brass alloy.

to minimize the energy losses to increase its functional performance and also increases aesthetic look of the parts by producing mirror like surface finish. In burnishing the hard and finished roller or ball is pressed against the pre machined surface. The process is done on almost all convectional machine tools and with simple tool thus making this process economically cheaper than convectional finishing processes.

The parameters selected in this work are feed, and force as they found more predominant in literature review. The main object of this work is to investigate the effect of burnishing speed and force on surface roughness and hardness of brass alloy.

Experimental Details

Burnishing tool

Roller burnishing is a surface finishing technique in which hardened; highly polished steel rollers are brought into pressure contact with a softer piece part. Burnishing processes are used in manufacturing to improve the size, shape, surface finish, or surface hardness of a workpiece. A 20 mm diameter hardened roller of 5 mm width was used for burnishing. The shanks of the burnishing tools are designed to be simply mounted or fixed onto the tool holder of a machine tool such as a lathe machine; as shown in Figures 2a and 2b.

Material

The work piece was received in the form of cylindrical bars of 20 mm diameter. The bars were turned to 18 mm diameter and 130 mm length. Brass alloy used in this experimental work, is made of copper and zinc and other element alloy, its chemical composition is shown in Table 1. The brass it is ideal for workmanship because it is has many properties like it is draw ability and its low hardness number in addition, it is easy to scribble and plating.

Burnishing procedure and parameters

In this experimental investigation five specimens were obtain by sawing operation of the rod brass. Each specimen was turned to length 130 mm and to diameter 18 mm by TRIUMPH 2000 lathe. After each specimen turning, the cutting tool was replaced by the roller burnishing tool. 4. Finally, the roughness measured roughness was measured by

means of roughness machine tester (SURFCORDER SE 3500 Surface Tester) and hardness was measured using Micro hardness machine tester (model HWDN-3). The burnishing condition considered in this work is shown in the Tables 2a and 2b.

Results and Discussion

These experiments were aimed to examine the effect of burnishing parameters on surface roughness and surface hardness.

Roughness results

The initial surface roughness before burnishing was equal to 0.99 μm. Figure 3 shows the effect of burnishing feed on the surface roughness at a constant force of 300 N.

The burnishing feed of specimen (1) feed was varied from 0.04 to 0.1 mm/rev, for specimen (2) the feed was varied from 0.12 to 0.30 mm/rev. and for specimen (3) feed from 0.35 to 1.00 mm/rev. It is clear that, as the burnishing feed rate increases, the surface roughness tends to be lowers. The optimum feed rate is an intermediate. Moreover, at at higher feed rates, fast rubbing, the roughness (Ra) value is slightly increased.

The effect of burnishing force on the surface roughness at a constant feed of 0.06 mm/rev is illustrated in Figure 4. For specimen (4) the force was varied force from 200 to 400 N and for specimen (5) the force varied from 450 to 650 N. It is noticed that the surface

Specimen Number	Pressing Force (N)	Zone Number	Feed (mm/rev)
1	300	1	0.04
		2	0.05
		3	0.06
		4	0.08
		5	0.1
2	300	1	0.12
		2	0.15
		3	0.20
		4	0.25
		5	0.30
3	300	1	0.35
		2	0.40
		3	0.45
		4	0.70
		5	1
Spindle Speed=350 rpm		Number of passes=3	

Table 2a: The burnishing condition considered in this work.

Specimen Number	Feed (mm/rev)	Zone Number	Pressing Force (N)
4	0.06	1	200
		2	250
		3	300
		4	350
		5	400
5	0.06	1	450
		2	500
		3	550
		4	600
		5	650
Spindle Speed=350 rpm		Number of passes=3	

Table 2b: The burnishing condition considered in this work.

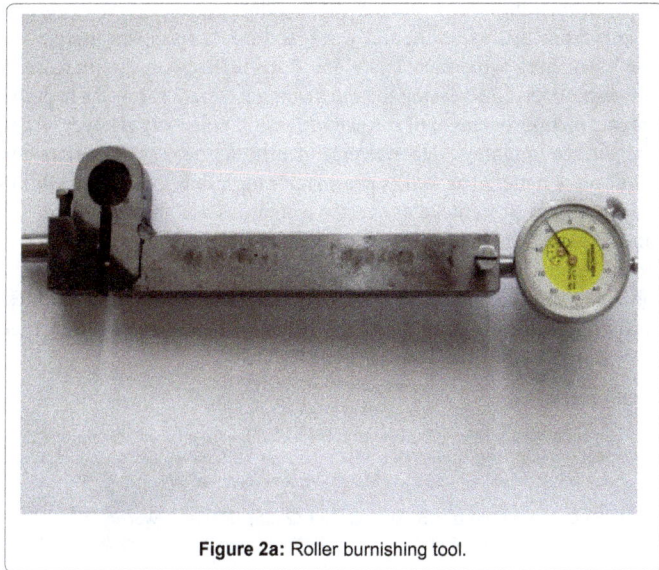

Figure 2a: Roller burnishing tool.

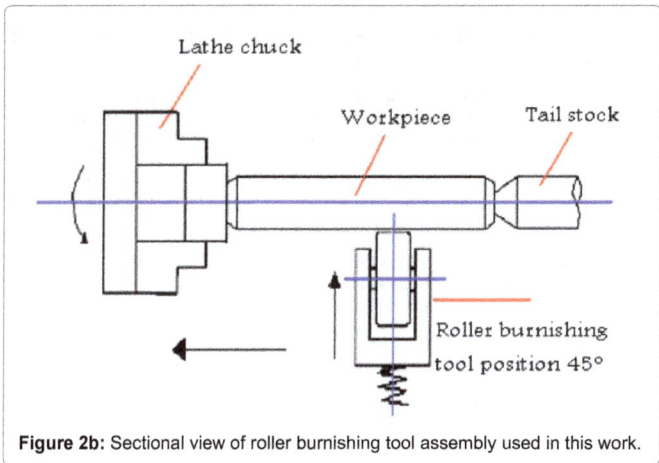

Figure 2b: Sectional view of roller burnishing tool assembly used in this work.

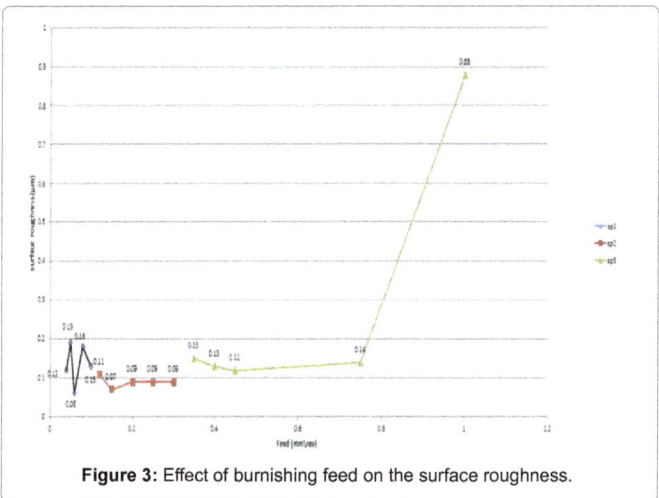

Figure 3: Effect of burnishing feed on the surface roughness.

when increasing the force beyond the mentioned optimum value, the swell of metal in front of the tool becomes large and the region of the plastic deformation widens which damages the already burnished surface, i.e., increase the surface roughness. Surface roughness is also found to be crucial. When the burnishing force is increased up to 300 N, the surface roughness is found to be decreased. Again when the force is increased above 300 N, the surface roughness is increased i.e., material get deteriorated surface roughness.

It can be seen that, the surface roughness decrease for the specimens 1 and 2, while for the specimen (3) the surface roughness increased when the feed increased, this is due to very short exposure time the specimen compared to other specimens.

For specimen (5) the surface roughness increases because the burnishing tool vibration was increased.

The optimum roughness value was at burnishing force equal 300 N and feed equal 0.06 mm/rev.

Surface microhardness t Results

The initial hardness before burnishing operation was equal 136 HV. Figure 5 shows effect of burnishing feed on the surface hardness at a constant force of 300 N. The burnishing feed for specimen (1) is varied between feed from 0.04 to 0.1 mm/rev and specimen (2) varied from 0.12 to 0.30 mm/rev and for specimen (3) feed varied from 0.35

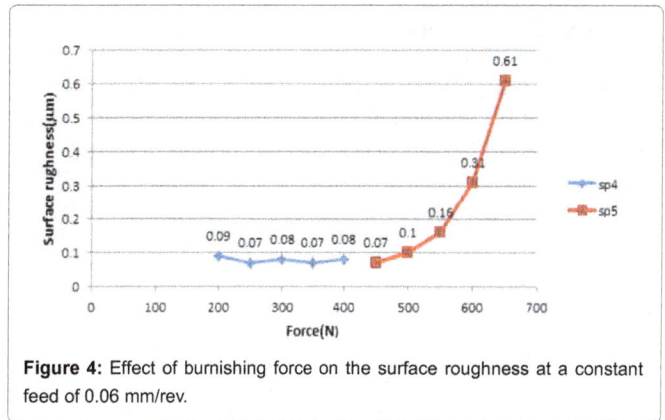

Figure 4: Effect of burnishing force on the surface roughness at a constant feed of 0.06 mm/rev.

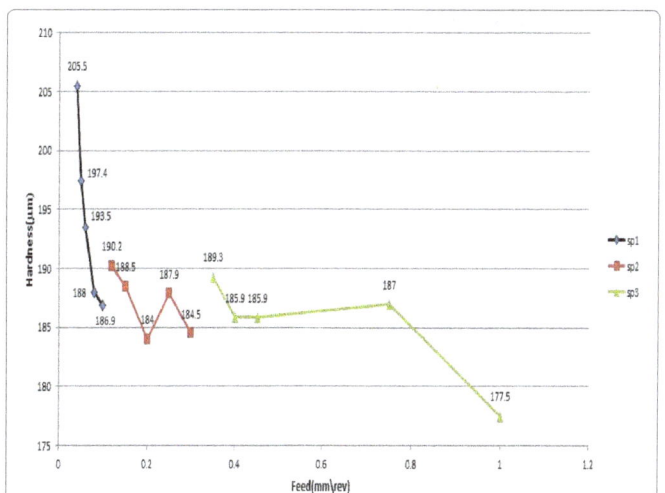

Figure 5: Effect of burnishing feed on the surface hardness at a constant force of 300 N.

roughness decreases with the increase of the applied force up to the maximum value of 300 N, after which the surface roughness increases with further increase of the force. This behavior can be explained that

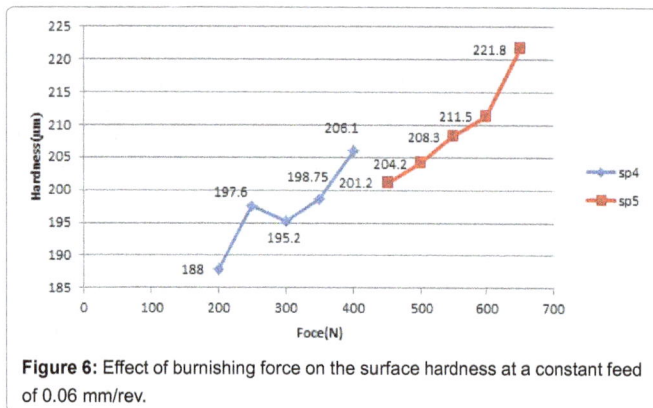

Figure 6: Effect of burnishing force on the surface hardness at a constant feed of 0.06 mm/rev.

to 1.00 mm/rev. It is obvious that can be seen that the relationship between the and when the force constant and equal and the feed for each specimen was like that the effect of burnishing force on the surface hardness at a constant feed of 0.06 mm/rev is depicted in Figure 6. The force for specimen (4) was varied from 200 to 400 N and for specimen (5) varied from 450 to 650 N.

It can be seen; from the above Figures 6 that the hardness increases as the force increased due to strain hardening when a metal is strained beyond the yield point. More stress is required to produce additional plastic deformation and the metal apparently becomes stronger and more difficult to deform. Whereas when the feed is increased the hardness was decreased. There are some irregularities in curves of micro hardness test valucs and since the pressure exerted by tool burnishing on the unit area along the specimens is not uniform and accurate in direction and magnitude. In addition there were non-homogeneities in the chemical elements and there are some impurities within the specimens which cause a hard spots on the surface. The maximum hardness value was obtained at a force of 650 N and feed of 0.06 mm/rev.

Conclusions

This article presents the results of the experimental investigation of fine machining conditions with roller burnishing of brass alloy. Roller burnishing process of cylindrical surface can be performed on standard performance machine tools. Surface quality improvement is obvious and can be visually detect when comparison of two parts before and

after burnishing is conducted. The following conclusions can be drawn from this experimental work, the micro hardness increase as the burnishing force increases while it decreases with the increase in burnishing feed, Increasing the burnishing feed rate showed a negative impact on improvement of surface roughness especially at lower forces, the surface roughness depends on burnishing process parameters and work piece material as well as pre-machining condition, the maximum improvement percentage of surface roughness was found to be 48.9% and the maximum improvement percentage of micro hardness was equal to 63.1%. Under the different roller burnishing parameters, the obtained surface roughness is dependent also on work piece hardness and pre-machining conditions. The roller burnishing tool can be used as a promising finishing method.

References

1. El-Tayeb NSM, Low KO, Brevern PV (2007) Influence of roller burnishing contact width and burnishing orientation on surface quality and tribological behavior of Aluminum 6061. J Mater Process Technol 186: 272-278.

2. Luca L, Neagu-Ventzel S, Marinescu I (2005) Effects of working parameters on surface finish in ball-burnishing of hardened steels. Pre Eng 29: 253-256.

3. Lin YC, Yan BH, Huang FY (2001) Surface improvement using a combination of electrical discharge machining with ball burnish machining based on the Taguchi Method. Int J Adv Manuf Technol 18: 673-682.

4. Low KO (2011) Surface characteristics modification of polyoxymethylene and polyurethane using burnishing. Tribol Trans 54: 96-103.

5. Wang L, Yu X (1999) Effect of various parameters on the surface roughness of an aluminum alloy burnished with a spherical surfaced polycrystalline diamond tool. Int J Mach Tools Manuf 39: 459-469.

6. Thamizhmnaii S, Bin Omar B, Saparudin S, Hassan S (2008) Surface roughness investigation and hardness by burnishing on titanium alloy. J Achiev in Materials and Manuf Engg 28: 139-142.

7. Hassan AM, Al- Bsharat AS (1997) Improvement in some properties of non-ferrous metals by the application of ball and roller burnishing process. Journal of Material Processing Technology 59: 250-256.

8. Siva Prasad T, Kotiveerachary B (1988) External Burnishing of Aluminum Components. Journal of Institution of Engineers (India) 69: 55-58.

9. Hassan AM, Sulieman AD (1999) Improvement in the wear resistance of brass components by the ball burnishing process. Journal of Materials Processing Technology 96: 73-80.

10. Shneider YG (1969) Classification of metal-burnished methods and tools. Mach Tool XL 1: 35.

11. El-Axir MH (2000) An investigation into roller burnishing. Int J Mach Tool Manuf 40: 1603-1617.

2

Poly(Itaconate) Derivatives, at the Air-Water Interface: Case of Poly(Monobenzyl) and Poly(Dibenzyl) Itaconate

Radic D*, Gargallo L and Leiva A

Pontificia Universidad Catolica de Chile Santiago, Chile

Abstract

The surface behaviors of monolayers of the poly(itaconate) derivatives, poly(dibenzyl itaconate) (PDBzI) and poly(monobenzyl itaconate) (PMBzI), at the air-water interface were investigated at 298 K on an aqueous subphase at pH 5.7 and 3.0. The monolayer characteristics of PDBzI and PMBzI were studied and compared in terms of surface pressure-area (-A) isotherms, surface compressional modulus-surface pressure (Cs-1-π) curves, static elasticity-surface concentration curves (-), hysteresis phenomena and phase images observed with a Brewster angle microscope (BAM.). The results showed that PMBzI and PDBzI gave rise to stable monolayers and that the isotherms presented pseudoplateau regions at different surface pressure values independent of pH. The PMBzI pseudoplateau region may because a change in the lateral packing of the chains. The PDBzI pseudoplateau region is attributed to a phase transition. The morphology of these monolayers was studied by Brewster angle microscopy (BAM). The surface pressure was expressed in terms of the scaling laws as function of surface concentrations. It can be concluded that the air-water interface was a poor solvent for both studied polymers. The degree of hydrophobicity of the polymers was estimated by determining the surface energy values based on wettability measurements.

Keywords: Poly(benzylitaconate)s; Pressure-area isotherms; Air-water interface; Monolayers; Static elasticity; Anphiphilicity isotherms

Introduction

Polymers derived from itaconic acid are of great interest from both the basic and industrial points of view [1-3]. These polymers, containing saturated rings as side chains, show significant mechanical and dielectric activity when they are affected by force fields [3,4]. This is partly due to the flexibility of the saturated rings which can flip between two conformational states, e.g. chair-to-chair [4]. In contrast, less activity can be expected with aromatic rings because of the planarity of the unsaturated ring. Monoesterification and diesterification of itaconic acid can be carried out to obtain monomers and polymers, with either one or two of the carboxyl groups esterified in each repeat unit [3,5]. In previous articles we have reported the relaxational behavior of poly(monobenzyl itaconate) (PMBzI) [4] and poly(dibenzyl itaconate) (PDBzI) [3] by means of dynamic mechanical and dielectric spectroscopy. From the comparison of the relaxational behavior of the two polymers it was concluded that the small differences in their chemical structures gave rise to significant differences in relaxational behavior. We also studied poly (itaconate)s monolayers at the air-water interface a decade ago [6]. Because of the peculiar relaxational behavior of poly (mono and dibenzyl itaconate)s [3,4] we felt it important to make a comparative study of the surface behavior of this kind of polymer at the air–water interface [6].

The aim of this work was to investigate the changes in surface activity and molecular organization of monolayers of PMBzI and PDBzI and to examine the effect of the chemical structure of these poly(itaconate) derivatives on the surface behavior at the air–water interface. Scheme 1 shows the chemical structures of the studied polymers.

Experimental Section

Synthesis and characterization of monomers and polymers

The monomers were prepared by esterification of itaconic acid with the corresponding benzyl alcohol according to the previously described method [3,7-10]. The polymers were obtained by radical polymerization in bulk at 323 K. using α,α'-azobisisobutyronitrile (AIBN) as catalyst (0.1% mol for PMBzI and 0.2-0.4% mol for PDBzI) [7-9].

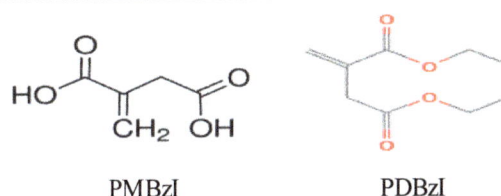

Scheme 1: Chemical structures of PMBzI and PDBzI

The polymers were characterized by light scattering and size exclusion chromatography measurements, as previously reported [8-10]. Samples used in the present work had a weight-average molecular weight of $M_w \approx 3.1 \times 10^4$ for PMBzI and $M_w \approx 3.7 \times 10^4$ for PDBzI corresponding to 141 and 119 repeat unit per molecule, respectively.

Surface pressure-area isotherms

A Langmuir film was obtained by depositing a polymer solution in toluene (1-1.5 mg mL^{-1}) dropwise on the air-water interface. Before compression, the film was allowed to equilibrate for about 15 minutes to ensure full evaporation of the spreading solvent and to allow the molecules to reach equilibrium.

The surface pressure-area (π-A) isotherms at the air-water interface were established in a Teflon Langmuir-Blodgett trough (NIMA-611 M) and a Langmuir film trough (Nima Technology-1232D1D2 and 622D2/D1, England.) equipped with two barriers and a Wilhelmy plate. The whole setup was enclosed in a transparent poly (methacrylate) box

*Corresponding author: Radic D, Pontificia Universidad Catolica de Chile Santiago, Chile, E-mail: dradic@puc.cl

to prevent environmental pollution. The polymer monolayers were compressed from 600 to 70 cm² with a compression speed of 5 cm² min⁻¹. The isotherms were determined in triplicate to ensure reproducibility. The temperature regulation of the monolayer was obtained by a thermostat-controlled water flow from an external bath, passing through the jacket at the bottom of the trough. All measurements were done at 298 K using ionized water purified with a Milli-Q system to 18.2 MΩ/cm resistivity as a subphase. The limiting surface area, A_0, was determined by extrapolation of the π-A isotherms to zero pressure as shown in Figures 1 and 2. The extrapolation was obtained by a straight line tangent to the isotherm as indicated in the Figure 1. The stability test of the monolayer was performed by following the variation of surface pressure and area against time, as shown in Figure 3. For the hysteresis experiments several compression-expansion cycles at 5 cm² min⁻¹ were performed to approximately 50% compression. The time interval between compression and expansion was 300 s and the temperature was fixed at 298 K (Figure 4).

Contact angle measurements of surface free energy

The surface free energies of the polymers were determined by contact angle measurements using diiodomethane and 1-bromonaphthalene. Polymers films were cast onto glass slides used for optical microscopy. The cast films were dried for 30 min at 60°C under vacuum. The

measurements were made using an optical contact angle system (OCA) (Dataphysics, England) with a conventional goniometer and high performance video camera controlled by the OCA20 software. A syringe connected to a Teflon capillary with an inner diameter of approximately 2 mm was used to supply liquid from above to measure the sessile drops, using drops with a radius of 0.4-0.5 cm. The contact angles were measured carefully from left to right of the drop and subsequently averaged. These procedures were repeated for 10 drops of each liquid on all six new surfaces, selected considering the homogeneity of the film and the reproducibility of the results. All readings were then averaged to give an average contact angle. All experiments were performed at room temperature.

Brewster angle microscopy

A Brewster angle microscope (BAM) (Nanofilm Technology GmbH, Germany) was mounted on the Langmuir trough (NIMA-1232D1D2 and 622D2/D1). Both devices were used to check the experimental results, which were in good agreement, in a perpendicular arrangement of the incidence plane with respect to the direction of barrier motion. For viewing and image storage the microscope was combined with a CCD camera and video system (Figures 5 and 6). A NIMA 702 Langmuir balance was used to follow the evolution of the thickness of the monolayer at various surface pressures, with BAM images recorded in several regions of the isotherms.

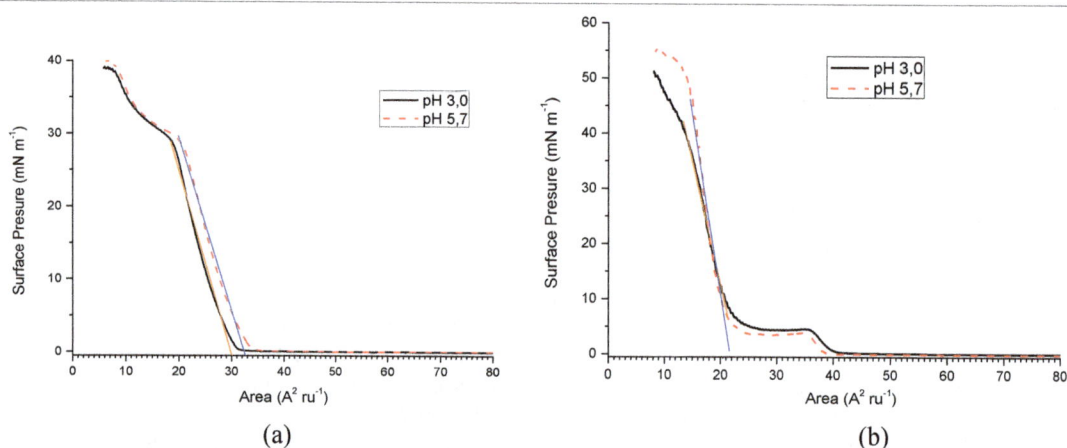

Figure 1: Surface pressure-area (π–A) isotherms of a) PMBzl and b) PDBzl at two pH levels of the water subphase (pH = 5.7 and 3.0).

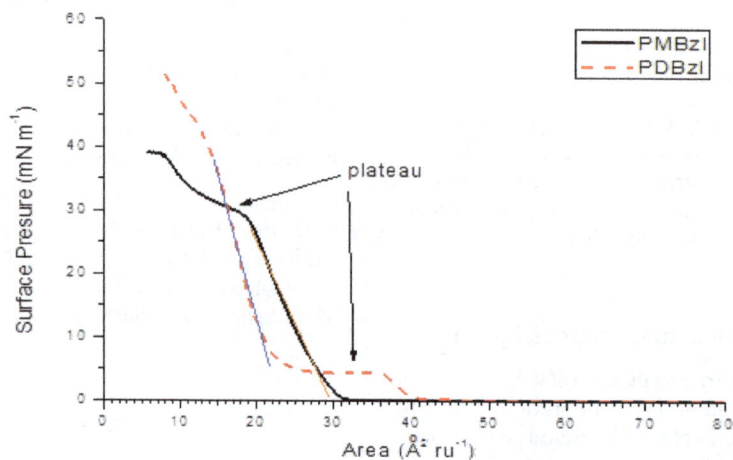

Figure 2: Surface pressure-area (π–A) isotherms of PMBzl and PDBzl at a pH level of 3.0 of the water subphase.

Poly(Itaconate) Derivatives, at the Air-Water Interface: Case of Poly(Monobenzyl)...

7

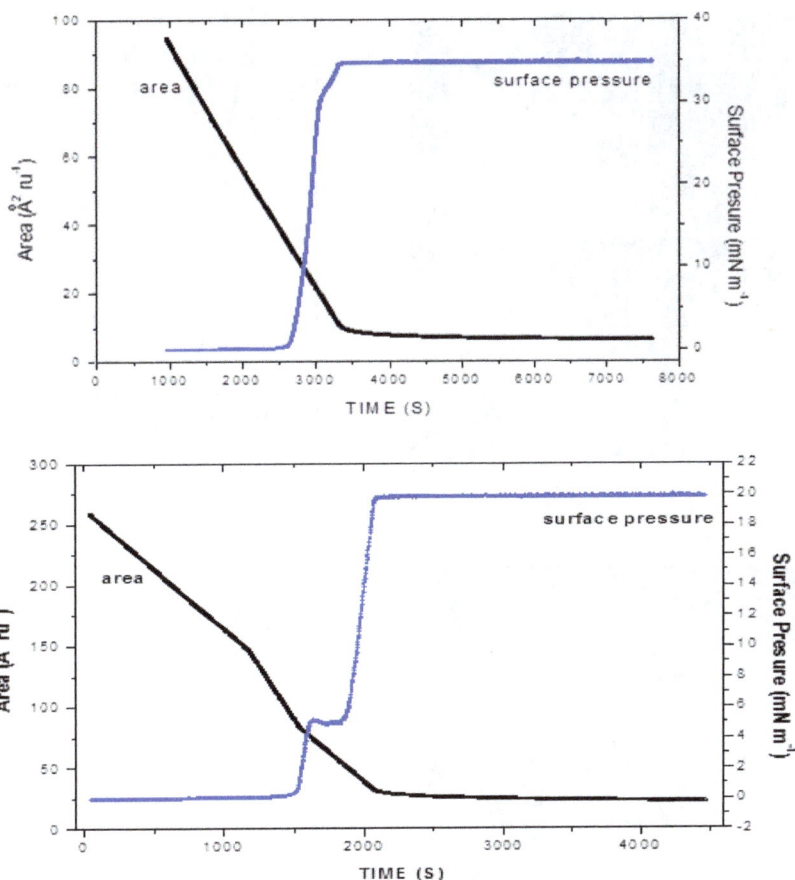

Figure 3: Stability test for PMBzI and PDBzI monolayers at the air-water interface.

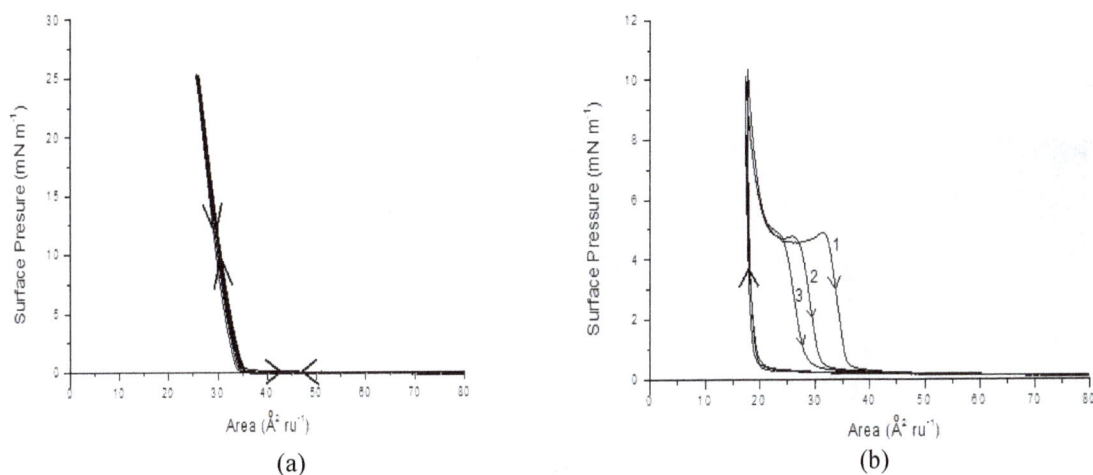

(a)

(b)

Figure 4: Hysteresis experiments with a) PMBzI and b) PDBzI using chloroform as a spreading solvent at a compression and expansion velocity of 10 cm2 min at 298 K.

Results and Discussion

The surface properties of the two polymers were studied by surface pressure measurements under compression. Figures 1a and 1b present the surface pressure-area (π-A) isotherms for the studied polymers on pure water acidified with HCl (pH: 5.7 and pH 3.0 with HCl at constant ionic strength) at 298 K. The limiting surface area, A_0, was determined by extrapolation of the π-A isotherms to zero pressure.

In general, three regions could be distinguished after compression in the surface area (π-A) plot, corresponding to the expanded liquid, condensed liquid and condensed-solid-like states in two dimensions.

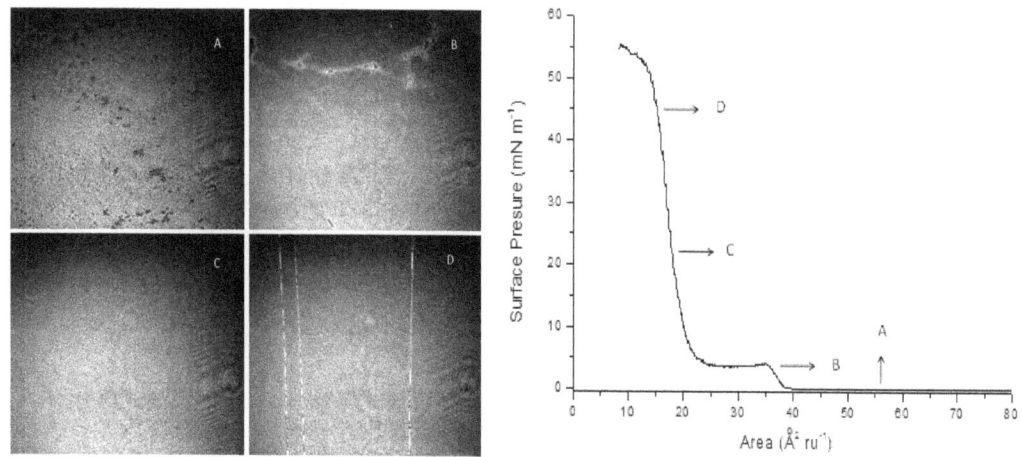

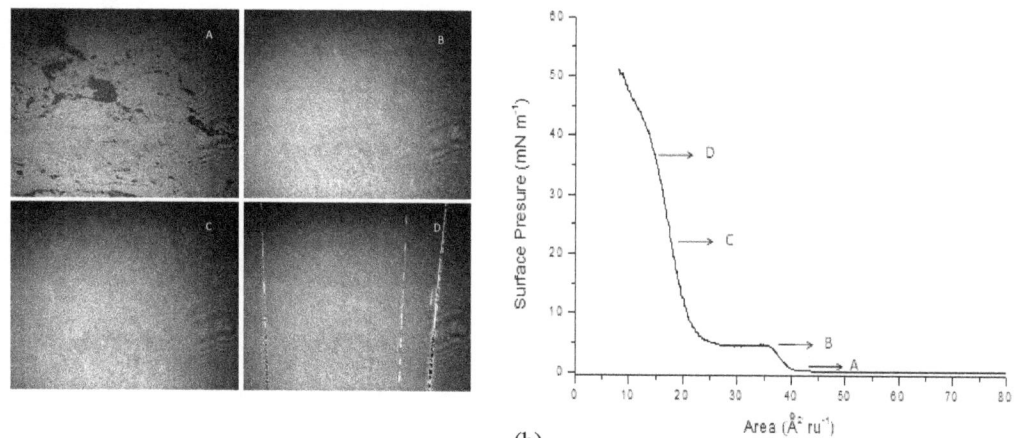

Figure 5: BAM images of the PDBzI monolayer corresponding to the points on the π−A isotherms of the water subphase at pH 5.7 (a) and pH 3.0 (b).

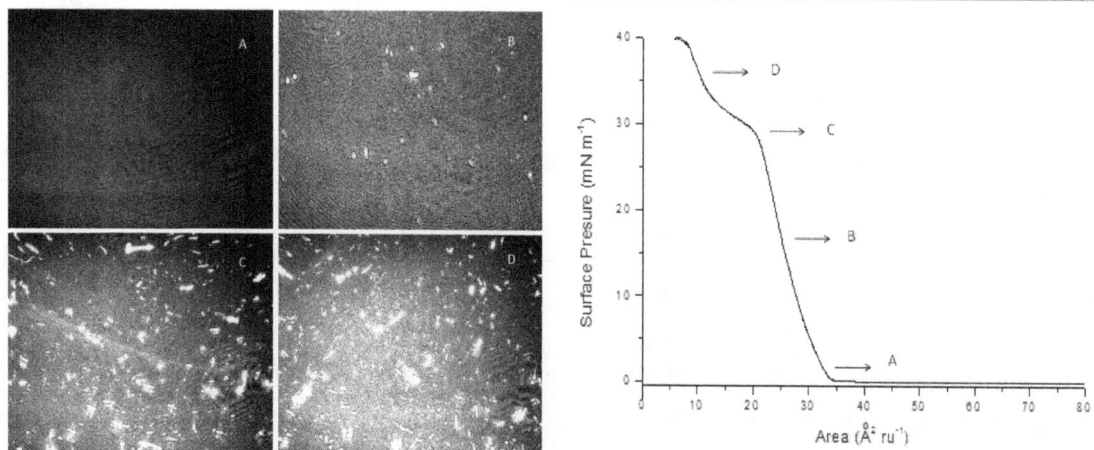

Figure 6: BAM images of the PMBzI monolayer corresponding to the points on the π−A isotherms of the water subphase at pH 5.7.

The last state can be identified as the lower area to zero pressure, corresponding to a projected area along axis of a hypothetical long organic chain from molecular models [11].

A pseudoplateau region can be observed in Figure 2 for PMBzI at low surface pressure, 8-20 Å^2 (ru^{-1}). At high surface pressures, at about 25-35 Å^2 (ru^{-1}) a pseudoplateau appeared for PDBzl that may represent

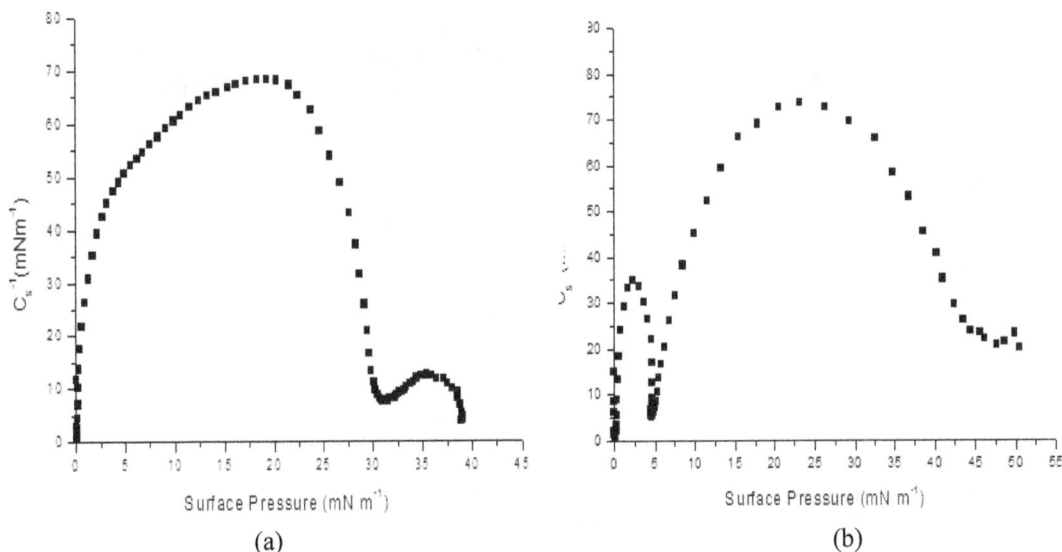

Figure 7: a) Compressional modulus values versus surface pressure plots for PMBzI and b) for PDBzI.

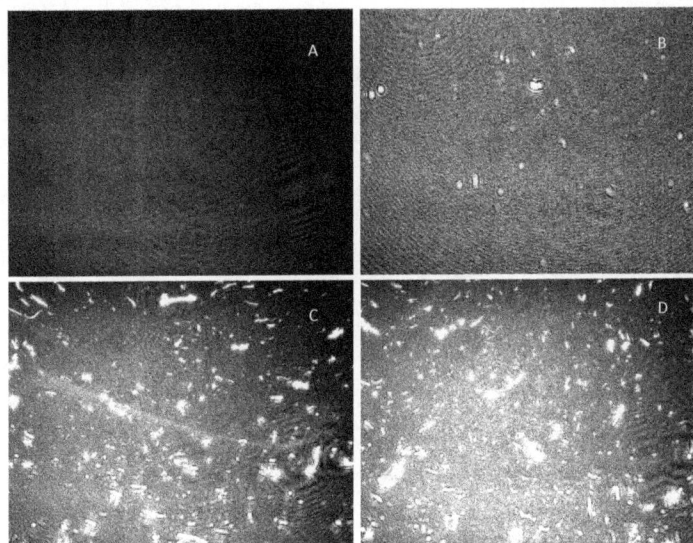

Figure 8: The BAM images of PMBzI monolayer corresponding to the points on the π-A isotherms on water subphase.

the transition between the condensed solid-like and liquid states at a low surface pressure of 5 mMm^{-1} (Figure 2). The reason for this behavior remains unknown, but we are currently running molecular dynamics simulations (MDS) to clarify if there was a change in the conformation of the polymer chains. A possible explanation is that the plateau is the result of repeated folding of the polymer segments, which are directed outwardly from the interface, forming loops and tails that reduce the contact points with the water subphase. The formation of these loops, resulting from monolayer compression, decreases the area occupied by the film. Table 1 summarizes the area per repeat unit at zero surface pressure; A0 values for both polymers. The A0 values for PDBzI and PMBzI are consistent with this explanation given that the area occupied by the hydrophobic film decreased at both pH levels.

The presence of the pseudoplateau in the isotherms implies a change in the surface organization of the polymer systems in this region due to

the collapse of the monolayer or phase transitions of the monolayers of both polymers. As compression continued, the surface pressure sharply increased, especially for PDBzI, apparently resulting in heterogeneous films, as shown in the BAM images in Figures 5 and 6. In conclusion, the surface isotherms of both polymers were similar, irrespective of the pH levels. However, they differed in the surface area and resulting surface pressure at which the monolayer presented pseudoplateau regions. The BAM images are in good agreement with the evolution of the shape of the isotherms, as shown in Figures 5 and 6. There is a difference in heterogeneity between PMBzI and PDBzI at pH 5.7, as shown in Figures 5 and 6, which could be attributed to differences in the hydrophobicity of the polymers.

Hysteresis experiments were conducted, consisting of three or four compression and expansion cycles (Figure 4), to evaluate the stability of the monolayers. The monolayers were subjected to four successive

Polymer	pH	A_0 (A^2 r.u.$^{-1}$)	π_c (mNm^{-1})
PMBzI	3.0	~30	~31
PMBzI	5.7	~33	~31
PDBzI	3.0	~23	≈45
PDBzI	5.7	~22	≈45

Table 1: Surface properties for PMBzI and PDBzI: Area per repeat unit at zero surface pressure, A_0 and collapse pressure, π_c, values.

cycles of compression-expansion at a rate of 5 cm^2/min. The first cycle was carried out by compressing the films up to a pressure of 1 mNm^{-1} in both Figures 5 and 6, so that the films were in a highly expanded state. Under these conditions the PMBzI decompression curves actually coincided with the compression curve, indicating the reversibility of the process (Figure 4a). The same occurred in the same area when the monolayer was recompressed in the second cycle. Gaines [12] called this behavior "reversible", characterized by the fact that the expansion and subsequent repeated compression curves reproduce the initial compression, although the process reaches higher pressures until 30 mNm^{-1} (Figure 4a). The results shown in Figure 4b suggested that the PDBzI monolayers adopted a different conformation after the first compression, that is compression up to 10 mNm^{-1}, involved some energy storage during the first expansion. It seems that the duration of the experiment was not sufficient to allow the full relaxation of the monolayer. However, hysteresis was much less pronounced in the subsequent cycles. The maximum surface pressure achieved in all the compressions, as shown in Figures 4a and 4b were constant, showing that no material loss was taking place during the experiments.

All the compression curves in Figure 4b are superposed, while the expansion curves showed changes that can be attributed to hysteresis and all the cycles were less pronounced. This fact may be due to entanglement of the chains during compression or a different arrangement in terms of hydration and conformation of the second cycle from that adopted after spreading. The expansion rate was apparently too rapid to allow the chains to adopt the same conformation they had prior to the compression when they decompress. Table 1 presents the zero pressure limiting areas per repeat units, A0, based on the π-A isotherms (Figures 4a and 4b). The A0 values for PDBzI were lower than those for PMBzI. This may be because in the case of PDBzI, which has two aromatic rings per repeat unit (ru), the area occupied by the aromatic hydrophobic residues per ru, must be smaller than it is for PMBzI; a hydrogen in PMBzI is replaced by a -CH$_2$- and an aromatic ring per repeat unit (Scheme 1). When the monolayer is compressed on the water surface, it may undergo several phase transformations. It is possible to make an analogy between the intervals of surface concentration and the solution concentrations often used in polymer science. By this analogy, the extrapolated area to zero surface pressure (A_0) and the collapse surface pressure, π_c correspond to the phenomena that occur in the concentrated region in solution in two dimensions. In the semi-dilute concentration zone, a phase transition, such as expanded liquid, condensed liquids and liquid-solid are observed [13,14]. Comparing surface and bulk concentration regimes commonly used in polymer science allow us to attribute the extrapolated area to zero surface, A_0, and the collapse pressure, π_c, that occur in the concentrated region.

The determination of collapse pressure is not a very clear topic in polymer science. Unlike small molecules, polymeric monolayers do not always collapse abruptly, making it difficult to identify the exact surface pressure at which collapse occurs. Consequently it is necessary to observe more than a single aspect to identify the collapse. A more precise way to determine the collapse surface pressure values

(compared to the π-A isotherms) is to plot the compressional modulus (C_s^{-1}) as a function of surface pressure (π) (Figure 7). This parameter, defined by Davies and Rideal [15] as the inverse of two-dimensional compressibility, is given by the relation:

$$C_S^{-1} = \varepsilon_0 = -A\left(\frac{\partial \pi}{\partial A}\right)_T = \Gamma\left(\frac{\partial \pi}{\partial \Gamma}\right)_T \tag{1}$$

where ε_o is the static elasticity and Γ is the surface concentration. C_s^{-1} values were obtained by numerical calculation of the first derivative from the isotherm data according to equation 1. By this way, after the determination of C_s^{-1}, (Figures 4a and 4b) in order to know the variation of the compressional modulus with surface pressure these values were plotted against surface pressure, as shown in Figure 7. The collapse pressures, π_c, for PMBzI and PDBzI, i.e., that is ~31 mNm^{-1} and ~45 mNm^{-1} respectively (Figure 1 and Table 1), are another difference between the two polymers that probably results from the difference in the degree of the hydrophobicity of the side chains.

Static elasticity, ε_o, of the polymer monolayers

The topography of the monolayer partially depends on the strength of the interfacial interactions with substrate molecules and the strength of the intersegmental polymeric interactions. The viscoelastic properties of polymer monolayers may also be dependent on these interactions. From the experimental π-A (Figures 1 and 2) and π-Γ curves (not shown for the sake of clarity), it was possible to calculate the classical static elasticity modulus ε_0 according to equation 1. However, this only accounts for hydrostatic compression.

Figure 8 shows the plot of the compressibility modulus or static elasticity, ε_o, calculated from the surface pressure isotherms π versus the surface concentration Γ. The greatest increase in elasticity occurred in the semi-diluted regime. The static elasticity ε_o values for the semidilute region represent normal behavior [11,16]. It is known that the maximum ε_o values for polymeric systems are found in diluted and semidiluted regions, since the chains are independent or are in contact with each other, but respond more or less individually to deformation. The plot of the compressibility modulus or static elasticity, ε_o, in Figure 8 shows that in the case of PDBzI the maximum was located in a more concentrated region where closer contact among chains can be expected, so that the response to deformation is like that of a polymer lattice. According to these results, areas of surface concentration have higher surface static elasticity values, with more sensitivity to changes than dilute regions.

As is generally known [17], the surface concentration of a polymer monolayer spread at the air-water interface can be easily regulated by compressing or expanding the monolayer. It is thus possible to make an analogy between the dilution ranges of superficial and bulk concentrations frequently used in polymer science. By this analogy, the extrapolated area at zero surface pressure, A_0, and the collapse pressure, π_C, take place in the semidilute region [18,19].

In the semidilute concentration region, surface pressure obeys a power law of surface concentration. Given the results obtained in the concentrated region, we considered it important to estimate the affinity of these systems for the air-water interface. We expected the interface to have different thermodynamic qualities as a solvent for the two polymers. To quantify this property we used the classic equation 2 of De Gennes [13].

$$\pi = \Gamma^{2v/2v-1} \tag{2}$$

According to this equation, the log π versus log Γ plot (Figure 6) in the semidilute region shows a linear variation with slope $2v/2v-1$. The v exponents for the two polymers (0.53 and 0.52 ± 0.03 for PMBzI and PDBzI, respectively) were obtained from the slope calculated for these plots. The theoretical prediction is a v value narrowly centered at 0.75 for two-dimensional polymer chains in good solvents [14]. Monte Carlo simulations predict a value of 0.753 [20], while the matrix-transfer method [21] predicts a value of 0.7503. The situation is not clear in the case of a poor solvent and predictions are less precise. Monte Carlo simulations [22] suggest $v_o \approx 0.51$, while matrix-transfer data suggest $v \sim 0.55$ [23]. The v values for PMBzI and PDBzI indicate that the two polymers have similar affinities for the interface, which is consistent with values for hydrophobicity (estimated by comparing the chemical structure of PMbzI and PDBzI). The surface free energy value for PDBzI, obtained by contact angle measurements, was lower than the value for PMBzI (27 mNm^{-1} and 45 mNm^{-1} respectively).

Because the collapse pressure for PDBzI at pH 5.7 was approximately 45 mNm^{-1}, Brewster angle microscopy experiments helped to clarify the collapsed areas of the isotherms of the studied polymers [24,25] The results obtained by this method were compared to those obtained from the isotherms. As we observed with the Brewster angle microscope (BAM), the images of point D of the isotherm for PDBzI at pH 5.7 reveal areas of local heterogeneity (Figure 5a) that can be attributed to the collapsed area. These results are consistent with other findings indicating that this area corresponds to collapse pressure π_c.

The results observed in Figure 6 are also consistent with previous findings and indicate that the collapse pressure for PMBzI, irrespective of the water subphase pH level, was located ~31 mNm^{-1} (Table 1). At low pressures, BAM images clearly reveal homogeneous morphology at point A, whereas a transition region appears at points B, C and D with areas of local heterogeneity that can be attributed to the collapsed regions.

Conclusions

PMBzI and PDBzI form stable monolayers at the air-water interface. The chemical structures of PDBzI and PMBzI significantly affect the shape of the surface isotherms irrespective of the pH level in the water subphase. The results indicate the existence of pseudoplateau regions at low surface pressures (5 mNm^{-1} for PDBzl and approximately 30 mNm^{-1} for PMBzI). This behavior has been interpreted as a phase transition in the first case and a collapse region in the second. The zero pressure limiting area per repeat unit (A0 values based on π-A isotherms) was lower for PDBzI than for PMBzI. In agreement with the concept of polymeric scales in semidilute regions, the air-water interface at 298 K is a bad solvent for these polymers, very close to the theta solvent. PDBzl is less hydrophilic than PMBzI according to level of surface free energy, which is consistent with the chemical structures of the polymers.

Acknowledgements

D.R and A.L. thanks to Fondecyt Grant 1120091 for financial support. L.G. Thanks to Universidad de Tarapacá, Arica for its unvaluable help.

References

1. Cowie JMG, McEwen I, Velickovic J (1975) Dynamic mechanical spectra of poly(itaconic acid esters) containing phenyl and cyclohexyl rings. Polymer 16: 869-872.

2. Gargallo L, Radić D, Yazdani-Pedram M, Horta A (1989) Properties of polyelectrolytes: poly(mono-methyl itaconate). conformational and viscometric behaviour in dilute solution. Eur Polym J 25: 1059-1063.

3. Diaz- Calleja R, Martinez-Piña F, Gargallo L, Radić D (2000) Dynamic mechanical and dielectric relaxations in poly(dibenzyl itaconate) and poly(diethylphenyl itaconate). Polymer 41: 1963-1969.

4. Diaz-Calleja R, Ribes-Greus A, Gargallo L, Radić D (1991) Dielectric relaxation in poly(monobenzyl itaconate). Polymer 32: 2331-2334.

5. Horta A, Hernández-Fuentes I, Gargallo L, Radić D (1987) Spectrophotometric Determination of Imatinib Mesylate using Charge Transfer Complexs in Pure Form and Pharmaceutical Formulation Chem Rapid Commun 8: 523.

6. Leiva A, Gargallo L, Gonzalez A, Radić D (2004) Poly(itaconates) monolayers behavior at the air/water interface. Study at different surface concentration. Eur Polym J 40: 2349-2355.

7. Radić D, Gargallo L (1996) CRC (in Press).

8. Yasdani-Pedram M, Gargallo L, Radić D (1985) Polymer conformation and viscometric behaviour-5: Synthesis, characterization and conformational studies for poly(monobenzyl itaconate). Eur Polym J 21: 707-710.

9. Leon A, Gargallo L, Radić D, Horta A (1990) Synthesis and solution properties of comb-like poly(mono-n-alkyl-itaconates): 2. Poly(monododecyl itaconate). Polymer 32: 761-763.

10. Gargallo L, Radić D, Leon A (1985) Polymer conformation and viscometric behaviour, 3. Synthesis, characterization and conformational studies in poly(mono-n-octyl itaconate). Makromol Chem 186: 1289-1296.

11. Gargallo L (2010) Polymer Behavior at the Air-Water Interface. MRS BULLETIN 35: 615-618.

12. Gaines JL (1991) Monolayers of polymers. Langmuir 7: 834-839.

13. de Gennes PG (1972) Exponents for the excluded volume problem as derived by the Wilson method. Phys Lett A 38: 339-340.

14. de Gennes PG (1979) Scaling Concepts in Polymer Physics. Cornell University Press, Ithaca, NY.

15. Davies JT, Rideal EK (1961) Interfacial Phenomena. Acade-mic Press, New York.

16. Miñones J, Miñones JJ, Count JM, Rodriguez Patino P, Latka D (2002) Mixed Monolayers of Amphotericin B-Dipalmitoyl Phosphatidyl Choline: Study of Complex Formation. Langmuir 18: 2817-2827.

17. Poupinet D, Vilanove R, Rondelez F (1989) Molecular weight dependence of the second virial coefficient for flexible polymer chains in two dimensions. Macromolecules 22: 2491-2496.

18. Vilanove R, Rondelez F (1980) Scaling Description of Two-Dimensional Chain Conformations in Polymer Monolayers. Phys Rev Lett 45: 1502-1508.

19. Cloizeaux JD, Jannick G (1990) Polymers in solution: Their modeling and structure. Clarendon Press, Oxford, UK

20. Havlin S, Ben Avraham D (1983) Corrections to scaling in self-avoiding walks. Phys Rev A 27: 2759.

21. Derrida B (1981) Phenomenological renormalisation of the self avoiding walk in two dimensions. J Phys A: Math Gen 14-L: 5.

22. Vilanove B, Poupinet D, Rondelez F (1988) A critical look at measurements of the .nu. exponent for polymer chains in two dimensions. Macromolecules 21: 2880-2887.

23. Derrida B, Saleur H (1985) Towards electrodynamical models of particles. J Phys A: Math Gen L-1075: 18.

24. Henon S, Meunier J (1991) Microscope at the Brewster angle: Direct observation of first-order phase transitions in monolayers. J Rev Sci Instrum 62: 936.

25. Hönig D, Möbius D (1992) The growth of organic thin films on silicon substrates studied by X-ray reflectometry. Thin Solid Films 64: 210.

Low-Cycle Fatigue Behaviors of Pre-Corroded Q345R Steel under Wet H$_2$S Environments

Luo Yun-Rong[1,3], Fu Lei[1,2]*, Zeng Tao[1], Lin Haibo[1,3] and Chen Yanqiang[1]

[1]College of Mechanical Engineering, Sichuan University of Science and Engineering, Zigong, 643000, China
[2]School of Aeronautics and Astronautics, Sichuan University, Chengdu 610065, PR China
[3]Key Lab in Sichuan Colleges on Industry Process Equipments and Control Engineering, Zigong, China

Abstract

Low cycle fatigue (LCF) behaviors of pre-corroded Q345R steel under different corrosion environments were investigated where both concentration of H$_2$S solution and pre-corrosion time served as two environmental factors. Low cycle fatigue tests were conducted at room temperature in air after the specimens were pre-corroded in H$_2$S solution. The low cycle fatigue behaviours of pre-corrosion test material, such as strain-life relationships and cyclic stress-strain responses were investigated and compared. Results show that environmental factors had little effect on the cyclic stress response while it had much effect on the fatigue life. And pre-corrosion time had more significant effect on low cycle fatigue life of test material than the solution concentration of H$_2$S. Furthermore, the fracture surfaces of the fatigue samples were characterized by scanning electron microscope (SEM) and the fracture mechanisms were discussed under different environmental conditions.

Keywords: Q345R; Pre-corroded; Wet H2S environments; Low cycle fatigue

Introduction

Q345R steel is one of the C-Mn steels widely used in pressure vessels in China, and almost one half of the pressure vessels for medium-low pressure are made of Q345R steel [1,2].

Q345R steel is a promising material for pressure vessels because of its favorable mechanical properties and corrosion resistance. Engineering components subjected to cyclic loading often fail due to fatigue. Its low cycle fatigue behaviors have attracted numerous interests during the past years. However, the pressure vessels in oil industry have to be exposed to severe conditions, where low-cycle fatigue is a predominant failure mode [3] and the surrounding environment contains H$_2$S corrosion solution caused by hydrogen sulfide [4]. During its service, pre-corroded low cycle fatigue is prone to occurring to the pressure vessels under alternate loading and wet H$_2$S corrosion environments. Therefore, it is of great significance to investigate the effect on the low-cycle fatigue behavior of pre-corroded Q345R steel.

There are some literatures on the fatigue properties of Q345R steel. For example, FAN Zhichao [5] investigated the LCF behavior of Q345R under stress control at elevated temperature, and the effects of temperature on fatigue properties and cyclic stress response were obtained. It has been found that Q345R steel exhibited the cyclic hardening and massing behavior at 300°C and 420°C. Ying et al. [6] and Zhuang et al. [7] conducted the fatigue crack growth tests of Q345R steel at different temperatures and strain rations (R), and investigated the fatigue crack growth laws. Guangxu, Zhiwen [8], Xu and Zhifang [9] have studied the LCF fatigue behaviour of the submerged-arc welding joint of Q345R steel. The mechanisms of stress corrosion of Q345R steel have been investigated under different concentrations of H$_2$S, temperature and pH values [10,11].

However, the research on the low-cycle fatigue behavior in wet H$_2$S environment is insufficient, yet. In comparison to low cycle fatigue, pre-corroded low cycle fatigue studies are relatively few, and very limited information on the pre-corroded low cycle fatigue of Q345R steel is available in the literature. The present study has been conducted to investigate the low cycle fatigue behavior of pre-corroded Q345R steel in different pre-corrosion environments, and then the effects of H2S solution concentration and pre-corrosion time on the LCF behavior of Q345R steel were evaluated. The fracture mechanism of Q345R steel in wet H2S environment was investigated as well.

Material and Testing Arrangements

Materials and test coupons

The specimens were machined from Q345R hot-rolled plate with the thickness of 20 mm. The chemical compositions and mechanical properties are given in Tables 1 and 2, respectively. The fatigue specimens used in the tests were cylindrical specimens with a gauge length of 20 mm and a central diameter of 7.25 mm as shown in Figure 1. The specimen dimensions were kept to avoid buckling phenomena under the highest compressive forces anticipated in the test program [12]. The surfaces of the specimens were carefully polished along the longitudinal direction to remove the scratches.

Experimental Details and Procedures

Pre-corroded tests were conducted with two environmental factors

Fe	C	Si	Mn	S	P
Bal.	0.12~0.16	0.35~0.50	1.20~1.60	0.005~0.05	0.015~0.025

Table 1: Chemical constant of test material (wt.%).

0.2% proof strength	Tensile strength	Elongation	Reduction of area
350 MPa	530 MPa	0.27	0.64

Table 2: Mechanical properties of test material.

*Corresponding author: Fu Lei, Lecturer, School of Aeronautics and Astronautics, College of Mechanical Engineering, Sichuan University of Science and Engineering, Zigong, 643000, PR China, E-mail: kunmingfulei@126.com

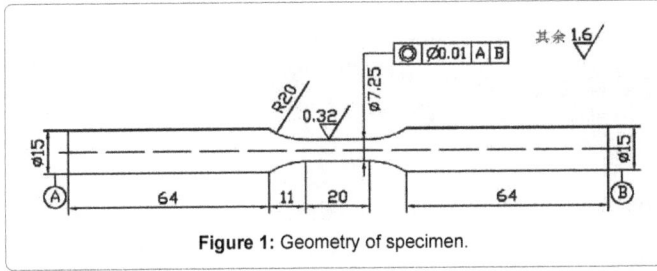

Figure 1: Geometry of specimen.

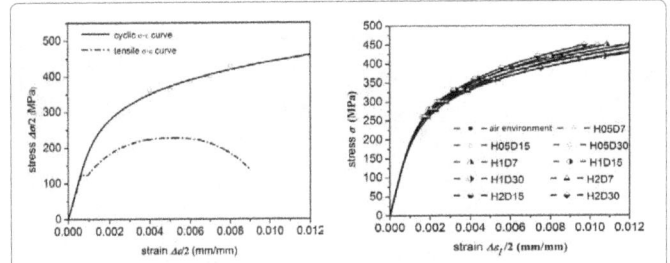

(a)   (b)

Figure 2: Cyclic stress-strain curves of Q345R steel in different test environments: (a) Comparison of cyclic stress-strain curve and uniaxial tensile curve, (b) Cyclic stress-strain curves in different environments.

Test arrangement		H$_2$S solution's concentration H		
		0.05% (wt. %)	0.1% (wt. %)	0.2% (wt. %)
Pre-corrosion time D	7 d	H05D7	H1D7	H2D7
	15 d	H05D15	H1D15	H2D15
	30 d	H05D30	H1D30	H2D30

Table 3: The pre-corroded experimental arrangement.

including corrosion solution's concentration and pre-corrosion time. Three levels have been selected for each factor according to the actual working condition [13]. The concentration of H$_2$S solution was set to 0.05%, 0.1% and 0.2% (wt, %), and the pre-corrosion time was set to 7 days (d), 15 days (d) and 30 days (d). Therefore, a total of 9 test groups were designed and the pre-corroded test arrangement is shown in Table 3. To maintain a stable corrosion solution's concentration, the solution was renewed every 7 days for those test groups of more than 7 days of corrosion time [14].

Low-cycle fatigue (LCF) tests were carried out on the MTS-809 fatigue test machine at room temperature in air. The LCF tests were conducted under total strain control at constant strain amplitude (ε_a) ranging from 0.4% to 0.8%. Triangular waveform was employed for all the fatigue tests at the frequency of 1 Hz. The cyclic loading was started from the tensile side. The total strain was measured by a dynamic extensometer with the span length of 10.0 mm which was attached to the specimen. The data was considered invalid for the tests with fracture outside the gage length of the dynamic extensometer. The fatigue life (N_f) was defined as the number of cycles when the peak tensile stress decreased to 75% of the maximum peak stress. The response at half of the fatigue life was used to obtain cyclic stress-strain curves in this study. After the low-cycle fatigue tests, the fatigue fracture surfaces of the specimens were observed by a scanning electron microscope (SEM) (TESCAN: VEGA 3 EasyProbe).

Results and Discussion

Cyclic stress response

Figure 2 shows the cyclic stress-strain curves of the test material in different test environments. Cyclic stress-strain curve is an important indication under cyclic loading, which could reflect the real stress-strain relationship of the test material. Figure 2a illustrates the difference between the cyclic stress-strain curve and uniaxial tensile curve of the uncorroded test material. It can be seen that the cyclic stress-strain curve coincide well with the uniaxial tensile curve in elastic deformation stage, which means that the low-cycle fatigue test system is under good control and the experimental data is of high reliability. In plastic deformation stage, the cyclic stress-strain curve surpassed the uniaxial tensile curve obviously, which means that Q345R steel exhibited cyclic hardening behavior. According to the regression of the experimental data, the cyclic stress-strain relationship of the test material was obtained as shown in eqn. (1). Figure 2b shows the cyclic stress-strain curves of the test material in different pre-corroded

LCF parameters	σ_f' (MPa)	b	ε_f'	c	K' (MPa)	n'
H05D7	904.68	-0.097	1.25	-0.63	1189.27	0.21
H05D15	985.89	-0.102	0.49	-0.54	1208.83	0.21
H05D30	893.51	-0.092	0.18	-0.44	1018.81	0.18
H1D7	1423.8	-0.14	1.02	-0.62	1228.49	0.21
H1D15	971.25	-0.105	0.43	-0.53	988.03	0.17
H1D30	734.16	-0.074	0.41	-0.53	940.29	0.17
H2D7	1200.57	-0.131	0.64	-0.57	1016.24	0.18
H2D15	796.95	-0.08	0.33	-0.51	1015.01	0.18
H2D30	796.53	-0.082	0.26	-0.49	1174.99	0.2

Table 4: LCF parameters of each test group.

environments, where little difference existed among each group and the distribution is irregular. It indicates that pre-corroded damage has little effect on cycling stress response of Q345R steel, which corresponds to the results reported [12].

$$\Delta\varepsilon_t/2 = \Delta\sigma/2E + (\Delta\sigma/2216.32)^{1/0.19} \quad (1)$$

Strain-life relationship

Cyclic strain-life relationship is an important indication to assess the fatigue properties and predict the fatigue life. For the low cycle fatigue test controlled by strain, cyclic strain-life relationship is generally expressed by *Manson-Coffin* as below.

$$\Delta\varepsilon_t/2 = \Delta\varepsilon_e/2 + \Delta\varepsilon_p/2 = \sigma'_f/E(2N_f)^b + \varepsilon'_f(2N_f)^c \quad (2)$$

where $\Delta\varepsilon_t$ is total strain range (%), $\Delta\varepsilon_e$ is elastic strain range (%), $\Delta\varepsilon_p$ is plastic strain range (%), σ_f' is fatigue strength coefficient, ε_f' is fatigue ductility coefficient, b is fatigue strength exponent, c is fatigue ductility exponent, E is the Young's modulus (Mpa) and N_f is the fatigue life (cycle). According to the tests results, the material constants of each test group were obtained as shown in Table 4. Particularly, the cyclic strain-life relationship in uncorroded test environment was shown in eqn. (3).

$$\Delta\varepsilon_t/2 = 0.00598(2N_f)^{-0.13} + 1.98(2N_f)^{-0.67} \quad (3)$$

Figure 3 shows the cyclic strain-life relationship of the test material in different pre-corroded environments according to eqn. (3). As it can be seen from Figure 3a, the fatigue life of the test material decreases dramatically with increasing strain amplitudes $\Delta\varepsilon_t/2$ (from 0.4% to 0.8%). Elastic strain amplitude increases from 0.16% to 0.19% while plastic strain amplitude increases from 0.24% to 0.61%, plastic strain amplitude is almost 2 or 3 times as much as elastic strain amplitude, which shows that plastic deformation plays a predominant role in total deformation under low-cycle fatigue loading. Figure 3b-3d show the cyclic strain-life curves of Q345R steel with three different

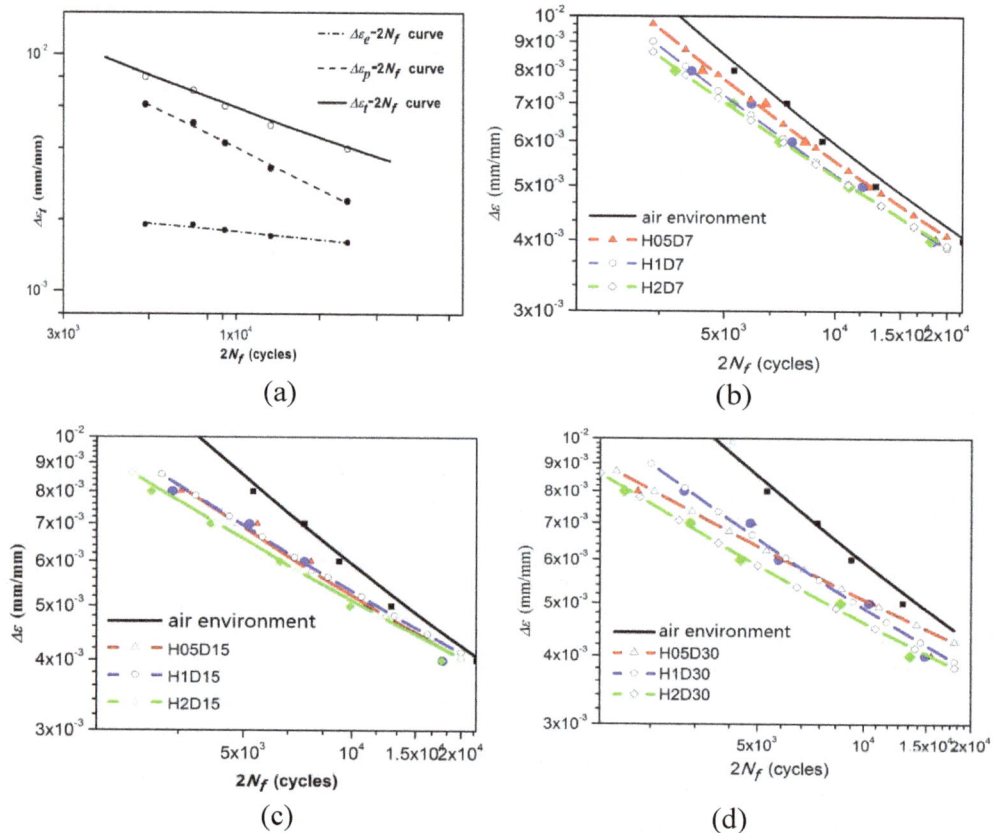

Figure 3: Strain-life curves of Q345R: (a) uncorroded, (b) in 7 days pre-corrosion environment, (c) in 15 days pre-corrosion environment, (d) in 30 days pre-corrosion environment.

H_2S solution's concentrations in 7, 15, 30 days of pre-corrosion environments, respectively. According to a series of observation and analysis, it is found that: 1. All cyclic strain-life curves in pre-corroded environments are below the curves of the uncorroded specimens, which shows a reduction in the fatigue life as a result of the corrosion of wet H_2S environment. 2. The distance between these cyclic strain-life curves with three different H_2S solution's concentrations for the same pre-corrosion time is extremely small, even overlap each other. It indicates the cyclic strain-life relationship was independent of the change of H_2S solution's concentration. 3. The fatigue life decreases sharply with increasing pre-corrosion time (from 7d to 30d) under the same strain amplitude. It can be deduced that the pre-corrosion time has a significant effect on the fatigue life of the test material. 4. The reduction rate of fatigue life increased with increasing total strain amplitude.

The fracture behaviors

In order to investigate the fracture mechanisms of Q345R steel in wet H_2S environment, fracture surfaces of samples were observed by SEM. Figure 4 shows the fracture of the uncorroded specimen fatigued, where the strain amplitude was 0.6% and the fatigue life was 4616 cycles. Figure 4a shows the overall fracture, which was typically consisted of crack, crack propagation and final fracture region. Multiple origins initiated at the specimen surface. The crack propagation region existed at the center of the specimen fracture, which approximately accounted for 50% of the total fracture area. Figures 4b and 4c show the low and high magnification of crack propagation region, respectively, where typical fatigue striations, discontinued river pattern and secondary

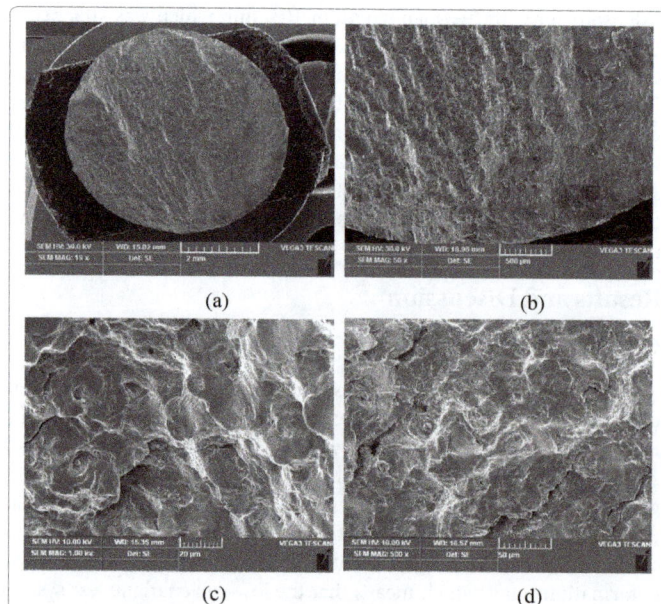

Figure 4: SEM micrographs of fracture surfaces fatigued in uncorroded environment.

cracks were observed. It indicates that the fracture behavior of the crack propagation region was quasi-cleavage fracture. Figure 4d shows the microstructure of the final fracture region, where shear lips occurred. The specimen was plastic fracture characterized by dimples with tyre-

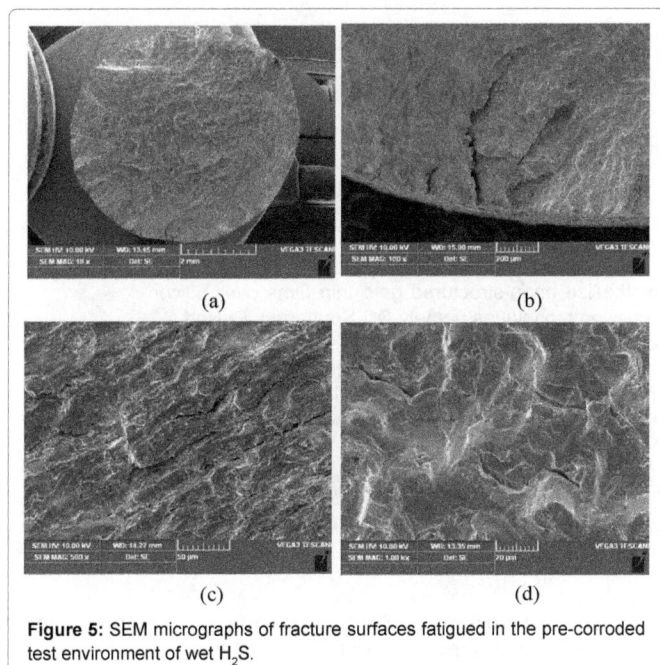

Figure 5: SEM micrographs of fracture surfaces fatigued in the pre-corroded test environment of wet H₂S.

like patterns and tiny secondary cracks. The fatigue striation with tyre-like patterns is typical fracture characteristics of LCF.

Figure 5 shows the SEM fractograph of the specimen fatigued in H2D30 test environment, where the strain amplitude was 0.6% and the fatigue life was 2242 cycles. As can be seen from Figure 5a, multiple crack origins initiated at the specimen surface. Figure 5b shows a typical crack origins initiated from the surface and the corrosion layer with the thickness of about 10 μm. Figure 5c shows the crack propagation region, where it exhibited the quasi-cleavage fracture with discontinued river pattern, typical fatigue striations and secondary cracks. Figure 4d shows the final fracture region characterized by tearing ridge, river pattern and secondary cracks. Secondary cracks existed in the whole fracture, and the length of the secondary cracks decreased while its amount increased.

Conclusions

The test results have provided useful and reliable data for the expected fatigue behavior of pre-corroded Q345R steel, which is most widely used in pressure vessel industry in China. The following conclusions can be described.

1.   Pre-corroded damage caused by wet H₂S environment has little effect on cycling stress response of Q345R steel. The test material exhibited cyclic hardening.

2.   The cyclic strain-life relationship was independent of the concentration of H₂S solution. However, the fatigue life of the test material reduced significantly with the increasing pre-corrosion time. The reduction rate of fatigue life increased with increasing total strain amplitudes.

3.   All fractures were characterized by LCF damage features, i.e. with fatigue striations, secondary cracks and tyre-like patterns. Multiple crack origins initiated at the specimen surface. The crack propagation region was quasi-cleavage fracture, and the final fracture in pre-corroded test environment of wet H₂S is mixture characteristics of plastic fracture and brittle quasi-cleavage fracture.

Acknowledgements

This research was financially supported by National Natural Science Foundation of China under grant No. 51301115, Key Science and Technology Support Program of Sichuan Province No. 2016GZ0294, Key Laboratory of Material Corrosion and Protection of Sichuan Province under grant (No. 2012CL10. No. 2016CL17), Project of Sichuan Province Department of Education No. 16ZB0255 and Talent Introduction Projection in Sichuan University of Science and Engineering under grant No. 2015RC34.

References

1. Li C, Chen G, Chen X, Zhang W (2012) Ratcheting strain and simulation of 16MnR steel under uniaxial cyclic loading. Computational Materials Science 57: 43-47.

2. Zengliang G, Kangda Z (1997) Comparison of the fracture and fatigue properties of 16MnR steel weld metal, the HAZ and the base metal. Journal of materials processing technology 63: 559-562.

3. Guangxu C (1994) The investigation of low-cycle fatigue life for pressure vessel by damage mechanics approach. Journal of Xi'an Jiaotong University 8.

4. Weizong W, Shiqun X, Penglin J (2001) Case of Corrosion Failure in Wet Hydrogen Sulfide Environment and Countermeasures. Petrochemical Corrosion and Protection 2: 002.

5. Fan ZC, Jiang JL (2004) Investigation of low cycle fatigue behavior of 16MnR steel at elevated temperature. Journal-Zhejiang University Engineering Science 38: 1190-1195.

6. Ying X, Bingbing C, Sanlong Z, Zengliang G (2009) Study on fatigue crack growth behavior of 16MnR steel under different conditions. Acta Metallurgica Sinica 45: 849-855.

7. Zhuang LJ, Gao ZL, Wang XG (2007) An experimental study and simulation of fatigue crack propagation of 16MnR steel. Pressure Vessel Technology 24: 1-7.

8. Guangxu C, Zhiwen L (1991) Low-cycle Fatigue Characteristic of Welding Joint for 16MnR Steel and Its Application to Fatigue Life Design in Pressure Vessel. Chemical Production and Technology 8(6).

9. Xu C, Zhifang Z (2000) Low Cycle Fatgue of 16MnR Steel Welded Joint[J]. Chemical Production and Technology 7(3).

10. Tang JQ, Gong JM, Zhang XC, Tu ST (2006) Comparison on the cracking susceptibility of different low alloy steel weldments exposed to the environment containing wet H₂S. Engineering Failure Analysis 13: 1057-1064.

11. Zhen L, Guodong L (2006) Investigation in damage from H₂S stress corrosion of 16MnR steel[J]. China Petroleum Machinery 34(9).

12. GB/T15248-2008. The Test Method for Axial Loading Constant-amplitude Low-cycle Fatigue of Metallic Materials. China Standards Institution: 2008.

13. Gang Y, Liang Z, Xiaohui Z (2004) The Investigation of H2S Corrosion of 16MnR Steel and the Rule of Hydrogen Permeation. Journal of Hunan University(Natural Sciences) 31(3).

14. JB/T 7901-1999, Metals materials - Uniform corrosion - Methods of laboratory immersion testing[S].

4

Observation of Optical Properties of Gold Thin Films Using Spectroscopic Ellipsometry

Pradhan SK*

Department of Physics, National Institute of Technology, Kurukshetra, Haryana, India

Abstract

In this research work, an attempt has been made to synthesize nano-structured gold thin films over Silicon dioxide (SiO_2) coated on Silicon(Si) Substrate using three deposition techniques namely DC Sputtering, Pulsed DC Sputtering, and Pulsed Laser Deposition(PLD). Optical measurements using spectroscopic ellipsometry showed that the dielectric constants of the films differed between films synthesized by different synthesis routes. This was expected because different synthesis routes yielded different microstructure. The difference in microstructures results in differences in electronic structure and therefore resulting in the differences on the optical response.

Keywords: Gold thin films; GIXRD; Spectroscopic ellipsometry; Pseudo dielectric functions

Introduction

Thin film science [1] and particularly nano-structured has received tremendous attention in recent years. They are being used in various fields of microelectronics, transistors, solar cells, radiation sources, magnetic memory devices, detectors, anti-reflection coating etc. The development of nano-technology [2] and nano-science has made inroads into new trends functional materials. As the size shrinks, the physical properties of the materials also change because of special importance is metallic thin films employed over a range of application areas, including macro electronics and microelectronics. The optical properties of the metallic films [3,4] depend mainly on the free and bound electrons. Most of the metals are transparent in UV and in far UV range except the case of the noble metals where they become transparent in the visible range itself. The difference between noble metals and other metals stems from the fact that the bound electron contribution plays a crucial role in the optical response [5]. Among the metallic thin films, gold has been of considerable interest for years owing to its better performance relative to other metal films [1] for optical applications. The main reason for the application of gold is that it is inertness to oxidation environment. Apart from this, it has some unique mechanical properties like strength, malleability and ductility, good electrical and thermal conductor. It is also known for its ability to endure extreme temperature changes. Nano particle films exhibiting strong Plasmon polarization [2,6] resonance are utilized as optical filters in the sensor applications. Moreover, the ultra thin films can be used in nano-scale electronics and as surface enhanced Raman spectroscopy. While there are several techniques to measure the linear optical properties of gold, spectroscopic ellipsometry [7,8] finds a special place owing to its potential advantages for study of thin films. The technique rests on the fact that, any linearly polarized light upon reflection form a sample becomes elliptically polarized [8,9]. Thus the change in polarization state of the incident light after reflection from the material of interest, their dielectric functions can be determined. It is mainly used in research fabrication to determine properties of layer stacks of thin films and interfaces between the layers. Metallic films in various architectural firms such as thin film, multi-layer, nano-composites, quantum wires, dots etc., find applications in diverse fields ranging from clinical to metallurgical applications [2]. Such a wide range of applications is possible owing to its amenability of being fabricated in such forms. Moreover, the physical properties of thin films are highly dependent on the thickness and dimensions of these films. The properties of materials evolve as one move on from confined structures to bulk forms. Thin films of gold have been deposited by transitional methods such as physical vapour deposition [1]. Recently, varieties of techniques such as electron beam MBE, cluster machines have been employed to deposit films so as to tailor certain specific properties. While sufficient literature is available on the structure and micro-structural aspects of deposited films, there is little literature on the correlation between microstructure and optical properties [5]. Noble metals such as Au, Ag and Cu have been investigated by many researchers owing to the fact that these metals are noble metals and show plasma resonance [2] in the UV-Visible region of the electromagnetic spectrum. Apart from this, these metals have been exploited for applications based on plasmonic resonances. The optical response in metals is explained by the free electron theory. Over and above the free electron behavior, the transition of d electrons to Fermi-level is known to play a vital role in influencing the dielectric response. Therefore, active research is underway to exploit and tune the optical response to various applications. In this direction, several publications are available in literature. According to Svetovoy et al., [10], the optical properties of the metallic thin films have not been measured accurately as it is commonly accepted. Invariably, these properties are taken from handbooks of tabulated data together with the Drude parameters which are necessary to extrapolate the data to low frequencies. The possible reasons may be based upon the fact that the optical properties of the deposited thin films [2,5] depend on the method of preparation, and can differ substantially from the bulk. Careful analysis of the data was performed using ellipsometry [7,8] to extract the values of the Drude parameters. It included joint fits of the real and imaginary parts of the dielectric function, or refractive index and extinction coefficient in the low frequency range, the Kramers-Kronig consistency of the dielectric function or complex refractive index performed at all

---

***Corresponding author:** Pradhan SK, Department of Physics, National Institute of Technology, Kurukshetra, Haryana-136119, India,
E-mail: sunilpradha@gmail.com

frequencies. The most important conclusions that followed from this analysis were that the dielectric response varied considerably with thicknesses of films. Recently, the interest in the preparation, characterization, and exploitation of the surface-Plasmon resonance (SPR) of the metal nanoparticles, and specifically of gold has grown exponentially to Losurdo et al., [11]. This interest has been driven both by technological applications spanning from biomedicine, sensing and bio sensing to catalysis and to nanophotonics and by the basic physics involved in the size-dependent structural, optical and electronic properties, which can be tailored to be suitable for device applications by the artificially designed geometries that enable the needed functionality for the nanostructures. Therefore, optical properties of gold nanoparticles are still widely investigated. Typically, optical investigations focus mainly on the SPR of Au nanoparticles, and its dependence on the extrinsic effects such as nanoparticles size and shape factor, and the internanoparticles coupling interactions. Indeed, the dielectric function, $\varepsilon(\omega)$, of metals, and of Au as well, is characterized by the Drude free-electron component, $\varepsilon_{Drude}(\omega)$, and by the interband transitions component, $\varepsilon_{inter}(\omega)$, involving transitions from d levels to an empty state above the Fermi level. i.e., $\varepsilon(\omega) = \varepsilon_{Drude}(\omega) + \varepsilon_{inter}(\omega)$. Although the impact of lattice defects within the nanoparticles as well as non- homogeneous distribution of size and shape of nanoparticles have to be also considered , the predominant size effects on the SPR mainly due to the Drude free electrons, $\varepsilon_{inter}(\omega)$, are extensively investigated. Indeed, the bound electron contribution to the dielectric function also plays an important role in determining Plasmon resonance wavelength and broadening. Nevertheless, there are two aspects that are still critical and interesting from both fundamental and technological points of views worthy of being duly investigated, which are the intrinsic size dependence of the dielectric function of the metal nanoparticles itself and the interconnection of the SPR to the Au nanoparticles interband transition contribution $\varepsilon_{inter}(\omega)$. In recent times research in nanoporous systems are increasing owing to the fact that they exhibit large free surface areas per unit volume compared to the bulk. This property concurs to make porous materials extremely appealing as high-efficiency catalysts and molecular sensors, and has accordingly fuelled huge research efforts in this direction in the past years. Among nanoporous materials, metals are receiving particular attention due to the possibility of exploiting the influence of infinite-size effects on their optical response (Plasmon resonances) for ultrasensitive optical detection. According to Bislo et al., [12] there are remarkable differences between the optical properties of porous materials and their bulk counterparts. After a detailed scan of the literature, we undertook preliminary work on understanding the role of deposition technique in influencing the optical properties of gold thin films.

Experimental

In our experiment, the substrate was taken in such a way that, the type of measurement and study could be intended to be carried out. In our case, for optical study using ellipsometry [7,8] in reflection mode SiO_2 coated on Si was chosen. The substrate was taken from the wafer of P type with <100> orientation with 525 +/- 25 μm thickness and 100 +/- 0.5 mm. diameter. Using diamond cutter the substrate was taken 20 mm × 20 mm. As good adhesion of the films to the substrate is extremely important, therefore we adopted standard procedures for cleaning up of the substrates. In this work, the standard RCA (Radio Corporation of America) procedure was followed, which involved cleaning of substrates with both acidic and basic baths to remove the basic and acidic impurities, respectively. The cleaned substrates were finally stored under isopropyl alcohol in closed container.

The thin films of gold of nominal thickness of ≈ 200 nm were then deposited over 200nm SiO2 coated on c-Si by DC Sputtering, Pulsed DC Sputtering, and Pulsed Laser Deposition (PLD), respectively. The measurement of ellipsometric parameters of the films were done by SOPRA ESVG rotating polarizer type spectroscopic ellipsometer in the energy range1.4–5.5 eV in intervals of 10 meV at an angle of incidence of 75°. In ellipsometry, the changes in the amplitude and the phase difference between the parallel(r_p) and perpendicular (r_s) components of the reflected light polarized with respect to the plane of incidence are measured. The change in polarization state of the incident light upon reflection from a sample is measured in terms of the ratios of their amplitudes and phase differences. The ratio, ρ is given by $\rho = r_{p/}r_s = \tan\Psi \cdot \exp(i\Delta)$ where Ψ is the ratio of the reflection coefficients of the parallel and perpendicular components, while Δ is their phase difference. From the ellipsometric parameters, the optical pseudo-dielectricfunction ε (E) is deduced using the following relation [13].

$$\varepsilon(E) = N_o^2 \left[\sin^2\theta + \left[\frac{1-\rho}{1+\rho} \right]^2 \sin^2\theta \tan^2\theta \right] \qquad (1)$$

Where θ is the angle of incidence, N_o is the refractive index of the ambient and ρ is the ratios of ellipsometric parameters.

Results and Discussion

Figure 1 shows the XRD patterns of Au thin films synthesized by PLD (5000 shots & 30000 shots) and DC Sputtering (deposited for 6 min) technique. The patterns were obtained from STOE diffractometer θ/θ mode at an angle of incidence of 0.5 degree in glancing incidence mode. Sharp diffraction peaks are observed in 2θ angles corresponding to the planes along (111), (200), (220) and (311) planes corresponding to glancing incidence of 0.5. The peaks are such that h+k, h+l and k+l are even and therefore characteristic of face centred cubic (fcc) structure. The experimental XRD pattern of Figure 1 matches nearly with JCPDS 04-0784. The JCPDS data show a 100% peak for grain oriented along 111 direction and 52% for (200) planes. However, our experimental data show that the almost equal population of grains oriented along (111) and (220) planes. Similarly there are small differences between the peak positions of the experimental data and the JCPDS data. The different in orientation and the shift in the peak positions are possible owing to the fact that thin films have different growth directions depending on the synthesis conditions. The diffraction peaks were fitted with gaussian profiles and the parameters are tabulated in Table 1.

The grain sizes are calculated by using the Debye-Scherer formula as $D = k \lambda / \beta \cos\theta$, where '$\lambda$ 'is the wavelength of X-ray (0.15405 nm), with Cu K_a radiation, 'W' is FWHM (full width at half maximum) in radian, 'θ' is the Bragg's angle, 'D' is the grain size (nm) and 'k' is taken to be 0.9

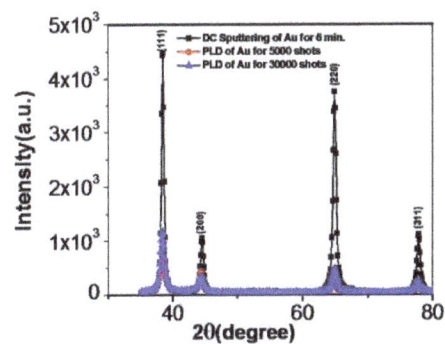

Figure 1: XRD pattern of Au on SiO₂/c-Si for DC Sputtering & PLD.

taken into consideration of spherical particles. In general, broadening of diffraction peaks different factors such as instrument, strain effects, and grain size [14]. Therefore, for an accurate determination of the grain size, one needs to take into account the above factors. The instrumental broadening was determined by obtaining the XRD pattern of standard Si powder whose parameters are very accurately known. The broadening sue to instrument was found to be 0.28126 (0.00381 radian). The XRD patterns were fitted with a gaussian profile in order to obtain the theta and FWHM values. Tables 2-4 show the grain size determination of DC Sputtering of Au on SiO_2 for 6 min and PLD of Au on SiO_2 for 5000 shots and 30000 shots, respectively.

DC sputtering of Au on SiO_2 for 6 min

Table 2 shows determination of grain size for DC Sputtering of Au on SiO_2 for 6 min.

PLD of Au on SiO_2 for 5000 shots

Table 3 represents determination of grain size for PLD of Au on SiO_2 for 5000 shots.

PLD of Au on SiO_2 for 30000 shots

Table 4 represents determination of grain size for PLD of Au on SiO_2 for 30000 shots.

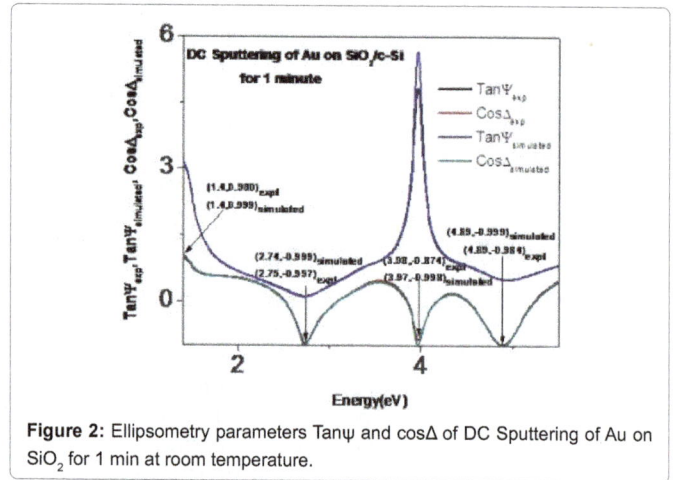

There seems to be wide variation in the grain sizes for different orientation. One possible reason for this could be due to the fact that the strain component has not been considered in fitting the XRD pattern. Thin films indeed how considerable strain depending on the synthesis route.

Figures 2-4 show the ellipsometry parameters tanψ and cosΔ of DC sputterd nanostrucured gold thin films. The tanψ spectrum for 1 and 2 min are very sharp and show maxima and minima. Concomitant

Figure 2: Ellipsometry parameters Tanψ and cosΔ of DC Sputtering of Au on SiO_2 for 1 min at room temperature.

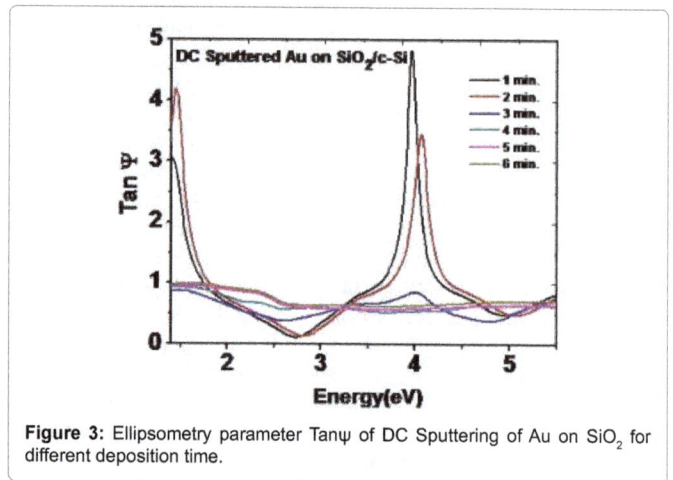

Figure 3: Ellipsometry parameter Tanψ of DC Sputtering of Au on SiO_2 for different deposition time.

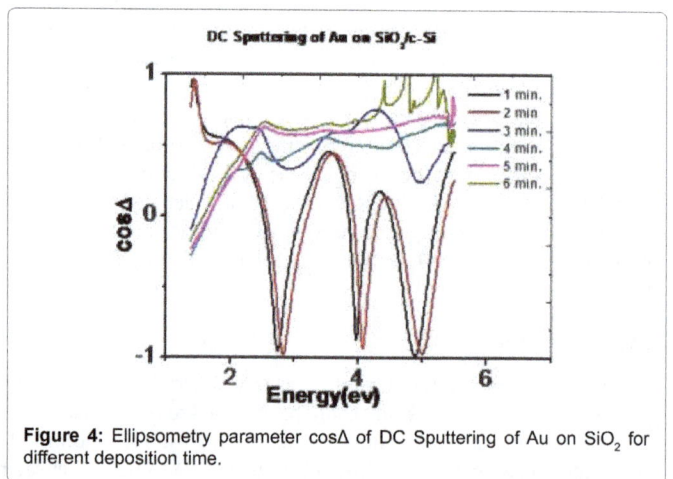

Figure 4: Ellipsometry parameter cosΔ of DC Sputtering of Au on SiO_2 for different deposition time.

Specimen	(111)	(200)	(220)	(311)
JCPDF	100%	52%	32%	36%
DC sputtering for 6 min	100%	23%	84%	24%
PLD for 5k shots	100%	48%	36.5%	27%
PLD for 30k shots	100%	28.63%	41.37%	19.45%

Table 1: Comparison between JCPDF and experimental data in terms of normalised counts per second.

Sl. No	2Θ	Θ	cos Θ	FWHM(rad.)	D(nm)
1	38.400	19.2	0.944	0.00292	50
2	44.600	22.3	0.925	0.004691	31
3	64.800	32.4	0.844	0.004375	37
4	77.800	38.9	0.778	0.00521	34

Table 2: Determination of grain size for DC Sputtering of Au on SiO_2 for 6 min.

Sl. No	2Θ	Θ	Cos Θ	FWHM(rad.)	D(nm)
1	38.2	19.1	0.944	0.00646	22
2	44.4	22.2	0.925	0.00308	48
3	64.6	32.3	0.845	0.00426	38
4	77.6	38.8	0.779	0.00618	28

Table 3: Determination of grain size for PLD of Au on SiO_2 for 5000 shots.

Sl. no	2Θ	Θ	Cos Θ	FWHM(rad.)	D(nm)
1	38.3	19.15	0.942	0.00772	19
2	44.3	22.15	0.926	0.01019	14
3	64.8	32.4	0.844	0.00819	20
4	77.5	38.75	0.779	0.00939	18

Table 4: Determination of grain size for PLD of Au on SiO_2 for 30000 shots.

peaks are found in the cosΔ values too. The value of cosΔ at +1 or -1 at some definite energies (1.4 eV, 2.75 eV, 3.97 eV, 4.89 eV) indicate the presence of interference. This implies that the films are very thin and the electromagnetic rays as shown in Figure 4 the underlying SiO_2 thereby causing interference. As the film thickness increases, we the decrease in the magnitude of these sharp peaks and therefore moving over to bulk like behavior. A detailed modeling of the optical data was undertaken for select specimens using the nonlinear least square fit method. A four layer model (Figure 5) was incorporated into the model. The refractive index of nano-structured gold was inferred based on a Drude Lorentz model. Two oscillators were used for this purpose. The Drude model was used to fit the data in the low energy region (i.e., below 2.2 eV) while the Lorentz model was used to account for transition of electrons from d-level to the Fermi level. The reference refractive index for gold and SiO_2 was taken from literature. Figure 4 shows the Ellipsometry parameter $\cos\Delta$ of DC Sputtering of Au on SiO_2 for different deposition time and Figures 6 and 7 show the ellipsometric parameters of gold thin films synthesized by PLD technique. The sharp structures near 4 eV and the zero value near 2.5 eV are clearly an indication of interference. The films are very thin and therefore the electromagnetic waves penetrate the gold and as shown in Figures 6 and 7 the underlying SiO_2 film. As the number of shots increase, we see that the effect due to interference comes down and there is a progressive move towards 'bulk like' feature. Figure 5 shows a representative fit using a four layer model (c-Si/200nm SiO_2/Au film/ambient). The model adopted was a Drude Lorentz combination

wherein the Drude model best fits for the free electron like behaviour and the Lorentz model takes care of electron transition from the d-level to Fermi level. Figure 8 shows the Ellipsometry parameters $Tan\psi$ and $\cos\Delta$ of Pulsed DC Sputtering of Au on SiO_2 for 1 min. It is clear from the figure that, there is no interference effect which signifies the film is thick.

Figures 9 and 10 show the variation of pseudo dielectric function at room temperature for DC Sputtering for 5 min, Pulsed DC Sputtering for 1 min, PLD for 30K shots. The variation of ε_r and ε_i in these films are typical as in the case of nano-structured thin film. Interference effects are seen in the case of PLD deposited samples. The films are not thin

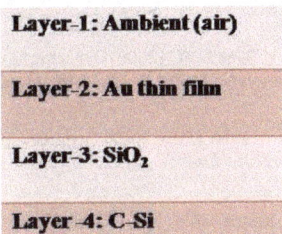

Figure 5: Four layer model.

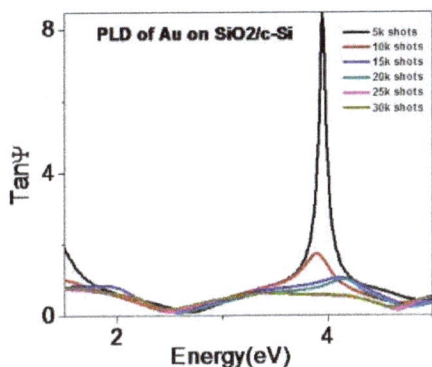

Figure 8: Ellipsometry parameters $Tan\psi$ and $\cos\Delta$ of Pulsed DC Sputtering of Au on SiO_2 for 1 min.

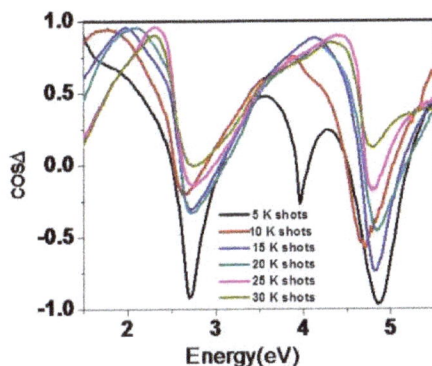

Figure 6: Ellipsometry parameter $Tan\psi$ of PLD of Au on SiO_2 for different shots.

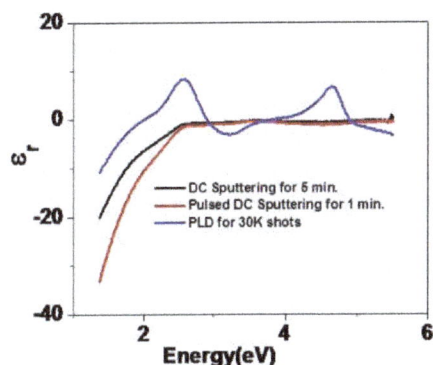

Figure 9: Real part of the experimental pseudo dielectric function of the specimens at room temperature.

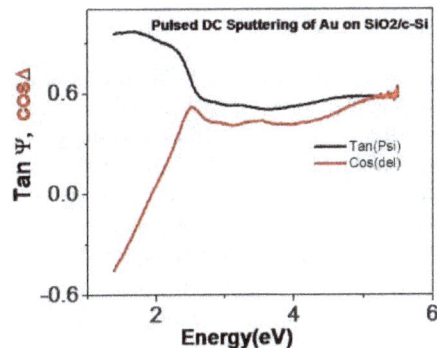

Figure 7: Ellipsometry parameter $\cos\Delta$ of PLD of Au on SiO_2 for different shots.

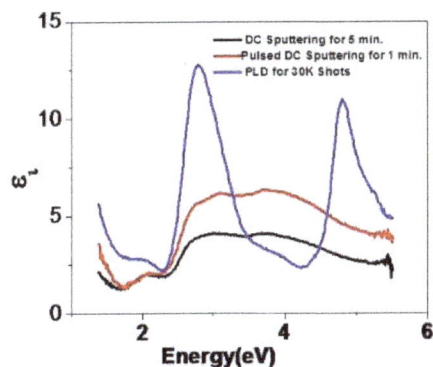

Figure 10: Imaginary part of the experimental pseudo dielectric function of the specimens at room temperature.

so that the SiO_2 is 'seen' by electromagnetic waves, but is thick so as to produce a discernible feature mimicking he bulk dielectric constants. As it is seen the thicker the films, the more negative is the ε_r value at low energy and larger is the value of ε_i at lower energy. The increase of ε_r with film thickness is consistent with the increase in reflection with film. The real and imaginary part of dielectric function shows a distinct change near 2.5 eV which happens to be the plasma frequency of the electrons. The typical negative value of the real part below 2.5 eV indicates that the free electrons are responsible for the optical properties. Below 2.5 eV, the specimens are reflecting while above 2.5 eV, there is a significant absorption (as seen in imaginary part). The increased absorption in this region is owing to the transition of electrons from d-level to the Fermi level.

Conclusion

Thin films were synthesized using different deposition techniques like DC sputtering, Pulsed DC Sputtering and Pulse Laser Deposition (PLD). Structural characterization of the films using glancing incidence X-ray diffraction showed that the films were *fcc* nature and compared well with star quality JCPDF data. The grain sizes computed from the Debye-Scherer formula showed that the grain sizes were more uniform for the specimens prepared by PLD technique using 30,000 shots. The difference in orientation and shift in the peaks were found owing to the fact that thin films have been adopted under different synthesis conditions. Optical measurements using spectroscopic ellipsometry showed that the dielectric constants of the films differed between films synthesized by different synthesis routes. This was expected because different synthesis routes yielded different microstructure. The difference in microstructures results in differences in electronic structure and therefore resulting in the differences on the optical response. Future direction in terms of the electrical resistivity measurements as a function of thickness of films can be performed in order to understand the evolution of the films from the nucleation and growth stage towards film formation. This will also throw light on the correlation between microstructure and the transport properties. Optical investigations at high and low temperatures can also be accomplished in order to discern the correlation between microstructure and the dielectric constants and to compute the thermo-optic coefficients. This quantity can be an important parameter for applications involving nanostructures at elevated temperatures.

Acknowledgement

The help and resources received from Department of Surface and Nano Science division at Indira Gandhi Centre for Atomic Research (IGCAR), Kalpakkam, India under the guidelines of Dr S Tripura Sundari is thankfully acknowledged.

References

1. Gan Q, Ding YJ, Bartoli FJ (2009) "Rainbow" trapping and releasing at telecommunication wavelengths. Phys Rev Lett 102: 056801.

2. Lazziri R, Jupille J (2011) Effect of hydrogen-switchable mirrors on the Casimir force. Nanotechnology 22: 445703.

3. Maaroof AI, Smith GB (2005) Halving the Casimir force with conductive oxides. Thin solid films 206: 485198.

4. Sun X (2007) Influence of random roughness on the Casimir force at small separations. Thin Solid Films. 6966: 5156962.

5. Wei J, Gan F (2003) Time-resolved thermal lens effect of Sb thin films induced by structural transformation near melting temperature. Opt Commun 219: 261-269.

6. Oates TWH (2006) Applied and Plasma Physics. University of Sydney, Australia.

7. Oates TWH, Ryves L, Bilek MMM (2008) Dielectric functions of a growing silver film determined using dynamic in situ spectroscopic ellipsometry. Opt Express 16: 2302-2314.

8. Oates TWH, Ryves L, Bilek MMM(2007) Dynamic spectroscopic ellipsometry determination of nanostructural changes in plasmonic silver films. Opt Expres 15: 15987-15998.

9. Born M, Wolf E, Bhatia AB (1980) Principles of Optics: Electromagnetic Theory of Propagation, Interference and Diffraction of Light. (7th Ed) Cambridge University Press, Oxford.

10. Svetovoy VB, Van Zwol PJ, Palasantzas G, De Hosson JTM (2008) Optical properties of gold films and the Casimir force. Phys Rev B 77: 035439.

11. Losurdo M (2010) Size dependence of the dielectric function of silicon-supported plasmonic gold nanoparticles. Phys Rev B 82: 155451.

12. Bisio F(2009) Optical properties of cluster-assembled nanoporous gold films. Physical Review B 80: 205428.

13. Azzam RMA, Bashara NM (1988) Ellipsometry and Polarized light. North-Holland, Amsterdam, New York.

14. Tripura Sundari S, Raut NC, Mathews T, Ajikumar PK, Dash S, et al.(2011) Ellipsometric studies on TiO_2 thin films synthesized by spray pyrolysis technique. Appl Surf Sci 257: 7399.

5

Optimization of Reaction Parameters for Silver Nanoparticles Synthesis from *Fusarium Oxysporum* and Determination of Silver Nanoparticles Concentration

Khan NT* and Jameel J

Department of Biotechnology, Faculty of Life Sciences and Informatics, Balochistan University of Information Technology, Engineering and Management Sciences, Balochistan, Pakistan

Abstract

A number of physical and chemical method available for the production of silver nanoparticles however these methods are quite costly and make use of poisonous chemicals. Thus use of biological organism as bionanofactories offers a clean and cost effective alternative process for the fabrication of silver nanoparticles. Extracellular synthesis of silver nanocrystals from *Fusarium oxysporum* was accomplished. Data obtained from Ultra violet visible spectrometry was used to calculate the concentration of siver nanoparticles at optimum conditions. The optimum conditions where the concentration of silver nanoaprticles were maximum was found to be 20 g of fungal biomass incubated at 40°C at pH 8.0 using substrate concentration of 2 mM.

Keywords: Mycosynthesis; *Fusarium oxysporum*; Silver nanoparticles; Ultra violet visible spectroscopy

Introduction

Biological organisms are known to involve in the synthesis of different metallic nanoparticles. For example extremophilic *Ureibacillus thermosphaericus* was explored by Juibari to have potential to produce AgNPs at raised temperatures and high Ag^+ ion concentrations [1]. Extracellular synthesis of AgNPs size ranging from 16-40 nm was produced by *Pseudomonas strutzeri* (bacterium) [2]. Mann reported the synthesis of magnetite (Fe_3O_4) or greigite (Fe_3S_4) nanoparticles by magnetotactic bacteria and the extracellular formation of siliceous material was documented in diatoms [3]. Another example reported by Ahmad et al. was the synthesis of CdS nanoparticles by the fungus *Fusarium oxysporum* extracellularly [4]. Not only microorganisms but plants can also be employed for nanoparticle synthesis as described by Shankar S that the synthesis of pure metallic silver and gold particles was achieved by the interaction between the Neem (*Azadirachta indica*) leaf broth with aqueous solution of silver nitrate or chloroauric respectively outside the plant cell [5]. Several species of *Fusarium oxysporum* [4,6,7] such as *Fusarium acuminatum* [8], *Fusarium solani* [9,10], *Fusarium semitectum* [11], *Penicillium brevicompactum* [12], *Penicillium fellutanum* [13], *Pleurotus sajorcaju* [14], *Phoma glomerata* [15], *Alternaria alternate* [16], *Aspergillus clavatus* [17] and *Aspergillus flavus* [18] have been known to synthesize AgNPs. In order to obtain silver nanoparticles of definite shape and size, optimization of the reaction parameters such as temperature, pH, silver nitrate concentration and fungal biomass was done.

Materials and Methods

The fungus culture of *Fusarium oxysporum* was obtained from Yeast and Fungal Biotechnology Lab, BUITEMS. Fungal biomass was obtained on one liter of CD (cezapex dox) broth. Cezapex dox broth consists of the following.

- Ferrous sulphate (0.01g)
- Calcium chloride (0.5 g)
- Magnesium sulphate (0.5 g)
- Sodium nitrate (2 g)
- Yeast extract (1 g)
- Glucose (10 g)
- zinc sulphate (0.01 g)
- Potassium dihydrogen phosphate (1 g)

The culture medium was autoclaved for 15 min at 121°C at 15 psi (pound/square inches) Inoculation was done in sterile air under Laminar flow cabinet. The cultured flasks were then incubated at room temperature on a rotatory shaker at 150 rpm for 120 hrs. Fungal mycelia were harvested after 120 hours of growth using Whatman's filter paper no.1 to obtain fungal filtrate. The filtrate was centrifuged at 15000 rpm for 15 min. 20 ml of centrifuged filtrate (supernatant) was brought in contact with 150 ml of $AgNO_3$ solution (1 mM) [19]. Control containing freshly prepared CD broth with aqueous $AgNO_3$ was run as standard.

Confirmation of silver nitrate formation

Silver nanoparticle formation was visually observed by the gradual change in color of the experimental flasks containing fungal filtrate with $AgNO_3$ solution incubated for a specific period of time. UV visible absorbtion analysis was done to obtain optimum wavelength. Optimization of external environment is important in order to control reaction parameters to achieve optimum conditions where maximum product yield could be obtained [20].

*Corresponding author: Khan NT, Department of Biotechnology, Faculty of Life Sciences & Informatics, Balochistan University of Information Technology, Engineering and Management Sciences, Balochistan, Pakistan
E-mail: nidatabassumkhan@yahoo.com

Effect of incubation temperature

Optimization was performed with temperature ranging from 20°C to 50°C with difference of 10°C on fungus, *Fusarium oxysporum* for AgNPs production. The sample was subjected to ultra violet visible spectrometry to study further effect of temperature on the rate of synthesis of silver nanoparticle.

Effect of pH

pH range from 5.0 to 8.0 was used with the difference of 1.0 to see the effect of pH on AgNPs formation. 1N Hydrochloric acid and 1N Sodium hydroxide was used to change the pH of the extracellular aqueous media.

Effect of AgNO$_3$ concentration

In this case different concentration of AgNO$_3$ from 0.5 to 2.0 mM was studied with a difference of 0.5 mM. The optimum concentration for the synthesis of nanosilver is confirmed by UV-visible absorption spectroscopy.

Effect of biomass (wet weight) concentration

Effect of fungal biomass concentration was studied by using 5 to 20 g of wet biomass with a difference of 5 of fungi *Fusarium oxysporum*. Biosynthesis of nanosilver particles at different biomass concentrations was characterized by UV-visible absorption spectroscopy.

AgNPs concentration calculation

Concentration of silver nanoparticles for each of the optimized parameter was calculated using UVvisible absorbtion data by applying Beer-lambert law:

A= $\varepsilon l c$

Where,

C=concentration of AgNPs

A=absorbance at specific wavelength

ε=molar absorptivity constant

L=path length

Results and Discussion

Silver nanoparticle formation was visually observed when the appearance of brownish black color was seen in the experimental (Figure 1). Presence of protein nitrate reductase in fungal filtrate is accountable for silver ion reduction causing the appearance of brownish black color [8,21,22]. UV visible absorbtion analysis

revealed a characteristic peak at 430 nm which is specific for silver nanoparticles (Figure 2). Silver nanoparticle yields extremely high absorption values within the UV/visible range [23,24]. Optimization studies with respect to temperature revealed that maximum synthesis of silver nanoparticles occurred at 40°C (Figure 3). At this high temperature enzymatic activity of nitrate reductases was maximum resulting in increased concentration of silver nanoparticles as confirmed by the measured concentration calculated from UV absorbtion data by using Beer-lambert law (Table 1). Mitra B et al. also stated that improved AgNPs synthesis occurred at high temperature [25]. Thus temperature greatly influences formation of silver nanoparticles in terms of concentration [26]. Optimum pH

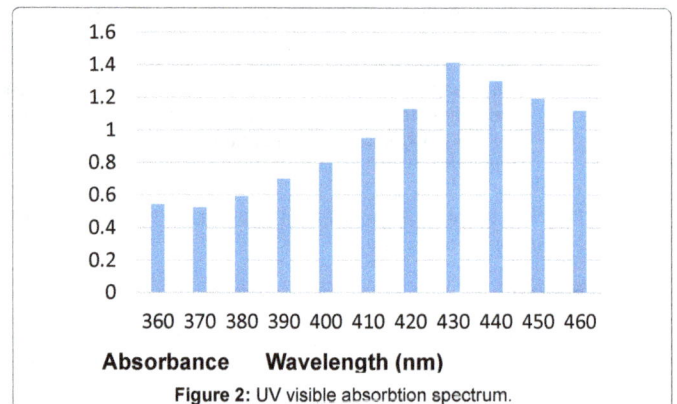

Figure 2: UV visible absorbtion spectrum.

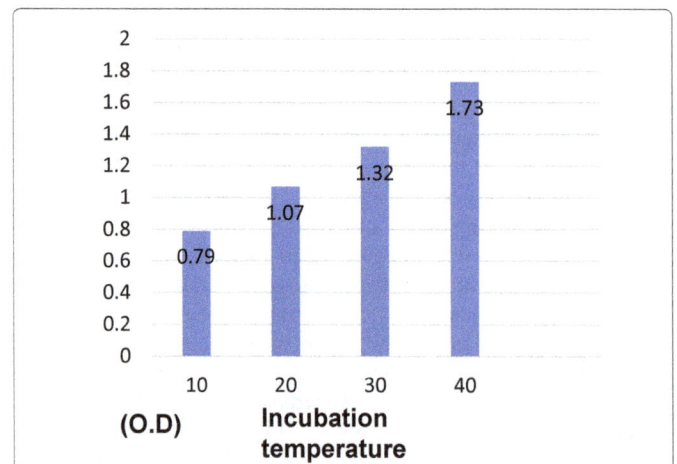

Figure 3: UV visible absorption spectrum showing optimum incubation temperature.

S.No	Incubation Temperature (°C)	Concentration (mol/lit)
1	10	9.46 *10^{-6}
2	20	1.28*10^{-5}
3	30	1.58*10^{-5}
4	40	2.07*10^{-5}

Table 1: Concentration of AgNPs at different incubation temperatures.

S.no	pH	Concentration (mol/lit)
1	5	1.44*10^{-5}
2	6	1.47*10^{-5}
3	7	1.4*10^{-5}
4	8	1.86*10^{-5}

Table 2: Concentration of AgNP$_s$ at different pH.

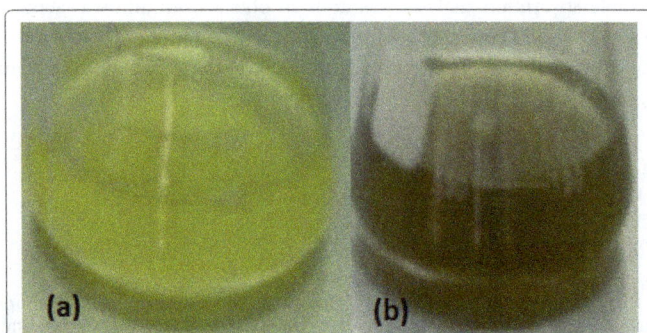

Figure 1: Color change (a) before AgNO$_3$ addition (b) after AgNO$_3$ addition.

was found to be 8.0. The results suggested that an alkaline medium is more suitable for the synthesis of silver nitrate than a low acidic pH as rate of reduction of silver ions were higher at pH 8.0 (Figure 4 and Table 2) .

Rate of bioreduction is directly proportional to the substrate concentration [27]. In this case the optimum concentration was found to be 2 mM of silver nitrate. Reaction kinetics and morphology of nanoparticle is affected by precursor solution (silver nitrate) [28-30] (Figure 5 and Table 3). Optimum fungal biomass was found to be 20 g. It seems that the biocatalysts were agents responsible for the amalgamation of nanoparticles. The enzyme reductase is an NADH dependent enzyme associated with the bioreduction of silver ions in case of fungi [8] (Figure 6 and Table 4).

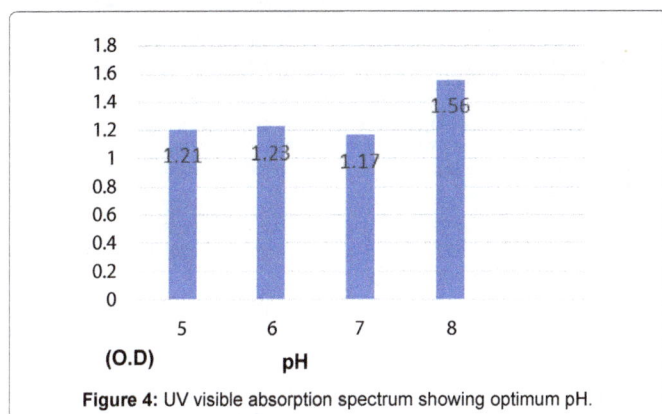

Figure 4: UV visible absorption spectrum showing optimum pH.

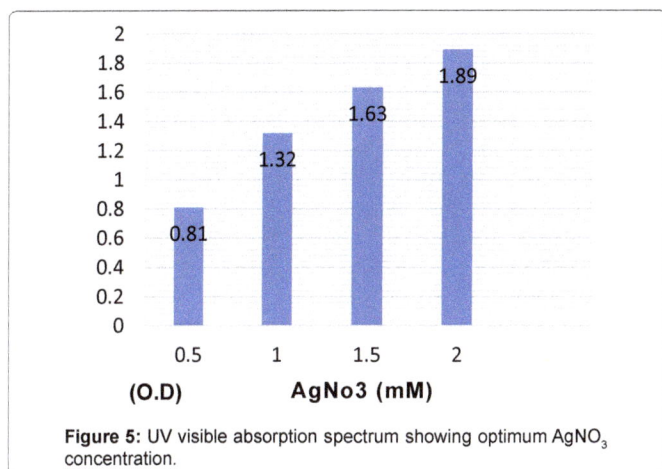

Figure 5: UV visible absorption spectrum showing optimum AgNO$_3$ concentration.

S.no	AgNO$_3$ Concentration	Concentration (mol/lit)
1	0.5	$2.85*10^{-6}$
2	1	$1.58*10^{-5}$
3	1.5	$2.5*10^{-5}$
4	2	$3.38*10^{-5}$

Table 3: Concentration of AgNPs at different AgNO$_3$ concentration.

S.no	Fungal Biomass (g)	Concentration (mol/lit)
1	5	$1.04*10^{-6}$
2	10	$1.44*10^{-5}$
3	15	$1.83*10^{-5}$
4	20	$2.33*10^{-5}$

Table 4: Concentration of AgNPs at different fungal biomass.

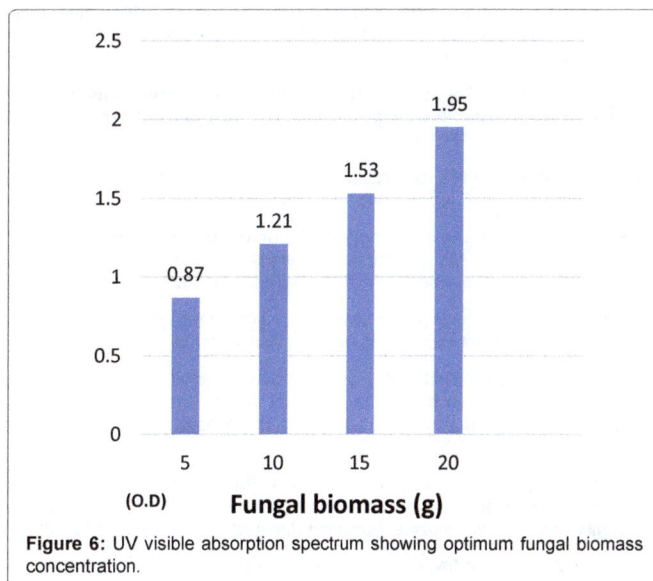

Figure 6: UV visible absorption spectrum showing optimum fungal biomass concentration.

Conclusion

Numerous synthetic methods were available for the synthesis of silver nanoparticles but these methods were quite cost ineffective and uses chemicals that were toxic in nature. Therefore employing biological organism such as fungi, plant etc. offers a clean and cheap alternative process for the amalgamation of silver nanoparticles. Among these microorganisms' fungi is the most suitable biological entity for nanoparticle fabrication because it not only offers simple downstream processing for product recovery but makes the handling of biomass quiet easy. Besides, optimization of the reaction parameters can easily be achieved to obtain maximum concentration of silver nanoparticles of unique morphology.

References

1. Juibari MM, Abbasalizadehb S, Jouzanib GS, Noruzic M (2011) Intensified biosynthesis of silver nanoparticles using a native extremophilic Ureibacillus thermosphaericus strain. Mater Lett 65: 1014-1017.

2. Jeorger K, Jeorger R, Granqvist O (2001) Bacteria as workers in the living factory: metal accumulating bacteria and their potential for material science. Trends Biotechnol 19: 15-20.

3. Mann S (2001) Biomineralization: principles and concepts in bioinorganic materials chemistry. Oxford University Press, Oxford.

4. Ahmad A, Mukherjee P, Mandal D, Senapati S, Khan MI, et al. (2002) Enzyme mediated extracellular biosynthesis of CdS nanoparticles by the fungus Fusarium oxysporum. J Am Chem Soc 124: 12108-12109.

5. Shankar SS, Rai A, Ahmad A, Sastry M (2004) Rapid synthesis of Au, Ag and bimetallic Au core–Ag shell nanoparticles using Neem (Azadirachta indica) leaf broth. J Colloid Interface Sci 275: 496-502.

6. Durán N, Marcato PD, Alves OL, de Souza GIH, Esposito E (2005) Mechanistic aspects of biosynthesis of silver nanoparticles by several Fusarium oxysporum strains. J Nanobiotechnology 3: 1-8.

7. Karbasian M, Atyabi SM, Siadat SD, Momem SB, Norouzian D (2008) Optimizing nano-silver formation by F. oxysporum (PTCC 5115) employing response surface methodology. Am J Agric Biol Sci 3: 433-437.

8. Ingle A, Gade A, Pierrat S, Sonnichsen C, Rai M (2008) Mycosynthesis of silver nanoparticles using the fungus F. acuminatum and its activity against some human pathogenic bacteria. Curr Nanosci 4: 141-144.

9. Ingle A, Gade A, Bawaskar M, Rai MK (2009) Fusarium solani: a novel biological agent for the extracellular synthesis of silver nanoparticles. J Nanopart Res 11: 2079-2085.

10. El-Rafie MH, Mohamed AA, Shaheen THI, Hebeish A (2010) Antimicrobial effect of silver nanoparticles produced by fungal process on cotton fabrics. Carbohydr Polym 80: 779-782.

11. Bhainsa KC, D Souza SF (2006) Extracellular biosynthesis of silver nanoparticles using the fungus Aspergillus fumigatus. Colloids Surf B Biointerfaces 47: 160-164.

12. Ahamed M, Alsalhi MS, Siddiqui MKJ (2010) Silver nanoparticle applications and human health. Clin Chim Acta 411: 1841-1848.

13. Kathiresan K, Manivannan S, Nabeel AM, Dhivya B (2009) Studies on silver nanoparticles synthesized by a marine fungus Penicillum fellutanum isolated from coastal mangrove sediment. Colloids Surf B 71: 133-137.

14. Nithya R, Ragunathan R (2009) Synthesis of silver nanoparticle using Pleurotus sajor caju and its antimicrobial study. Dig J Nanomater Bios 4: 623-629.

15. Birla SS, Tiwari VV, Gade AK, Ingle AP, Yadav AP, et al. (2009) Fabrication of silver nanoparticles by Phoma glomerata and its combined effect against Escherichia coli, Pseudomonas aeruginosa and Staphylococcus aureus. Lett Appl Microbiol 27: 76-83.

16. Gajbhiye M, Kesharwani J, Ingle A, Gade A, Rai M (2009) Fungus mediated synthesis of silver nanoparticles and their activity against pathogenic fungi in combination with fluconazole. Nanomed Nanotechnol Biol Med 5: 382-386.

17. Verma VC, Kharwar RN, Gange AC (2010) Biosynthesis of antimicrobial silver nanoparticles by the endophytic fungus Aspergillus clavatus. Nanomedicine 5: 33-40.

18. Vigneshwaran N, Ashtaputre M, Nachane RP, Paralikar KM, Balasubramanya H (2007) Biological synthesis of silver nanoparticles using the fungus Aspergillus flavus. Mater Lett 61: 1413-1418.

19. Basavaraja S, Balaji SD, Lagashetty A, Rajasab AH, Venkataraman A (2007) Extracellular biosynthesis of silver nanoparticles using the fungus Fusarium semitectum. Mater Res Bull 43: 1164-1170.

20. Gurunathan S, Kalishwaralal K, Vaidyanathan R, Venkataraman D, Pandian SR, et al. (2009) Biosynthesis, purification and characterization of silver nanoparticles using Escherichia coli. Colloids Surf B Biointerfaces 74: 328-335.

21. Hamedi S, Shojaosadati SA, Shokrollahzadeh S, Hashemi-Najafabadi S (2014) Extracellular biosynthesis of silver nanoparticles using a novel and non-pathogenic fungus, Neurospora intermedia: controlled synthesis and antibacterial activity. World J Microbiol Biotechnol 30: 693-704.

22. Gholami-Shabani M, Akbarzadeh A, Norouzian D, Amini A, Gholami-Shabani Z, et al. (2014) Antimicrobial activity and physical characterization of silver nanoparticles green synthesized using nitrate reductase from F. oxysporum. Appl Biochem Biotechnol 172: 4084-4098.

23. Wilcoxon J (2009) Optical absorption properties of dispersed gold and silver alloy nanoparticles. J Phys Chem B 113: 2647- 2656.

24. Jain PK, Lee KS, El-Sayed IH, El-Sayed MA (2006) Calculated absorption and scattering properties of gold nanoparticles of different size, shape, and composition: Applications in biological imaging and biomedicine. J Phys Chem B 110: 7238-7248.

25. Mitra B, Vishnudas D, Sant SB, Annamalai A (2012) Green synthesis and characterization of silver nanoparticles by aqueous leaf extracts of Cardiospermum helicacabum leaves. Drug invent today 4: 340-344.

26. Chen JC, Lin ZH, Ma XX (2003) Evidence of the production of silver nanoparticles via pretreatment of Phoma sp 32883 with silver nitrate. Lett Appl Microbiol 37: 105-108.

27. Christensen L, Vivekanandhan S, Misra M, Mohanty AK (2011) Biosynthesis of silver nanoparticles using Murraya koenigii (curry leaf): An investigation on the effect of broth concentration in reduction mechanism and particle size. Adv Mat Lett 2: 429-434.

28. Khan M, Khan M, Adil SF, Tahir MN, Tremel W, et al. (2013) Green synthesis of silver nanoparticles mediated by Pulicaria glutinosa extract. Int J Nanomedicine 8: 1507-1516.

29. Chandran SP, Chaudhary M, Pasricha R, Ahmad A, Sastry M (2006) Synthesis of gold nanotriangles and silver nanoparticles using Aloe Vera plant extract. Biotechnol Prog 22: 577-583.

30. Singh AK, Talat M, Singh DP, Srivastava ON (2010) Biosynthesis of gold and silver nanoparticles by natural precursor clove and their functionalization with amine group. J Nanopart Res 12: 1667-1675.

Influence of Dopants on Mechanical Properties of Steel: A Spin-Polarized Pseudopotential Study

Zavodinsky V[1]* and Kabaldin Y[2]

[1]*Institute for Materials Science of the Russian Academy of Sciences 153 Tikhookeanskaya str., Khabarovsk, 680042, Russia*
[2]*Nizhny Novgorod State Technical University, 24 Minin str. Nizhny Novgorod, 603013, Russia*

Abstract

Density functional theory and pseudopotentials were used to study reaction of the ferrite grains interface (doped with C, P, N, Ti, Ti+N and Ti+C) on deformations. It was shown that impurities could increase or decrease the tensile strength and the elongation limit. The best effect was demonstrated for cases when Ti presents simultaneous with C; the worst case is doping iron with P. As for the shift modulus, effect is not significant.

Keywords: Ab initio simulation; α-Fe; Tensile strength; Shift modulus; Dopants influence

Introduction

Low carbonized steel is a well-known Fe allow used as a material for various engineering details and constructions. Usually it contains up to 0.2 per cent of carbon and some small amounts of other dopants. There are some works where influence of dopants on the mechanical properties of steel is studied on the atomic level [1-5]. However, in the most of them only some energetic characteristics are studied, while engineers are interested in technical parameters. Here we present results of the tensile strength and the shift modulus calculations for the doped Fe.

Methods and Models

As real steel is a polycrystalline material consisted mainly of α-Fe grains, its durability depends mainly on the durability of grain interfaces (borders). For this reason our investigation is limited with studying of the interface between two grains of α-Fe.

It is well know that α-Fe is a ferromagnetic with magnetic polarization of 2.2 μ_B per atom, *bcc*-lattice with period a=2.867 Å, and cohesion energy of 5.48 eV. In this work we used the spin-polarized version of DFT realized within the FHI96md package [6] based ont he density functional theory (DFT) [7,8] and the pseudo-potential method [9]. This package was previously used with advantage for many systems, including transition metal compositions [10-14]. In all cases, the generalized gradient approximation [15] to description of the exchange-correlation interactions has been chosen and the optimization of the atomic geometry has been performed. All pseudopotentials were constructed with the FHI98PP package [16]. They were checked for the absent of the so-called 'ghost' states and were used for founding the equilibrium lattice parameters, magnetic and the cohesion energy of α-Fe.

To model the bulk *bcc* α-Fe we used a cubic cell with two iron atoms: one atom was placed in the lattice cell corner; the second atom was situated in the center of the cell. The energy cut-off for the pane wave set was equal to 40 Ry and the special k-point (0.25; 0.25; 0.25) was used. The self-consistence convergence was provided by stabilizing the total energy with an accuracy of 0.005 eV per atom. We have found the equilibrium lattice constant of 2.88 Å, magnetic moment of 2.0 μ_B, and cohesion energy of 4.5 eV, in satisfactory accordance with experimental data.

An interface between two α-Fe grains was modeled as a contact of two thin crystalline slabs having infinity dimensions in X and Y directions. Thickness of the each slab was made of three atomic layers. The interacting slabs approached among themselves before the minimum of their total energy was achieved. The scheme of such system is given in Figure 1.

In order to test our approach we calculated the interface energy (the energy gain of iron grains bonding) and have obtained the value of 1.33 J/m². According to refs. [5,17] published values of the grain interface energy of Fe lie in the interval of 0.47-1.63 J/m².

After leading of the system in a state with the minimum energy we made computer experiments of two types (Figure 2). The top panel: investigation of the reaction of the interface to break; the down panel: investigation of the reaction of the interface to shift.

In the stretch case we found the variations of the total energy ΔE as functions of elongation ΔZ, and then the applied strength P was

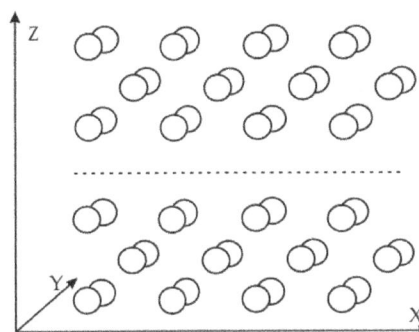

Figure 1: The scheme of the interface between two crystallites of alpha ferrite. The dotted line shows the defective plane in which there are no atoms inside of cubic cells.

*Corresponding author: Victor Zavodinsky, Institute for Materials Science of the Russian Academy of Sciences 153 Tikhookeanskaya str., Khabarovsk, 680042, Russia, E-mail: vzavod@mail.ru

calculated according the formulae $P = \dfrac{\Delta E}{\Delta Z} \dfrac{1}{S_{XY}}$, where S_{XY} is the system square in the XY plane. In the shift case we studied the dependence of ΔE on $tg\varphi$, where φ is the shift angle, and calculated the shift module G: $G = \dfrac{\Delta E}{\Delta X} \dfrac{1}{S_{XY}} \dfrac{1}{tg\phi}$.

Results and Discussion

Undoped α-Fe

First of all we studied reaction of the undoped iron system on stretching along Z direction and on shift along X direction. Atoms of upper and down planes were moved step-by-step and fixed at the each step. Other atoms were able to relax up and to find equilibrium positions. The dependences of tension on elongation and energy on tangent of the shift angle are presented in Figure 3.

The top panel of this figure demonstrates the strength limit of 18 GPa for breaking the undoped α-Fe grains interface. The down panel shows that the shift process has two stages. Firstly, the total energy grows sharply with the shift modulus of 160 GPa; secondly, the shift modulus decreased to 80 GPa. Comparatively, experimental value of

tensile strength is 60 MPa for the low carbonized steel and 1.3 GPa for single-crystalline whiskers. The published shift module value is 14 GPa [3].

Dopants influence

We placed dopant atoms as it is shown in Figure 2B-2D. Namely, for single-atomic cases we replaced one atom of the interface with C, P, N or Ti atom. For two-atomic cases (TiN or TiC) dopant atoms were placed on both sides of the interface. Results of calculations are presented in Figure 4 and Table 1.

It is clear (Figure 4) that presence of carbon increases the tensile strength by 1.5 times in comparison with undoped ferrite; N and Ti+N decrease it approximately by 20 percent. The Ti and Ti+C cases demonstrate the tensile strength very close to the case of undoped iron. At another hand addition of carbon reduces the limit of elongation to 1 percent (for undoped α-Fe it is 10 percent); nitrogen reduces it to 8 percent; while Ti, Ti+N and Ti+C increase its value up to 14-18 percent. As for phosphorus it decreases strongly both the tensile strength and the limit of elongation. In more details results are collected in Table 1.

Influence of dopants on the shift behavior of the Fe grains interface is much less significant than their effect for the stretch characteristics. Our modeling shows that in all doped cases the dependence of the shift module on the shift angle tangent looks like that of the undoped Fe case, and its value varies between 70 and 170 Gpa.

Conclusion

Quantum-mechanics modeling of the α-Fe grains interface as a contact of two nano thin slabs lets us to obtain adequate data of dopants influence (C, P, N, Ti, Ti+C and Ti+N) on mechanical properties of steel. Summarize all results we can conclude that addition of Ti (single or with C) increases significantly the elongation limit keeping approximately the level of the tensile strength of un-doped iron. Thus, addition of Ti in the C doped α-Fe can improve elasticity of steel details and constructions at low temperatures. Phosphorus is the worst impurity in steel.

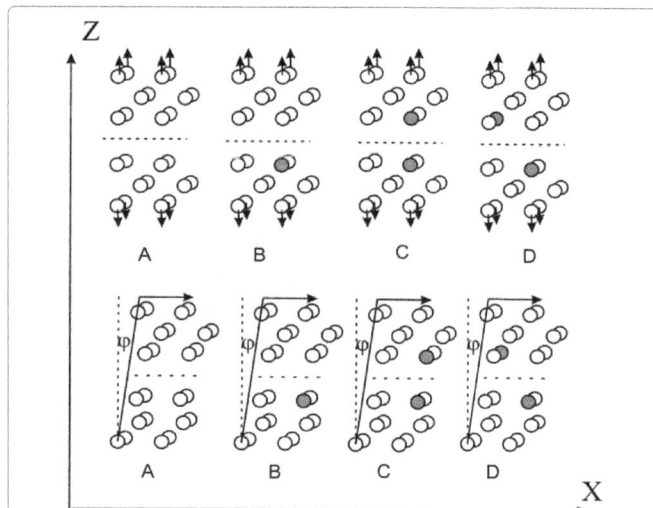

Figure 2: Schemes of computer experiments: A) undoped α-Fe grains interface; B) one dopant atom; C and D) two dopant atoms. Dopant atoms are shown as grey balls, φ is a shift angle.

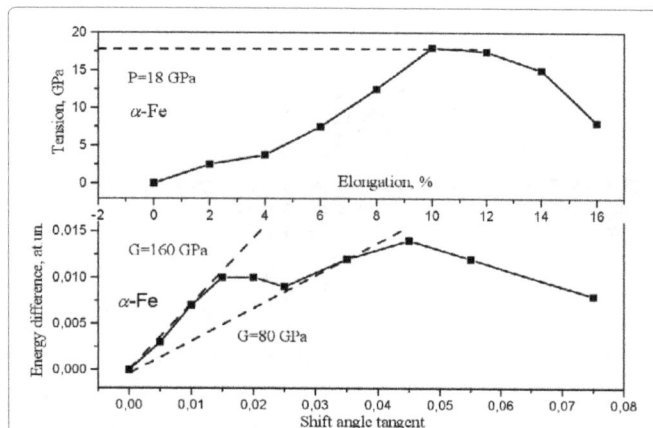

Figure 3: Stretch and shift study of the undoped α-Fe grains interface.

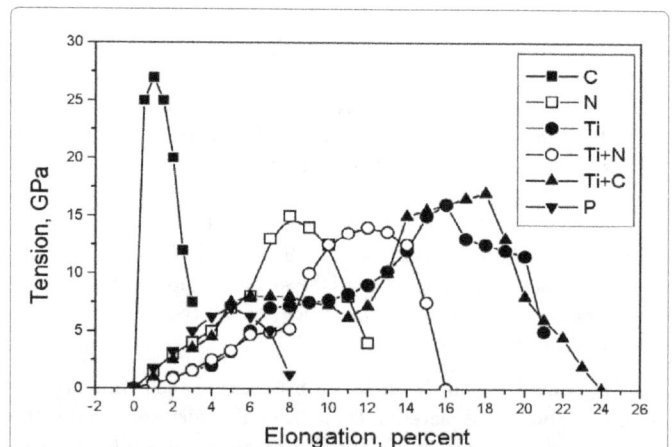

Figure 4: Stretch study of the dopant influence on durability of the Fe grains interface.

System	Undoped Fe	Fe+C	Fe+N	Fe+Ti	Fe+Ti+C	Fe+Ti+N	Fe+P
P, GPa	18	27	14.8	15.5	16.5	14.5	7.1
L, percent	10	1	8	16	18	12	5

Table 1: Tensile strength P and elongation limit L for the Fe grains interface.

References

1. Kolesnikov V, Myasnikova N, Sidashov A, Myasnikov P, Kravchenko J, et al. (2010) Multilayered antifriction nanostraction covering for lubrication in the tribocoupling "wheel-rail". Transport problems 5: 71-79.

2. Ridny YaM, Mirzoev AA, Mirzaev DA (2014) Ab initio modelirovanie azota v GTsK-reshotke zheleza. Vestink Yuzhnogo Gosudarstvennogo Universiteta. Ser. Metallurgia (in Russian) 14: 59-63.

3. Hu SY, Ludwig M, Kizler P, Schmauder S (1998) Atomistic simulations of deformation and fracture of. Modelling and Simulation in Materials Science and Engineering 6: 567.

4. Oila A, Bull SJ (2009) Atomistic simulation of Fe-C austenite. Computational Materials Science 45: 235-239.

5. Verkhovykh AV, Mirzoev AA (2013) Ab intio modelirovanie energii formirovaniya granitsy zerna v OTsK-zheleze. Vestink Yuzhno-Uralskogo Gosudarstvennogo Universiteta. Ser. Metallurgia, Mekhanika, Fizika (in Russian) 5: 76-81.

6. Bockstedte M, Kley A, Neugebauer J, Scheffler M (1997) Density-functional theory calculations for poly-atomic systems: electronic structure, static and elastic properties and ab initio molecular dynamics. Computer physics communications 107: 187-222.

7. Hohenberg P, Kohn W (1964) Inhomogeneous electron gas. Physical Review 136: B864.

8. Kohn W, Sham LJ (1965) Self-consistent equations including exchange and correlation effects. Physical review 140: A1133.

9. Heine V, Weaire D (1970) Pseudopotential theory of cohesion and structure. Solid State Physics 24: 249-463.

10. Dabrowski J, Müssig HJ, Zavodinsky V, Baierle R, Caldas MJ (2002) Mechanism of dopant segregation to SiO2/Si (001) interfaces. Physical Review B 65: 245305.

11. Zavodinsky VG (2004) The mechanism of ionic conductivity in stabilized cubic zirconia. Physics of the Solid State 46: 453-457.

12. Zavodinsky VG, Chibisov AN (2006) Zirconia nanoparticles and nanostructured systems. Journal of Physics: Conference Series 29: 173). IOP Publishing.

13. Zavodinsky VG (2010) Small tungsten carbide nanoparticles: Simulation of structure, energetics, and tensile strength. International Journal of Refractory Metals and Hard Materials 28: 446-450.

14. Zavodinsky VG (2011) Cobalt layers crystallized on the WC (100) surface: Spin-polarized ab initio study. International Journal of Refractory Metals and Hard Materials 29: 184-187.

15. Perdew JP, Burke K, Wang Y (1996) Generalized gradient approximation for the exchange-correlation hole of a many-electron system. Physical Review B 54: 16533.

16. Fuchs M, Scheffler M (1999) Ab initio pseudopotentials for electronic structure calculations of poly-atomic systems using density-functional theory. Computer Physics Communications 119: 67-98.

17. Chicago Zybin IN, Bulychev VV, Latypov RA (2015) Analiz energii granits zeren v metallakh primenitelno k protsesu soedinenia metallov pri svarke davleniem. Sovremennye Problemy Nauki i Obrazovania 1: 18-23.

Polymer Electrode Material for Microbial Bioelectrochemical Systems

Moutcine A[1], Akhramez S[2], Maallah R[1], Hafid A[2] and Chtaini A[1]*

[1]Equipe Electrochemistry and Molecular Inorganic Materials, University Sultan Moulay Slimane, Faculty of Science and Technology of Beni Mellal, Morocco
[2]Laboratoirre of Organic Chemistry and Analytical, Sultan Moulay Slimane University, Faculty of Science and Technology of Beni Mellal, Morocco

Abstract

Bioelectrochemical systems based on polymer-bacteria thin film modified electrode were explored. The prepared polymer-bacteria modified copper electrode was characterized with voltametric methods, as cyclic voltammetry (CV) and electrochemical impedance spectroscopy (EIS). The proposed electrode indicated a definite redox response, high conductivity and electrochemical stability. The experimental results revealed that the prepared electrode could be a feasible for degradation of hazardous phenol pollutants.

Keywords: *Staphylococcus aurous*; Cyclic voltammetry; EIS; Phenol; Oxidation

Introduction

The chemical contamination of the environment is a serious problem across the world. Phenol is considered as one of the most products rejected by industrial plants such as pharmaceuticals, formaldehyde resins, pesticides, textiles, petroleum refineries, chemical industry and agricultural activities. Phenol is one of 129 chemical compounds considered important pollutants listed by the Agency for Environmental Protection (EPA) [1,2]. Due to severe legislative laws, prohibiting the discharge of toxic products into the environment, various methods for phenol treatment have been studied, such as wet air oxidation [3], adsorption [4], chemical oxidation [5], photo catalyst [6], biological treatment [7], and ozone oxidation [8]. But there are few sufficiently efficient processes for the removal of phenol. Microbial bioelectrochemical systems have evolved as a potential alternative technology for phenol treatment. The electrochemical oxidation methods has become a promising method for the toxic, bio refractory and highly concentrated organic wastewater treatment because of its simplicity, easy control, strong oxidation performance and environmental compatibility [9]. For anodic oxidation reactions, that the organic matter electro oxidation process, hydroxyl radicals ($\cdot$OH), produced by water discharge through the retardation of oxygen at high electrode over potential on metal/particle electrode surface, are required for organic matter elimination [10]. The combination of electrochemical and biological method can bring added value to the degradation of toxic materials such as phenol. The function of microbial bioelectrochemical systems is based on operating the microorganisms to catalyze an electrochemical reaction. The system allows for different tasks, such as, biodegradation, and electrochemical processes [11]. The aims of this work were to examine the new electrode, based on polymer-bacteria modified copper electrode for simultaneous production of electricity and degradation of phenol.

Experimental

Reagents and apparatus

All chemicals were analytical grade and used without further purification. The monomer ε-caprolactone was purcharged from Sigma-Aldrich. All solutions were prepared with distilled water. Voltammetric experiments were performed using a voltalab potentiostat (modelPGSTAT 100, Eco Chemie B.V., Utrecht, The Netherlands) driven by the general purpose electrochemical systems data processing software (voltalab master 4 software. The three electrode system consisted of a polymer-bacteria modified copper electrode as the working electrode a saturated calomel electrode (SCE) serving as reference electrode, and platinum as an auxiliary electrode. Prior to its modification the copper plate was polished with 0.05 μm alumina slurry for 2 min, rinsed with doubly-distilled water and sonicated in a water bath for 5 min.

Bacterial cultivation

The bacterial strain used in this study was Staphylococcus aureus ATCC 25923. The strain was cultured in Luria Burtani broth at 37°C for 24 h after culture; the cells were harvested by centrifugation for 15 min at 8400xg and were washed twice with and resuspended in KNO$_3$ solution with ionic strength 0.1 M. The physicochemical properties of this strain were measured by contact angle measurements.

Provisions were made for oxygen removal by bubbling the solution with azotes gas for about 5 min then the solution was blanketed with azotes gas while the experiment was in progress. For reproducible results, a fresh solution was made for each experiment. The re-suspended bacteria suspension was diluted with water to obtain needed suspension of different concentration before use.

Electrode preparation

The bacteria suspension was versed onto the copper electrode plate (1 cm × 1 cm × 1 mm), after immobilizing the bacteria on the surface of copper plate, we launch a scan cyclic voltammetry, 10 cycles, in the potential range between -3 V and 2 V at 50 mV/s in a solution of 1 M NaCl.

Results and Discussion

Electrodeposition of polymer and bacteria

Electrochemical polymerization of the polymers can be carried out by either potential step or potential sweep methods, using a typical coating solution 0.1 M monomers in 1M NaCl solution (pH 5). In the

*Corresponding author: Chtaini A, Equipe Electrochemistry and Molecular Inorganic Materials, University Sultan Moulay Slimane, Faculty of Science and Technology of Beni Mellal, Morocco, E-mail: a.chtaini@usms.ma

cyclic voltammetric (CV) curves for the polymerization, the anodic current is even bigger during the early reverse scan that during the forward scan, leading to a cross-over. These features indicate [12,13] that deposition of the polymer proceeds through a nucleation and growth mechanism as reported for other conducting polymers.

The CV (Figure 1) exhibits three picks, the first one (Ia), in the anodic scan, at about 0 V attributed to monomers oxidation, the second one at 0.5 V, corresponds to the evolution of the polymer, the third peak appears in the direction of cathodic scan may correspond to the polymer reduction.

The proposed mechanism of polymerization is as follows:

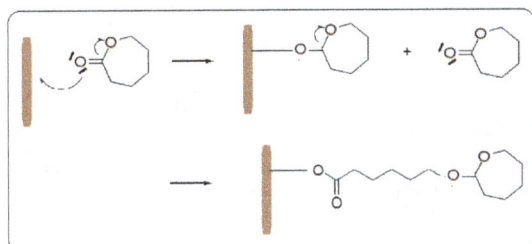

The EIS experiments were carried out in 1.0 mol L^{-1} NaCl in order to evaluate the effectiveness of the polymer modified copper electrode to oxidize the phenol. Figure 2 shows the Nyquist plot for phenol-free polymer-modified electrode and polymer-Cu/phenol system at different concentrations of phenol. In all cases (presence or absence of phenol in the electrolytic solution) curves included a semicircle at higher frequencies corresponding to the electron transfer limited process and the linear part at lower frequencies corresponding to the diffusion process. It appears clearly from these data that the resistance became smaller in presence of phenol in electrolytical solution. This means that the modified electrode becomes more conductive, which can be explained by the presence of phenol on the electrode surface.

In the presence of bacteria in the polymer matrix, oxidation of phenol has been investigated by impedance spectroscopy (EIS). The same behavior in the absence of bacteria, the impedance diagrams have the shape of a semi circle, indicating that the mechanism of the phenol oxidation is not affected by bacteria (Figure 3).

In Figure 4, we illustrate the impedance diagrams recorded in the presence of phenol in the electrolytic medium, respectively for the electrodes, polymer modified copper electrode (curve a) and polymer-bacteria modified electrode (curve b), in the both cases EIS curves have the same shape, the mechanism of phenol oxidation has not changed. However, the electrochemical parameters deduced from these diagrams (Table 1) reveal that, in presence of phenol in electrolytical solution, the electron transfer resistance, Rt, recorded for polymer modified electrode, at the metal/solution interface is about 71.69 ohm. cm^2, while in the case of the polymer-bacteria electrode, Rt is of the

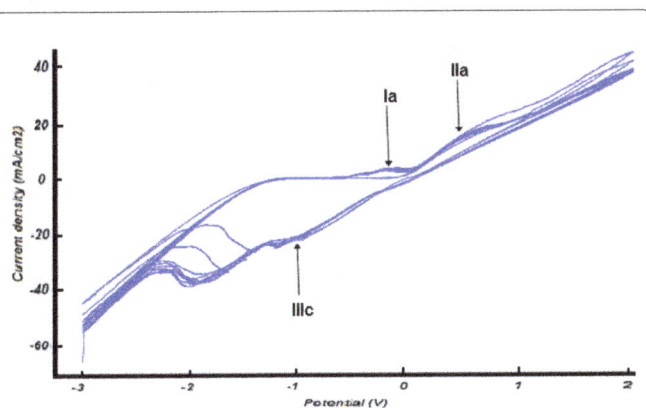

Figure 1: Cyclic voltammogram for electrochemical polymerization of 0.1 M monomer in 0.1 M NaCl, at copper electrode, at 50 mv/s, 10 cycles.

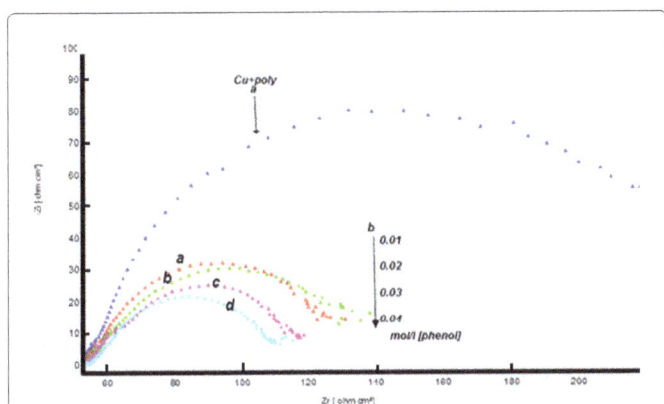

Figure 2: Impedance spectra in 1.0 mol L^{-1} NaCl, at polymer modified copper electrode in 1 M NaCl solution containing different concentrations of phenol.

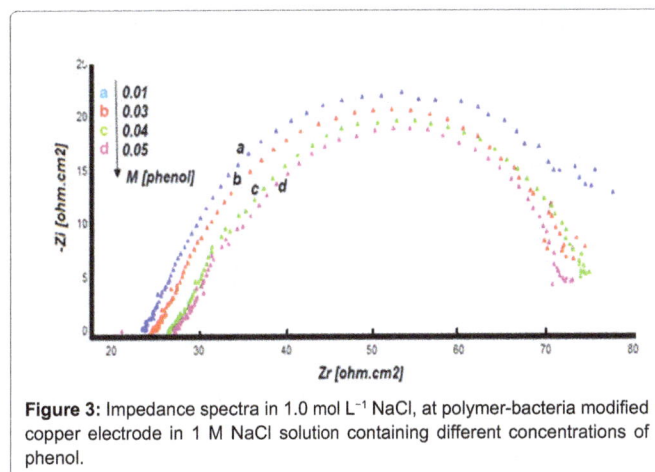

Figure 3: Impedance spectra in 1.0 mol L^{-1} NaCl, at polymer-bacteria modified copper electrode in 1 M NaCl solution containing different concentrations of phenol.

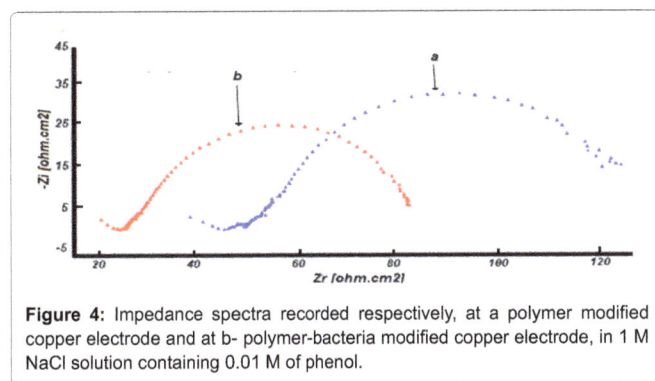

Figure 4: Impedance spectra recorded respectively, at a polymer modified copper electrode and at b- polymer-bacteria modified copper electrode, in 1 M NaCl solution containing 0.01 M of phenol.

Electrode	R₁ (ohm.cm²)	R₂ (ohm.cm²)	Diameter (ohm.cm²)
Polymer-Copper	55.86	70.24	71.69
Polymer-bacteria-Copper	30	53.25	53.48

Table 1: Parameters of the electrochemical spectra of impedance.

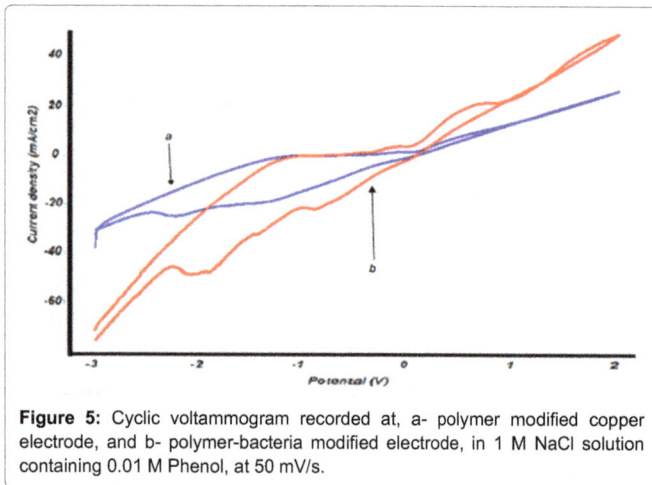

Figure 5: Cyclic voltammogram recorded at, a- polymer modified copper electrode, and b- polymer-bacteria modified electrode, in 1 M NaCl solution containing 0.01 M Phenol, at 50 mV/s.

order of 53.48 ohm.cm^2. Which shows that the electrode polymer-bacteria modified copper electrode is more active for the oxidation of phenol that polymer modified copper.

Figure 5 illustrates the CV's registered respectively at (curve a) and polymer-bacteria modified copper electrode, in electrolyte medium containing 0.01 M phenol. We note that, the oxidation current densities are very important in the case of the electrode modified by bacteria, confirming the previous results; bacteria catalyze the oxidation of phenol connection witch facilitate the adsorption of the molecule of phenol.

Conclusion

In this study, we experimentally prepared two electrodes, polymer modified copper electrode and polymer-bacteria modified copper electrode. The presence of bacteria onto polymer matrix improves sensitivity towards phenol and provides a rapid response.

Oxidation of phenol was invested by cyclic voltammetry and EIS. EIS diagrams resulted in separate time constants; the oxidation of phenol is mostly represented by half a circle, whose diameter corresponds to the electron transfer resistance. Electron transfer resistance produced by polymer-bacteria modified copper electrode is less than that obtained by polymer modified copper electrode. The presence of bacteria in the polymer matrix increases the activity of the electrode and catalyzes the oxidation of phenol.

References

1. Keith LH, Telliand WA (1979) ES & T special priority pollutants. I A perspective view. Environmental Science Technology 13: 416-423.

2. Zhang F, Li M, Li WQ, Feng CP, Guo YX (2011) Degradation of phenol by a combined independent photocatalytic and electrochemical process. Chemical Engineering Journal 175: 349-355.

3. Levévre S, Boutin O, Ferrasse JH, Miand I (2011) Thermodynamic and kinetic study of phenol degradation by a non-catalytic wet air oxidation process. Chemosphere 84: 1208-1215.

4. Su J, Lin H, Wang Q, Xie Z, Chen Z (2011) Adsorption of phenol from aqueous solutions by organo montmorillonite. Desalination 269: 163-169.

5. Manojlovic D, Ostojic DR, Obradovic BM, Kuraica MM, Krsmanovic VD, et al. (2007) Removal of phenol and chlorophenols from water by new ozone generator. Desalination 213: 116-122.

6. Xu X, Yi Z, Chen D, Duan X, Zhou Z, et al. (2012) Evaluation of photocatalytic production of active oxygen and decomposition of phenol in ZnO suspensions. Rare metals 28: 347-348.

7. Ahmad SA, Shamaan NA, Arif NM, Koon GB, Shukor MYA, et al. (2012) Enhanced phenol degradation by immobilized Acinetobacter sp. Strain AQ5NOL 1. World J Microbiol Biotechnol 28: 347-352.

8. Joshi MJ, Shambaugh RL (1982) The kinetics of ozone phenol reaction in aqueous solution. Water Res 16: 933-938.

9. Zhu X, Trong M, Shi S, Zhao H, Ni J (2008) Essential explanation of the strong mineralization performance of boron doped diamond electrodes. Environ Sci Technol 42: 4914-4920.

10. Liu HI, Liu Y, Zhang C, Chen RS (2008) Electrocatalytic oxidation of nitrophenols in aqueous solution using modified PbO$_2$ electrodes. J Appl Electrochem 38: 101-108.

11. Venkata M, Veer R, Dinakar P, Sarma P (2009) Integrated function of microbial fuel cell (MFC) as bio-electrochemical treatment systemassociated with bioelectricity generation under higher substrate load. Biosens Bioelectrons 24: 2021-2027.

12. Gunawardena G, Hille G, Montenegro I, Scharifker B (1982) Nucleation and growth of Cu onto polycrystalline Pt electrode. J Electroanal Chem 138: 225-228.

13. Asavapiriyanont S, Chandler G, Gunawardena G, Pletcher D (1984) The electrodeposition of poly-N-methylpyrrole films from aqueous solutions. J Electroanal Chem 177: 245-251.

8

Obtainment of Sorbitol on Ferroalloy Promoted Nickel Catalysts

Kedelbaev BS*, Korazbekova KU and Kudasova DE

M. Auezov South Kazakhstan State University, Kazakhstan

Abstract

The systematic study of the activity of the catalyst suspended with ferroalloys additives in catalytic hydrogenation of glucose over a wide variation of process parameters is given in this article. Since nickel catalysts were studied sufficiently, we limited to the data of the phase composition and structure; specific surface of alloys and catalysts based on aluminum-nickel alloys, modified ferroalloys. Results of the study of phase, chemical, particle size distribution and structure of nickel alloys and catalysts have shown that modifying metals affect to the ratio of $NiAl_3/Ni_2Al_3$ in the alloys, crush crystals, increase catalyst particle size, surface area and large size pore volume and simultaneously increase the micro- and supermicro pore ratio. Highly active, stable and selective catalysts based on nickel for the sorbitol synthesis process was developed.

Keywords: Catalysts; Nickel; Autoclave; Hydrogenation; Glucose; Sorbitol; Physico-chemical properties; Ferroalloys; Kinetics

Introduction

Liquid-phase hydrogenation of unsaturated compounds is a complex process consisting of several successive stages: transport of reactants to the catalyst surface and their subsequent adsorption, catalytic conversion on the surface, and finally desorption of the reaction products from the catalyst surface. The most complex of these is the adsorption stage and acts on the reaction surface, having a chemical nature. At the same time it is impossible to calculate the rate constants of all these stages of the process, so it is assumed that the overall reaction rate should be determined by the speed of the slowest (limiting) of these stages.

As known, hydrogenation of the unsaturated compound can also flow through one or other mechanism depending on the nature of the catalyst, solvent and reaction conditions [1,2]. In this connection, we studied the effect of glucose concentration of the solution, the hydrogen pressure and temperature on the rate of hydrogenation of glucose.

It is known that the initial concentration of the hydrogenated compound in the reaction medium is one of the key factors for ensuring optimal process in the kinetic region. The hydrogenation rate is continuously increased with increasing glucose concentration up to 30% in glucose hydrogenation in the presence of skeletal nickel catalyst was found by Bizhanov [3]. A further increase in the concentration of glucose has practically no effect on the rate of reaction.

Experimental Part

The objects of the study were the following multicomponent alloys and catalysts: skeletal aluminum-nickel catalyst with modifying additives of ferrosilicon (FSi), ferromanganese (FMn) and ferrosiliconmanganese (FSiMn).

In this paper we studied the catalytic properties of alloy copper catalysts, modified with not pure metals, only with ferroalloys. The following ferroalloys were used as additives: ferromanganese (FMn), ferromolybdenum (FMO) and ferromanganesemolybdenum (FMnMo). The alloys were prepared in a high-melting furnace. The quartz crucible was placed a calculated amount of Al in the form of ingots and it was gradually heated up to 1000-1100°C, and then the necessary amount of Ni and ferroalloy additive in the form of chips or powder was added.

As a result of the exothermic reaction, the melting temperature was raised to 1700-1800°C, stirring by the induction field was lasted 3-5 min. The alloy was air-cooled in the graphite molds and was crushed to grains of 0.25 mm. To activate the alloy, 1.0 g of alloy was leached with 20% sodium hydroxide solution (in volume 40 cm³) in a boiling water bath for 1 h, after which the catalyst was washed from alkali with water until neutral reaction to phenolphthalein.

The study of kinetics was carried out in a modified batch reactor system (capacity of 0.5 l). The device was equipped with a hermetic drive capacity of 0.6 kW, the number of turnovers of the agitator was 2800 rev/min. Complete analysis of the reaction products was consisted of determining the reducing sugars by Machen Shoorlya method and polyhydric alcohols by paper chromatography. The experimental results of hydrogenation of different glucose batches on the ferroalloys modified nickel catalysts under a wide variation of process parameters are given in Table 1. It is seen that with increasing the concentrations of an aqueous glucose solution from 5 to 20%, sorbitol yield was decreased due to surface blocking by molecules of hydrogenated substances, and the reaction rate was remained constant or gradually increased. Independence of reaction rate on the concentration of glucose on less active skeletal nickel catalyst is maintained in areas of 80-100°C and 8 MPa and 120°C and 8 MPa and 60-120°C and 6-8 MPa and 100°C and 6 MPa and 40-80⁰C and 4-6 MPa and a gradual increase was observed for the remaining conditions. The rate constancy in change of the glucose concentration indicates zero order in unsaturated compounds, i.e., in these conditions, the hydrogenation is carried out at full saturation of catalyst surface by molecules of the starting material. Increasing the reaction rate with increasing glucose concentration in a relatively high temperature and hydrogen pressure in investigated catalysts indicates fractional reaction hydrogenated substance. The latter circumstance is due to apparently lack of the unsaturated compound on the surface

---

*Corresponding author: Kedelbaev BS, M. Auezov South Kazakhstan State University, Kazakhstan, E-mail: ortaev@mail.ru

Top, °C	P_{H2}, MPa	C_{main}, %	Sorbitol yield (%) in time (min)		W□10^4 mol/g$_{ct}$□min	$n_{кс}$
			20	60		
1	2	3	4	5	6	7
Ni-FMn						
80	8	5	70.1	86.9	24.8	
		10	44.0	67.9	25.5	0.0
		20	25.2	39.3	26.0	
100	10	5	74.0	89.1	35.1	
		10	55.9	75.2	38.9	0.3
		20	33.3	58.3	42.6	
120	8	5	92.3	97.3	37.2	
		10	60.7	77.1	36.8	0.0
		20	30.9	52.7	35.8	
Ni-FMnMo						
60	8	5	60.7	80.9	22.0	
		10	35.4	61.3	21.8	0.0
		20	18.5	26.0	21.9	
100	8	5	81.1	96.2	32.4	
		10	57.6	74.5	34.4	0.2
		20	30.5	43.7	37.0	
100	6	5	74.9	96.2	31.8	
		10	58.3	73.8	31.7	0.0
		20	25.3	41.3	31.9	
Ni-FMo						
40	6	5	57.4	78.8	20.4	
		10	34.0	51.3	20.3	0.0
		20	16.3	25.0	20.1	
80	6	5	83.7	97.0	34.8	
		10	52.9	71.4	35.1	0.0
		20	22.7	34.4	34.8	
80	12	5	85.1	98.1	35.6	
		10	63.8	83.8	39.7	0.3
		20	36.1	51.7	43.8	

Note: Conditions; 1.0 g of catalyst (ct).

Table 1: Effect of concentration of the glucose aqueous solution on the rate of hydrogenation.

resulting in a high process speed with relatively harsh conditions. Bizhanov studied in detail the effect of hydrogen pressure on the kinetics and mechanism of hydrogenation of sugars in the presence of nickel catalysts. It was noted, that the rate of hydrogenation is proportionally increases to a certain limit with increasing hydrogen pressure. The limiting pressure depends on the nature of the hydrogenated compound, solvent, catalyst, and temperature of the experiment. Reaction order with respect to hydrogen is changed from the first to zero, and with respect to hydrogenated substance from zero to fractional depending on the process conditions.

However, the kinetics of the hydrogenation of glucose in the presence of promoted ferroalloys of skeletal nickel catalyst was not investigated. In this connection it is of great interest to track the effect of the simultaneous change of hydrogen pressure and temperature of the experiment on glucose kinetics on promoted nickel catalysts.

The results of our research on the hydrogenation of glucose on nickel catalysts with ferroalloys additives showed that with increasing hydrogen pressure from 2 to 10 MPa and experimental temperatures in the range of 40-120°C, hydrogenation rate of glucose is increased from 5.4 to 54.7 mol/kg min. The most beneficial effect of hydrogen pressure is affected the activity of relatively passive contacts at low temperatures. Raising the test temperature in the range 40-120°C and in 2 and 10 MPa leads to a sharp increase in the process rate. The highest value of the temperature coefficient is occurred in the least active catalysts under

relatively low hydrogen pressures [4,5].

Experiments on the effect of the concentration of glucose and hydrogen show that the order of the reaction of hydrogenated substance is varied from zero to fractional and the order of hydrogen is fractional. Effect of hydrogen pressure on the kinetics and mechanism of monosaccharide hydrogenation in the presence of nickel catalysts was studied in detail by Bizhanov. He showed that the rate of hydrogenation is proportionally increased to a certain limit with increasing hydrogen pressure. The limiting pressure depends on the nature of the compound to be hydrogenated, solvent, catalyst, and the experimental temperature. Reaction order with respect to hydrogen is changed from the first to zero, and with respect to the hydrogenated substance from zero to fractional depending on the conditions of the process [6].

Modifying influence of ferroalloys consistent with the study of physical-chemical and adsorption properties of the starting alloys and catalysts and it is explained due to the formation of new and additional phases and changes in the available phases. Results of the effect of imposed additives on the activity of the multi-component nickel catalyst at a hydrogen pressure of 2-10 MPa are shown in Table 2. The results of our numerous studies carried out earlier [7] showed that the ferroalloys are semi-metallurgical plants, and they may be used as modifiers to increase the activity of alloyed nickel catalysts of xylose hydrogenation.

| Top. °C | P_{H2}, MPa | Xylitol yield (%) in time (min) | | $W \cdot 10^4$ mol/g$_{ct}$·min | n_{H2} | n_{main} |
		20	60			
1	2	3	4	5	6	7
			Ni – FMn			
40	2	4.9	8.5	13.0	1.0	0.0
	6	11.6	20.0	15.9	1.0	0.0
	10	15.0	26.5	17.8	1.0	0.0
80	2	11.1	16.7	20.8	0.9	0.0
	6	21.4	37.9	24.9	0.9	0.0
	10	31.6	53.7	25.8	0.9	0.0
120	2	23.7	31.1	40.9	0.7	0.0
	6	34.4	55.3	48.6	0.7	0.0
	10	48.4	81.5	54.7	0.7	0.3
			Ni-FMo			
40	2	5.5	9.3	5.4	1.0	0.0
	6	10.9	21.3	8.8	1.0	0.0
	10	20.8	33.0	11.7	1.0	0.0
80	2	9.1	18.9	8.8	0.9	0.0
	6	22.9	35.7	16.0	0.9	0.0
	10	30.3	53.2	20.2	0.9	0.0
120	2	15.1	28.6	18.8	0.7	0.0
	6	35.0	56.0	29.5	0.7	0.0
	10	49.4	84.8	9.8	0.3	
Note: Conditions: 200 cm³ of 15% glucose solution 1.0 g catalyst.						

Table 2: Influence of hydrogen pressure and experimental temperature on the rate of glucose hydrogenation.

This work is a continuation of previous research and devoted to the study of physical and chemical properties of the most active nickel alloys and catalysts modified by ferromanganese, ferromolybdenum and ferromanganese molybdenum (FMn, FMo and FMnMo).

Therefore, we studied the effect of ferroalloys additives on the phase composition, the pore structure and specific surface area of nickel catalysts. It is known that the skeletal nickel Ni- (50% Al) alloy, which consists essentially of aluminides NiAl$_3$, Ni$_2$Al$_3$ is used in the industry for the monosaccharide hydrogenation. Introduction of various additives of metals to Ni-Al alloys is the most effective way to obtain high-modified skeletal nickel catalysts having a high activity, selectivity and stability in the hydrogenation process [2,3]. Introduction of modifying metals to the nickel alloys generates new additional active centers, in particular aluminides changes phase composition of leached alloy, i.e., skeletal catalysts. The majority of the alloying metal (Ti, Mo, Ta, Zr, Fe) in the leaching of the starting alloys are oxidized to form oxides [4]. Studies have shown that in the received catalyst is presented gamma oxides of these metals in various degrees of valency in the stages of the adsorption and activation of reagents. These metal oxides represent refractory phase, which are located at the mouths of the pores and at the boundaries between the grains of a catalyst, which is reflected on its catalytic activity. Despite the fact that the last time a sufficient quantity of studies on the catalytic activity in the recovery of monosaccharide of ferroalloys modified nickel catalysts is presented, but there are insufficient attention is paid to the correlation of activity with physico-chemical characteristics.

It should be also noted that the literature has not matter about influence of ferroalloys additives on the physico-chemical properties of alloyed aluminum-nickel catalyst. Therefore we investigated the influence of FMn, FMo and FMnMo on the phase composition and structure of aluminum-nickel alloys and catalysts.

The results are shown in Table 3, which shows that the modifying metals have a significant impact on the qualitative and quantitative composition and structure of the starting alloys and catalysts. The test additives of metals form except conventional for Ni-Al (50-50) alloy phases – NiAl$_3$, Ni$_2$Al$_3$ and eutectic (NiAl$_3$ + Al), a new Fx phase, which is not yet deciphered.

Square of NiAl$_3$ and Ni$_2$Al$_3$ phases are fluctuated within 36-52%, and 29-40%, and it is decreased with increasing concentrations of metals in alloys. The content of the eutectic mixture and Fx is advantageously increased to 18-20% and 12-15% respectively with increasing amounts of additives in the alloys. Ratio of NiAl$_3$/Ni$_2$Al$_3$ in the promoted alloys is more (1.28-1.37) than in the Ni-Al (50-50) alloy without additive (1.25); decreased with increasing concentration of alloying metals or increased from ferromanganese containing alloys to ferromanganesemolybdenum containing.

The studies show that the catalysts are consisted of a skeletal nickel, -Al$_2$O$_3$, Ni$_2$Al$_3$ and Fx. Modifying additives do not affect the crystal lattice parameter of nickel, but it significantly pulverize its crystals (from 5.4 to 3.4 nm); increase the surface area of the catalyst within 100-112.5 m²/g.

We studied the porous skeletal structure of nickel catalysts with ferroalloys additives. Argon adsorption isotherms show that the shape of hysteretic loops for most modified nickel catalysts are characterized by a parallel arrangement of the adsorption and desorption branches in the middle region of the relative pressures and belong to the A-type according to De Boer classification, that suggests the predominance of cylindrical pores. The maximum of the pore distribution is not allocated, but you can see that they are in a close area.

Table 4 shows the parameters of a porous structure of skeletal nickel (50% Al) catalysts with additions of ferroalloys. From the data of Table 4, it is shown that the modifier metal generally increases S$_{BET}$, respectively S$_{CUM}$ to 110-130.5 and 85-98 m²/g; pore volume to 1.14-1.38 times;

Modifying additives	Alloys					Catalysts		
	Phase square, %			F$_x$	$\dfrac{NiAl_3}{Ni_2Al_3}$	Crystal lattice parameter (a), nm	Crystal size (L), nm	Specific surface area (S), m²/g
	NiAl$_3$	Ni$_2$Al$_3$	eutectic Al + NiAl$_3$					
Ni-Al = 50-50								
-	50	40	10	-	1.25	0.353	5.4	15
Ni-50% Al-FMn								
3-10.0	50	39	7	3	1.28	0.353	4.7	110
Ni-50% Al-FMo								
3-10.0	48	44	12	6	1.33	0.353	4.6	130
Ni-50% Al-FMnMo								
3-10.0	45	33	11	10	1.36	0.353	3.4	112.5

Table 3: Physicochemical properties of modified nickel alloys and catalysts.

Catalyst	S$_{BET}$*, m²/r	S$_{CUM}$**, m²/r	$\dfrac{S_{BET} - S_{KUM}}{S_{BET}} 100\%$	V pore, cm³/r	R$_{EFF}$***, A	Type of isotherm
Ni (50% Al)	105	75	28.5	0.105	30	A
Ni-3-10% ФMn	110	85	22.7	0.120	34	A
Ni-3-10% ФMo	130.5	98	24.9	0.138	36	A
Ni-3-10%ФMnMo	112.5	86	23.5	0.145	37	A

Note: *Bayer- Emmett-Teylor, **Cumulative, ***Effective.

Table 4: Parameters of the porous structure of skeletal aluminum-nickel (50% Al) catalysts with additions of ferroalloys.

effective pore radius R$_{EFF}$, to 1.13-1.23 times. The simultaneous increase in the specific surface area and pore volume at a relatively high effective radius is occurred apparently due to the nickel phase dispersion of catalyst of modifying metal.

Conclusion

Thus, the introduction of ferroalloys additives to Ni-50% Al alloy significantly effects on the phase structure, the porous structure and the specific surface area of the skeletal nickel catalyst, which ultimately affects the catalytic activity in the reduction of monosaccharide. Results of physico-chemical studies outlined in this article can be successfully used in predicting the catalytic activity of alloyed nickel catalysts modified by ferroalloys in hydrogenation reactions of various unsaturated compounds in nature. The experimental data are corresponds to the certain data by others authors, working in the field of preparation and testing of alloyed catalysts not only based on nickel, but also on copper and cobalt.

References

1. Paquette LA, Crich D, Fuchs PL, Molander GA (2009) Encyclopedia of reagents for organic synthesis. John Wiley, New York.

2. Solomons TWG, Fryhle CB (2004) Organic Chemistry. (8thedn). Wiley International, USA.

3. Bizhanov FB (1977) Method of producing polyhydric alcohols. United States Patent 4018835.

4. Turabdzhanov SM, Kedelbaev B, Tashkaraev RA (2013) Hydrogenation of benzene on nickel catalysts promoted by ferro-alloys. Theoretical Foundations of Chemical Enqineering 47: 633-636.

5. Turtabaev SK, Kedelbaev BS, Shalabaeva GS, Sarbaeva KT (2015) Synthesis and research of the nickel catalysts of liquid-phase hydrogenation of benzene promoted by ferro-alloys. Contemporary Engineering Sciences 8: 127-135.

6. Aytmuhanbetov GB, Kuatbekov AM, Kedelbaev BS, Dauylbaev AD (2015) Development of the promoted floatable nickel catalysts of hydrogenation of benzene to cyclohexane. International Journal of Applied and Fundamental Research 2: 251-255.

7. Kedelbaev BS (2004) Production of xylitol on modified catalysts. Works of International Conference, Shymkent.

9

Polysulfone/Cellulose Acetate Butyrate Environmentally Friendly Blend to minimize the Impact of UV Radiation

Raouf RM[1,2]*, Wahab ZA[1], Ibrahim NA[3,4] and Talib ZA[1]

[1]Department Of Physics, Universiti Putra Malaysia, 43400 UPM Serdang, Selangor, Malaysia
[2]Materials Engineering Department, College of Engineering, Al-Mustansiriyah University, Baghdad, Iraq
[3]Department Of Chemistry, Faculty of Science, Universiti Putra Malaysia, 43400 UPM Serdang, Selangor, Malaysia
[4]Materials Processing and Technology Laboratory, Institute of Advanced Technology, Universiti Putra Malaysia, 43400 UPM Serdang, Selangor, Malaysia

Abstract

An eco-friendly transparent blend for ultraviolet and visible light rays from polysulfone (PSF) and cellulose acetate butyrate (CAB) was prepared by melting and re-molding. Some optical, mechanical, thermal and morphological properties of the blend were studied by means of UV-Vis spectroscopy, dynamic mechanical analysis (DMA), thermogravimetric analysis (TGA), scanning electron microscope (SEM) and X-ray diffraction (XRD). The UV-Vis spectroscopy results showed that the blend became more transparent in the ultraviolet region with an increase in CAB concentration, especially at the damage threshold (268 nm). The preferable blend sample contained 0.2% w/w CAB in the PSF/CAB blend, signifying low ultraviolet light absorption while preserving the transparency of the blend. The results also demonstrated that the amount of 0.2% CAB in PSF increased the modulus and thermal stability while decreasing the value of the glass transition temperature.

Keywords: Blend; Cellulose acetate butyrate (CAB); Polysulfone (PSF); Ultraviolet

Introduction

Polymer blend is a combination of two polymers or composites at least. The purpose of polymer blending is to obtain a new material with different physical properties to meet performance requirements that cannot be satisfied by the currently available polymer. There are two ways to blend polymers. One is by blending the compound component in the molten state; the other is to blend them in common solution [1]. Melting by using the twin screw extruder (used in this work) is the other method. The twin screw extruder is preferred for melt-blending polymers with added materials instead of using a single screw extruder because of several issues, including its extraordinary mixing ability, the high degree of process resilience, better control of process parameters, and higher process productivity. Therefore, mixing and melting two materials for the purpose of forming a homogeneous blend or composite will be easier and more accurate by using the twin screw extruder [2,3].

In recent years, much attention has been focused on environmentally friendly materials as a renewable source to reduce the amount of waste that accumulates on the Earth. Therefore, inserting renewable polymeric materials in the blend has become the focus of many academic studies and industrial undertakings [4-7]. Renewable polymeric materials carry a higher grade of complexity than artificial ones. This is the result of nature's long growth in terms of material design, which is of great importance to many applications. Natural polymers have numerous positive properties compared to artificial polymers that make them ideal applicants for different applications.

Conversely, there are numerous disadvantages of biodegradable polymers obtained from renewable sources, such as unacceptable mechanical properties especially under wet conditions, solubility in water, and the rapid degradation rate. Although there is no ideal polymeric blend or composite from renewable resources, the properties of polymer blends can notably improve by blending synthetic polymers with natural polymers [8,9].

Bio-based cellulose acetate butyrate (CAB) is a thermoplastic polymer, obtained from the esterification of acetyl and butyryl group.

This polymer has showed its significance in different applications in terms of transparency and weathering resistance when used as a coating to repair or re-finish, and especially for ultraviolet protection, in addition to the high usability of casting and molding. Therefore, CAB is used as biodegradable inhibitor with polymers. The biodegradation inhibitor has postulated that the change in the biodegradability of the blend with CAB was caused by the new structural phase formed when CAB is added to the blend [10-13].

Polysulfone (PSF) is also a thermoplastic polymer like cellulose acetate butyrate (CAB). It is well known for its transparency, toughness, high strength and high thermal stability. In some applications, it is used as a substitute for polycarbonate because of its unique properties, but its low resistance to a number of solvents and rapid weathering restricts its applications [14,15].

Studies have confirmed that the microstructure and the performance of PSF are affected by ultraviolet irradiation within a very short time of exposure, leading to simultaneous chain scission and crosslinking [16]. PSF photooxidation in an oxygen atmosphere leads to an oxidized surface with an increase in the relative amount of sulphur on the surface. This change is attributed to rapid oxidation at carbon sites in the polymer [17]. At temperatures below 140°C, long heating times cause little changes in the PSF properties, but exposure to ultraviolet light results in greater changes, as indicated by broad absorption in both the carbonyl and hydroxyl areas [18].

The purpose of this work is to make the PSF/CAB blend more

*Corresponding author: Raouf RM, Faculty of Science, Department Of Physics, Universiti Putra Malaysia, 43400 UPM Serdang, Selangor, Malaysia
E-mail: raoufmahmood@yahoo.com

transparent in the UV region in order to reduce the damage that is caused from trapped radiation inside the material, and to withstand weather conditions.

Experimental

Materials

Transparent pellets of polysulfone (PSF; average M_w ~35,000 by LS, average M_n ~16,000 by MO) and white powder cellulose acetate butyrate (CAB; average M_n ~12,000) were supplied by Sigma-Aldrich(USA).

Methods

The samples were prepared using twin screw (three zones, electric heating and air cooling) Thermo-Haake Poly Drive Internal Mixer (Germany) (D=19.05 mm) and Hsin-Chi Machinery Co. Ltd. hot press (Taiwan). The materials were dried in a vacuum oven at 50°C for 4 h before mixing. The CAB/PSF blends were prepared by adding a fixed weight of PSF to different weight ratio of CAB. Firstly, the polymer pellets was melt-kneaded in the extruder at a rotation rate of 50 rpm at 225°C for 10 min. Then, variable percentage weights of CAB (0.1%, 0.2%, 0.3% and 0.4%) were added to the molten of PSF. Mixing continued until reached to constant torque, which took about 10-15 min. The samples were transparent and homogeneous in their outward appearance. After that, each blended sample was pressed in a hot press at 110 KPa and 130°C to form a sheet 70 mm × 90 mm, 1 mm thick.

Measurements

The character of the transparency and absorbance of the samples was determined using ultraviolet-visible (UV-VIS) spectroscopy (Shimadzu UV-3600 spectrophotometer, Japan). Surface analysis was obtained out by scanning electron microscopy (SEM) studies on a Hitachi S-3400N (Japan) microscope. Sample morphology and crystallinity identification were determined by XRD analysis using a X-ray diffractometer (Philips/X'Pert Pro Panalytical - PW 3040/60 MPD, Netherlands). The diffractometer data were obtained from 2Θ=20°C to 80° at a scanning speed of 5°C/min. Dynamic mechanical analysis (DMA) was made on a PerkinElmer Pyris Diamond (USA) apparatus in tension mode at a frequency of 1 Hz and a heating rate of 5°C/min in a liquid nitrogen atmosphere. Thermal gravimetric analysis (TGA) was undertaken on a TGA/DSC1 STAR System (USA) at a heating rate of 20°C/min from room temperature to 1000°C in a continuous highly pure nitrogen atmosphere.

Characterization

Complete scans over the ultraviolet and visible spectra were made from 220 nm to 800 nm for highly transparent pure PSF samples using a Shimadzu UV-3600 spectrophotometer. The results show that the absorbance peak in the ultraviolet range for pure PSF was at 268 nm, which represents the UV damage peak for the polymers [19-21], whilst the transmittance peak in the visible range for pure PSF was 712.5 nm. The transmittance peak indicates the real effectiveness for pure PSF [21]. The absorbance and transmittance peak values were adopted for all subsequent measurements on the blended samples under study, as well as the sample transparency.

Results and Discussion

UV-VIS spectroscopy

The absorbance and transmittance curves, in addition to the spectra and the visual appearance of the PSF and PSF-CAB blends are shown in Figure 1.

The CAB concentration in PSF was 0.1, 0.2, 0.3 and 0.4%. Figure 1a shows the relationship between absorbance and concentration for PSF at 268 nm. It is clear that an increase in the CAB concentration in PSF reduced the optical absorbance in the ultraviolet area, especially at the damage threshold (268 nm). The retention of ultraviolet radiation within PSF molecules leads to the breakdown of bonds in the polymeric chain and disintegration over time. Therefore, a change in physical properties will take place, which reduces the lifetime of utilization [18]. On the other hand, the optical transmittance in the visible region, especially at a wavelength of 712.5 nm, was relatively stable at the first two concentrations, then declined from 0.2% to 0.4%; this is evident from the foggy samples shown in Figure 1d. Therefore, 0.2% CAB was the best concentration of PSF in terms of the lack of absorption in the mid-ultraviolet region and the maintenance of optical transmittance with the required transparency.

In the PSF-CAB blends, it was noted that the samples were transparent down to 0.3%; after that, the samples started to become hazy with an increasing CAB concentration in PSF. Samples with concentrations more than 0.4% were discarded because they did not meet the required purpose.

Scanning electron microscopy

The SEM images taken for the internal morphological study of pure PSF and PSF/0.2% CAB blend are shown in Figure 2. The SEM images for each sample are shown at three different magnifications (2, 5, 10 μm). As can be seen in the pure PSF images, there were uniform morphological features, indicating a single material (Figures 2a-2c) [22].

The morphology of PSF/0.2% CAB blend shown in Figure 2d-2f. It is easy to see the size reduction of the CAB components and separation from the continuous phase of PSF. Figure 2f shows that the size of CAB sphere not more than (99.5 nm). In polymer blends, if the concentration of one components is small, it tend to present in spherical droplets dispersed all over the matrix and the polymer nature determined the size [23]. The clear difference in the size of the spherical shapes may be due to variation in the interaction between functional components that led to an increase or decrease the surface tension [24,25]. As a result foggy appeared as early as in low concentrations of CAB in PSF.

X-Ray diffraction

The X-ray diffraction pattern of pure PSF and PSF/0.2% CAB blend are shown in Figure 3. The patterns in Figures 3a and 3b show that there were no clear peaks. The broadened background scattering areas of pure PSF indicate their amorphous nature and non-crystalline structure [26-28]. So, change absence that observed in the pure polymer to the blends indicates that CAB did not change the essential random compositional structure of the original polymer. This corresponds with previous data indicating that the spherical clusters were no larger than 99.5 nm in PSF/0.2% CAB, i.e., are not crystalline regions.

Dynamic mechanical analysis

The dynamic mechanical analysis is a technique involves studying the properties of materials as they are deformed under periodic stress by applying a variable sinusoidal stress under thermal conditions. Most polymers are viscoelastic and exhibit a phase difference between applied stress and the resultant sinusoidal strain. This phase difference, together with the amplitudes of the stress and strain waves, is used to determine a variety of parameters, including storage modulus E' (the ability of the material to store potential energy), loss modulus E" (energy

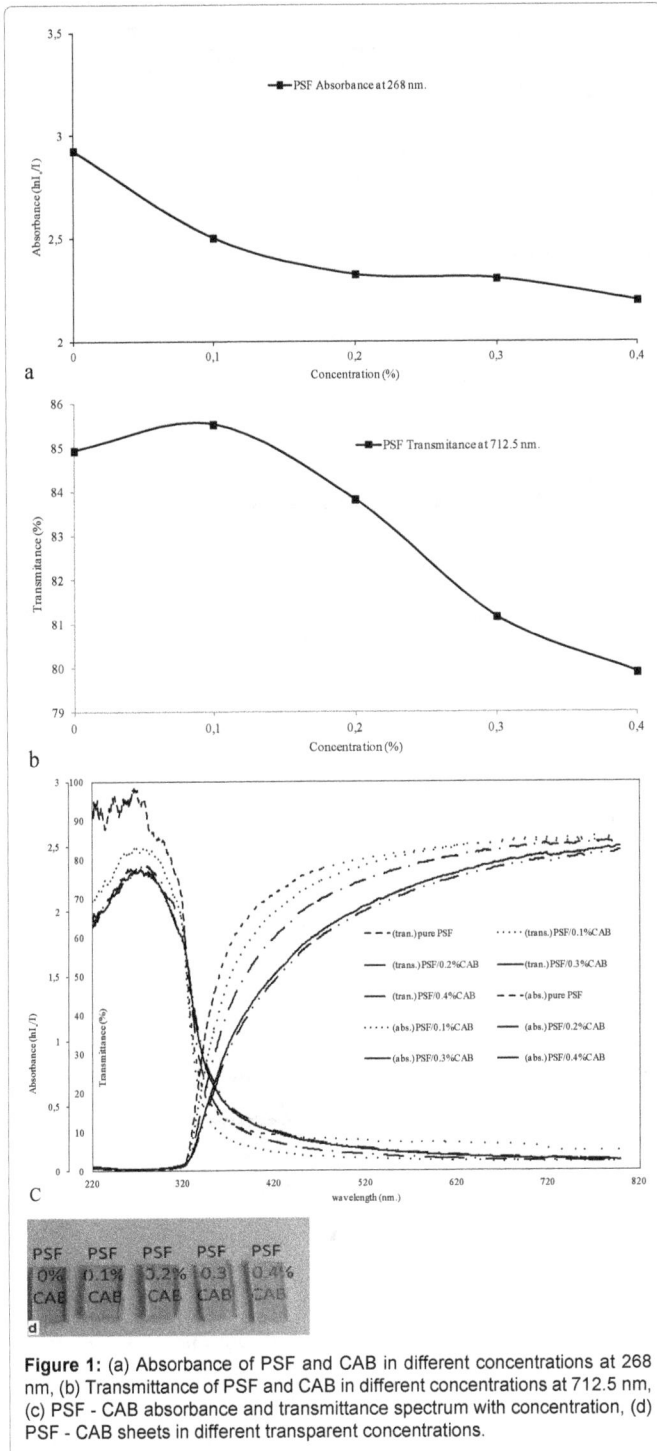

Figure 1: (a) Absorbance of PSF and CAB in different concentrations at 268 nm, (b) Transmittance of PSF and CAB in different concentrations at 712.5 nm, (c) PSF - CAB absorbance and transmittance spectrum with concentration, (d) PSF - CAB sheets in different transparent concentrations.

dissipation in the form of heat upon deformation) and phase angle tanδ (the mechanical damping or internal friction in a viscoelastic system). The dynamic modulus indicates the intrinsic stiffness of the material under dynamic loading conditions. Some polymeric blends are well-suited to this test due to their single phase. Nevertheless, most polymer blends form two phases because of incompatibility between the blend elements [29].

The curves of the storage modulus E', loss modulus E" and phase angle tanδ vs. temperature for pure PSF and PSF/0.2% CAB are shown in Figure 4a-4c. The storage modulus E' curve for PSF and the PSF/0.2% CAB blend showed three main regions, but in the glassy region there is a significant appearance of the three deformation phases (secondary dispersion at 82.6/23.9°C, γ-relaxation at 90.4/30-40°C and β-relaxation at 187/164°C) for pure PSF and the PSF/0.2% CAB blend, respectively. It was clear that adding CAB to PSF led to significant changes in molecular motion, so that two γ-relaxation processes were observed (30-40°C) because the substituent in the CAB component introduced phenylene rings [30]. At 187/164°C, the E' curve for pure PSF and PSF/0.2% CAB went into the transition region with a steep slope in order to meet with the tanδ curve at the glass transition temperature (T_g) region for each curve, after that went into the rubbery region. The melt temperature of PSF was 220°C, which has been observed in several experiments. The sulfone with two neighboring benzene rings in PSF contains a highly conjugated diphenyl structure, with significant rigidity of the molecular chains of pure PSF, which led to a high T_g at 177.3°C, with the peak of tanδ curve within the rubbery region [31].

The increase in the modulus value for the PSF/0.2% CAB blend indicated improved storage of potential energy but decrease in T_g value, so the addition of CAB increased stiffness and made the blend technologically compatible to some extent even though molecular level

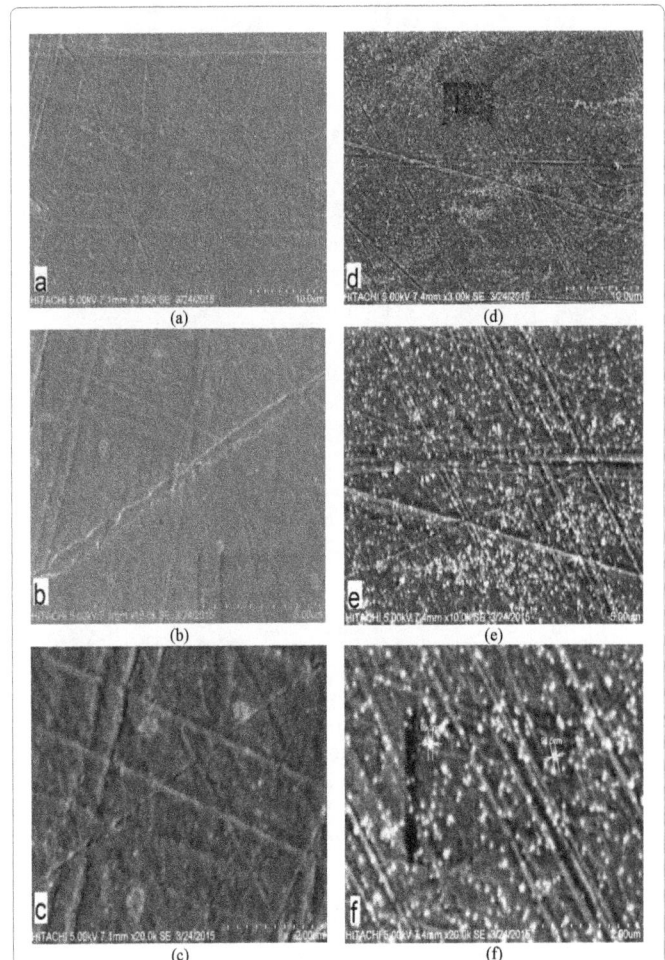

Figure 2: SEM images for PSF and PSF-CAB 0.2% blend with different scale bar (a) PSF/10 μm. (b) PSF/5 μm. (c) PSF/2 μm. (d) PSF-CAB0.2%/10 μm. (e) PSF-CAB0.2% /5 μm. (f) PSF-CAB0.2%/2 μm (particle measure).

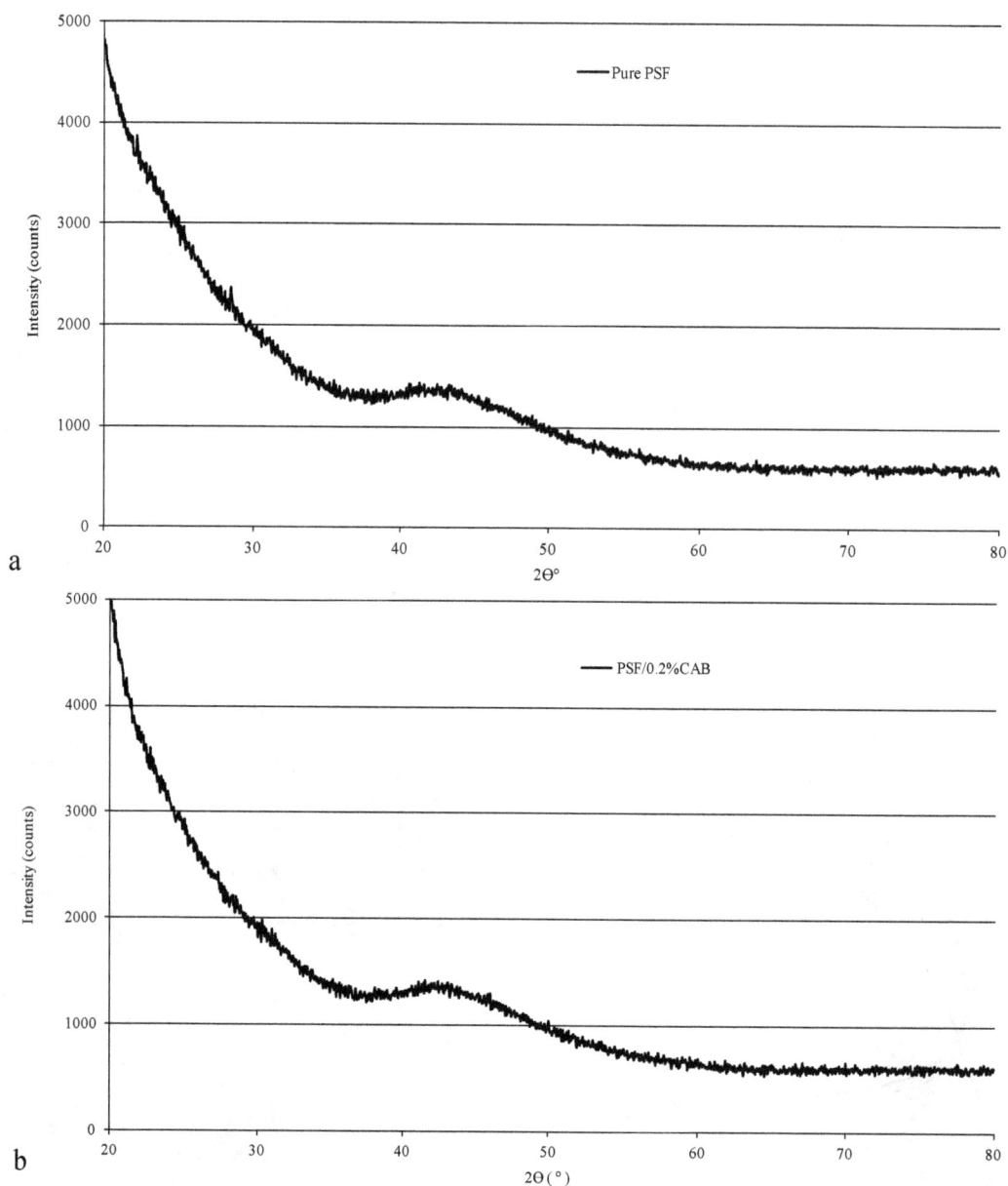

Figure 3: XRD pattern for; (a) pure PSF (b) PSF/0.2%CAB.

miscibility could not be achieved [28,32].

The position of a peak maximum in the tan δ versus temperature curve can be attributed to Tg. It was seen that the addition of 0.2% CAB shifted Tg of pure PSF towards a lower value by about 15°C Figure 4c. This result suggested that the presence of CAB molecules caused an increase of molecular mobility in PSF at high temperatures near to T_g due to decrease in van der Waals bonding forces between PSF and CAB chains [33].

Thermogravimetric analysis

The thermogravimetric analysis (TGA) and derivative thermogravimetry (DTG) diagrams for pure PSF and the PSF/0.2% CAB blend are shown in Figures 5a and 5b. All the DTG peaks represent degradation point for each stage. The total weight loss for pure PSF

amounted to ~27% from 89°C-245°C and for the PSF/0.2% CAB blend was ~5% from 70 to 430°C, due to the residual water bound to the hydrophilic imidazole moieties [33]. Pure PSF (Figure 5a), showed a lower degradation temperature at 389°C, because of the loss in sulfonic acid groups and this represents the most important thermal degradation in polymer chains. The second at 527.8°C as a result of carbonization of the degraded residuals (polymeric backbone degradation) [34,35]. The high degradation temperature at 527.46°C for PSF/0.2% CAB blend was attributed to decomposition of the imidazole group without phase inversion components [33,36].

It has been observed that there is thermal stability due to CAB addition therefore CAB contributed to the improvement of thermal stability of polymer although not change the degree of decomposition of the main chains temperature. The break in each thermogram

Polysulfone/Cellulose Acetate Butyrate Environmentally Friendly Blend to minimize the Impact...

39

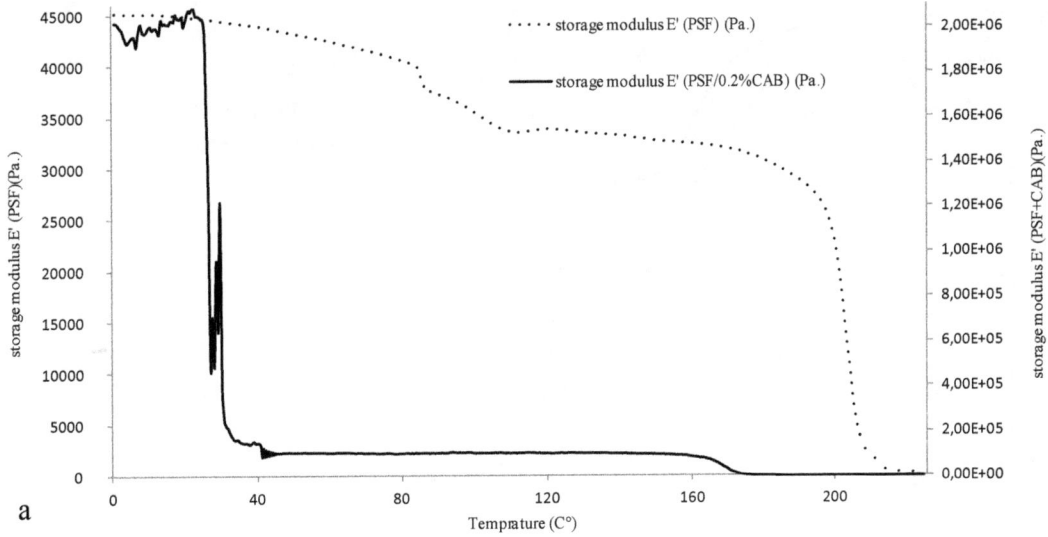

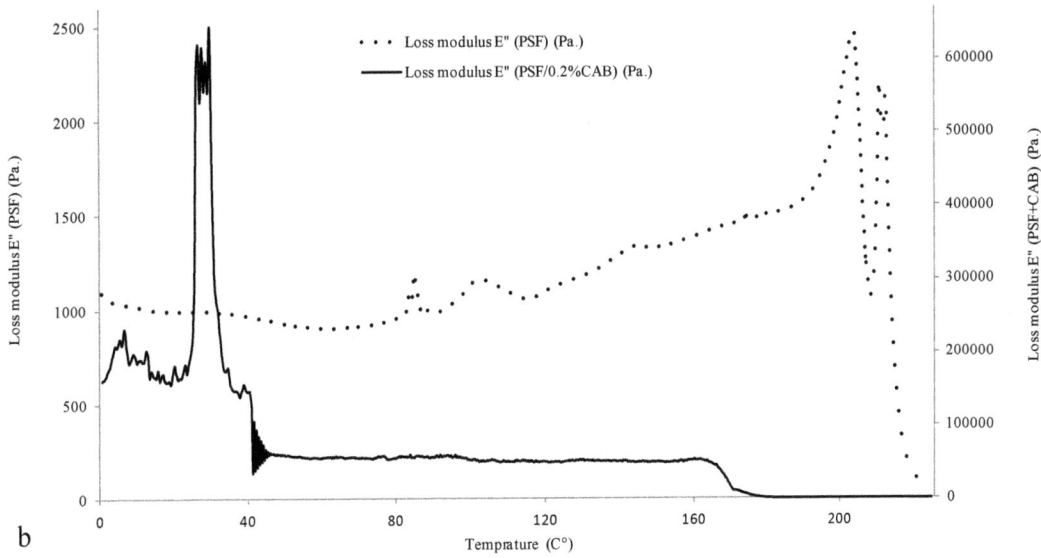

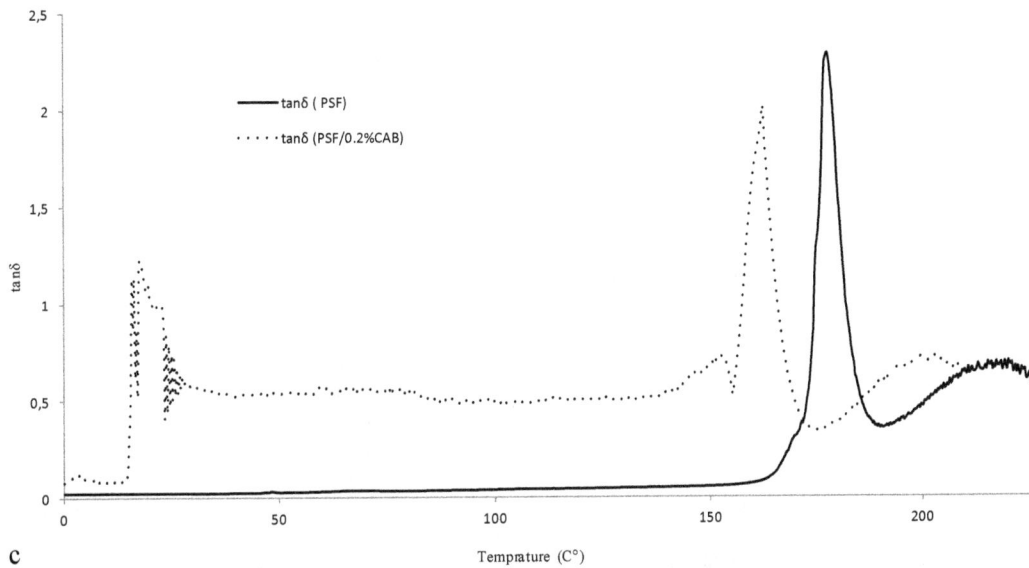

Figure 4: (a) Storage modulus E', (b) loss modulus E" and (c) tan δ trace for pure PSF and PSF-CAB 0.2% as a function of temperature by DMA.

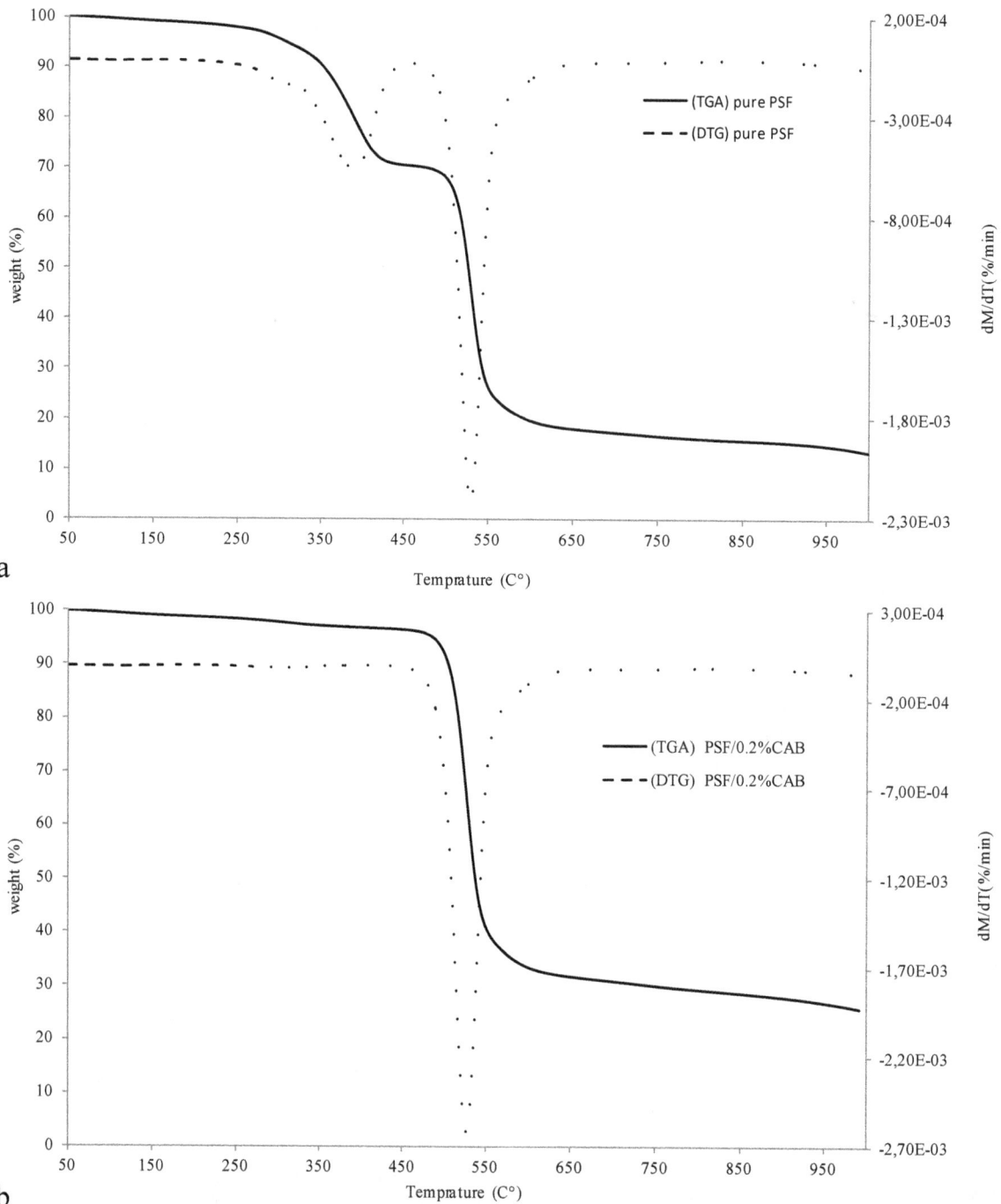

Figure 5: TGA- trace for (a) pure PSF (b) PSF/0.2% CAB.

indicates the onset of the decomposition process involving a rapid loss in weight [37].

Conclusions

In this paper, different concentrations of cellulose acetate butyrate (CAB) were mixed with polysulfone (PSF) using Internal Mixer for transparent blend of ultraviolet-visible area, and investigations of some optical, morphological, mechanical and thermal properties were carried out. The study revealed that PSF has absorbance peak at 268 nm and a transmittance peak at 712.5 nm. The value of these peaks decreases with increasing CAB concentration in PSF. The addition of CAB to PSF does not change the random structure of the blend, and this seems clear in the sample transparency. It was found that PSF/0.2% CAB blend representing low ultraviolet light absorption and high transparency, for this purpose, was chosen as the best concentration. The PSF/0.2% CAB blend showed an obvious thermal stability with an increase in the value of the modulus, but the value of the glass transition temperature decreased after adding CAB. The properties of the blends were a function of CAB concentration in PSF. Additionally; environmentally-friendly CAB improved the PSF properties while maintaining the transparency of the samples, at least to some extent.

References

1. Rauwendaal C (2014) Polymer extrusion. Carl Hanser Verlag GmbH Co KG, Germany.

2. Rauwendaal CJ (1981) Analysis and experimental evaluation of twin screw extruders. Polymer Engineering & Science 21: 1092-1100.

3. White JL (1991)Twin Screw Extrusion: Technology and Principles.Hanser Publishers, New York.

4. Armentano I, Fortunati E, Burgos N, Dominici F, Luzi F, et al. (2015) Processing and characterization of plasticized PLA/PHB blends for biodegradable multiphase systems. Express Polymer Letters 9: 583-596.

5. Imre B, Renner K, Pukánszky B (2014) Interactions, structure and properties in poly (lactic acid)/thermoplastic polymer blends. Express Polymer Letters 8: 2-14.

6. Chen Y, Yuan D, Xu C (2014) Dynamically vulcanized biobased polylactide/ natural rubber blend material with continuous cross-linked rubber phase. ACS applied materials & interfaces 6: 3811-3816.

7. Arrieta MP, Lopez J, Hernández A, Rayón E (2014) Ternary PLA-PHB-Limonene blends intended for biodegradable food packaging applications. European Polymer Journal 50: 255-270.

8. Yu L, Dean K, Li L (2006) Polymer blends and composites from renewable resources. Progress in polymer science 31: 576-602.

9. Yu L (2009) Biodegradable polymer blends and composites from renewable resources. John Wiley & Sons, USA.

10. Meyer LWA, Gearhart WM (1951) Ultraviolet Inhibitors for Cellulose Acetate-Butyrate Plastics. Phenyl Hydroxybenzoates, Hydroxyphenyl Benzoates and their Methyl Ethers. Industrial & Engineering Chemistry 43: 1585-1591.

11. Meyer WC (1979) Refinishing automobile and truck bodies, Google Patents.

12. Tachibana Y, Truong Gianga NT, Ninomiyaa F, Funabashia M, Kuniokaa M (2010) Cellulose acetate butyrate as multifunctional additive for poly (butylene succinate) by melt blending: Mechanical properties, biomass carbon ratio, and control of biodegradability. Polymer Degradation and Stability 95: 1406-1413.

13. Laskar J, Vidala F, Ficheta O, Gauthier C, Teyssié D (2004) Synthesis and characterization of interpenetrating networks from polycarbonate and cellulose acetate butyrate. Polymer 45: 5047-5055.

14. Noshay A, McGrath JE (2013) Block copolymers: Overview and critical survey. Elsevier Science, Burlington.

15. Lane SL, Lindstrom RL, Cameron JD, Thomas RH, Mindrup EA, et al. (1986) Polysulfone corneal lenses. Journal of Cataract & Refractive Surgery 12: 50-60.

16. Rupiasih NN, Suyanto H, Sumadiyasa M,Wendri N (2013) Study of Effects of Low Doses UV Radiation on Microporous Polysulfone Membranes in Sterilization Process. Open Journal of Organic Polymer Materials 3: 12.

17. Peeling J, Clark DT (1981) Photooxidation of the surfaces of polyphenylene oxide and polysulfone. Journal of Applied Polymer Science 26: 3761-3772.

18. Gesner B, Kelleher P(1968) Thermal and photo-oxidation of polysulfone. Journal of Applied Polymer Science 12: 1199-1208.

19. Abouelezz M, Waters PF (1978) Studies on the Photodegradation of Poly (methyl methacrylate) DTIC Document.

20. Charlesby A, Thomas DK (1962) Comparison of the Effects of Ultra-Violet and Gamma Radiation in Polymethylmethacrylate. Proceedings of the Royal Society of London. Mathematical and Physical Sciences 269: 104-124.

21. Michelson J, Werner L, Ollerton A, Leishman L, Bodnar Z (2012) Light scattering and light transmittance in intraocular lenses explanted because of optic opacification. Journal of Cataract & Refractive Surgery 38: 1476-1485.

22. Silverstein RM, Bassler GC, Morrill TC (1991) Spectrometric identification of organic compounds. John Wiley & Sons, USA.

23. Bower DI (2002) An introduction to polymer physics. Cambridge University Press, India.

24. Guo HF, Packirisamy S, Mani RS, Aronson CL, Gvozdic NV, et al. (1998) Compatibilizing effects of block copolymers in low-density polyethylene/ polystyrene blends. Polymer 39: 2495-2505.

25. Robeson L (2014) Historical Perspective of Advances in the Science and Technology of Polymer Blends. Polymers 6: 1251-1265.

26. Elizalde-Pena EA, Flores-Ramirezd N, Luna-Barcenasa G,Vásquez-Garcíab SR, Arámbula-Villaa G, et al. (2007) Synthesis and characterization of chitosan-g-glycidyl methacrylate with methyl methacrylate. European Polymer Journal 43: 3963-3969.

27. Silvestre C, Cimmino S, Martuscelli E (1987) Poly (ethylene oxide)/poly (methyl methacrylate) blends: influence of tacticity of poly (methyl methacrylate) on blend structure and miscibility. Polymer 28: 1190-1199.

28. Lufrano F, Squadrito G, Patti A, Passalacqua E (2000) Sulfonated polysulfone as promising membranes for polymer electrolyte fuel cells. Journal of Applied Polymer Science 77: 1250-1256.

29. Menard KP (2008) Dynamic mechanical analysis: A practical introduction. (2nd eds.) CRC press, USA.

30. Lee KJ, Jho JY, Kang YS, Won J, Dai Y, et al. (2003) Gas transport and dynamic mechanical behavior in modified polysulfones with trimethylsilyl groups: Effect of degree of substitution. Journal of membrane science 223: 1-10.

31. Zhang J, He J (2002) Interfacial compatibilization for PSF/TLCP blends by a modified polysulfone. Polymer 43: 1437-1446.

32. Merenga AS, Katana GA (2010) Katana Dynamic mechanical analysis of PMMA-cellulose blends. International Journal of Polymeric Materials and Polymeric Biomaterials 60: 115-123.

33. Said KA, Amiinu IS, Zhang H, Pan M (2014) Functionalized Polysulfones as an Alternative Material to Improve Proton Conductivity at Low Relative Humidity Fuel Cell Applications. Chemistry and Materials Research 6: 19-29.

34. Krishnaswamy RK, Wadud SEB, Baird DG (1999) Influence of a reactive terpolymer on the properties of in situ composites based on polyamides and thermotropic liquid crystalline polyesters. Polymer 40: 701-716.

35. Momeni S, Pakizeh M (2013) Preparation, characterization and gas permeation study of PSf/MgO nanocomposite membrane. Brazilian Journal of Chemical Engineering 30: 589-597.

36. Subrahmanyan S (2003) An Investigation of Pore Collapse in Asymmetric Polysulfone Membranes. Virginia Polytechnic Institute and State University.

37. Tosh B (2011) Thermal analysis of cellulose esters prepared from different molecular weight fractions of high a-cellulose pulp. Indian Journal of Chemical Technology 18: 451-457.

10

Mechanical Behavior of Long Carbon Fiber Reinforced Polyarylamide at Elevated Temperature

Wang Q[1], Ning H[1]*, Vaidya U[2] and Pillay S[1]

[1]*Department of Materials Science and Engineering, University of Alabama at Birmingham, Birmingham, USA*
[2]*Department of Mechanical, Aerospace and Biomedical Engineering, University of Tennessee, Knoxville, USA*

Abstract

Long fiber reinforced thermoplastic (LFT) composites have recently found increasing use in transportation, military and aerospace applications and become well established as high volume and low cost materials with high specific modulus and strength, superior damage tolerance, and excellent fracture toughness. This study is conducted to evaluate the performance of long fiber reinforced thermoplastic composite at elevated high temperature. Long carbon fiber reinforced polyarylamide (CF/PAA) composites containing 20 wt% and 30 wt% carbon fibers are used and processed using extrusion compression molding. Flexural and tensile samples are tested at three temperatures, room temperature, medium temperature (MD 65°C) and glass transition temperature (TG 80°C). Samples in both longitudinal and transverse directions are prepared to show the effect of the orientation on mechanical properties at different temperatures. The testing results show that as temperature increases, both of the flexural and tensile properties of the CF/PAA decrease as expected. Both of the flexural and tensile modulus reduce more dramatically than the flexural and tensile strength, indicating that the temperature has more pronounced effect on modulus than strength. The transversely oriented samples generally show larger reduction in properties than the longitudinally oriented samples. Temperature significantly affects flexural strength at the elevated temperature section between MD and TG temperature.

Keywords: Long carbon fibers; Thermoplastic; Polyarylamide; PA-MXD6; Mechanical properties; Elevated temperature

Introduction

Long fiber reinforced thermoplastic (LFT) is a class of composite material comprised of reinforcement fibers (carbon and glass fibers etc.) and thermoplastic polymers such as polypropylene (PP), polyamide (PA), and polyphenylene sulfide (PPS), etc. The reinforcement fibers are typically 5-25 mm in length with resultant high fiber aspect ratio, compared to 0.5-1.0 mm in short fiber reinforced thermoplastic (SFT). The long fibers in LFT composites provide several property advantages, such as high impact strength, improved modulus and better dimensional stability, over SFT composites. LFT is manufactured by pulling continuous fiber tows through a thermoplastic polymer melt in a processing die. The ratio of fiber to resin is controlled by a metering orifice. The resulting rods are cut into pellets, 5-25 mm in length, which can be injection molded or compression molded to form a part. LFT composites are now used in numerous high volume commercial applications in transportation, military, sporting goods and aerospace due to their lightweight characteristics, high specific modulus and strength, ease of processability, and recyclability [1-3].

In spite of their general use at ambient temperature, LFTs have also found their use in applications at elevated temperature. Glass fiber LFT (GF/nylon 66) has been used as a tailcone for an XM-1002 training round [4,5]. The LFT composite tailcone made of 40 wt% glass fiber reinforced nylon 66 was manufactured to replace aluminum counterpart using compression molding process. Skin temperature of the tailcone can reach up to 270°C for over 5 seconds during its flight. The flexural creep behaviors of GF/nylon 66 LFT, glass fiber reinforced polypropylene LFT, and glass fiber reinforced high density polyethylene LFT at elevated temperatures have been studied [6]. It is found that the long fiber composites exhibit non-linear viscoelastic response at moderate to high applied stresses and the addition of the long reinforcement fibers enhance the creep resistance [6]. Glass fiber reinforced PP LFT has also been used in the automotive industries as

underbody panel and dashboard [3,7] where temperature can rise up to over 60°C in hot summer days. Thin-walled shell made of CF/PAA LFT has to stand elevated temperature more than 100°C for durations up to 90 seconds [8].

In this study, long carbon fiber reinforced polyarylamide (CF/PAA) is used for the testing at elevated temperatures and consequent failure analysis is conducted. PAA, also known as PA-MXD6 which is a kind of nylon and produced from m-xylylenediamine and adipic acid through polycondensation reaction, is a semi-crystalline aromatic polyamide [9,10]. Currently there is no publication available that is related to the mechanical behavior of CF/PAA at elevated temperature. This research effort will be focused on systematically studying the effect of elevated temperature on the mechanical properties (flexural and tensile properties) of CF/PAA LFT samples in different orientations with different fiber contents and their failure mechanisms are evaluated.

Material and Processing

20% and 30 wt% long carbon fiber PAA pellets supplied by Celanese were used to produce the plates from which the mechanical testing samples were prepared. All of the pellets have a length of 25 mm. In order to determine the proper processing and testing temperature of samples, it is necessary to understand the thermal properties and

---

*Corresponding author: Ning H, Department of Materials Science and Engineering, University of Alabama at Birmingham, Birmingham, USA, E-mail: ning@uab.edu

stability of the CF/PAA using differential scanning calorimetry (DSC) and thermogravimetric analysis (TGA) before processing.

CF/PAA samples were analyzed using Q100 DSC (TA Instruments) for the melting temperature and glass transition temperature of PAA matrix. The CF/PAA sample was scanned from room temperature to 450°C at a rate of 10°C per minute. The resultant curve is illustrated in Figure 1. The melting of the PAA in the sample starts from approximately 270°C. The valley beginning at that temperature represents the melting endotherm for the CF/PAA sample. The sudden change of the heat capacity at the segment of 70-80°C corresponds to the glass transition temperature Tg. The peak at approximately 120°C is exothermic and represents crystallization of the PAA.

TGA (TA Instrument DSC Q100) was used to measure the degradation temperature of the PAA matrix. The sample was run in air environment and the temperature was ramped up at a rate of 10°C per minute. TGA was run to mainly evaluate the mass loss as a function of temperature. Figure 2 illustrates the mass loss as a function of temperature for CF/PAA and the mass loss dramatically increases at approximately 440°C.

Extrusion-compression molding process was used to manufacture the testing samples. Before processing, the pellets were dried for 24 hours at 90°C in a desiccant dryer to eliminate any moisture which could affect material processing and properties. Processing starts with

feeding the pellets into the hopper of a plasticator which has single screw and low shear. The pellets were metered down a barrel that was heated above the melting point and extruded in low shear. A temperature which is 30-40°C higher than the melting temperature of the PAA matrix was used for melting the PAA matrix to ensure adequate melting in the plasticator and flowing in the mold while avoiding any polymer degradation. The molten material was accumulated at the front end of the barrel and a specific amount of material was extruded and cut to a cylindrical-shaped charge. The extruded charge was transferred in a closed-cavity compression mold with the dimension of 305 mm by 305 mm housed within a heated hydraulic press under 60 tons pressure. The CF/PAA plate was then demolded after 5-minute dwell in the mold.

Multiple plates were used to prepare large number of samples needed for the flexural and tensile testing. In order to make sure that all the fibers have consistent orientation from plate to plate, the charge was placed at the same position and in the same orientation in the mold for every run. The direction along the axis of the cylindrical charge is defined as transverse orientation (90°), and the direction perpendicular to the axis of the charge (parallel to the charge flow direction) is defined as longitudinal orientation (0°). Figure 3 shows the schematic of the positioning of the charge and flow directions in relation to the orientation of the samples to be prepared for the mechanical testing.

Flexural samples were prepared according to ASTM D790-Standard Test Methods for Flexural Properties of Unreinforced and Reinforced Plastics and Electrical Insulating Materials. Tensile testing samples were prepared by cutting rectangular sample into dog-boned shape based on ASTM-D638 Standard Test Method for Tensile Properties of Plastics. Figure 4 shows representative flexural and tensile testing samples.

Results and Discussion

Three different temperatures were selected for the mechanical testing: room temperature (RT), glass transition temperature (TG), and medium temperature 65°C (MD) between room temperature and Tg. In addition, different fiber content (20 and 30 wt%) of the LFT and orientation (longitudinal and transverse) of the molded samples were also taken into account in this study. Three samples were tested for each material category.

A servo-hydraulic universal testing frame with a heating chamber was used for flexural testing at different temperatures. All of the CF/PAA samples were 180 mm in length, 15 mm in width and 4.5 mm in thickness. Support span length was 120 mm and the loading rate was 2 mm/min.

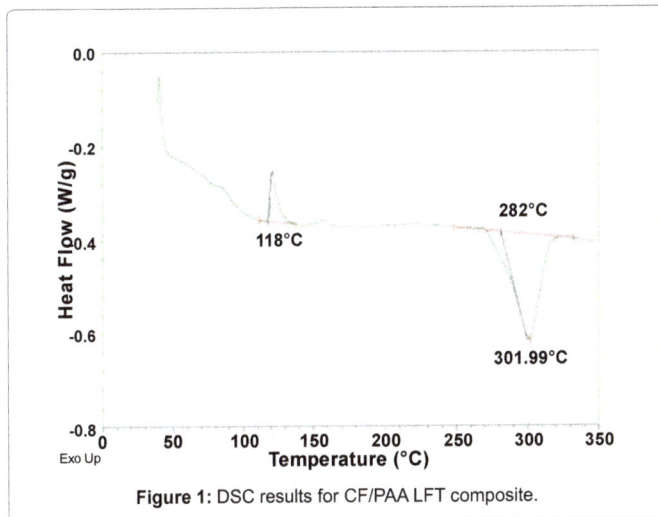

Figure 1: DSC results for CF/PAA LFT composite.

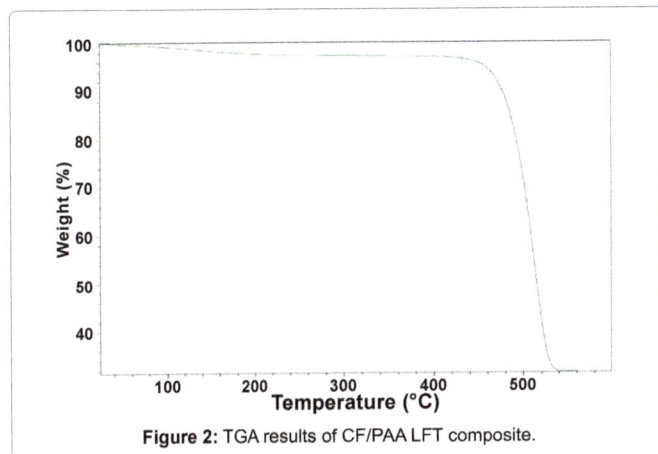

Figure 2: TGA results of CF/PAA LFT composite.

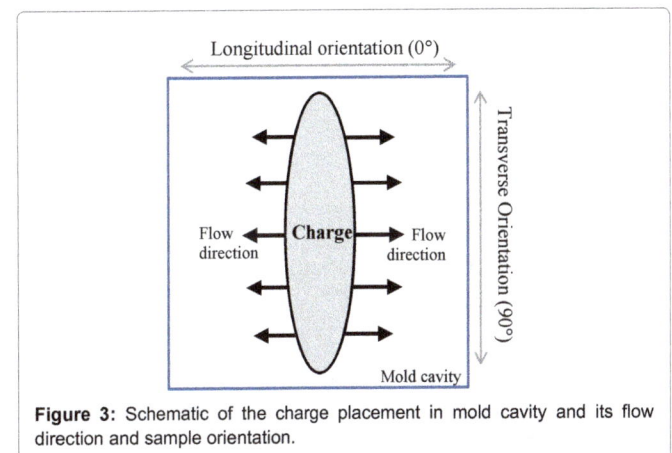

Figure 3: Schematic of the charge placement in mold cavity and its flow direction and sample orientation.

Tensile testing was conducted using the same testing frame and an extensometer was attached to the sample for obtaining the strain data during tensile testing, which was used to calculate tensile modulus. The sample was loaded at a loading rate of 2 mm/min.

Typical flexural load-displacement curves of 20 wt% CF/PAA samples in longitudinal (0°) orientation are presented in Figure 5. It is seen that the RT samples show linear load-displacement curve until catastrophic fracture when peak load is reached, indicating the rigid and brittle characteristic of the CF/PAA LFT composite at ambient temperature. However, the samples tested at the other two temperatures show nonlinearity, especially the ones tested at Tg, which indicates that their matrix softens and the interface between fiber and matrix weakens. The curves at glass transition temperature (Tg) exhibit obvious ductile material characteristic fracture with higher strain to failure. Zig-zag patterns on the Tg curves indicates that there is continuous debonding between fiber and matrix due to weakened fiber-matrix interface at elevated temperature. The slope of the load-displacement curve at the RT temperature possesses the highest value while the lowest at TG temperature. There is a change in the shape of the curves as a result of increasing testing temperature. The slopes of the curves decrease with increasing temperatures, indicating lower stiffness at higher temperatures. Peak load is reduced with increasing temperatures. Below Tg, PAA is in a glassy state which results in the highest stiffness and strength of the CF/PAA LFT composite.

Effect of temperature on the flexural strength of both 20 and 30 wt% samples is presented in Figure 6. The average flexural strength values

are plotted and compared along with the standard deviation added as the error bar. Temperature significantly affects flexural strength at the elevated temperature section between MD and TG temperature. The samples in 0° orientation show that flexural strength drops 10% at 65°C for 20% CF/PAA and 5% for 30% CF/PAA, respectively. However, when temperature reaches Tg, flexural strength drops 26% and 64% for 20% CF/PAA and 30% CF/PAA, respectively. For the samples in 90° orientation, the reductions are 11% and 8% for 20% CF/PAA and 30% CF/PAA at 65°C, respectively, and dramatically increase to 43% and 68% at 80°C. The change in the shape of the curves as a result of increasing the temperature does not follow a linear trend. The property decreases exponentially when the temperature increases from 65 to 80°C.

Effect of temperature on the flexural modulus of the samples is presented in Figure 7. It is seen that flexural modulus is also largely affected by temperature and its reduction is much more as compared with that of flexural strength. In 0° orientation, flexural modulus drops 21% and 17% at 65°C for 20% CF/PAA and 30% CF/PAA, respectively. However, when temperature reaches Tg, flexural modulus drops 42% and 75%, respectively. In 90° orientation, the reductions are 40% and 40% for 20% CF/PAA and 30% CF/PAA at 65°C, respectively, and dramatically increase to 71% and 68% at 80°C. This again shows the non-linearity for the effect of temperature on flexural modulus.

A similar behavior to that of flexural load-deflection curves is

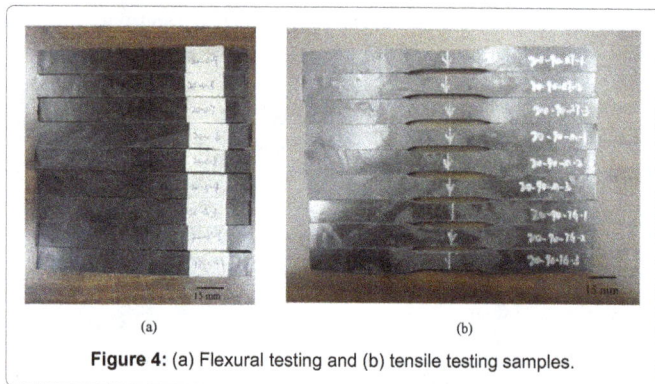

Figure 4: (a) Flexural testing and (b) tensile testing samples.

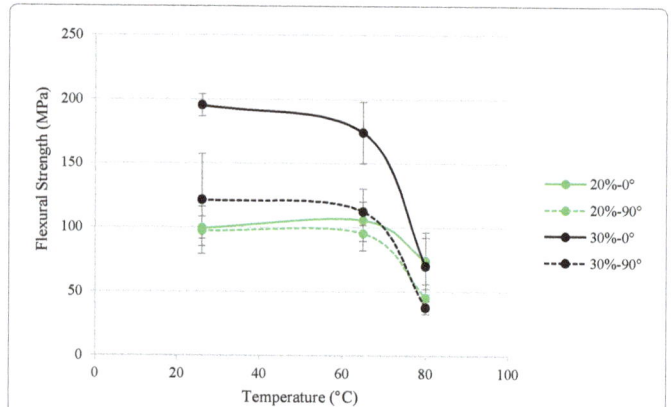

Figure 6: The flexural strength comparison at different temperatures for 20 and 30 wt% samples in longitudinal (0°) and transverse (90°) orientation.

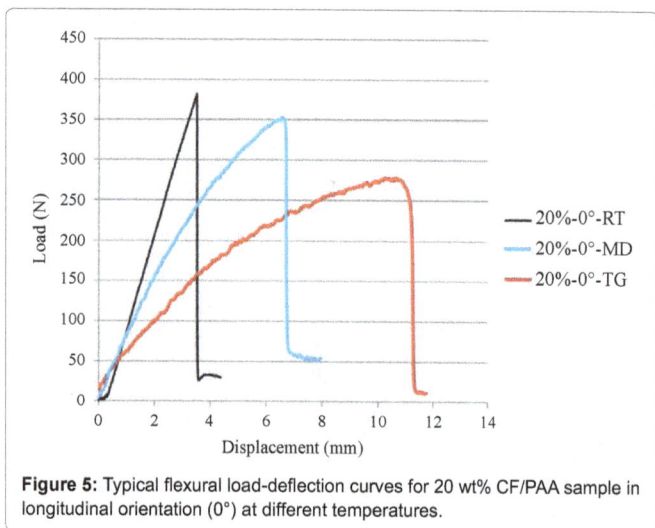

Figure 5: Typical flexural load-deflection curves for 20 wt% CF/PAA sample in longitudinal orientation (0°) at different temperatures.

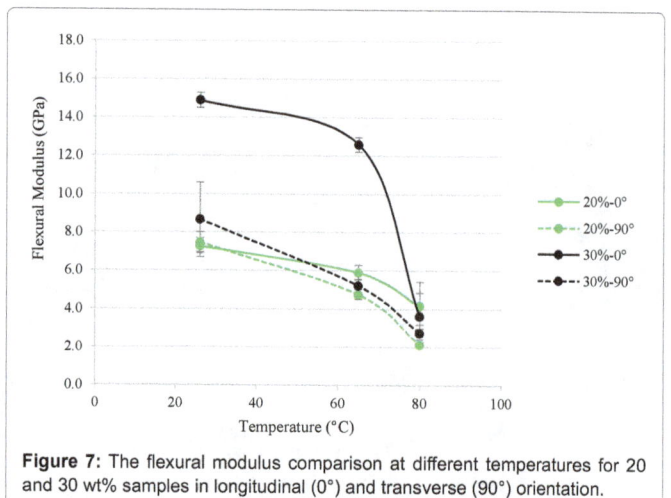

Figure 7: The flexural modulus comparison at different temperatures for 20 and 30 wt% samples in longitudinal (0°) and transverse (90°) orientation.

noticed for tensile testing results. As expected, the modulus of elasticity of the samples was reduced with increasing temperature. Figure 8 exhibits the effect of temperature on the tensile strength of the samples in both orientations. Tensile strength drops around 7 and 12% at MD temperature for samples containing 20 and 30 wt% carbon fibers in 0° orientation, respectively, and drops 10 and 15% in 90° orientations, which are very close to the reduction values of flexural strength. When the temperature is reached to Tg, the corresponding reductions are 30% and 40% for 20% CF/PAA and 30% CF/PAA at 65°C, respectively, and still remain similar at 40% and 50% at 80°C, indicating that tensile strength is less affected by temperature compared to its flexural counterpart. It is understood that tensile modulus and strength are fiber-dominated property [11]. Tensile property is not as much affected as flexural properties which are matrix-dominated properties at elevated temperature, especially when the matrix PAA softens significantly at Tg.

Effect of temperature on the tensile modulus of the samples is illustrated in Figure 9. Comparing with the tensile strength, tensile modulus is more affected by temperature. Tensile modulus reduces by approximately 20 wt% and 30 wt% samples at MD temperature (around 24% and 26%, respectively). At Tg, the corresponding reductions are 43% and 51%, respectively.

Figures 6-9 illustrate that both strength and modulus show higher value in longitudinal orientation compared to transverse orientation, although that trend is more obvious for 30 wt% fiber loading samples.

More fibers were oriented in the longitudinal direction because of the charge flow during compression molding which induces preferred fiber alignment in the flow direction. Thus, the properties of the samples in longitudinal orientation show higher mechanical properties. The samples in the transverse orientation have fewer fibers oriented in the loading direction and therefore lower property is resulted.

Generally it is found that the effect of temperature was more pronounced on the stiffness of composites (both flexural and tensile) as compared with that of strength values because of the fact that stiffness of the composite is directly related to the stiffness of the constituents which heavily depends on temperature. Strength is mainly governed by interfacial adhesion which is less influenced by temperature. Both of the strength and modulus have the highest decrease when the temperature is raised from MD temperature to Tg temperature. Below Tg, the interfacial bonding between the fiber and matrix is not much affected and PAA matrix is glassy below glass transition temperature. When temperature is reached to Tg, the weakened interfacial bonding and softened matrix account for the large reduction in material properties. Similar trend of unidirectional fiber composite has been reported by other research groups. It was found that the longitudinal strength and modulus of a unidirectional laminate remain virtually unaltered with increasing temperature, but its transverse and off-axis properties are significantly reduced as the temperature approaches the Tg of the polymer matrix [12].

Stereoscope was used to analyze the failure mechanisms of the samples tested at different temperatures. Figure 10 shows typical fracture surface of the flexural sample tested at RT and TG temperature (30% CF/PAA longitudinal samples). The fractured sample at RT temperature shows a typical brittle failure and there is no obvious fiber pull-out. However, the sample tested at Tg has distinct fiber pull-out and delamination.

Scanning electron microscope (SEM) was also used to study the fracture surfaces of flexural test specimens. Figure 11 shows the SEM images of the same fracture surfaces shown in Figure 10. Both fracture surface of the specimens (30% CF/PAA samples in longitudinal orientation) were studied using a Quanta 650 FEG field emission scanning electron microscope. Figure 11a shows the fracture surface image of the sample tested at RT. It is noticed that the main failure mechanism is fiber fracture and there is also some fiber pull-out with polymer adhering to the fibers, indicating a good fiber-matrix bonding. However, the fracture surface of the flexural specimen tested at TG temperature (Figure 11b) show that fiber pull-out phenomenon

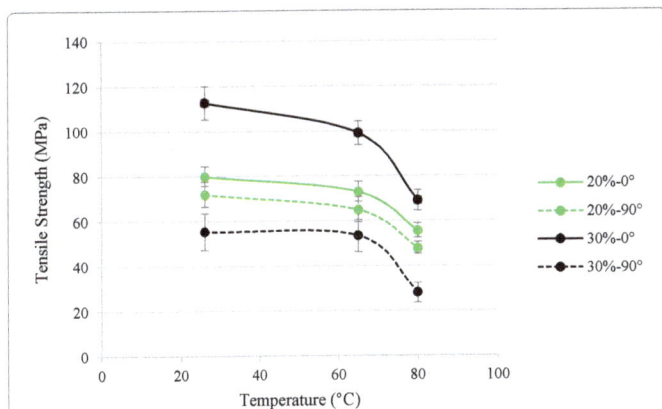

Figure 8: The tensile strength comparison at different temperatures for 20 and 30 wt% samples in longitudinal (0°) and transverse (90°) orientation.

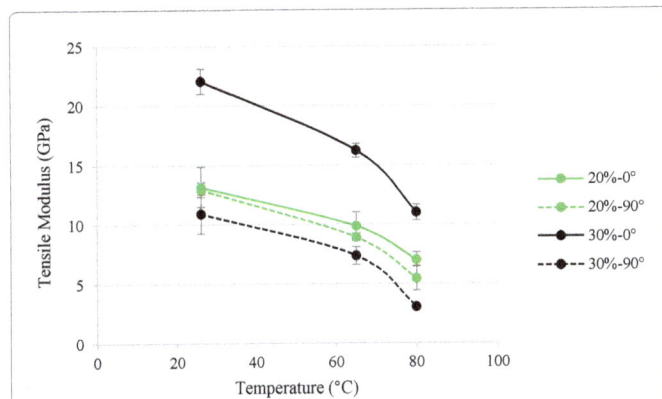

Figure 9: The effect of the temperature on the tensile modulus for 20 and 30 wt% samples in longitudinal (0°) and transverse (90°) orientation.

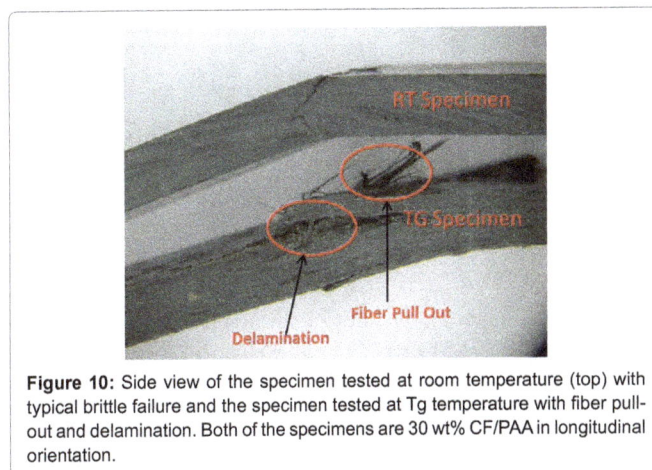

Figure 10: Side view of the specimen tested at room temperature (top) with typical brittle failure and the specimen tested at Tg temperature with fiber pull-out and delamination. Both of the specimens are 30 wt% CF/PAA in longitudinal orientation.

Figure 11: (a) Typical SEM of fracture surface of flexure specimen tested at RT and (b) Typical *SEM of fracture surface of flexure specimen tested at TG.* Both of the specimens are 30 wt% CF/PAA in longitudinal orientation.

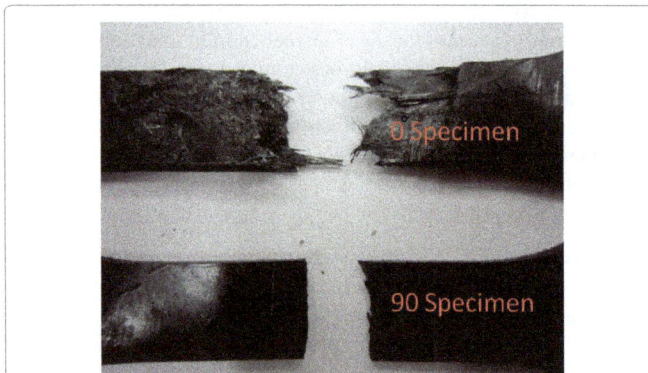

Figure 12: Typical fracture surface of tested tensile specimen in 0° (top) and in 90° (bottom) orientation. Both of the specimen have fiber loading of 30 wt% and are tested at MD temperature.

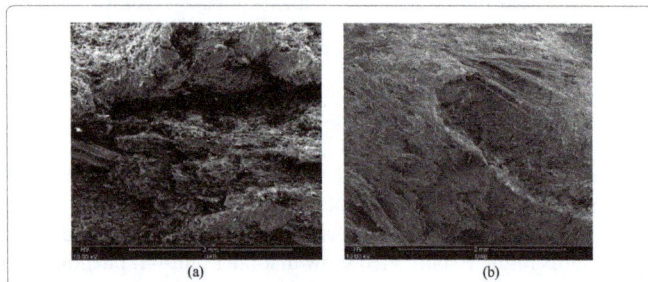

Figure 13: (a) SEM image of fracture surface of tensile specimen in 0° orientation; and (b) SEM image of fracture surface of tensile specimen in 90° orientation. Both of the specimens have 30 wt% fiber loading and are tested at MD temperature.

is dominant and there is minimum fiber fracture. The length of the fiber pulled out of the matrix is much longer than the fibers pulled out as shown in Figure 11a. It indicates that the bonding strength between fiber and matrix at elevated temperature was much weaker as compared to that at RT temperature and the pull-out is the main failure mechanism at elevated temperature.

Typical stereoscopic pictures of the tested tensile samples are shown in Figure 12 (30% CF/PAA). It is seen that these two samples have totally different morphology at the fracture surface. For the sample in 0° orientation, the fracture surface is not straight and there are obvious fibers exposed at the surface. However, the sample in 90° orientation has nearly straight fracture surface and no noticeable fibers

exposed. This indicates that there are a large number of fibers involved in the tensile test for the 0° orientation sample. The involvement of large number of fibers in the 0° samples results in higher mechanical properties than the 90° samples. Figure 13 shows the SEM images of the same tensile samples. Figure 13a shows the fracture surface of the sample in 0° orientation and it is noticed that a large number of fibers are exposed on the fracture surface as compared to the fracture surface of the sample in 90° orientation as shown in Figure 13b. This indicates that the main failure mechanism is fiber breakage for the sample in 0° orientation due to the fact that more fibers are oriented in 0°. During tensile testing, the fibers oriented in 0° mainly bear the tensile loading and therefore these 0° oriented samples show higher tensile strength and modulus. The fracture surface of 90° orientated samples is much smoother compared to 0° orientation sample. There is fewer fiber breakage compare to the 0° oriented sample and the main failure mechanism for the sample in 90° orientation is matrix fracture. In this case, fibers do not play a load-bearing role since few fibers are aligned in loading directions. As a result, the samples in 90° orientation show less strength and modulus.

Conclusions

The effect of temperature on the mechanical properties of the carbon fiber reinforced polyarylamide was studied in this research. The CF/PAA samples were processed using extrusion-compression molding and tested in flexural and tension. The effect of sample orientation and weight percentage of carbon fibers on the flexural and tension properties were evaluated. The following conclusions are drawn from the results and discussion presented above: 1) Flexural strength, flexural modulus, tensile strength, and tensile modulus of the CF/PAA LFT composites reduce exponentially from medium temperature 65°C to glass transition temperature, 80°C; 2) The effect of temperature on both of flexural and tensile modulus of the composite is more pronounced compared with flexural and tensile strength; 3) The reduction of both strength and modulus in 90° orientation samples is larger than in 0° orientation samples; 4) Flexural properties of the composites were influenced by temperature to a higher degree than tensile properties.

Acknowledgements

The authors would like to thank Department of Energy (DOE) Graduate Automotive Technology Education (GATE) program for funding this research.

References

1. Hartness T, Husman G, Koenig J, Dyksterhouse J (2001) The characterization of low cost fiber reinforced thermoplastic composites produced by the DRIFT™ process. Compos Part A: Appl S 32: 1155-1160.

2. Steffens M, Himmel N, Maier M (1998) Design and analysis of discontinuous long fiber reinforced thermoplastic structures for car seat applications. Transactions on Engineering Sciences 21: 35-44.

3. Thattaiparthasarathy KB, Pillay S, Ning H, Vaidya U (2008) Process simulation, design and manufacturing of a long fiber thermoplastic composite for mass transit application. Compos Part A: Appl S 39: 1512-1521.

4. Sands JM, Vaidya U, Husman G, Serrano J, Brannon R (2008) Manufacturing of a composite tailcone for an XM-1002 training round.

5. Vaidya UK, Serrano JC, Villalobos A, Sands J, Garner J (2008) Design and analysis of a long fiber thermoplastic composite tailcone for a tank gun training round. Mater Des 29: 305-318.

6. Chevali VS, Dean DR, Janowski GM (2009) Flexural creep behavior of discontinuous thermoplastic composites: Non-linear viscoelastic modeling and time–temperature–stress superposition. Compos Part A: Appl S 40: 870-877.

7. Henning F, Ernst H, Brüssel R (2005) LFTs for automotive applications. Reinf Plast 49: 24-33.

8. Ning H, Pillay S, Vaidya U, Andrews JB (2008) Processing and characterization of thin-walled long fiber reinforced thermoplastic (LFT) composites, in International SAMPE Symposium.

9. Xi Z, Chen L, Zhao Y, Zhao L (2013) Experimental and modeling study of melt polycondensation process of PA-MXD6, in Macromolecular Symposia. Molecular Symposia 333: 172-179.

10. Doudou B, Dargent E, Grenet J (2006) Crystallization and melting behaviour of poly (m-xylene adipamide). J Therm Anal Calorim 85: 409-415.

11. Kumar BG, Singh RP, Nakamura T (2002) Degradation of carbon fiber-reinforced epoxy composites by ultraviolet radiation and condensation. J Compos Mater 36: 2713-2733.

12. Mallick PK (2007) Fiber-reinforced composites: materials, manufacturing, and design. CRC press, USA.

11

Fourier Transform Infrared Spectroscopy and Liquid Chromatography – Mass Spectrometry Study of Extracellular Polymer Substances Produced on Secondary Sludge Fortified with Crude Glycerol

Nouha K[1], Hoang NV[2] and Tyagi RD[1*]

[1]*Université du Québec, Institut national de la recherche scientifique, Centre Eau, Terre and Environnement, 490 de la Couronne, Québec, G1K 9A9, Canada*
[2]*Institute of Environmental Technology, Vietnam Academy of Science and Technology, 18 Hoang Quoc Viet, Hanoi, Vietnam*

Abstract

This study was conducted to characterize the extracellular polymeric substances (EPS) produced by growing *Cloacibacterium normanense* in wastewater sludge alone and fortified crude glycerol. The EPS highest concentration of 17.5 g/L was produced using 25 g/L sludge suspended solids supplemented with 25 g/L of crude glycerol. Galactose and Glucose was the main compound of EPS produced with or without crude glycerol, respectively. EPS FTIR spectra revealed a variation in the different functional groups such as amines, carboxyl, and hydroxyl groups for EPS produced. Different functional groups were observed in the EPS produced with or without glycerol. The EPS exhibited kaolin flocculation activity up to 95% and was stable at high temperature. The high viscosity bioflocculant properties of EPS make it suitable for potential industrial applications

Keywords: Cloacibacterium normanense; EPS; Activated sludge; Crude glycerol; FTIR; Rheology; Bioflocculation

Abbreviations: CAN: Acetonitrile; ANOVA: Analysis of Variance; B-EPS: Broth EPS; BSA: Bovine Serum Albumin; C-EPS: Capsular EPS; CDW: Cell Dry Weight; CHNS: Carbon, Hydrogen, Nitrogen and Sulphur; CFU: Colony Forming Units; C/N: Carbon/Nitrogen Ratio; CUQ: Communauté Urbaine du Québec; DLVO: Derjaguin-Landau-Verwey-Overbeek; EPS: Extracellular Polymeric Substances; FA: Flocculation Activity; FTIR: Fourier Transform Infrared Spectroscopy; KBr: Potassium Bromide; LC/MS: Liquid Chromatography/Mass Spectrometry; mg EPS/g of kaolin: mg of EPS added per gram of kaolin suspension in water; NTU: Nephelometric Turbidity Unit; S-EPS: Slime EPS; SEM: Scanning Electron Microscopy; SS: Suspended Solids; TSB: Tryptic Soy Broth; (v/v): Volume per Volume; ζ-potential: Zeta potential

Introduction

Microbial polysaccharides have been demonstrated with a vast range of functional properties and applications including food products, pharmaceuticals, bioemulsifiers [1], bioflocculants [2], chemical products [3] and the biosorption of heavy metals [4]. The chemical compositions of EPSs produced by different bacterial strains are very diverse. The main constituents of EPS are sugars such as galactose, glucose and rhamnose. Acetate and puryvate are characteristic substituents. Whilst the structures are very complex, consist of branched repeating units and different linkages that exist between the monosaccharides [5].

Depending on their specific composition, EPSs have different cellular functions, including accumulation of nutrients, diffusion barrier for toxins and heavy metals, cell motility, attachment to surfaces, protection against desiccation. Thus, several researchers have discussed recent advancements in the understanding of the structure–function relationships, i.e., to relate EPSs structure with their properties (bioflocculant, bioemulsfiers) in order to improve EPSs synthesis and applications [6]. Therefore, there is a need to develop an understanding of structure–function relationships to relate EPS structure, chemical composition and molecular weight with bioflocculation properties, especially for the EPS produced in wastewater sludge by new isolated bacterial strain.

In this work, a novel microbial biopolymer produced by *Cloacibacterium normanense* is described. Along with the fermentation process for the extracellular polysaccharides (EPSs) production from sludge supplemented with crude glycerol, a preliminary polymer characterization in terms of its chemical composition and structure is presented. Glycosyl composition analysis was performed by LC/MS assigning the different polysaccharide monomers. EPS was also characterized by quantification of the content of proteins and carbohydrates, its structure by Fourier transform infrared spectroscopy (FT-IR), and its morphology and surface attachment by scanning electron microscopy (SEM).

Methods

EPS production

Cloacibacterium normanense was grown on 25 g/L suspended solids (SS) of activated sludge collected from bio-filtration unit of Communauté Urbaine du Québec (Municipal wastewater treatment plant, CUQ, Québec, Canada). For inoculum preparation, *Cloacibacterium normanense* (NK6, accession number KF675204) was inoculated in Tryptic soy broth (TSB) and incubated at 30°C, 180 rpm for 24 h. The pre-culture was added (5% v/v) to each experimental flask containing 150 ml sterilized sludge (25 g/L SS, pH 7). Crude glycerol 25 g/L was added at 24 h to obtain the desired C/N concentration ratio 25. The C/N ratio was calculated taking into account the initial concentration of nitrogen in sludge and carbon content of glycerol. The flasks were then incubated in a shaker at 180 rpm and 30°C for 96 h. The control experiments were also performed without adding crude

***Corresponding author:** Tyagi RD, Université du Québec, Institut national de la recherche scientifique, Centre Eau, Terre & Environnement, 490 de la Couronne, Québec, G1K 9A9, Canada, E-mail: tyagi@ete.inrs.ca*

glycerol. During the experiments, samples of the broth were collected every 24 h to determine EPS concentration, glycerol concentration and cell population. The crude glycerol solution also contains other components (soap and methanol), which can act as carbon source. Therefore, consumption of soap and methanol was determined during the fermentation [7].

EPS extraction and dry weight

The centrifugation method was used to extract S-EPS (slime EPS) and C-EPS (capsular EPS). Broth EPS (B-EPS) consists of both S-EPS and C-EPS, or the fermented broth is named as B-EPS. The fermented broth was centrifuged at 9000 g, 4°C for 20 min to obtain supernatant (containing S-EPS). The biomass pellet was re-suspended in deionized water equal to the initial volume and then heated at 60°C for 20 min followed by centrifugation at 9000 g, 4°C for 20 min to extract capsular EPS [8].

For measuring dry weight, one volume of supernatant obtained after centrifugation (crude S-EPS) was mixed with two volumes of chilled ethanol (95% v/v) and precipitated at -20°C overnight. After precipitation, the sample was centrifuged at 6000 g for 15 min to obtain the precipitated pellets. The pellet was dried at 60°C until constant weight.

The EPS concentration was estimated by the following formula:

$$[EPS](g/L) = \frac{W_2 - W_1}{V}$$

Where, W_1: Initial dry weight of the empty aluminium dish without a sample (g)

W_2: Dry weight of the aluminum dish with dried sample (g)

V: volume of the sample (L)

The total EPS (B-EPS) contained in the broth was calculated as sum of S-EPS and C-EPS. All the measurements were carried out in triplicates and the average was presented

Analytical methods

Glycerol concentration in the cell-free supernatant, methanol and soap content was determined according to Hu et al. [7]. The growth was measured based on the dry weight per volume of the culture. The cell dry weight (CDW) or biomass was determined by centrifugation (8000 g, 4°C, 15 min) and after the C-EPS extraction, followed by overnight drying the sample to a constant weight in an oven at 60°C. The cell concentration of all the samples (diluted with saline solution), was measured as CFU employing standard agar-plate technique. Total nitrogen and organic carbon in the samples collected at various times of fermentation were measured by the CHNS analyzer (Shimadzu VCPH).

EPS characterization

Chemical composition of the EPS: Glycosyl composition analysis was performed by combining Liquid Chromatography/Mass Spectrometry (LC/MS) with Hypersil Gold column (100*2.1 mm ID), using 85% Water, 0.1% formic acid (Phase A)/ 15% acetonitrile (ACN), 0.1% formic acid (Phase B), as eluent, at a flow rate of 0.4 mL/min and a temperature of 30°C [9]. EPS sample 20 µl was used for the identification and quantification of acyl group and monosaccharides present in the purified EPS (the ethanol precipitated EPS). The monosaccharides were identified by their retention times in comparison to standards.

FT-IR spectroscopy: Precipitated EPS were collected by centrifugation at 4000 g, 30 min at 4°C and dried at 60°C. The purified and dry S-EPS (0.1-0.2 mg) and 100 mg of potassium bromide (KBr) were mixed and pressed in a die (at five tons and one minute) to form a pellet. Afterwards, the pellet was immediately put into the sample holder and FT-IR spectra were recorded. The transmission FT-IR spectra were obtained using a Perkin Elmer 2000 FT-IR spectrometer. FT-IR scanning was conducted in ambient conditions. The resolution was set to 4 cm^{-1} and the operating range was 400 to 4000 cm^{-1} [10].

EPS properties

Rheology: To investigate the stability of the EPS at different temperatures, the crude EPS solution was incubated for 10 min at 80°C to 200°C in an incubator. The viscosity of the above EPS solution (20 ml) was measured using a ULA S 34 spindle (Digital Viscometer, DV-II+ Pro, Brookfield), at 60 rpm and room temperature.

Enzymatic digestions test: Proteinase K was used as the proteolytic enzyme and 80 units of the enzyme were added to 200 ml sample. Proteinase K is a non specific enzyme that hydrolyses proteins at a number of cleavage sites. Cellulase (β glucosidase) was used as an extracellular polysaccharide degrading enzyme and 150 active units were added to 200 mL sample of B-EPS. Cellulase hydrolyses the polysaccharides present in EPS [1]. Enzymes were added to each 200 mL sample of the fermented broth collected at 48h and the enzymatic digestion was conducted for 36h at 30°C. The samples, each 20 ml, were collected at every 12 h to measure ζ-potential, viscosity and turbidity index. All the measurements were carried out in duplicates and the average result was presented.

Scanning electron microscopy (SEM): Samples of enzymatic digestion test were collected 6 hr after enzyme addition. Broth sample of 10 µl was taken on the glass slide and the cells were fixed with 3% glutaraldehyde. After fixing, the cells were washed with ethanol solution of different concentration (30–100% v/v) and air dried between each wash to completely remove the water adhered to the cells. After final washing, cells were subjected to overnight drying. The dried samples were mounted on conventional 12.7 mm or 25.4 mm diameter aluminium stubs using double sided adhesive carbon discs and coated with gold film to a thickness of 10-20 nm using a sputter coater (SPI™ sputter coater module) to examine their morphology under scanning electron microscopy (Model: Carl Zeiss EVO®50 smart SEM) [11].

Zeta potential: The charge of B-EPS after enzymatic digestion was determined by adding 50-1000 µL of individual EPS sample to 100 ml of deoinzed water. Characterization of charge (zeta potential) was implemented using Zetaphoremeter (Zetaphoremeter IV, Zetacompact Z8000, CAD Instrumentation, France) with the application of the Smoluckowski equation. Surface charge of the wastewater sludge was also measured. The zeta potential values were obtained from an average of around 10 measurements, the average values are presented with its half-width confidence interval at 95% confidence level.

Measurement of turbidity index: Kaolin clay was used as a test material to measure the flocculation activity or the turbidity index of EPS samples digested by enzymes. Flocculation activity or turbidity index of EPS was carried out through the jar test method [8]. Kaolin with concentration of 5 g/L was suspended in distilled water, 150 mg/L of Ca^{2+} was added to the kaolin suspension and pH was adjusted to 7.5 after addition of Ca^{2+}. The samples collected at different times of (6 h, 12 h, 24 h and 36 h) after adding enzymes, were added in different volumes (corresponding to desired concentrations of EPS, which was calculated through dry weight and the volume of EPS solution required) to kaolin suspension and rapidly mixed at 100 rpm for an initial 5 min then slowly mixed for an additional 30 min at 70 rpm. After 35 min of mixing, samples were transferred to a 500 ml cylinder and allowed to settle for 1 h. The supernatants of the settled samples were collected to measure the turbidity using turbidimeter (Micro 100 turbidimeter, Scientific Inc.). The flocculation activity (FA) was determined using the formula [100*(B-A)/B]; where 'A' is the turbidity of the sample (treated with EPS; Slime, Capsular or Broth-EPS) and 'B' is the turbidity of the control (in which equal volume of EPS solution was replaced with distilled water).

Statistical analysis

All analysis reported in this manuscript was performed in triplicate, and the results are presented as the mean values. The results were analysed by analysis of variance (ANOVA), using Excel's Analysis ToolPak.

Results and Discussion

EPS production

The variation of biomass, nitrogen concentrations and C/N ratio (with and without adding crude glycerol) during the fermentation is presented in Figure 1 and Table 1. Cell count, glycerol and EPS concentrations are depicted in Figure 2. The C/N ratio was calculated as the concentration ratio (i.e. concentration of organic carbon/ concentration of total nitrogen). Different C/N ratio (C/N 10, C/N 25 and C/N50) have been studied before to determine the optimum C/N ratio (Data not shown). Thus, C/N 25 was chosen in present study.

The crude glycerol was added at 24 h fermentation as an extra carbon source to aid EPS production by *Cloacibacterium normanense*, which increased the C/N ratio to 25 compared to 11 in the control or without glycerol (Table 1). The maximum EPS production (about 17.5 g/L) was observed at 72 h of fermentation with glycerol fortification (C/N 22, Table 1), whereas EPS concentration of 13.3 g/L was observed at 48 h (C/N reached 14) fermentation time without glycerol fortification. Without glycerol, the C/N ratio initially increased (0 to 48 h) and was due to higher consumption of nitrogen than carbon, both required for growth (Figure 2). Further, decrease in the C/N ratio after 48 h could be due to higher consumption of carbon required for maintenance of cells, because EPS did not increase during this time. Corresponding to results found C/N we suggest that C/N 22 is considered to be the most favourable for EPS production by *Cloacibacterium normanense* in present study. However, Miqueleto et al. [12] have reported that C/N 10 was recommended for high EPS production in an anaerobic sequencing batch biofilm reactor. In the study of Liu et al. [13], the suggested C/N ratio was 12 for maximum EPS concentration (8.90 g/L) produced by *Zunongwangia profunda* SM-A87. Nevertheless, for most EPS-synthesizing microorganisms, the highest polymer productivities are usually achieved at high C/N ratio and is microbial strain dependent.

With glycerol addition, EPS concentration steadily increased until 72 h, whereas C/N ratio first increased to 25 (at 24 h due to the addition of extra glycerol) and then decreased to 22 at 72 h. Most of the added glycerol was consumed at 48 h (residual glycerol 2.0 g/L, Figure 2); however, EPS substantially increased from 48 to 72 h. Therefore, the microbe used carbon from sludge for synthesis of EPS, thus decreased the C/N ratio (even in spite of nitrogen decrease, Figure 1).

The CFU (colony forming units) count increased during the first 48 hours of fermentation (with the addition of glycerol) and nitrogen used was 37 mg/L, whereas between 48 and 72 h, the increase in CFU was relatively less but nitrogen consumed was almost 100 mg/L (Figure 1

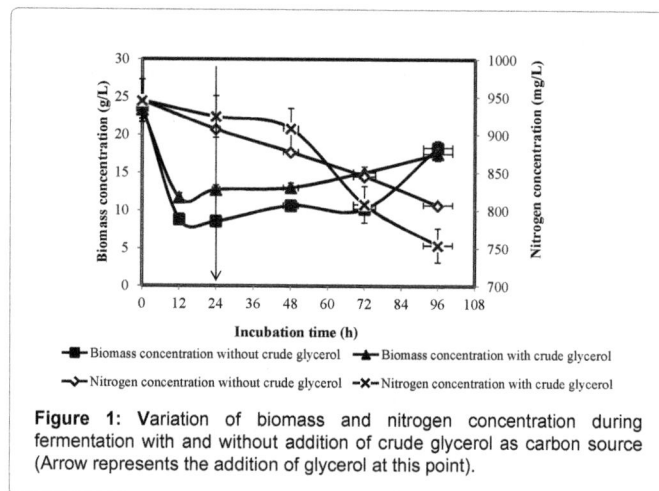

Figure 1: Variation of biomass and nitrogen concentration during fermentation with and without addition of crude glycerol as carbon source (Arrow represents the addition of glycerol at this point).

Experiments with crude glycerol addition	Incubation time (h)	0	24	48	72	96
	C/N ratio	3	25	21	22	16
	S-EPS (g/L)	1.4 ± 0.3	4.6± 0.3	13.9 ± 0.4	15.6 ± 0.5	4.3± 0.3
	B-EPS (g/L)	1.8 ± 0.3	5.2 ± 0.2	14.9± 0.2	17.5± 0.4	15.4± 0.6
	EPS/biomass ratio (g/g)	0.08 ± 0.01	0.4 ± 0.2	1.1 ± 0.3	1.1 ± 0.1	0.9 ± 0.4
Experiments without crude glycerol addition	C/N ratio	3	11	14	12	11
	S-EPS (g/L)	1.5 ± 0.3	3.9 ± 0.3	12.9 ± 0.5	12.2 ± 0.3	10.4± 0.2
	B-EPS (g/L)	1.9± 0.3	4.7± 0.3	13.3± 0.2	12.5± 0.1	10.8± 0.4
	EPS/biomass ratio (g/g)	0.08 ± 0.3	0.55 ± 0.2	1.25 ± 0.3	1.27 ± 0.1	0.60 ± 0.4

Table 1: Production of biomass and EPS produced with and without crude glycerol fortification during fermentation.

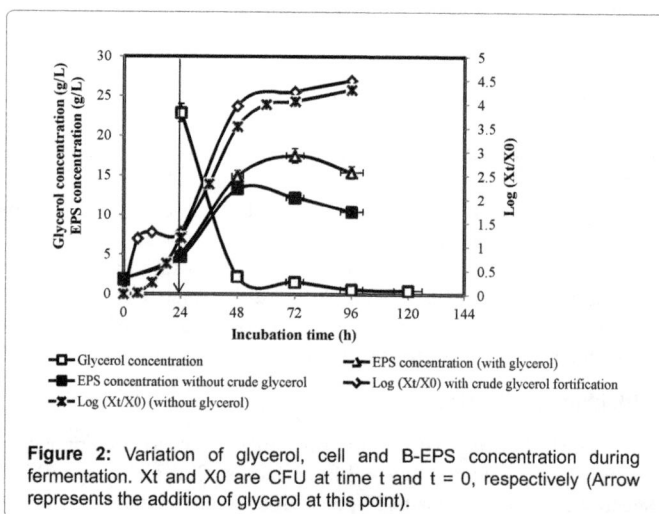

Figure 2: Variation of glycerol, cell and B-EPS concentration during fermentation. Xt and X0 are CFU at time t and t = 0, respectively (Arrow represents the addition of glycerol at this point).

and 2). The excess consumption of nitrogen (48-72 h) was routed toward formation of other unknown metabolites [14]. In this context, Duenas et al. [15] reported that EPS production by *Pediococcus damnosus* was mainly enhanced with an increase in glucose concentration, but not by an increase in nitrogen concentration.

In addition, the crude glycerol is source of soap and methanol, which can be used as substrate for the EPS production. Figure 2 shows the decrease of crude glycerol content during the fermentation concomitant with the growth of biomass on semi log scale ($X0$ is the initial CFU and Xt is CFU at time 't'). Figure 3 illustrates the decrease of methanol and soap concentration during the fermentation process indicating that soap and methanol are degraded by the strain for growth and to generate EPS. Kumar et al. [16] have also reported that non sugar carbon sources (like methanol) could contribute to produce microbial EPS.

The carbon source was studied for the EPS and biomass production of *Cloacibacterium normanense*, determining the relative significance of two variables (crude glycerol added and sludge media without crude glycerol addition), using ANOVA Excel's Analysis ToolPak. The ANOVA analysis for EPS production and biomass is shown in Table 2. The value of regression coefficient R^2 was 0.9890 for biomass and 0.9723 for EPS production. The p-values of the models were 0.0002 and 1.3632E-05, in case of crude glycerol fortification, 4.27E-06 and 3.4637E-05 without crude glycerol for biomass and EPS production, respectively, indicating that the models were significant. Usually, a model term is considered to be significant when its value of "p-value" is less than 0.05 [17].

EPS physico-chemical characterization

The B-EPS obtained in the experiments were analysed for their sugar and acyl group composition (Figure 4). Five main constituent sugar residues were identified by the glycosyl composition analysis, namely, glucose, galactose, lactose, sucrose and xylose. Galactose was the most abundant monosaccharide, accounting for 67 mol % of the total carbohydrate content of the B-EPS. Glucose represented 13 mol %, but lactose, sucrose and xylose was present in only minor amounts i.e. 3, 8 and 9 mol %, respectively.

EPS synthesized without the addition of crude glycerol was distinct from the EPS produced with glycerol addition. The EPS without glycerol addition was characterized by only one type of monosaccharide, i.e. eighty mol % of glucose. The appearance of galactose and lactose monosaccharide after adding crude glycerol was demonstrated. This can be explained by the fact that the strain *Cloacibacterium normamnense* converts glucose to galactose or/and glucose to lactose in the presence of glycerol. We can, therefore, anticipate that the addition of glycerol has influenced the level of activation of the enzymes necessary for the different metabolic and involved in the synthesis and assembling of the sugar nucleotides, thus resulting in the production of polymers with such a diverse sugar composition. Similar observations have been recorded by Yolunda et al. [18]. Polysaccharides containing galactose are produced by several bacteria, such as *Pseudomonas, Lactobacillus* and *Streptococcus* [9]. The data obtained in our study are in agreement with the findings reported by Freitas et al. [9]. They demonstrated high galactose content of the EPS produced by *Pseudomonas oleovarans* grown on pure glycerol.

The present results have also shown that the EPS contained (with or without glycerol addition) non-saccharide components, namely, acyl groups. Two different acyl groups were identified in small quantity. Pyruvate (0.014 wt%) appeared after addition of glycerol; however, succinate (0.057wt.%) was present in the absence of glycerol (Figure 4c and 4d). These components are frequently present in microbial EPS and notably influence polymer's properties, namely, solubility and rheology [19]. According to Freitas et al. [9], three acyl groups were identified using media containing pure glycerol, i.e. pyruvate (3.35 wt%), succinate (1.04 wt%) and acetate (0.38 wt%).

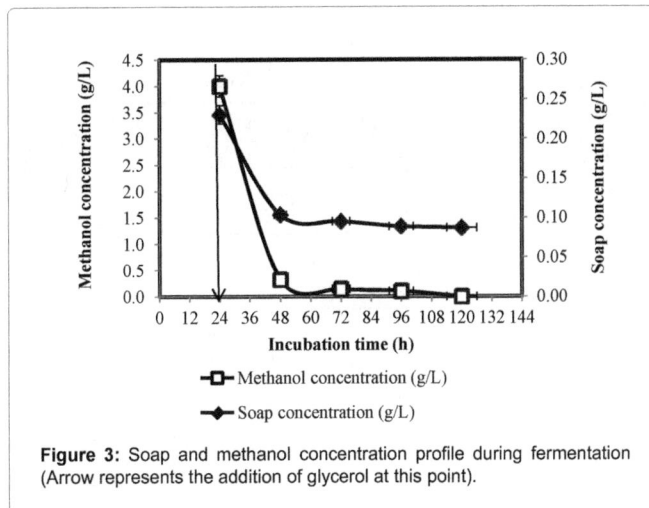

Figure 3: Soap and methanol concentration profile during fermentation (Arrow represents the addition of glycerol at this point).

Variables	Biomass[a]			EPS production[b]		
	coefficient estimate	F value	p value	coefficient estimate	F value	p value
Sludge + crude glycerol	0.93	0.07	0.0002[c]	1.01	0.013	1.3632E-05[c]
Sludge without crude glycerol	0.99	0.01	4.2769E-06[c]	1.0087	0.016	3.4637E-05[c]

[a]R^2 = 0.9890;
[b]R^2 = 0.9723;
[c]Model terms are significant

Table 2: Identification of significant variables for EPS production and Biomass of *Cloacibacterium normanense* using ANOVA of Excel.

FT-IR spectroscopy

The FTIR analysis of EPS (produced with or without crude glycerol) was presented in Figure 5a and 5b. The FT-IR spectrum of the exopolysaccharide is in agreement with the chemical analysis (LC/MS) described above. The main functional groups are hydroxyl, carboxyl, acyl and amino, which corresponds mostly to the presence of carbohydrates and proteins in the EPS.

In the two cases (with and without crude glycerol addition), it revealed a broad stretched peak at 3455 cm⁻¹ (range 3600-3200 cm⁻¹), corresponding to the hydroxyl group. A weak absorption at 2925 cm⁻¹ (range 300-2500 cm⁻¹) was assigned to an asymmetrical l C-H stretching vibration of the aliphatic CH_2 group, which represents the presence of organic substances like sugars and proteins. The amide II band at 1729 cm⁻¹ originates from N-H bonding and C-N stretching vibrations in -CO-NH of proteins. Presence of an asymmetric stretching peak or vibration at 1643 cm⁻¹ may correspond to the ring stretching of galactose as the LC/MS analysis indicated in the last section. Another peak at 1404 or 1455 cm⁻¹ could be attributed to the symmetric stretching of the COO– group. The absorption peaks (1271 or 1233 cm⁻¹) ranging from (1500-1200 cm⁻¹) were designated to C-O-C and C=O, which indicates the occurrence of carbohydrates. A peak at 1058 cm⁻¹ (1000-1125 cm⁻¹ range) may be attributed to O-acetyl ester linkage bond of uronic acid. Absorption peak approximately in the range of 781-522 cm⁻¹ corresponded to stretching of alkyl-halides. These findings are in agreement with the studies of [9,20,21].

However, the EPS produced with sludge supplemented with crude glycerol differs from the EPS produced without crude glycerol by having

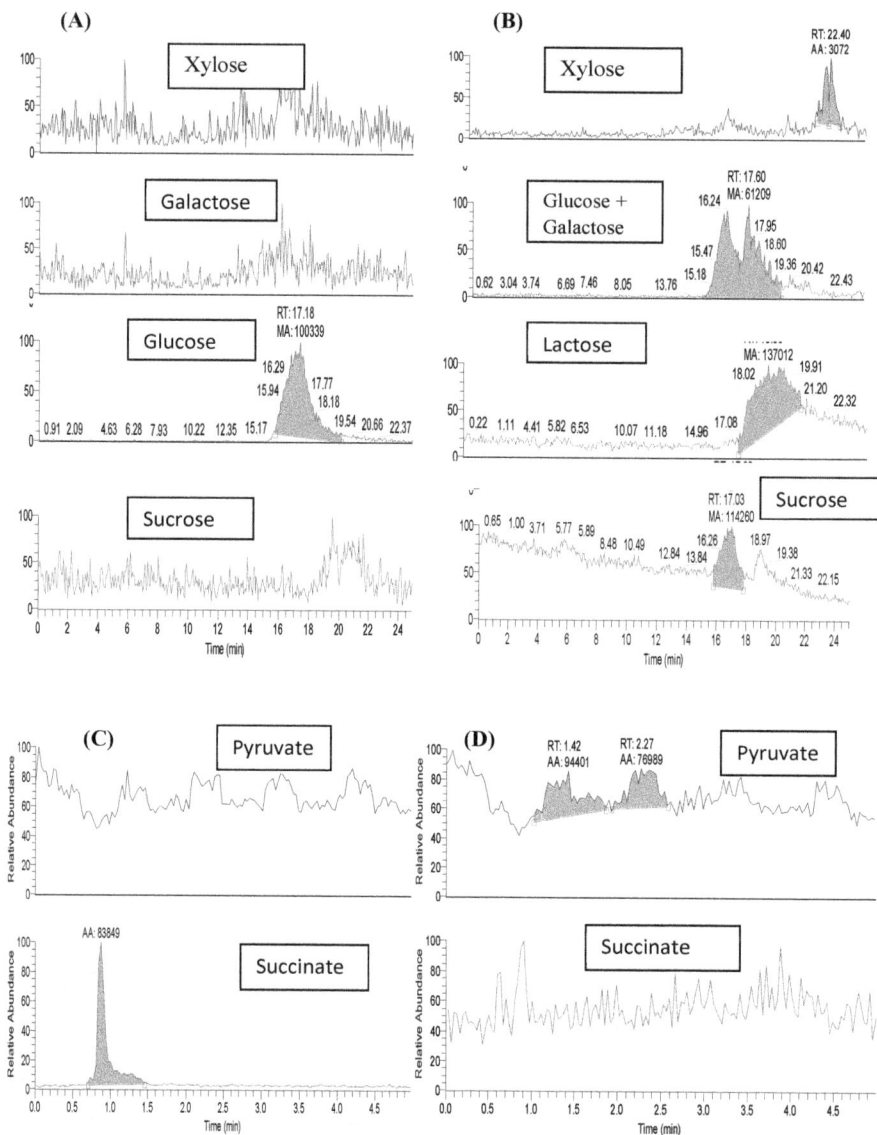

Figure. 4. LC/MS analysis of the EPS produced by Cloacibacterium normanense : (A) Monosaccharide analysis without crude glycerol, (B) Monosaccharide analysis with crude glycerol, (C) Acyl group analysis without crude glycerol, (D) Acyl group analysis with crude glycerol.

additional peaks at different regions. IR spectra of EPS produced in the case of crude glycerol addition (Figure 5a) show particular bands, which do not appear for EPS produced in the absence of crude glycerol (Figure 5b). Three bands around 2855 cm^{-1} (symmetric stretching vibration of CH$_2$, C = O stretching), 1678 cm^{-1} (deformation vibration of N-H or Amide I and C-N stretching) and 1578 cm^{-1} (deformation vibration of N-H or Amide II) appeared when crude glycerol was added to the production medium. These characteristic bands can be attributed to protein and polysaccharide functional groups. By the presence of these peaks, FTIR of EPS (with crude glycerol) demonstrates relatively higher quantity of hydroxyl (-OH), amide (-CO-NH), carboxyl (-COO-), and primary amine (-NH$_2$) groups compared with those observed without crude glycerol. The abundance of these groups in EPS produced with supplementation of crude glycerol in the medium may contribute to the difference in flocculation activity (93.4%) compared to flocculation activity (90.2%) of EPS produced without crude glycerol fortification.

Further, the presence of carboxyl, hydroxyl, and amine groups is very important for bioflocculation with relatively quantity, providing surface charges, which serve as the binding sites for suspended particles causing aggregation or floc formation as also discussed by Li et al. [22] and Kavita et al. [20]. However, the excess of these groups may increase the negative charge causing a repulsion of particles [20].

The difference in the occurrence, position and frequency of these groups could affect the flocculation activity and that could be the reason of high flocculation activity of the EPS obtained in this study (with or without fortification of crude glycerol) compared to those reported by other researchers. The flocculation activity of EPS synthesised by *Pseudomonas* SM9913 was 49.3% [22] and 40% of the EPS produced by *Oceanobacillus iheyensis*. These groups (C-O-C; O-H; C = O…etc.) were either absent or present in insufficient quantity in EPSs as described in Table 3. Table 3 recaps the functional groups corresponding to bands observed in the IR spectra of B-EPS produced by the strain with or

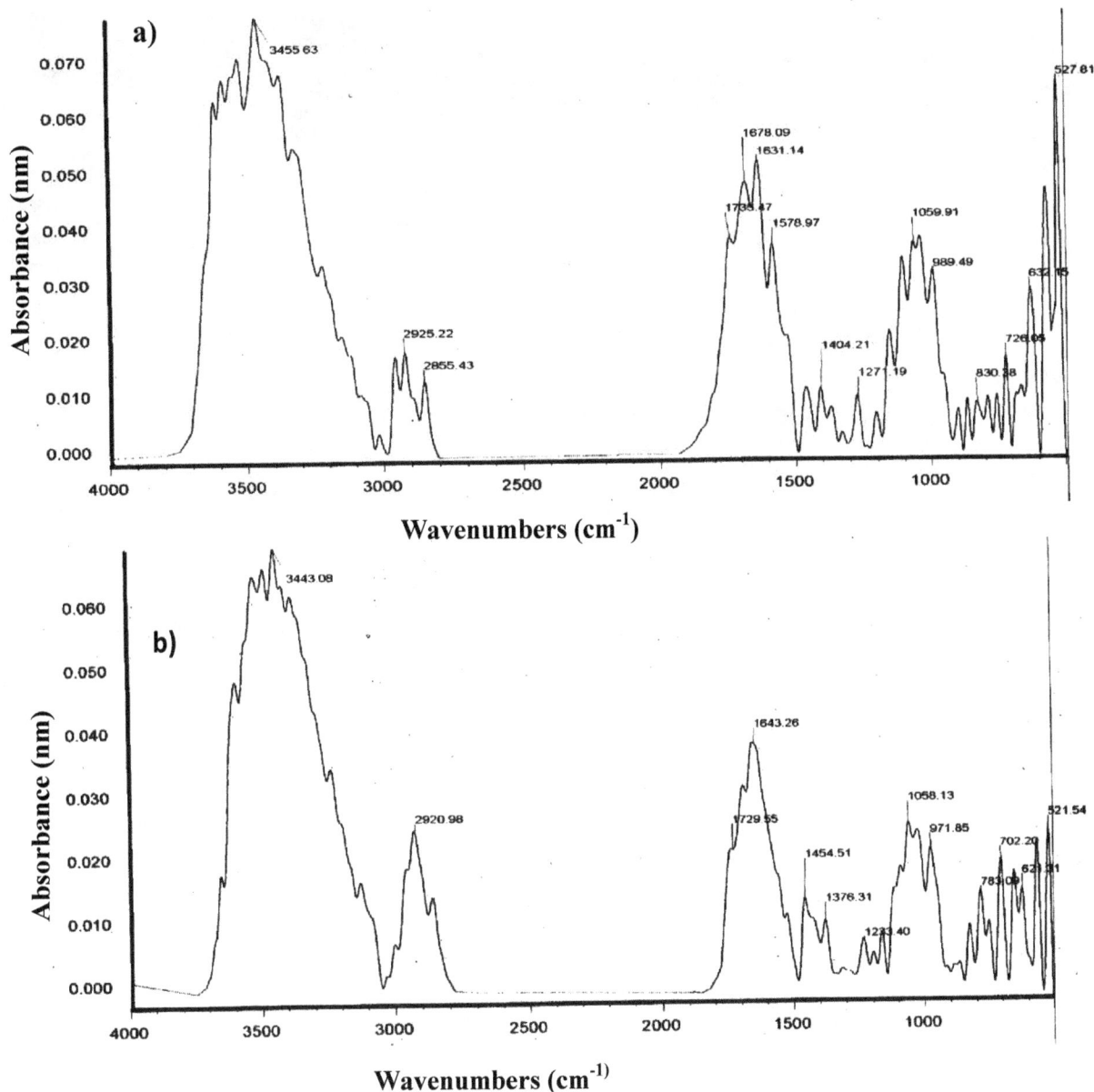

Figure 5: IR spectra of the S-EPS produced in sludge by pure culture of *Cloacibacteriun normanense* grown: a) with crude glycerol and b) without crude glycerol addition.

without glycerol fortification of sludge as well as those available in the literature.

In general, flocculation activity is due to the interaction between the cations and the EPS functional groups and the particles to be flocculated. Several researchers [23] have performed experiments that support the DLVO theory, as London force, for the mechanism of bioflocculation. London forces are present between all chemical groups. The source of this type of interaction is the spontaneous formation of transient dipoles due to fluctuations in the electron distribution within the molecule. This temporary dipole polarises the molecule and thus creating dipolar attraction forces. It represents the main cohesive force between hydrocarbon chains.

The carboxyl functional group (–COOH) prefers to lose the H+

ion in to the solution and the –COO⁻ ion can be formed. These results were confirmed by Li et al. [22]. They proposed that a large number of carboxyl groups of EPS can also serve as binding sites for divalent cations (Ca^{2+}). When bridged by cations, the negatively charged EPS combines the flocs together. This enables EPS to serve a key role in the flocculation of activated sludge. The 1678 cm^{-1} band appeared in the IR spectrum of EPS produced in the presence of crude glycerol, whereas it was absent in the IR spectrum of EPS without glycerol (Figure 5a and 5b). This band represents the carboxyl group, which could bind to cations and form flocs by bridging mechanism. In relation to the carboxyl group, the EPS produced by *Pseudoalteromonas* sp. SM991 [22] exhibited low FA of 37.7% compared to 93.4% FA of EPS produced in this study with crude glycerol supplementation. The low activity can be related to a relatively smaller strength (15.4%) of the band at 1678

Wave number (cm⁻¹)	Absorbance (nm)							Vibration type	Functional type
	EPS with glycerol (present study)	EPS without glycerol (present study)	EPS by *Pseudoaltermonas* SM9913 (Li et al. [22])	EPS by *Oceanobacillis iheyensis* BK6 (Kavita et al. [20])	EPS by *Klebsiella pneumoniae* (Nie et al. [24])	*EPS by Cobetia* sp. and *Bacillus* sp. (Ugbenyen et al. [25])	EPS by *Bacillus megaterium* TF10 (Yuan et al. [26])		
3200-3420	0.05	0.065	0.3	0.4	0.5	0.6	0.3	Stretching vibration of OH	OH into polymeric compounds
2930-2935	0.025	0.02	0.3	0.3	0.4	0.5	0.2	Asymmetric stretching vibration of CH$_2$	Proteins (peptidic bond)
2850-2865	0.015	-			-	-	-	Symmetric stretching vibration of CH$_2$	Proteins (peptidic bond)
1678	0.05	-		-	-	-	-	Stretching vibration of C = O and C-N (Amide I)	Proteins (peptidic bond)
1630-1660	0.06	0.04	0.8	0.3	0.4	0.3	0.3	Stretching vibration of C = O and C-N (Amide I)	Proteins (peptidic bond)
1550-1580	0.04	-	-	-	-	-	0.02	Stretching vibration of C-N and deformation vibration of N-H (Amide II)	Proteins (peptidic bond)
1450-1460	-	0.01	-	-	-	0.6		Deformation vibration of CH$_2$	phenols
1400-1410		-	0.7	0.3	0.2	-	0.2	Stretching vibration of C = O / Deformation vibration of OH	Carboxylates
1235-1245	0.015	0.01	-	-	0.3	-	0.09	Deformation vibration of CH$_2$/ Stretching vibration OH	Alcohols and phenols
1130-1160	0.04	0.04	-	0.3	-	-	-	Stretching vibration C-O-C	phenols
1060-1100	0.03	0.03	0.01	-	0.4	0.6	0.4	Stretching vibration of OH	polysaccharides
< 1000	0.06	0.06	-	0.2	-	0.2	0.02	Several bands visible	Phosphorus or sulphur functional group

Table 3: Main functional group observed from IR spectra of broth EPS (B-EPS) with and without crude glycerol fortification

cm⁻¹ in EPS produced by Li et al. than 40% strength in EPS obtained in the present study (with crude glycerol).

The bands (1678, 1630-1660, and 1550-1580) are characteristic of C-N stretching and deformation vibration of N-H (Amide II). Hydrogen bonds may contribute to the active reaction between water molecules and amide group (N-H), which is present frequently in protein moiety of EPS (with crude glycerol). The researchers also suggested that higher content of protein moiety in EPS, as indicated by the presence of peptide bond in the IR spectrum of EPS obtained with supplementation of crude glycerol (corresponding to wave numbers 1678, 1630-1660, and 1550-1580 in Table 3 and Figure 5a), could bring more negatively charged amino groups, thus strengthening electrostatic interaction with cations. This plays an important role in flocculation. Further, the large band of Amide II (wave number 1550-1580, Table 3) is present in the EPS (produced with glycerol) and absent in the EPS produced without glycerol. This implies a high concentration of protein moiety in EPS with glycerol, which leads to a high flocculation activity. Furthermore, the EPS obtained in the present study revealed better flocculation comparing to those produced by *Klebsiella pneumoniae*

strain NY1 (85.3%) or the consortium of *Cobetia* sp. and *Bacillus* sp. MAYA (90.2%) due to the absence of the amide band (in the latter), where the bioflocculant structure was a polysaccharide (Table 3) [24,25].

The symmetric vibration of CH$_2$ group was observed in the EPS obtained in the present study (with glycerol case, corresponding to wavenumber 2850-2865, Table 3). This type of band was absent in the other IR spectrum of EPS obtained by other studies [22,24-26] as indicated in Table 3. This band could offer the covalent C = C bond, which is more effective in playing an important role in the aggregation of flocs compared to any other type of main interactions (such as hydrogen bond, London force...) [26].

EPS properties

Effect of temperature on EPS viscosity and FA: The high temperature resistance of the polymers or EPS is important for two reasons: Firstly, the EPS will be produced as powder product using atomiser, which is prepared by spray drying at high temperature (more than 80°C). Secondly, there are many other applications of EPS (chemical,

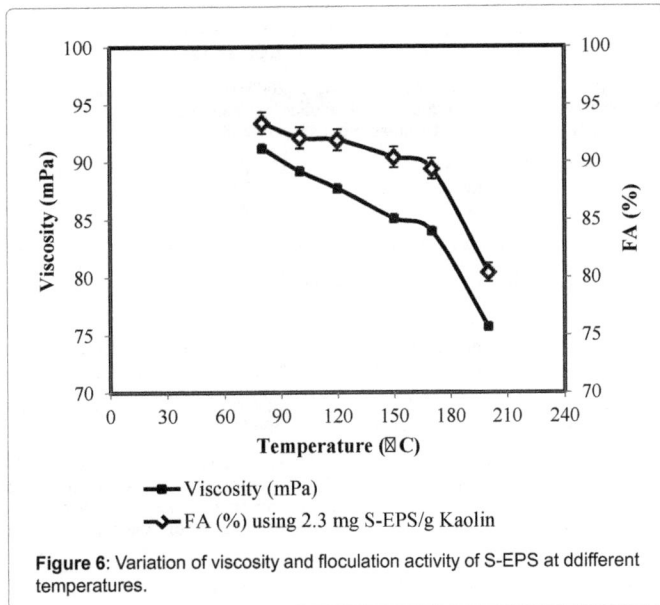

Figure 6: Variation of viscosity and floculation activity of S-EPS at ddifferent temperatures.

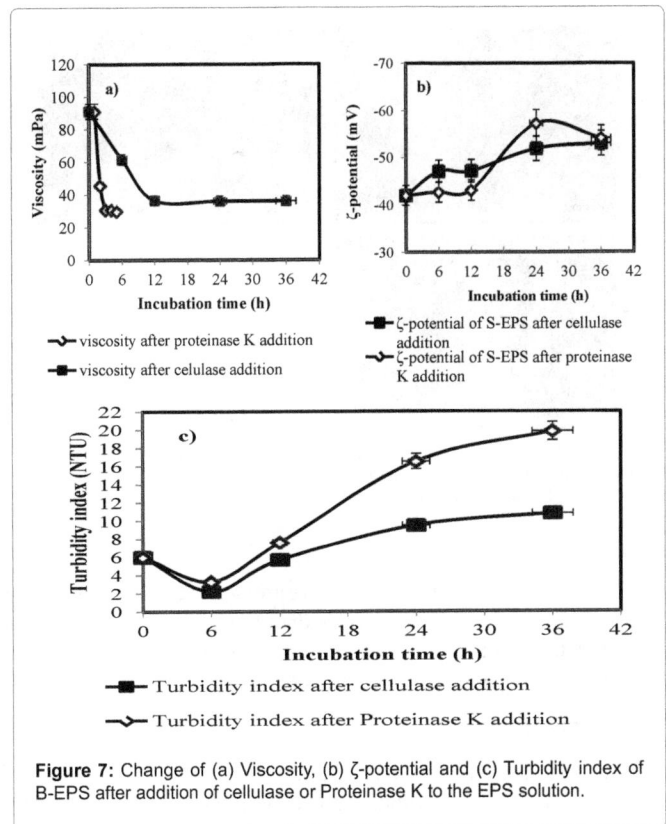

Figure 7: Change of (a) Viscosity, (b) ζ-potential and (c) Turbidity index of B-EPS after addition of cellulase or Proteinase K to the EPS solution.

food and medical industries) where high temperature resisting polymers are used [27,28]. For food and medical requirement where high purity of EPS is required, the pure EPS product could be produced by growing high EPS yielding pure strain (*Cloacibactérium normanense* NK6) in a synthetic medium.

In the present study, the stability of S-EPS produced by *Cloacibactérium normanense* (NK6) was investigated by exposing the EPS for 10 minutes at different temperatures (from 80 to 200°C) and then measuring their viscosity. The S-EPS of the sample collected at 72 h of fermentation with glycerol addition to sludge and possessing highest EPS concentration was used. Figure 6 presents the impact of temperature on the flocculation activity and viscosity of the EPS, which revealed that the S-EPS from *Cloacibacterium normanense* strain exhibited a good stability at high temperature. The viscosity starts to decrease slowly from 80°C until 150°C and then decreased rapidly at 200°C. The flocculation activity decreased in a similar way as the viscosity (Figure 6). The decrease in viscosity could be attributed to the polymer degradation due to the cleavage of glycosidic bonds within the polysaccharide structure [9].

Past studies have discussed the degradation of polymers in aqueous and organic solutions, which was accelerated by strong acids, certain oxidizing agents, ultraviolet light and temperature [29]. They observed various reaction mechanisms and revealed that redox reactions involving free radicals were probably the cause of polymer degradation and concomitant viscosity losses, which could affect their ability to flocculate.

Enzymatic study of S-EPS: Protein and carbohydrate are the main components of EPS. Many previous reports proposed that protein was more important than carbohydrate to floc formation and demonstrated that activated sludge deflocculated after incubation with a proteolytic enzyme [30,31]. In order to better understand the role of physicochemical properties, as well as the role of protein and carbohydrate in bioflocculation, enzymes were used in this study to degrade biopolymers of S-EPS produced by *Cloacibacterium normanense*. Viscosity, surface charge and turbidity index were measured at different incubation time. Proteinase K was used to degrade extracellular proteins moiety and cellulase was used to degrade extracellular polysaccharide moiety of the EPS. As shown in Figure 7,

the viscosity decreased after the addition of both enzymes. Treatment of EPS with cellulase caused the viscosity decrease by 60% (decreased from 91.2 mPa to 36.4 mPa) of the original value after 12 h; after that the viscosity remained stable.

Proteinase K treatment caused a rapid viscosity drop by 67% (decreased from 91.2 mPa to 30 mPa) in three hours. This clearly demonstrated that protein in S-EPS is more important to viscosity drop than carbohydrate. With the degradation of the carbohydrate and protein in S-EPS, the cell surface charge decreased (Figure 7b). The ζ-potential values decreased from –41 mV to around –57.3 mV and –47.1 mV after adding cellulose and Proteinase K, respectively. Increased negative surface charge would cause deterioration in flocculation activity of the EPS [32].

Six hours after the Proteinase K treatment (Figure 7c), turbidity index was measured by adding EPS to kaolin solution (2.3 mg S-EPS/g Kaolin). During the first six hrs, the turbidity decreased and then increased rapidly reaching 19.8 NTU after 36 hours. There was no sign of turbidity recovery with an increase in incubation time. Cellulase addition gave almost the same trend. Turbidity decreased, followed by a relatively slow increase (compared with proteinase K results) with incubation time. These findings suggested that hydrolysis of carbohydrate and protein moieties in S-EPS decreased the viscosity and increased the surface charge, which caused deflocculation. Thus, these results established that both protein and carbohydrate moiety of the EPS play an important role in floc formation. Moreover, the increase in turbidity (or deterioration of FA) was higher and rapid in case of proteinase treatment of EPS, which establishes major role of the protein moiety of EPS in determining flocculation activity.

Figure 8 demonstrates SEM photos presenting the state of bacterial

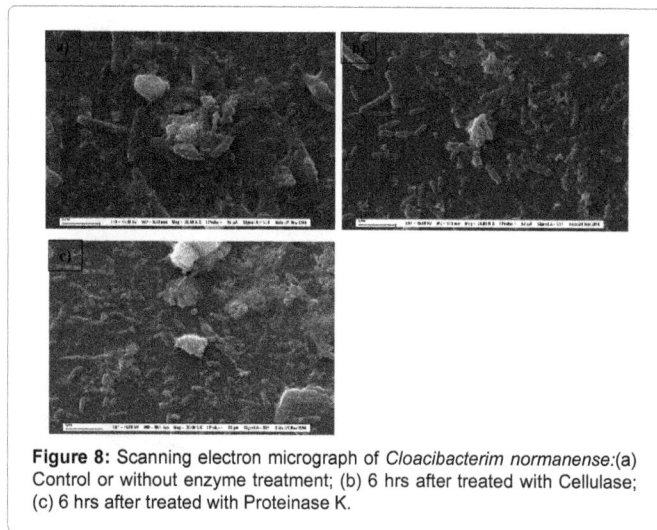

Figure 8: Scanning electron micrograph of *Cloacibacterim normanense:*(a) Control or without enzyme treatment; (b) 6 hrs after treated with Cellulase; (c) 6 hrs after treated with Proteinase K.

aggregate before and after enzyme treatment. Bacterial flocs before enzymes digestion were more visible than those treated with enzymes for six hours (Figure 8a-8c). This suggested that the enzymes destroyed the EPS structure and the cells were dispersed.

Conclusion

This research investigated the carbon and nitrogen content influence on EPS production, the chemical characterisation of EPS produced and their degree of stability as bioflocculant. The following conclusions could be drawn from the foregoing research:

- 17.5 g/L EPS was produced by *Cloacibacterium normannese*.

- LC/MS/MS results analysis demonstrated that the EPS obtained had higher galactose contents (67 mol%) and lower content of glucose (13 mol%), xylose (9 mol%), sucrose (8 mol%) and lactose (3 mol%).

- IR spectra revealed distinct functional groups in EPS produced with or without fortification of glycerol.

- The degree of EPS stability decreased under high temperature (150°C).

- Deflocculation was induced due to digestion of protein and carbohydrate moieties of the EPS by proteinase K and cellulase enzymes, respectively.

- Detailed structural analysis using SEM photos revealed that both protein and carbohydrate moieties of EPS are important factors to impact surface properties and bioflocculation ability of the EPS.

Acknowledgements

Authors would like to acknowledge the Natural Sciences and Engineering Research Council of Canada (grant A4984, Canada Research Chair) for financial support. Sincere thanks are also due to Ministry of Science and Technology of Vietnam (Project No. 134/2013/HD-NDT) for the financial support. We are grateful to technical staffs of INRS-ETE, for their timely help to analyse the samples on LC/MS, FTIR.

References

1. Xie P, Hao X, Mohamad OA, Liang J, Wei G (2013) Comparative study of chromium biosorption by mesorhizobium amorphae strain CCNWGS0123 in single and binary mixtures. Applied Biochem Biotechnol 169: 570-587.

2. Sathiyanarayanan G, Kiran GS, Selvin J (2013) Synthesis of silver nanoparticles by polysaccharide bioflocculant produced from marine Bacillus subtilis MSBN17. Colloids and Surfaces B: Biointerfaces 102: 13-20.

3. Wang Y, Ahmed Z, Feng W, Li C, Song S (2008) Physicochemical propertiesof exopolysaccharide produced by Lactobacillus kefiranofaciens ZW3 isolated from Tibet kefir. Int J of Biol Macromol 43: 283-288.

4. Mohamad OA, Hao X, Xie P, Hatab S, Lin Y, et al. (2012) Biosorption of copper(II) from aqueous solution using non-living Mesorhizobium amorphae strain CCNWGS0123. Microbes Environ 27: 234-241.

5. Klock JH, Andrea W, Richard S, Walter M (2007) Extracellular polymeric substances (EPS) from cyanobacteriamats: characterisation and isolation method optimisation. Mar Biol 152: 1077-1085.

6. Pepi M, Cesaro A, Liut G, Baldi F (2005) An antarctic psychrotrophic bacterium Halomonas sp. ANT-3b, growing on n-hexadecane, produces a new emulsyfying glycolipid. FEMS Microbiol Ecol 53: 157-166.

7. Hu S, Luo X, Wan C, and Li Y (2012) Characterization of crude glycerol from biodiesel plants. J Agric Food Chem 60: 5915-5921.

8. Bala Subramanian S, Yan S, Tyagi RD, Surampalli RY (2010) Extracellular polymeric substances (EPS) producing bacterial strains of municipal wastewater sludge: isolation, molecular identification, EPS characterization and performance for sludge settling and dewatering. Water Res 44: 2253-2266.

9. Freitas F, Vitor DA, Joana P, Nuno C, Cristin O, et al. (2009) Characterization of an extracellular polysaccharide produced by a Pseudomonas strain grown on glycerol. Bioresour Technol 100: 859-865.

10. Madejová J (2003) FTIR techniques in clay mineral studies. Vib Spectrosc 31: 1-10.

11. Ajila CM, Prasada Rao UJS (2008) Protection against hydrogen peroxide induced oxidative damage in rat erythroctyes by Mangifera indica L peel extract. Food Chem Toxicol 46: 303-309.

12. Miqueleto AP, Dolosic CC, Pozzi E, Foresti E, Zaiat M (2010) Influence of carbon sources and C/N ratio on EPS production in anaerobic sequencing batch biofilm reactors for wastewater treatment. Bioresour Technol 101: 1324-1330.

13. Liu SB, Qiao LP, He HL, Zhang Q, Chen XL, et al. (2011) Optimization of Fermentation Conditions and Rheological Properties of Exopolysaccharide Produced by Deep-Sea Bacterium Zunongwangia profunda SM-A87. PLoS One 6: e26825.

14. Lee BC, Bae JT, Pyo HB, Choe TB, Kim SW, et al. (2004) Submerged culture conditions for the production of mycelial biomass and exopolysaccharides by the edible basidiomycete Grifolafrondosa. Enzyme Microbiol Technol 35: 369-376.

15. Duenas M, Munduate A, Perea A, Irastorza A (2003) Exopolysaccharide production by Pediococcus damnosus 2.6 inasemi defined medium under different growth conditions. Int J Food Microbiol 87: 113-120.

16. Kumar AS, Mody KJ, Jha B (2007) Bacterial exopolysaccharides–a perception. J Basic Microbiol 47: 103-117.

17. Zhou WW, He YL, Niu TG, Zhong JJ (2010) Optimization of fermentation conditions for production of anti-TMV extracellular ribonuclease by Bacillus cereus using response surface methodology. Bioprocess Biosyst Eng 33: 657-663.

18. Yolanda A, Inmaculada L, Fernando MC, Montserrat A, Emilia Q (2005) Eps ABCJ genes are involved in the biosynthesis of the exopolysaccharide mauran produced by Halomonas maura. Microbiol 151: 2841-2851.

19. Rinaudo M (2004) Role of substituents on the properties of some polysaccharides. Biomacromol 5: 1155-1165.

20. Kavita K, Singh VK, Mishra A, Jha B (2014) Characterisation and anti-biofilm activity of extracellular polymeric substances from Oceanobacillus iheyensis. Carbohydr Polym 101: 29-35.

21. Nwodo UU, Okoh AI (2012) Characterization and flocculation properties of biopolymeric flocculant (Glycosaminoglycan) produced by Cellulomonas sp. J Appl Microbiol 114: 1325-1337.

22. Li WW, Zhou WZ, Zhang YZ, Wang J, Zhu XB (2008) Flocculation behavior and mechanism of an exopolysaccharide from the deep-sea psychrophilic bacterium Pseudoalteromonas sp. SM9913. Bioresour technol 99: 6893-6899.

23. Sobeck David C, Matthew JH (2002) Examination of three theories for mechanisms of cation-induced bioflocculation. Water Res 36: 527-538.

24. Nie M, Yin X, Jia J, Wang A, Liu S (2011) Production of a novel bioflocculant MNXY1 by Klebsiella pneumoniae strain NY1 and application in precipitation of cyanobacteria and municipal wastewater treatment. J Appl Microbiol 111: 547-558.

25. Ugbenyen AM, Okoh AI (2014) Characteristics of a bioflocculant produced by a consortium of Cobetia and Bacillus species and its application in the treatment of wastewaters. Appl Biochem and Microbiol 50: 49-54.

26. Yuan SJ, Sheing GP, Li Y, Li WW, Yao RS, et al. (2011) Identification of Key Constituents and Structure of the Extracellular Polymeric Substances Excreted by Bacillus megaterium TF10 for their Flocculation. Environ Sci Technol 45: 1152-1157.

27. Patil SV, Salunkhe RB, Patil CD, Patil DM (2011) Bioflocculant exopolysaccharide production by Azotobacterinducus using flower extract of Madhucalatifolia L. Appl Biochem Biotechnol 162: 1095-1108.

28. Carneiro-da-Cunhaa MG, Miguel AC, Bartolomeu WS, Sandra C, Mafalda AC, et al. (2010) Physical and thermal properties of a chitosan/alginate nanolayered PET film. Carbohydr Polym 82: 153-159.

29. Salehizadeh H, Shojaosadati SA (2001) Extracellular biopolymeric flocculants. Recent trends and biotechnological importance. Biotechnol Adv 19: 371-385.

30. Liao BQ, Allen DG, Droppo IG, Leppard GG, Liss SN (2001) Surface properties of sludge and their role in bioflocculation and settleability. Water Res 35: 339-350.

31. Wilen BM, Jin B, Lant P (2003) The influence of key chemical constituents in activated sludge on surface and flocculating properties. Water Res 37: 2127-2139.

32. Liu H, Fang HHP (2002) Extraction of extracellular polymeric substances (EPS) of sludge. J Bacteriol 95: 249-256.

12

Preparation of a Photoactive 3D Polymer Pillared with Metalloporphyrin

Zargari S[1], Rahimi R[1]* and Rahimi A[2]

[1]Department of Chemistry, Iran University of Science and Technology, Iran
[2]Department of Civil Engineering, Iran University of Science and Technology, Iran

Abstract

Among the very few efforts for the preparation of stable pillared graphene nanostructures, there is no report of tin porphyrin intercalated between TiO_2-graphene (TG) nanosheets. In this study, a nanostructure material of pillared graphene made of tin porphyrin functionalized graphene-TiO_2 composite (TG) was successfully synthesized. The prepared compound showed high activity in the photodegradation reaction under irradiation of visible light.

Photocatalytic results showed that the composite of graphene-TiO_2 containing 3% graphene had the highest photoactivity. The photoactivity of TG (3%) was about 1.5 times higher than that of the pure TiO_2.

Besides, tin porphyrin-pillared TG composite (TGSP) material exhibited an excellent visible light photocatalytic performance in degradation of methyl orange dye. This compound could destruct 100% of methyl orange dye in 180 min. Such pillared carbon nanostructure exhibited unique photoactivity due to the synergistic effect between the graphene sheets and the SnTCPP pillars. It is found that the highly efficient light-harvesting structure of the SnTCPP pillared TG composite can be attributed to densely embedded porphyrin chromophores with high visible absorptivity within the framework. The investigation of photocatalytic mechanism determined that hydroxyl radical is a main species in photodegradation process of methyl orange over TGSP compound.

Keywords: Nanostructure; Material; Photocatalytic; Semiconductors

Introduction

Developing high efficiency semiconductors photocatalysts [1-4] have become the research hotspot to decompose effectively harmful organic contaminants for solving environment problems. Among the various photocatalysts, the polymeric graphitic carbon nanostructures have been more and more widely concerned due to its unique inherent characteristic. Defect-free graphene has an infinite number of repetitive elements, with the smallest being any of its sp²-hybridized carbon atoms, whose one p orbital and three sp² orbitals are filled with one electron each. These carbon atoms correspond to the smallest repetitive chain segments representing the repeating units of common linear polymers [5]. Graphene is an excellent electron-transport material in the process of photocatalysis [6]. Therefore, graphene can create a two dimensional conductive support path for charge transfer and collection. This can be used to enhance electron transport properties of the nanostructured TiO_2 and the dye sensitized photocatalytic system. Graphene matrix increases the active surface area as well as modulating the electronic structure of the TiO_2 support [7,8].

Photocatalysts, which accelerate light-driven chemical reactions, are currently under intensive research and development [9,10]. Photocatalytic materials, especially TiO_2 are currently of great interest for application in photocatalysis [11,12]. The efficiency of this application is based on generating electron-hole pair on the catalyst surface and inducing the formation of radicals [13]. Furthermore, the efficiency of photocatalysis relies on the recombination dynamics of the electron and the hole on the surface of TiO_2 nanomaterials [14,15]. The photosensitization of TiO_2 via dye molecules is an effective approach toward conversion of light to electricity in dye-sensitized solar cells [16]. As presented in our previous investigations [17,18], photosensitization of TiO_2 with porphyrin dyes was also considered to be an effective approach to enlarge the light absorption to the visible light range and photocatalytic activity under visible light irradiation.

In this study, a new graphitic nanostructure named pillared graphene-TiO_2 composite (TGSP) was prepared, which is a composite of graphene-TiO_2 layers joined through relatively spaced tin porphyrin

($SnTCPP.Cl_2$). Using this chemical functionalization, the graphene nanosheets were kept separate, and new electronic and optical properties were obtained.

The principal objective of the present research is to prepare an efficient visible-light photocatalyst to significantly use the visible light in photocatalysis reactions. In the present investigation, TiO_2 was composed with various amounts of graphene (TG) to reduce the recombination of the electron and the hole. The photocatalytic activity of these composites was investigated in degradation of Methyl Orange (MO). The TG composite with the optimum content of graphene was photosensitized by a tin complex of tetrakis(4-carboxyphenyl) porphyrin (SnTCPP). In this synthesis, SnTCPP pillared the TG composite (TGSP) was obtained. The photocatalytic activity of the prepared TGSP in the degradation of MO aqueous solution and the photoelectrochemical properties of the prepared TGSP in the generation of photocurrent were investigated. Furthermore, the mechanism of the photocatalytic and the charge transfer process was discussed in details.

Experimental

Synthesis of TiO_2-graphene composites

Graphite oxide was synthesized by the Hummers method and exfoliated to give a brown dispersion of graphene oxide (GO) under

---

*Corresponding author: Rahimi R, Department of Chemistry, Iran Universty of Science and Technology, 16846-13114, Iran
E-mail: rahimi_rah@iust.ac.ir

ultrasonication [19]. A facile one-step solvothermal method was used to synthesize the TiO_2-graphene nanocomposite (TG) by dissolving different mass ratios of graphene oxide in 30 mL of a mixture solution of ethanol and H_2O under ultrasound irradiation. Then, 0.2 g commercial TiO_2 nanoparticles were added to the calculated amount of the graphene oxide solution to prepare the TiO_2-graphene nanocomposites with 1, 3, 5 W% of graphene. The solutions were stirred completely to obtain homogeneous suspensions. Then, the suspensions were transferred into a 40 mL Teflon-sealed autoclave and kept at 120°C for 24 h. During the solvothermal reaction, both graphene oxide reduction and TiO_2 loading were achieved. The resulting products were recovered by filtration, rinsed with DI water, and dried at 60°C for 12 h. The samples were donated as TG (1%), TG (3%), and TG (5%).

Synthesis of TGSP

Tetrakis(4-carboxyphenyl)porphyrin (TCPP) was prepared according to our previous reports [20]. To the synthesis of its tin complex, TCPP was solved in pyridine and refluxed with the excess amount of $SnCl_2.2H_2O$ salt for 6 h. The final product was obtained after evaporating off the solvent and washing with water.

The TiO_2-graphene nanocomposite intercalated by the tin porphyrin molecule was synthesized by refluxing TG (3%) and $SnTCPP.Cl_2$ with a 1:10 ratio in DMF. The excess amount of $SnTCPP.Cl_2$ was washed with DMF. The material formed by the functionalized SnTCPP of pillared graphene-TiO_2 was named TGSP. The proposed schematic illustration of the prepared TGSP is shown in Figure 1.

Photocatalytic performance of the prepared samples

The photocatalytic activity of the pure TiO_2, TG composites, and

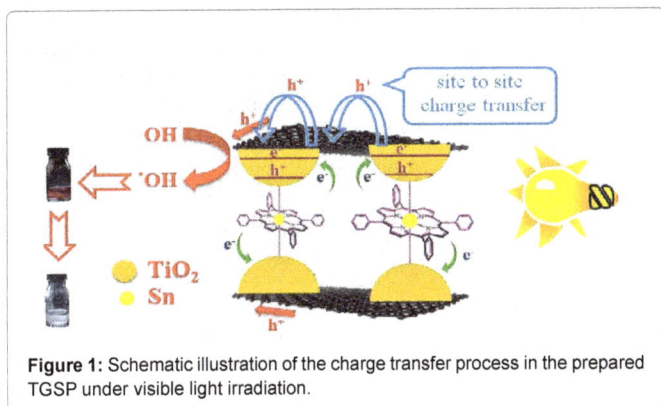

Figure 1: Schematic illustration of the charge transfer process in the prepared TGSP under visible light irradiation.

the TGSP compound was measured at 25°C to degrade the 10 mgL^{-1} MO aqueous solution under visible light irradiation. Figure 2a shows the temporal absorption spectral changes of MO in the photodegradation process over the prepared TGSP photocatalyst. Control experiments on the photodegradation of MO indicated that a negligible degradation of MO molecules could be observed in the absence of irradiation or photocatalyst. This suggests that the self-sensitized photodegradation of MO could hardly occur under the experimental conditions of this study. As shown in Figure 2a, the main absorption peak of the MO molecule decreases with extension of the exposure time and completely disappears after 180 min, which indicates the total decomposition of MO.

To demonstrate the enhancement in the photocatalytic performance of the prepared TGSP, comparative experiments were conducted for the degradation of MO by the pure TiO_2 under the same experimental conditions. Based on the photodegradation results, the photoactivity of TG (3%) was about 1.5 times higher than that of the pure TiO_2 (Figure 2b). To the best of our knowledge, this result represents the highest photoactivity recorded for the graphene-TiO_2 composites. Compared to the pure TiO_2, the TG composites with 1% and 3% content of graphene showed an enhanced photodegradation efficiency of MO. In addition, the TG (3%) composite exhibited the highest photoactivity among the synthesized TG composites. Further increase of the mass ratio of graphene led to a decrease in photoactivity of the composites. These results are correspondence with the photocurrent responses of the prepared samples.

Results and Discussion

Description of photocatalytic activity

In the photocatalysis process, electron-hole pairs are produced on photocatalysts. The photogenerated electrons and holes migrate to the surface of the photocatalysts and then can be trapped generally by the oxygen and surface hydroxyls to produce radical species which react with the adsorbed reactants [21]. It is well known that the photodegradation efficiency is strongly related to the electron-hole pairs produced by light irradiation on photocatalysts [22].

In the prepared TG composites, the photoactivity enhancement of TiO_2 after composition with graphene could be attributed to the higher separation efficiency of the electron-hole pair caused by the rapid photoinduced charge separation. Furthermore, the inhibition of the electron-hole pair recombination increases the number of the holes participating in the photooxidation reaction. The photogenerated

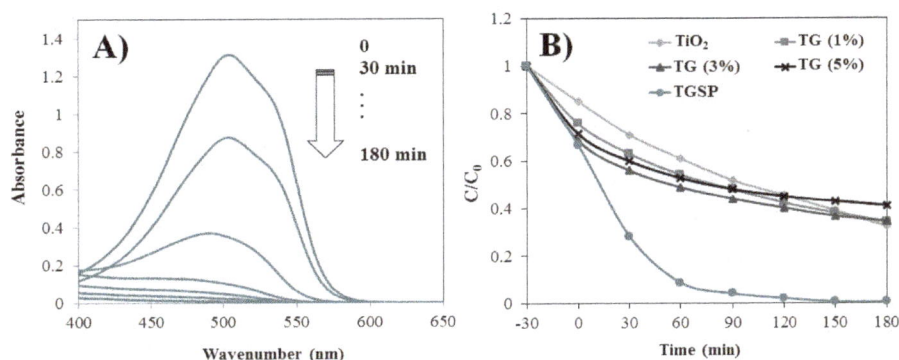

Figure 2: (a) Temporal absorption spectral changes of MO aqueous solution over TGSP, (b) Photodegradation results of MO over the prepared photocatalysts (0.6 gL^{-1}) under visible light irradiation.

electrons and holes would transfer from TiO$_2$ to the graphene [4]. The graphene with delocalized conjugated π structure and superior electrical conductivity rapidly transfers the electrons and effectively inhibits the charge recombination. Hence, this would result in a higher photocurrent density and more charge carriers to form reactive hydroxyl radicals promoting the photodegradation of MO dyes. In addition to improved charge separation, the graphene sheets can overcome the mass transfer limitation by increasing the availability of the MO molecules near the photocatalyst surface. Furthermore, the improved photoactivity of the TG composites is attributed to the ability of the graphene sheets in capturing and transferring the photogenerated charges.

With regard to Figure 2b, it is determined that the photoactivity of the TG (3%) composite was effectively enhanced after photosensitization by tin porphyrin dyes (SnTCPP). This pillared graphene-TiO$_2$ (TGSP) exhibited significant activity for the complete photocatalytic degradation of MO aqueous solution in the period of three hours. Such pillared carbon nanostructure exhibited unique photoactivity due to the synergistic effect between the graphene sheets and the SnTCPP pillars. In this photocatalyst, the photoinduced electron transfers from the excited SnTCPP to deep surface of the TG layers in the TGSP structure. Then, in the TGSP system, the effectively separated electrons and holes are formed upon illumination. The schematic illustration of charge migration in the prepared TGSP is shown in Figure 1.

In TGSP, the SnTCPP molecules are excited under visible light irradiation. According to our finding about porphyrin based MOF in the previous work and as can be seen in the DRS spectra of TGSP, it is found that the highly efficient light-harvesting structure of the SnTCPP pillared TG composite can be attributed to densely embedded porphyrin chromophores with high visible absorptivity within the framework [23]. The excited electrons of SnTCPP transfer from SnTCPP to TiO$_2$ through the bond between the tin atom and the hydroxyl group of TiO$_2$ surface. Moreover, graphene due to its high charge mobility prevents the recombination of photoinduced electron-hole pairs. The axial ligand increases the electronic coupling between SnTCPP and TiO$_2$. On the other hand, the metal ion, among other factors, plays an important role in the photocatalytic activity of TGSP. As shown in Figure 3 and via the electrochemical investigations, the excited potential of tin porphyrin matches well with the conduction bond potential of TiO$_2$. The schematic illustration of the charge transfer in TGSP is shown in Figure 3.

As shown in Figure 3, in the TGSP compound the SnTCPP molecules acts as a light harvesting agent. SnTCPP can generate

photoinduced electrons and holes, leading to a larger shift in the visible light range. The enhanced photoactivity of the prepared TGSP is due to the facile intra-nanostructure site-to-site charge migration to the TGSP surface, and efficient charge transfer at the TGSP-solution interface. Furthermore, such aggregate type of SnTCPP might be beneficial to inducing a cascade of photoinduced charges from the parallel TG composite sheets without photodamaging the SnTCPP molecules in the TGSP structure.

Photocatalytic mechanism

O^{2-} and OH· are the main reactive agents which produce, in the photocatalytic process. Therefore, to investigate the photodegradation mechanism of methyl orange, the photoreaction was investigated in two conditions: (1) in the absence of O2, and (2) in the presence of hydroxyl radical scavenger.

Figure 4 exhibits the relevant photodegradation results of the TG (3%) and TGSP photocatalysis reactions under visible light irradiation in the absence of O$_2$ as well as in the presence of tert-butanol (1 mM) as an OH radical scavenger (Figure 4 Photodegradation of MO over TGSP A) in the oxygen free solution, B) in the presence of t-BuOH (1 mM) as radical scavenger and over TG (3%) C) in the oxygen free solution, D) in the presence of t-BuOH (1 mM) as radical scavenger

The absence of O$_2$ (under N$_2$ saturation condition) severely inhibited the photocatalytic activity of the TG composite. This could be resulted from the generation of O^{2-} in the photocatalytic reaction of the TG composite. Wenfeng Shangguan and coworkers [24] determined that the photocatalytic hydrogen evolution rate of the TiO$_2$-graphene composite increased significantly after injection small amount of air into the reaction vessel under UV light irradiation. TiO$_2$ is an n-type semiconductor because of its donor-like oxygen vacancies. Graphene is a p-type semiconductor due to the residual oxygen containing groups. Therefore, a p-n junction at their interface forms in the TG composition. Figure 4c shows that the photoactivity of the TG (3%) composite decreased significantly after the injection of N$_2$ into the sealed reaction cell. Therefore, It is determined that the O^{2-} produced by the reaction of the photoinduced electrons can moderately enhance the oxygen groups on the graphene sheets. Thus, the p-doping concentration of the residual oxygen containing groups on the graphene sheets is directly affected. Additionally, it would have an important influence on the separation efficiency of the photogenerated electrons and holes in the TG composite [25]. Hence, the photocatalytic activity of the TG composite increased.

Figure 4b and 4d show the photodegradation of MO with the addition of tert-butanol as an OH radical scavenger in the photoreaction of the TGSP and TG (3%) photocatalysts. Adding tert-butanol caused a significant decrease in the photocatalytic activity of the TGSP photocatalyst. Thus, the free OH radical formation is the main active oxidation species in the TGSP photocatalytic mechanism. However, the TG photoactivity relatively decreased after the addition of the OH radical scavenger. This result indicates that a dual mechanism involving both the OH· and the O^{2-} radical oxidation is expected in the photocatalysis process of the TG (3%) composite.

Conclusion

In summary, TiO$_2$-graphene sheets were pillared with SnTCPP to prepare graphene-based materials with good electron conductivity. The complete photodegradation of MO over TGSP was observed within 180 min. This is because the intercalation of SnTCPP between the graphene sheets led to high efficiency in the photocatalytic activity.

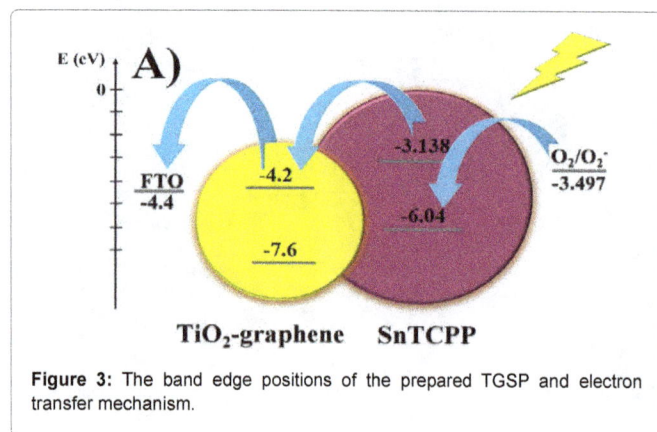

Figure 3: The band edge positions of the prepared TGSP and electron transfer mechanism.

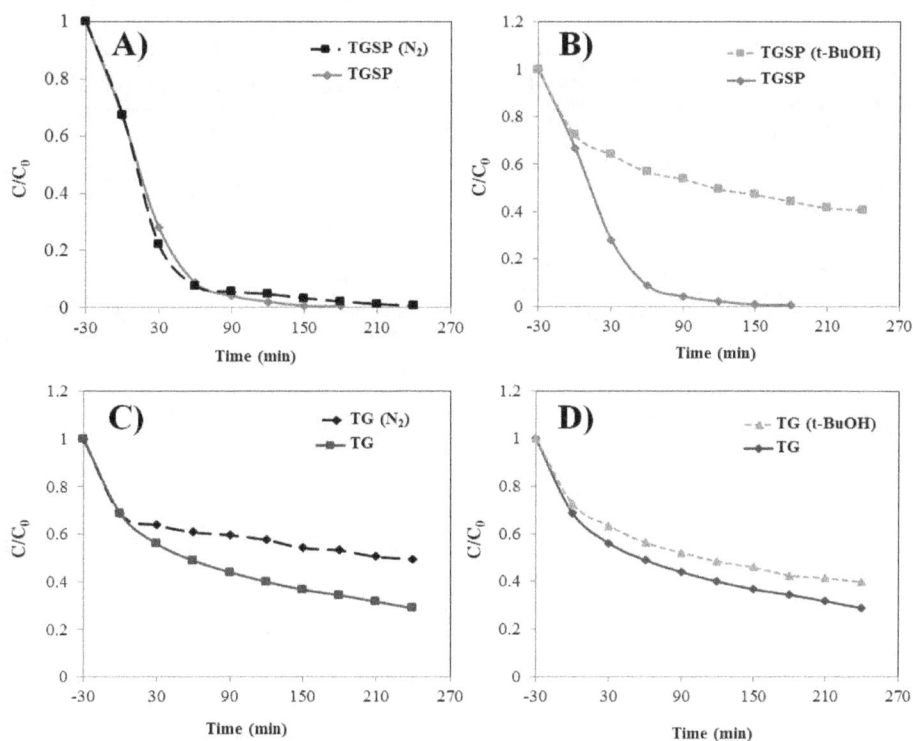

Figure 4: Photodegradation of MO over TGSP (a) in the oxygen free solution, (b) in the presence of t-BuOH (1 mM) as radical scavenger and over TG (3%) (c) in the oxygen free solution, (d) in the presence of t-BuOH (1 mM) as radical scavenger.

This intercalation not only increases the adsorption quantity of porphyrin photosensitizer in the TGSP structure, but also acts as a charge transfer channel, which facilitates electron injection through the axial ligand of the excited SnTCPP into the conduction band of TiO_2. This process is similar to facilitating site-to-site intra-nanostructure charge migration through the SnTCPP photosensitizer to the TGSP surface. Furthermore, the charge recombination is decreased because of the excellent conductivity of the pillared graphene. Thus, the TGSP was observed to have highly efficient photoactivity with high stability under visible light irradiation.

Furthermore, it was determined that the OH radical formation is the main active oxidation species in the TGSP photocatalytic mechanism. Overall, pillared graphene-TiO_2 materials exhibits sufficient potential for further utilization of solar energy.

References

1. Zargari S, Rahimi R, Yousefi A (2016) An efficient visible light photocatalyst based on tin porphyrin intercalated between TiO2-graphene nanosheets for inactivation of E. coli and investigation of charge transfer mechanism. RSC Advances 6: 24218-24228.

2. Zargari S, Rahimi R, Ghaffarinejad A, Morsali A (2016) Enhanced visible light photocurrent response and photodegradation efficiency over TiO2-graphene nanocomposite pillared with tin porphyrin. Journal of colloid and interface science 466: 310-321.

3. Rahimi R, Zargari S, Ghaffarinejad A, Morsali A (2015) Investigation of the synergistic effect of porphyrin photosensitizer on graphene-TiO2 nanocomposite for visible light photoactivity improvement. Environmental Progress & Sustainable Energy 35: 642-652.

4. Rahimi R, Zargari S, Yousefi A, Berijani MY, Ghaffarinejad A, et al. (2015) Visible light photocatalytic disinfection of E. coli with TiO2-graphene nanocomposite sensitized with tetrakis (4-carboxyphenyl) porphyrin. Applied Surface Science 355: 1098-1106.

5. Sakamoto J, van Heijst J, Lukin O, Schlüter AD (2009) Two-Dimensional Polymers: Just a Dream of Synthetic Chemists? Angewandte Chemie International Edition 48: 1030-1069.

6. Wang X, Yuan B, Xie Z, Wang D, Zhang R (2015) ZnS-CdS/Graphene oxide heterostructures prepared by a light irradiation-assisted method for effective photocatalytic hydrogen generation. Journal of colloid and interface science 446: 150-154.

7. Kim CH, Kim BH, Yang KS (2012) TiO2 nanoparticles loaded on graphene/ carbon composite nanofibers by electrospinning for increased photocatalysis. Carbon 50: 2472-2481.

8. Hu C, Chen F, Lu T, Lian C, Zheng S, et al. (2014) Water-phase strategy for synthesis of TiO2-graphene composites with tunable structure for high performance photocatalysts. Applied Surface Science 317: 648-656.

9. Das D, Dutta RK (2015) A novel method of synthesis of small band gap SnS nanorods and its efficient photocatalytic dye degradation. Journal of colloid and interface science 457: 339-344.

10. Saravanan R, Khan MM, Gupta VK, Mosquera E, Gracia F, et al. (2015) ZnO/ Ag/CdO nanocomposite for visible light-induced photocatalytic degradation of industrial textile effluents. Journal of colloid and interface science 452: 126-133.

11. Shao Y, Cao C, Chen S, He M, Fang J, et al. (2015) Investigation of nitrogen doped and carbon species decorated TiO2 with enhanced visible light photocatalytic activity by using chitosan. Applied Catalysis B: Environmental 179: 344-351.

12. Li C, Wang J, Guo H, Ding S (2015) Low temperature synthesis of polyaniline-crystalline TiO2-halloysite composite nanotubes with enhanced visible light photocatalytic activity. Journal of colloid and interface science 458: 1-13.

13. Rabbani M, Rahimi R, Bozorgpour M, Shokraiyan J, Moghaddam SS (2014) Photocatalytic application of hollow CuO microspheres with hierarchical dandelion-like structures synthesized by a simple template free approach. Materials Letters 119: 39-42.

14. Zheng X, Li D, Li X, Chen J, Cao C, et al. (2015) Construction of ZnO/TiO2 photonic crystal heterostructures for enhanced photocatalytic properties. Applied Catalysis B: Environmental 168: 408-415.

15. Li X, Chen X, Niu H, Han X, Zhang T, et al. (2015) The synthesis of CdS/TiO2 hetero-nanofibers with enhanced visible photocatalytic activity. Journal of colloid and interface science 452: 89-97.

16. Batniji AY, Morjan R, Abdel-Latif MS, El-Agez TM, Taya SA, et al. (2014) Aldimine derivatives as photosensitizers for dye-sensitized solar cells. Turkish Journal of Physics 38: 86-90.

17. Rahimi R, Moghaddas MM, Zargari S (2013) Investigation of the anchoring silane coupling reagent effect in porphyrin sensitized mesoporous V-TiO$_2$ on the photodegradation efficiency of methyl orange under visible light irradiation. Journal of sol-gel science and technology 65: 420-429.

18. Rahimi R, Fard EH, Saadati S, Rabbani M (2012) Degradation of methylene blue via Co-TiO2 nano powders modified by meso-tetra (carboxyphenyl) porphyrin. Journal of Sol-Gel science and technology 62: 351-357.

19. Hummers WS, Offeman RE (1958) Preparation of graphitic oxide. J Am Chem Soc 80: 1339-1339.

20. Rahimi R, Mahjoub Moghaddas M, Zargari S (2013) Synthesis and exfoliation of isocyanate-treated graphene oxide nanoplatelets. Sol-Gel Sci Technol 65: 420-429.

21. Dutta RK, Nenavathu BP, Talukdar S (2014) Anomalous antibacterial activity and dye degradation by selenium doped ZnO nanoparticles. Colloids and Surfaces B: Biointerfaces 114: 218-224.

22. Hu C, Zheng S, Lian C, Chen F, Lu T, et al. (2014) One-step synthesis of a sulfur-graphene composite with enhanced photocatalytic performance. Applied Surface Science 314: 266-272.

23. Rahimi R, Shariatinia S, Zargari S, Berijani MY, Ghaffarinejad A, et al. (2015) Synthesis, characterization, and photocurrent generation of a new nanocomposite based Cu-TCPP MOF and ZnO nanorod. RSC Advances 5: 46624-46631.

24. Gao H, Chen W, Yuan J, Jiang Z, Hu G, et al. (2013) Controllable O2•-oxidization graphene in TiO2/graphene composite and its effect on photocatalytic hydrogen evolution. International journal of hydrogen energy 38: 13110-13116.

25. Chang J, Yang J, Ma P, Wu D, Tian L, Gao Z, et al. (2013) Hierarchical titania mesoporous sphere/graphene composite, synthesis and application as photoanode in dye sensitized solar cells. Journal of colloid and interface science 394: 231-236.

13

Microstructure Evolution and Mangnetique Proprieties of Nanocrystalline Fe₆₀ Cu₃₀Al₁₀ Prepared by Combustion Processes

Hafs A[1]*, Benaldijia A[2] and Aitbara A[1]

[1]University of El Tarf, B.P. 73, El Tarf, 36000, Algeria
[2]University of Annaba, B.P. 12, Annaba, 23000 Algeria

Abstract

Nanostructured disordered iron-aluminium-copper alloy of Fe-30% Al-10% Cu composition was prepared by Self-propagating high-temperature synthesis (SHS) and thermel explosin (TE) techniques of mechanically activated mixture of Fe, Al, Cu powders. The transformations occurring in the material during combustion were studied with the use of X-ray diffraction. Auger spectroscopy and atomic emission spectroscopy (AES) was used to determine the phase composition of the phase formation. Finally, the Magnetic properties were also investigated, for combution processers the magnetic behavior slightly softened becoming a semi hard ferromagnetic.

Keywords: Combustion processes; Noncrystalline; Fe₆₀ Cu₃₀Al₁₀; AES; Magnetic

Introduction

Recently a great deal of interest has been generated in the synthesis of advanced engineering materials in their stable, metastable, nanocrystalline and amorphous phases and making their scope much wider. For that, nanocrystalline alloys are relatively new functionally important materials that demonstrate superior properties in a wide range of conditions and have a number of important applications. Mechanical alloying (MA) is one of the most promising and rapidly developing methods for nanocrystalline materials on an industrial scale [1,2]. Some properties, such as high strength and hardness [3,4] ductilization of brittle materials [5,6] and enhanced diffusivity [7], that are superior to those of the conventional materials, may result from the nanocrystalline structure. In comparison with the solidification route, MA is an expensive and simpler technique to process the material into nanocrystalline or amorphous state. Furthermore, it is easier to produce the nanocrystalline phase in a wider composition range by MA than by solidification methods. Additionally, since MA processing is carried out in the solid state, phase diagram restrictions do not seem to apply to the phases produced by the technique [1,8,9]. Further, it appears that the mechanism by which a nanocrystalline or an amorphous phase is formed is different between solid-state processed and liquid-state processed alloys. The Fe-Al alloy systems are attractive for their potential magnetic and mechanical applications [10,11]. Moreover, they have n excellent corrosion resistance at elevated temperatures, and a low cost and relatively low density compared to Ni-, Co-, and Fe-based superalloys [12-15]. FeAl has a B2 (CsCl) structure and exists over a wide range of Al concentration at room temperature (34-52% Al) [16].

In recent years, a number of studies have been reported on mechanical alloying of Fe-Al and Fe-Cu binary systems. It is established that the formation of the supersaturated solid solution (SSS) α-Fe(Al) as a final product of MA takes place with x ≤ 60 at.% Al though in a number of papers the formation of SSS was found with x = 75 at.% Al [17-22]. The Fe-Cu system has negligible mutual solid solution solubility in equilibrium at temperatures below 700°C (a miscibility of approximately 3%) [23,24] because of the large positive enthalpy of mixing (ΔH_{mix} = 13 kJ mol⁻¹ for Fe-Cu system of 50: 50 in mol % according to the Miedema model) [25]. Eckert et al. [26] have reported that single phase FCC alloys and single phase BCC alloys are formed by MA in the FexCu100-x system with x b 60 and with x>80, respectively.

Moreover, Majumdar et al. [27] and Gaffet et al. [28] have deduced that the formation of the BCC Fe-Cu solid solution is restricted to 0-20 wt.% Cu. According to their investigation, this process occurs in two steps: a nanocrystallization step and an Fe (Cu) and/or Cu(Fe) formation step. Knowledge of ternary additions to B2 alloy systems is one of the key points in understanding and controlling the mechanical proprieties of these materials and is critical to alloy design. Moreover, successful application of Fe-Al based intermetallics at elevated temperatures is critically dependent on improvement of their creep resistance [29].

Strengthening by second phase particles represents a potential method in this direction [30,31]. Disordering of FeAl intermetallics has been a subject of numerous theoretical [32,33] and experimental investigations [34-37] over the past two decades. Their results indicated that crushing creates, through plastic deformation, a large number of antiphase boundaries leading to antiphase domains with average linear dimensions of about six atomic distances. The influence of structural disorder on the magnetic properties has been evidenced, in FeAl, in different types of microstructures such as cold worked single crystals [33,38-40] quenched or cold worked polycrystalline materials [33,40,41] or ball milled and mechanically alloyed nanoparticles [33,42-44]. It is well established that magnetic properties of FeAl alloy, show a strong dependence on the degree of atomic order [38]. Disordering the alloy induces paramagnetic → ferromagnetic transitions. It was actually argued [33] that the origin of the magnetic interactions in disordered FeAl may not arise solely from nearest-neighbor magnetism but also from changes in the band structure of the material induced by Δa (increase in the lattice parameter). Noguès et al. [33], have shown, besides the effects of the local environment of the magnetic ions, disorder-induced lattice changes and they demonstrate experimentally and theoretically that about 35-45% of the magnetic moment of Fe₆₀Al₄₀ alloy arises from lattice expansion effects induced during

*Corresponding author: Hafs A, University of El Tarf, B.P. 73, El Tarf, 36000, Algeria, E-mail: hafsali2006@yahoo.fr

the disordering process. MA of magnetic Fe-based alloys leads to the formation of supersaturated solid solution, multiphase or possibly amorphous structure. Increasing milling time is accompanied by both a decrease of grain size and an increase of internal microstrain, which is a common behavior to all metallic systems prepared by mechanical alloying. In general, some magnetic properties can be improved when the grain size is reduced to the nanoscale, while the presence of stresses and defects introduced by MA impairs the magnetic behavior; the overall magnetic property is a competition between decreasing grain size and increasing strain [45].

In the present work, $Fe_{60}Al_{30}Cu_{10}$ powders were mechanically alloyed by high energy ball milling. The fabrication of $Fe_{60}Al_{30}Cu_{10}$ by different combustion techniques such as SHS [46-48] and electrothermal explosion (ETE). The advantage of this two methods is shown by comparing structural properties of $Fe_{60}Al_{30}Cu_{10}$ synthesized both by SHS and the ETE.

The crystalline phases of this material are examined magnetic properties of the bulk such as Hc, Js, Br are studied by a vibrating sample magnetometer (VSM) under the applied field of 1500 kA/m.

Experimental

Elemental powders of Fe, Al and Cu with purity of 99.99% of the nominal composition $Fe_{60}Cu_{30}Al_{10}$ powder (in wt.%) were mixed and milled for 10 h. The mechanical alloying experiments were performed in a high energy planetary ball mill (Fritsch Pulverisette 6) in argon atmosphere using stainless steel vials and balls with diameter of about 12 mm. The rotation speed could be varied within the range 400–600 rpm. In order to avoid oxidation during alloying, the ball mill was filled with high purity argon gas. The vial was opened in mill for 60 and 120 min to assure high homogenization. The powders were compacted into small discs (2-4) × 13 mm in size at the compacting pressure P = 7000 psi and then subjected to SHS (Self-propagating high-temperature synthesis), thermal explosion, and annealing upon heating with a heavy current (400 A) for 1 to 4 min under an Ar pressure of 10 atm (Figure 1).

The structural properties were determined by XRD using a Philips diffractometer (Co-Kα radiation). Further structural characterizations were carried out by energy dispersive X-ray microanalysis (EDX), scanning electron microscopy (SEM). AES and depth profiling measurements were carried out in the UHV cell of a scanning Auger spectrometer (Microlab VG MKII) equipped with a hemispherical analyzer. In addition we have used the using a vibrating magnetometer (VSM) to explore the magnetic properties and domains of the $Fe_{60}Al_{30}Cu_{10}$ phase after synthesis.

Results and Discussion

Figures 2a-2c shows the SEM pictures of raw Fe, Cu and Al powders. The powder particles of Cu were found to be nearly spherical shaped with an average size in the range of 5-50 μm. the Al and Fe elemental particle powders with irregular shapes. For the powders subject to 10 h of milling, it is very clear from Figure 3 that all the initial shape of the powders was changed and powders having composite structure started to form. An increase in the particle size is noticed, indicating a "primary" welding of very small particles to the surface.

Figure 4 shows the X-ray diffraction patterns of the $Fe_{60}Cu_{30}Al_{10}$ mixture at milling time of 10 h of the combustion products formed in SHS (Self-propagating high-temperature synthesis) and ETE reaction. Clearly, in Figures 4a and 4b after a combustion of SHS products and ETE reaction, the peaks on phase formation of Nanocrystalline Fe(Al, Cu) appeared in all the samples (this is can be explained par the formation of the BCC Fe(Al, Cu)) and though a little FeAl phase. Its mean crystallite size are refined as the milling time increases and reached a final value of about 20 nm. The mean crystallite size at 10 h of milling is on the same level in the two coexists, Fe(Al) and Fe (Al, Cu) phases.

According to this model, when the grain size is smaller than magnetic exchange length L_{ex}, the origin of the soft magnetic properties in the nanocrystalline materials is ascribed to average out the magnetocrystalline anisotropy. Theoretically, as the L_{ex} for Fe-based alloys is of the order of 20-30 nm that corresponds approximately to D value of the Fe (Al, Cu) alloy in the present study at earlier stage of processing. Further milling does not reduce the grain size appreciably. For this reason the coercivity does not show great changes after 4 h of milling.

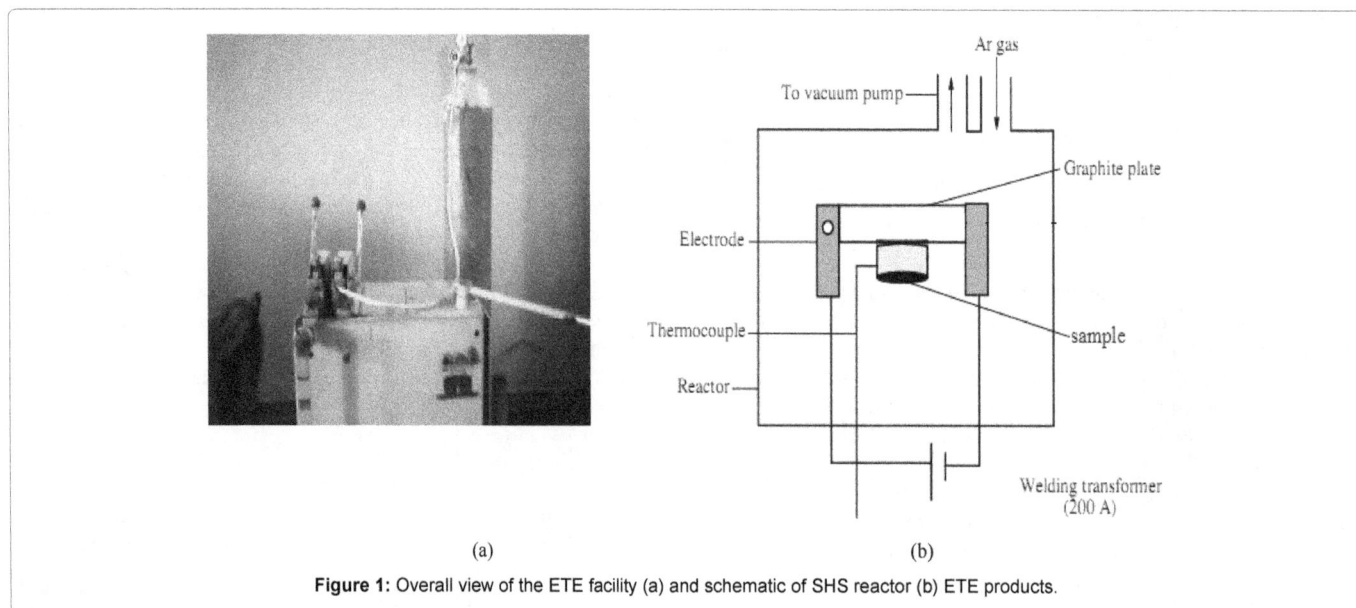

(a)   (b)

Figure 1: Overall view of the ETE facility (a) and schematic of SHS reactor (b) ETE products.

Figure 2: SEM images of the powders (a) Fe (b) Cu, (c) Al.

Figure 3: SEM micrographs of the powder mixture of Fe, Al and Cu milled for 10 h.

It suggests that SHS and ETE is an effective methods to accelerate the formation of BCC Fe(Al, Cu) phase. X-ray diffraction is a convenient method for determining the mean size of nanocrystallites in nanocrystalline bulk materials.

This can be attributed to the fact that "crystallite size" is not synonymous with "particle size", while X-Ray diffraction is sensitive to the crystallite size inside the particles. From the well-known Scherrer formula the average crystallite size, L, is:

$$L = \frac{K\lambda}{\beta . \cos\theta} \tag{1}$$

Where λ is the X-ray wavelength in nanometer (nm), β is the peak width of the diffraction peak profile at half maximum height resulting from small crystallite size in radians and K is a constant related to crystallite shape, normally taken as 0.9. The value of β in 2θ axis of diffraction profile must be in radians. The θ can be in degrees or radians, since the $\cos\theta$ corresponds to the same number.

Table 1 shows the crystallite size values for different peaks at combustion of SHS products and ETE reaction. The XRD patterns are observed in Figures 4a and 4b with gradual sharpness of the peaks as the soaking combustion processes (SHS,ETE), indicating the growth and increase of crystallite size. Relatively gradual decrease in $\beta.\cos\theta$ and almost increase in L values L= (const/ $\beta.\cos\theta$) is observed with the increase of 2θ. The AES data in Figures 5a and 5b (corelevel excitation) show the peaks around the kinetic energies E_k = 505, 918, 1392 and 1833 eV corresponding to O, Cu, Al and Fe respectively, according to the energy Atlas of AES.

In Figures 6a and 6b it is shown the hysteresis curves at 5 K and 300 K of $Fe_{60}Cu_{30}Al_{10}$ alloy produced by SHS (Self-propagating high-temperature synthesis) and ETE(electrothermal explosion). It can be seen that increasing the temperature, all the magnetic parameters studied (Ms, Hc, Mr and (Mr/Ms)) as expected, decrease due to the thermal excitation (The rapid decrease in Ms is mainly due to the early interaction between the Fe atoms which are ferromagnetic with Al and Cu atoms that are non-ferromagnetic in nature. Al decreases the magnetic moment of individual Fe sites due to a decrease in the direct ferromagnetic interaction between Fe-Fe sites, and also an antiferromagnetic super-exchange interaction between Fe sites mediated by Al atoms as suggested by Plascak [49]. Increasing temperature in a ferromagnetic material causes an increase of the thermal vibrations of atoms and therefore the magnetic moments are free to rotate and are then arranged randomly. For this reason the magnetic parameters are affected by temperature, reducing them quantitatively and significantly. According to studies carried out by Hu et al. [50] and Amils et al. [51] among others, it has been confirmed the transition from paramagnetic to ferromagnetic state in $48Fe_{32}Al_{20}Cu$ at % alloy by MA even at room temperature.

Conclusions

Nanocrystalline Fe(Al, Cu) powder with crystallite of 10 nm has been successfully synthesized from SHS (Self-propagating high-temperature synthesis) and ETE (electrothermal explosion). The single

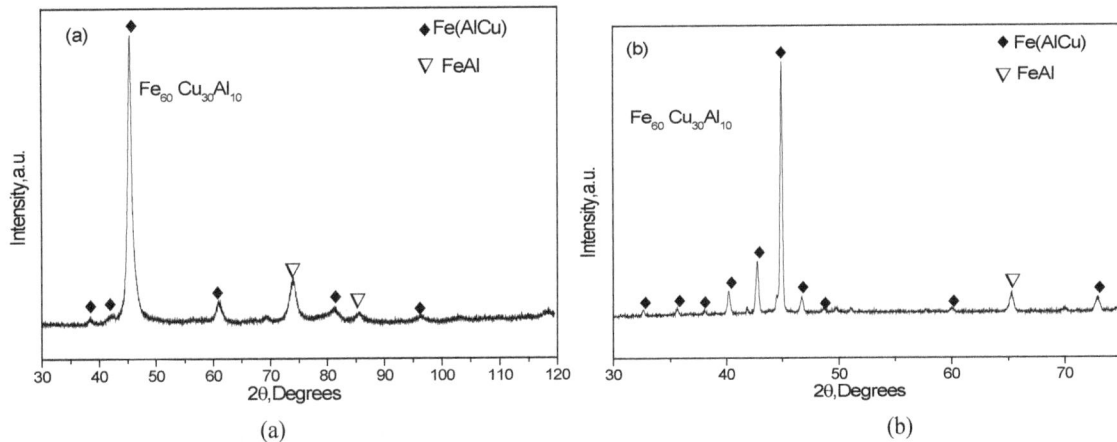

(a)

(b)

Figure 4: Diffraction patterns of mixture powders with the milling time of 10 h (a) obtained by SHS products (b) ETE products.

$2\theta_{hkl}$ Degrees	Values of $\beta.\cos\theta$ for different peaks		Calculated L (nm)
	SHS products	ETE products	
35,341	-	$3,84\times10^{-3}$	35,93
38,155	$4,6\times10^{-3}$	$4,6\times10^{-3}$	30
40,218	$5,58\times10^{-3}$	$5,58\times10^{-3}$	24,73
42,761	$7,4\times10^{-3}$	$7,4\times10^{-3}$	18,65
45,011	$5,50\times10^{-3}$	$5,50\times10^{-3}$	25,09
46,75	-	$12,31\times10^{-3}$	11,21
48,66	-	$3,36\times10^{-3}$	38,33
60,59	$7,1\times10^{-3}$	$7,1\times10^{-3}$	19,43
73,01	-	8×10^{-3}	17.25
81,20	$6,5\times10^{-3}$	-	21.23
96,20	$6,7\times10^{-3}$	-	20,59

Table 1: Calculation of the crystallite sizes of Fe-30% Al-10% Cu composition was prepared by Self-propagating high-temperature synthesis (SHS) and thermel explosin (TE) techniques of mechanically activated mixture of Fe, Al, Cu powders.

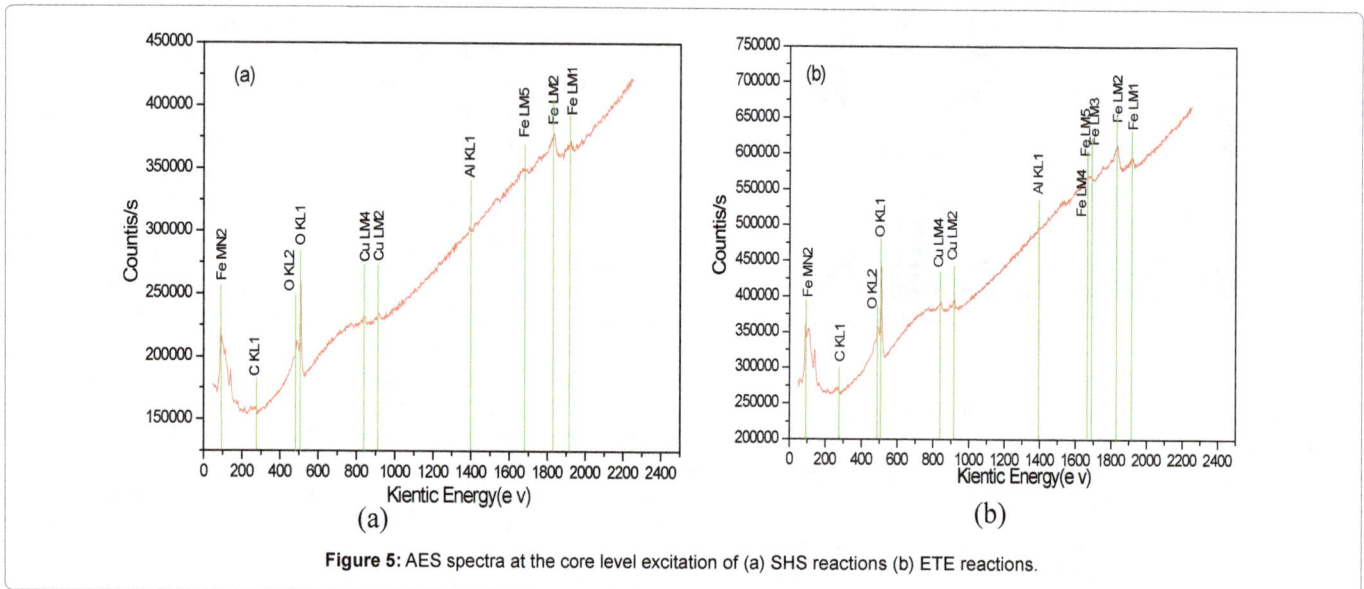

Figure 5: AES spectra at the core level excitation of (a) SHS reactions (b) ETE reactions.

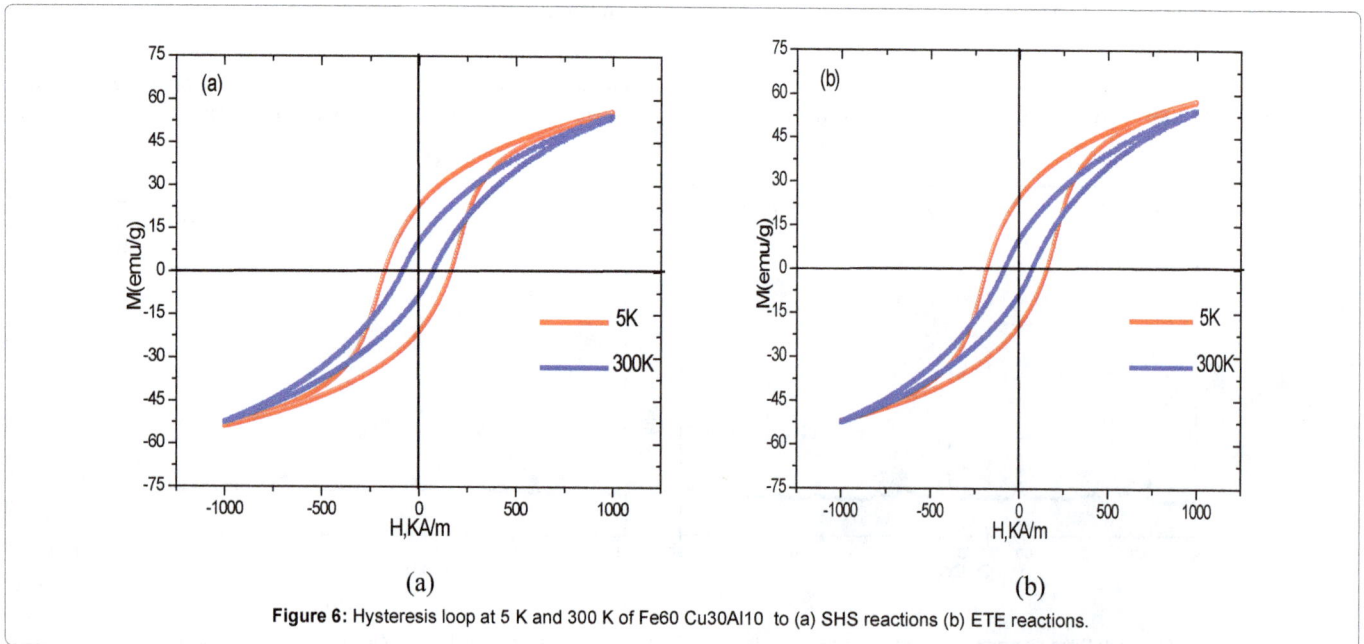

Figure 6: Hysteresis loop at 5 K and 300 K of Fe60 Cu30Al10 to (a) SHS reactions (b) ETE reactions.

phase of BCC Fe (Al,Cu) was formed. It suggests that SHS and ETE is an effective method to accelerate the formation of Fe (Al,Cu) phase. From the magnetic point of view there is a correspondence between microstructural evolution and magnetic behavior in the form of that decreasing the crystallite size increases the magnetism by increasing coercivity and squareness ratio. Consequently as low crystallites size is, stronger hard ferromagnetic material results. Due to thermal vibrations higher temperature produces a decrease in magnetic properties.

Acknowledgments

We are grateful to Prof A. Benaldijia for the synthesis of the samples and to the members of Institut Jean Lamour-Prof. Jamal Bougdira, Dr. Christine Bellouard, Dr. Ghouti Medjaldi, and also to Pascal and Damien Geneve-for their kind help and assistance in XRD, SEM, EDX, and AES measurements.

References

1. Fougere GE, Weertman JR, Siegel RW (1993) On the hardening and softening of nanocrystalline materials. Nanostructured Materials 3: 379-384.

2. Siegel RW, Fougere GE (1995) Mechanical properties of nanocrystalline materials. Nanostructured Materials 6: 205-216.

3. Karch J, Birringer R, Gleiter H (1987) Ceramics ductile at low temperature. Nature 330: 556-558.

4. McFadden SX, Mishra RS, Valiev RZ, Zhilyaev AP, Mukherjee AK (1999) Low temperature superplasticity in nanostructured nickel and metal alloys. Nature 398: 684-686.

5. Birringer R (1989) Nanocrystalline materials. Materials Science and Engineering A 117: 33-43.

6. Suryanarayana C (2001) Mechanical alloying and milling. Progress in Materials Science 46: 1-184.

7. Baghbaderani HA, Sharafi S, Chermahini MD (2012) Investigation of nanostructure formation mechanism and magnetic properties in $Fe_{45}Co_{45}Ni_{10}$ system synthesized by mechanical alloying. Powder Technology 230: 241-246.

8. Nowroozi MA, Janghorban K, Hashemi B, Taghvaei AH, Maleksaeedi S (2013) Microstructural evolution and amorphous phase formation of $Fe_{42}Ni_{28}Zr_8Ta_2B_{10}C_{10}$ alloy produced by mechanical alloying. Powder Technology 233: 255-260.

9. Suryanarayana C (2004) Mechanical Alloying and Milling. Marcel Dekker, New York.

10. Mhadhbi M, Khitouni M, Escoda L, Suñol JJ, Dammak M (2011) Microstructure evolution and magnetic properties of nanocrystalline FeAl obtained by mechanical alloying. Journal of Alloys and Compounds 509: 3293-3298.

11. Salimon AI, Korsunsky AM, Shelekhov EV, Sviridova TA, Kaloshkin D (2001) Crystallochemical aspects of solid state reactions in mechanically alloyed Al-Cu-Fe quasicrystalline powders. Acta Materialia 49: 1821-1823.

12. Run-hua F, Jia-tao S, Hong-Yu G, Kang-ning S, Wei-min W (2005) Structural evolution of mechanically alloyed nanocrystalline Fe-28Al powders. Powder Technology 149: 121-126.

13. Baker I, George EP (1997) Proceedings of the International Symposium on Nickel and Iron Aluminides. The Minerals, Metals and Materials Society, Warrendale.

14. Tortorelli PF, DeVan JH (1992) Behavior of iron aluminides in oxidizing and oxidizing sulfidizing environments. Materials Science and Engineering A 153: 573-577.

15. Haraguchi T, Yoshimi K, Yoo MH, Kato H, Hanada S, et al. (2005) Vacancy clustering and relaxation behavior in rapidly solidified B_2FeAl ribbons. Acta Materialia 53: 3751-3764.

16. Colas D (2004) PhD Thesis, Université Jean Monnet de Saint-Etienne, France.

17. Yelsukov EP, Dorofeev GA (2004) Mechanical alloying in binary Fe-M (M=C, B, Al, Si,Ge, Sn) systems. Journal of Materials Science 39: 5071-5079.

18. Oleszak D, Shingu PH (1994) Mechanical alloying in the FeAl system. Materials Science and Engineering A 181-182: 1217-1221.

19. Enzo S, Frattini R, Gupta R (1996) X-ray powder diffraction and Mossbauer study of nanocrystalline Fe-Al prepared by mechanical alloying. Acta Materialia 44: 3105-3113.

20. Enzo S, Mulas G, Frattini R (1998) The structure of mechanically alloyed AlFe end-products after annealing. Materials Science Forum 385: 269.

21. Enzo S, Frattini R, Mulas G, Delogu F (1998) Structural evolution of AlFe and AlFe powders prepared by mechanical alloying. Materials Science Forum 391: 269.

22. Zeng Q, Baker I (2006) Magnetic properties and thermal ordering of mechanically alloyed Fe-40 at.% Al. Intermetallics 14: 396-405.

23. In: D.P.M. Hansen (Ed.), Constitution of Binary Alloys, Genium Publishing Corporation p. 580.

24. Xie YP, Zhao SJ (2011) The energetic and structural properties of bcc NiCu, FeCu alloys: A first-principles study. Computational Materials Science 50: 2586-2591.

25. Huang X, Mashimo T (1999) Metastable BCC and FCC alloy bulk bodies in Fe-Cu system prepared by mechanical alloying and shock compression. Journal of Alloys and Compounds 288: 299-305.

26. Eckert J, Holzer JC, Johnson WL (1993) Thermal stability and grain growth behavior of mechanically alloyed nanocrystalline Fe-Cu alloys. Journal of Applied Physics 73:131.

27. Majumdar B, Manivel Raja M, Narayanasamy A, Chattopadhyay K (1997) Structural and magnetic investigations on the metastable phases of the mechanically alloyed Fe-Cu system. Journal of Alloys and Compounds 248: 192-200.

28. Gaffet E, Harmelin M, Faudot F (1993) Far-from-equilibrium phase transition induced by mechanical alloying in the Cu-Fe system. Journal of Alloys and Compounds 194: 23-30.

29. Stoloff NS (1998) Iron aluminides: present status and future. Materials Science and Engineering A 258: 1-14.

30. Palm M (2005) Concepts derived from phase diagram studies for the strengthening of Fe-Al based alloys. Intermetallics 13: 1286-1295.

31. Morris DG, Muñoz-Morris MA (2007) Materials Science and Engineering A 462: 45-52.

32. Apiñaniz E, Garitaonandia JS, Plazaola F (2001) Influence of the disorder on themagnetic properties of FeAl alloys: theory. Journal of Non-Crystalline Solids 287: 302-307.

33. Nogués J, Apiñaniz E, Sort J, Amboage M, d'Astuto M, et al. (2006) Volume expansion contribution to the magnetism of atomically disordered intermetallic alloys. Physical Review B 74: 024407.

34. Gialanella S (1995) FeAl alloy disordered by ball milling. Intermetallics 3: 73-76.

35. Collins GS, Peng LSJ (1996) Point defects in FeAl, II, Nuovo Cimento D 18: 329-336.

36. Meyer M, Mendoza-Zélis L, Sánchez FH, Mora H, Clavaguera MT, et al. (1999) Mechanical milling of the intermetallic compound AlFe. Physical Review B 60: 3206.

37. Collins GS, Peng LSJ (1997) Disordering of FeAl by mechanical milling. Materials Science Forum 235: 535-540.

38. Guilemany JM, Cinca N,Casas L, Molins E (2009) Ordering and disordering processes in MA and Mintermetallic iron aluminide powders. Journal of Materials Science 44: 2152-2161.

39. Jonhson CE, Cranshaw TE, Ridout MS (1963) Mössbauer effect in iron alloys. Proceedings of the Physical Society of London 81: 1079-1090.

40. Takahashi S, Umakoshi Y (1990) The influence of plastic deformation on the magnetic properties of Fe-Al alloys. Journal of Physics: Condensed Matter 2: 4007-4011.

41. Yang Y, Baker I, Martin P (1999) On the mechanism of the paramagnetic to ferromagnetic transition in Fe-Al. Philosophical Magazine B 79: 449-461.

42. Varin RA, Czujko T, Bystrzycki J, Calka A (2002) Cold-work induced phenomena in B_2FeAl intermetallics. Materials Science and Engineering A 329: 213-221.

43. Hu WY, Kato T, Fukumoto M (2003) Synthesis and characterization of nanocrystalline iron aluminide intermetallic compounds. Materials Transactions 45: 2678-2687.

44. Amils X, Nogués J, Suriñach S, Muñoz JS, Baró MD, et al. (2003) Quantized topological magnetoelectric effect of the zero-plateau quantum anomalous Hall state. Physical Review B 92: 081107.

45. Apiñaniz E, Plazaola F, Garitaonandia JS, Martin D, Jiménez JA (2003) Study of the enhancement of the magnetic properties of $Fe_{70}Al_{30}$ in the order–disorder transition. Journal of Applied Physics 93: 7649-7651.

46. Benaldjia A, Guellati O, Bounour W, Guerioune M, Ali-Rachedi M, et al. (2008) Titanium Carbide by the SHS Process Ignited with Aluminothermic Reaction. International Journal of Self-Propagating High-Temperature Synthesis 17: 54-57.

47. Guerioune M, Amiour Y, Bounour W, Guellati O, Benaldjia A, et al. (2008) SHS of Shape Memory CuZnAl Alloys. International Journal of Self-Propagating High-Temperature Synthesis 17: 41-48.

48. Bendjemil B (2010) SHS Produced Ni-Mn-Al Magnetic Shape Memory Alloy. International Journal of Self Propagating High Temperature Synthesis19: 110-113.

49. Pearson WB (1967) Handbook of Lattice Spacing and Structures of Metals and Alloys. Pergamon, Oxford.

50. Hu WY, Kato T, Fukumoto M (2003) The Formation and Characterization of the Primary Mg2Si Dendritic Phase in Hypereutectic Al-Mg2Si Alloys. Materials Transactions 44: 2681-2686.

51. Amils X, Nogués J, Suriñach S, Muñoz JS, Baró MD, et al. (2001) Tailoring of paramagnetic (structurally ordered) nanometric grains separated by ferromagnetic (structurally disordered) grain boundaries: Isolating grain-boundary magnetic effects. Phys Rev B 63: 052402.

14

Investigation of the Flexoelectric Coupling Effect on the 180° Domain Wall Structure and Interaction with Defects

Mbarki R[1]*, Borvayeh L[2] and Sabati M[3,4]

[1]*Department of Mechanical Engineering, The Australian College of Kuwait, Meshraf, Kuwait*
[2]*Department of Mathematics, The Australian College of Kuwait, Meshraf, Kuwait*
[3]*Department of Electrical Engineering, The Australian College of Kuwait, Meshraf, Kuwait*
[4]*The Dr. John T. Macdonald Foundation Biomedical Nanotechnology Institute, Department of Radiology, University of Miami, USA*

Abstract

A new theory for 180° domain wall in ferroelectric perovskite material is presented in this work. The effect of flexoelectric coupling on the domain structure is analyzed. We show that the 180° domain wall has a mixed character of Ising and Bloch type wall and that the polarization perpendicular to the domain wall is not zero though it is very small compared to the spontaneous polarization in the case of tetragonal Barium Titanate. Finally, we present the effect of the new finding on the domain wall interaction with defects in the material.

Keywords: Flexoelectric; Coupling; Dielectric; Piezoelectric; Ferroelectric materials

Introduction

Perovskite ferroelectric materials are known for a superior dielectric and piezoelectric response. They have attracted attention for applications in technologies like capacitors, microelectromechanical systems (MEMS), nonvolatile memories as well as nonlinear optical applications. Ferroelectric materials usually display domains consisting of regions that are either ordered along the same axis but with opposite polarity also called 180° domain walls (Figure 1) or along different orthogonal axis. The existence of these domain walls has a significant influence on the material properties. The measured material properties have two contributions, an intrinsic response which is the property of a single domain material and the extrinsic response which consists of domain wall contribution. For example, it has been shown that for Lead zirconate titanate (PZT) at composition near the morphotropic phase boundary, the domain wall contribution represents more than half of the dielectric and piezoelectric response at room temperature [1]. Ghosh et al. [2] studied experimentally the effect of domain wall motion on the piezoelectric properties of ferroelectric material. They showed that the motion of the domain wall is responsible for the high permittivity and piezoelectric coefficient in Barium Titanate.

The 180° domain wall formed in ferroelectric materials is generally assumed to be an Ising type domain. However, it can generally be either an Ising or a Bloch type domain. In the Ising type, the polarization tends to switch sign by decreasing in magnitude in the plane of the domain wall. While in the Bloch type, the polarization tends to switch sign by rotating outside the plane of the wall and maintaining a constant magnitude (Figure 2). A mixture of Ising and Bloch types is possible where the polarization changes sign by rotating outside of the wall plane and decreasing in magnitude as it gets closer to the wall.

Huang et al. [3] investigated the type of the domain that can form in the tetragonal phase of a ferroelectric perovskite. They showed theoretically that upon cooling material, the domain wall can switch from an Ising type to a Bloch type. Lee et al. [4] performed an ab *initio* calculation and showed that the 180° domain wall in ferroelectric perovskite shows a mixed character.

Ferroelectric domain walls have been studied by many researchers [5-9]. The work started by the Landau Devonshire theory where the Gibbs free energy density is expanded to sixth power of the polarization including a polarization gradient term. However, this formulation did not take into consideration the electromechanical coupling between strain and polarization. Further expansion of the Gibbs free energy is required to take into consideration the ferroelastic nature of this material. Later studies included this coupling e.g., Huang et al. [3] and Cao et al. [9].

In the last decade, ferroelectric materials have also been exploited and studied at small scales. Researchers have shown that at the nanoscale flexoelectricity becomes very important especially for perovskite materials which are known for their high flexoelectric coefficient [10,11]. Recent work by Yudin et al. [8] presented a new formulation for the domain wall energy in which, they included the flexoelectric coupling. They assumed that the polarization in the plane

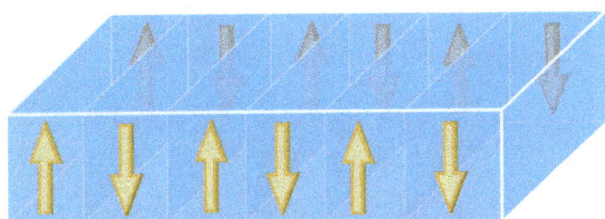

Figure 2: Different types of domain wall: a) Ising type b) Bloch type.

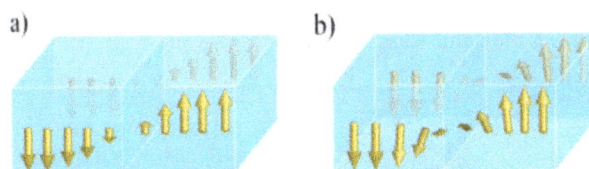

Figure 1: Example of 180° domain wall.

*Corresponding author: Mbarki R, Department of Mechanical Engineering, The Australian College of Kuwait, Meshraf, Kuwait, E-mail: rmbarki@uh.edu

of the domain wall is not zero and that the polarization perpendicular to the wall vanishes because of the depoling field across the wall. Such assumption can reduce the accuracy of the calculation especially with the existence of a high strain gradient across the domain which induces polarization in the perpendicular direction due to the flexoelectric coupling. In this work, we develop a theory for perovskite ferroelectrics in the tetragonal phase. We present numerical results for 180° domain wall in Barium Titanate material with a random defects distribution. We show that the flexoelectric effect transforms the 180° domain wall to a mixture of Bloch and Ising type. Then, we show the effect of the flexoelectricity on the defects interaction with the domain wall.

Model Formulation

Let us consider a thin film of perovskite material with a 4 *mm* symmetry in the tetragonal phase occupying a volume in the space (Figure 3). $[x_1; x_2; x_3]$ is the cubic crystallographic direction. The 180° domain wall lies in the (100) plane. Two platinum electrodes are fixed on each surface of the thin film. Upon cooling to temperature below the Curie temperature T_c, the material tends to reduce its total free energy by creating the domain structure. The creation of the domain structure is accompanied by the distortion of the unit cell which induces a spontaneous strain ε^s and the generation of spontaneous polarization P^s. Far away from the domain wall, the strain and the polarization are defined by these vectors. Near the domain wall, it is reasonable to assume that the polarization varies significantly although the elastic strain remains small. The equilibrium state is defined by the spontaneous polarization P^s, the spontaneous strain ε^s and the spontaneous strain gradient η^s. The polarization gradient is zero at equilibrium. The free energy density can be expanded around the equilibrium state

$$W(P,\varepsilon,\nabla P,\nabla\varepsilon)=\frac{1}{2}\nabla P.\frac{\partial^2 W}{\partial(\nabla P)^2}\nabla P+\frac{1}{2}(\nabla\varepsilon-\eta^s).\frac{\partial^2 W}{\partial(\nabla\varepsilon)^2}(\nabla\varepsilon-\eta^s)$$
$$+\frac{1}{2}(\varepsilon-\varepsilon^s).\frac{\partial^2 W}{\partial\varepsilon^2}(\varepsilon-\varepsilon^s)+\frac{1}{2}\nabla P.\frac{\partial^2 W}{\partial\nabla P\partial\varepsilon}(\varepsilon-\varepsilon^s)+W_L(Ps,\varepsilon^s,0,\eta^s) \quad (1)$$

where $W_L(P, \varepsilon^s, 0, \eta^s)$ is the Landau free energy given by the following expression

$$W_L(P)=\frac{a_1}{2}(P_1^2+P_2^2)+\frac{a_2}{4}(P_1^4+P_2^4)+\frac{a_3}{2}(P_1^2+P_2^2)+\frac{a_4}{6}(P_1^6+P_2^6)+\frac{a_5}{4}(P_1^4+P_2^4) \quad (2)$$

During the phase transition from cubic to tetragonal phase, the free energy given by equation 1 can describe the limit transition. If we match Equation 1 with the free energy in the cubic phase, it is possible to determine an expressions for the spontaneous strain and the spontaneous strain gradient. The free energy density in the cubic phase Wc is given by the following expression

$$W_c(P,\varepsilon,\nabla P,\nabla\varepsilon)=W_L(P)+\frac{1}{2}(\varepsilon-\varepsilon^s).C(\varepsilon-\varepsilon^s)+\frac{1}{2}(\nabla\varepsilon-\eta^s).G(\nabla\varepsilon-\eta^s)+\frac{1}{2}\nabla P.a_0\nabla P+\frac{1}{2}\nabla P.d(\varepsilon-\varepsilon^s)+Pf\nabla\varepsilon \quad (3)$$

where C is the elastic tensor given by the following expression

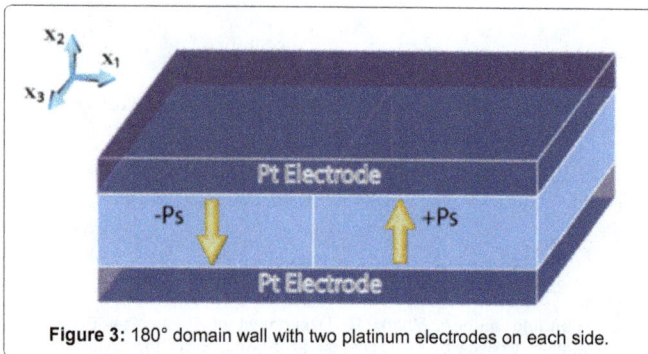

Figure 3: 180° domain wall with two platinum electrodes on each side.

$$C_{ijkl}=\frac{\partial^2 W}{\partial\varepsilon_{ij}\partial\varepsilon_{kl}} \quad (4)$$

d is the fourth order tensor introduced by Mindlin in the polarization gradient theory [12]

$$d_{ijkl}=\frac{\partial^2 W}{\partial P_{i,j}\partial\varepsilon_{kl}} \quad (5)$$

and f is the fourth order flexoelectric tensor

$$f_{ijkl}=\frac{\partial^2 W}{\partial P_l\,\partial\varepsilon_{ij,k}} \quad (6)$$

It is obvious that the spontaneous polarization P^s, the spontaneous strain ε^s and the spontaneous strain gradient η^s are function of the polarization. In order to determine the expressions of these tensors, a matching between the equations 1 and 3 is necessary.

The term coupling the polarization and the strain in equation 1 is the piezoelectric tensor e given by the following expression

$$e=\frac{\partial^2 W}{\partial P\partial\varepsilon} \quad (7)$$

In the cubic phase, the material is not piezoelectric so there is no coupling between polarization and strain. However, at the transition limit between cubic and tetragonal phases, deriving the free energy in the cubic phase given by Equation (3) with respect to the polarization and the strain, we should be able to find an equivalent to this coupling.

$$e=\frac{\partial^2 W_c}{\partial P\partial\varepsilon}=-C\frac{d\varepsilon^s}{dP} \quad (8)$$

The integration of the equation 8 gives the following result

$$\varepsilon^s=\varepsilon^s(P^s)-S:e(P-P^s) \quad (9)$$

where $S = C^{-1}$ is the elastic compliance.

Same procedure used to determine the spontaneous strain gradient η^s gives the following result

$$\eta^s=-G^{-1}:f(P-P^s) \quad (10)$$

Equation 1 can be rewritten as follow:

$$W(P,\varepsilon,\nabla P,\nabla\varepsilon)=\frac{1}{2}(\varepsilon-\varepsilon^s).C(\varepsilon-\varepsilon^s)+\frac{1}{2}(\nabla\varepsilon-\eta^s).G(\nabla\varepsilon-\eta^s)+\frac{1}{2}\nabla P.a_0\nabla P+\frac{1}{2}\nabla P.d(\varepsilon-\varepsilon^s)+W_L(P) \quad (11)$$

The governing equations of polarization and strain evolution are obtained as the gradient flow associated with the Gibbs free energy. For further information on how to obtain these equations, the reader is advised to refer to Zhang et al. [13]:

$$\mu\frac{dP}{dt}=-\frac{\partial W}{\partial P}+\nabla.(\frac{\partial W}{\partial P\nabla})-\nabla\phi \qquad in\,\Omega \quad (12)$$

$$v\frac{du}{dt}=\tilde{N}.(\frac{\partial W}{\partial\varepsilon})-\nabla\nabla.(\frac{\partial W}{\partial\nabla\varepsilon})=0 \qquad in\,\Omega \quad (13)$$

Subjected to the corresponding boundary conditions on $\partial\Omega$:

$$\frac{\partial W}{\partial\nabla P}.n=0 \qquad on\,\partial\Omega\,and \quad (14)$$

$$(\frac{\partial W}{\partial\varepsilon}-\nabla.(\frac{\partial W}{\partial\nabla\varepsilon})).n=0 \qquad on\,\partial\Omega, \quad (15)$$

where n is the outward normal vector to the surface $\partial\Omega$, Φ is the electrostatic potential, μ and v are respectively, the inverse of the polarization and strain mobility.

The 180° domain wall can be described as 1D problem where all the variables are function of one independent variable x_1. There are no assumptions on the polarization vector. The spontaneous polarization can be described by the following equations:

$$P_1^s = \frac{P_0}{2}(tanh(\frac{P_1 + P_2}{l}) + tanh(\frac{P_1 - P_2}{l})) \tag{16}$$

$$P_2^s = \frac{P_0}{2}(tanh(\frac{P_1 + P_2}{l}) - tanh(\frac{P_1 - P_2}{l}))$$

where P_0 is the magnitude of the spontaneous polarization at large distance from the domain wall, and l is a shape factor that controls how fast the spontaneous polarization changes across the domain wall. The component of the polarization P_2 is parallel to the domain wall and the component P_1 is normal to the wall. We consider two cases: the first case is a perovskite crystal without defects and the second case is a doped perovskite crystal with random oxygen vacancies. The electrostatic equation for the first case is the solution of the Poisson equation:

$$\nabla.(-\varepsilon_0\nabla\Phi + P) = 0 \tag{17}$$

$$n.(-\varepsilon_0\nabla\Phi + P) = 0 \tag{18}$$

In the presence of defects in crystal, the electrostatic potential Φ, the polarization P and the charge density ρ are related by the Maxwell equation

$$\nabla.(-\varepsilon_0\nabla\Phi + P_{X\Omega}) = \rho X\Omega \tag{19}$$

where $X\Omega$ is a characteristic function

$$\times_\Omega = \begin{cases} 1, & if \ x \in \Omega \\ 0, & otherwise \end{cases} \tag{20}$$

The charge density at any point is

$$\rho = q(N_d^+ - n_c + n_v) \tag{21}$$

where q is the electron charge, N_d^+ is the density of ionized donors, n_c is the density of electron in conduction band and n_v is the density of holes in the valence band. N_d^+, n_c and n_v are given by the Fermi-Dirac Distribution for fermions:

$$n_c = \frac{N_c}{1 + e^{\beta(E_c - q\phi - \mu_{nc})}} \tag{22}$$

$$n_v = \frac{N_v}{1 + e^{\beta(-E_v + q\phi - \mu_{nv})}} \tag{23}$$

$$N_d^+ = N_d(1 - \frac{1}{1 + \frac{1}{2}e^{\beta(E_d - q\phi - (\mu_{N_d^0} - \mu_{N_d^+}))}}) \tag{24}$$

$\mu_{n_c}, \mu_{n_v}, \mu_{N_d^0}$ and $\mu_{N_d^+}$ are the electrochemical potentials. E_c is the energy at the bottom of the conduction band, E_v is the energy on the top of the valence band and E_d is the donors level and $\beta = (K_bT)^{-1}$.

The space charge evolution in Ω is given by:

$$\frac{dn_c}{dt} = k_1N_d^0(1 - e^{\beta(\mu_{nc} + \mu_{N_d^+} - \mu_{N_d^0})}) + k_2N_v(1 - e^{\beta(\mu_{nc} + \mu_{pv})}) + \nabla.(K_1\nabla\mu_{n_c}) \tag{25}$$

$$\frac{dn_v}{dt} = k_3N_d^+(1 - e^{\beta(\mu_{nc} - \mu_{N_d^+} + \mu_{N_d^0})}) + k_2N_v(1 - e^{\beta(\mu_{nc} + \mu_{pv})}) + \nabla.(K_2\nabla\mu_{n_c}) \tag{26}$$

$$\frac{dN_d^+}{dt} = k_1N_d^0(1 - e^{\beta(\mu_{nc} + \mu_{N_d^+} - \mu_{N_d^0})}) - k_3N_d^+(1 - e^{\beta(\mu_{nc} - \mu_{N_d^+} + \mu_{N_d^0})}) \tag{27}$$

$$N_d^0 = N_d - N_d^+ \tag{28}$$

With the corresponding boundary conditions:

$$K_1\nabla\mu_{n_c}.n = -k_4(e^{\beta(\mu_{nc} + \Phi_e)} - 1) \quad on \ \partial\Omega \tag{29}$$

$$\nabla\mu_{n_v}.n = 0 \quad on \ \partial\Omega \tag{30}$$

Where k_1, k_2, k_3 are rate constants for the species interconversion reactions, K_1, K_2 are diffusion constants and Φ_e is the work function of the electrodes.

The expression of the spontaneous strain is

$$\varepsilon_{11}^s = \alpha^t - 1 - (S_{11}e_{12} + S_{12}e_{22})(P_2 - P_2^s)$$
$$= \alpha^t - 1 - K_{11}(P_2 - P_2^s)$$

$$\varepsilon_{22}^s = \beta^t - 1 - (S_{12}e_{12} + S_{11}e_{22})(P_2 - P_2^s) \tag{31}$$
$$= \beta^t - 1 - K_{22}(P_2 - P_2^s)$$

$$\varepsilon_{12}^s = -S_{44}e_{61}(P_1 - P_1^s)$$
$$= K_{12}(P_1 - P_1^s)$$

where $\alpha^t = 0.9958$ and $\beta^t = 1.0067$ are the two stretch components for tetragonal distortion [14-16]. The equation 12 becomes

$$\mu\frac{\partial P_1}{\partial t} = -a_1P_1 - a_2P_1^3 - a_3P_1P_2^3 - a_4P_1^5 - a_5P_1^3P_2^2 + (d_{11} - f_{11}(1 - \frac{\partial P_1^s}{\partial P_1}))\frac{\partial^2 u_1}{\partial x_1^2}$$
$$+ e_{12}\frac{\partial P_1^s}{\partial P_2}\frac{\partial u_1}{\partial x_1} - \frac{d\phi}{dx_1} + (K_{11}e_{12} + K_{22}e_{22} + f_{12}\Gamma_{12} + f_{11}\Gamma_{11})\frac{\partial P_1^s}{\partial P_2}(P_2 - P_2^s)$$
$$- (K_{12}e_{61} + f_{11}\Gamma_{11} + f_{12}\Gamma_{12})(1 - \frac{\partial P_1^s}{\partial P_1})(P_1 - P_1^s) + e_{12}(1 - \alpha^t)\frac{\partial P_1^s}{\partial P_2} + a_0\frac{d_2P_1}{dx_1^2}$$
$$+ (d_{11}K_{11} + d_{12}K_{12} - 2d_{44}K_{12})(1 - \frac{\partial P_1^s}{\partial P_1})\frac{dP_2}{dx_1} \tag{32}$$

$$\mu\frac{\partial P_2}{\partial t} = -a_1P_2 - a_2P_2^3 - a_3P_2P_1^2 - a_4P_2^5 - a_5P_2^3P_1^4 + f_{11}\frac{\partial P_1^s}{\partial P_2}\frac{\partial^2 u_1}{\partial x_1^2} + a_0\frac{d^2P_2}{dx_1^2}$$
$$- e_{12}(\frac{\partial u_1}{\partial x_1} + 1 - \alpha^t)(1 - \frac{\partial P_1^s}{\partial P_1}) - (K_{11}e_{12} + K_{22}e_{22})(1 - \frac{\partial P_1^s}{\partial P_1})(P_2 - P_2^s)$$
$$- (f_{12}\Gamma_{12} + f_{11}\Gamma_{11})(1 - \frac{\partial P_1^s}{\partial P_1})(P_2 - P_2^s) + (K_{12}e_{61} + f_{11}\Gamma_{11} + f_{12}\Gamma_{12}\frac{\partial P_1^0}{\partial P_2}(P_1 - P_1^s) \tag{33}$$
$$+ (d_{11}K_{11} + d_{12}K_{12} - 2d_{44}K_{12})(1 - \frac{\partial P_1^s}{\partial P_1})P_{1,1}$$

The equation 13 becomes:

$$v\frac{\partial u_1}{\partial t} = C_{11}\frac{\partial^2 u_1}{\partial x_1^2} + e_{12}(\frac{\partial P_2}{\partial x_1} - \frac{\partial P_2^s}{\partial x_1}) + d_{11}\frac{\partial^2 P_1}{\partial x_1^2} - G_{11}\frac{\partial^4 u_1}{\partial x_1^4} - f_{11}(\frac{\partial^2 P_1}{\partial x_1^2} - \frac{\partial^2 P_2^s}{\partial x_1^2}) \tag{34}$$

$$v\frac{\partial u_2}{\partial t} = e_{61}(\frac{\partial P_1}{\partial x_1} - \frac{\partial P_1^s}{\partial x_1}) - f_{44}(\frac{\partial^2 P_2}{\partial x_1^2} - \frac{\partial^2 P_2^s}{\partial x_1^2}) + d_{44}\frac{\partial^2 P_2}{\partial x_1^2} \tag{35}$$

The Boundary conditions become:

$$\frac{dP_1}{dx_1} + d_{11}\frac{du_1}{dx_1} + (d_{11}K_{11} + d_{12}K_{22})(P_2 - P_2^s) = d_{11}(\alpha^t - 1) \tag{36}$$

$$\frac{dP_2}{dx_1} + d_{44}K_{12}(P_1 - P_1^s) = 0 \tag{37}$$

$$C_{11}\frac{du_1}{dx_1} - (\alpha^t + 1) - G_{11}\frac{d^3u_1}{dx_1^3} + (d_{11} - f_{11})\frac{dP_1}{dx_1} + e_{12}(P_2 - P_2^s) + f_{11}P_1^s = 0 \tag{38}$$

$$e_{61}(P_1 - P_1^s) + (d_{44} - f_{44})\frac{dP_2}{dx_1} + f_{44}\frac{dP_2^s}{dx_1} = 0 \tag{39}$$

Numerical Simulation

The problem is defined in bidirectional space with one independent spatial variable x_1. The size of simulation cell is taken as twice bigger than the domain wall transition distance. In the case of our simulations, the domain wall was transiting in 120 nm for P_2 and much smaller for P_1. Thus the simulation cell size was chosen to be 250 nm to guarantee that the boundary conditions will not induce any artificial effect. All boundary conditions in terms of strain are given in both direction x_1 and x_2 to accommodate the spontaneous strain geometrical constraint. Equations 32-39 are discretized through finite differences on a 2500 grid with a constant grid size $\Delta x = 0.1$ nm. The grid size was chosen small enough to provide good resolution in order to allow a better observation of the transition phenomena. All variable are assumed to be zero in the start of the simulation then the strain at $\pm\infty$ is imposed to be equal to the spontaneous strains. Equations 32-35 are explicitly integrated from time t^n to $t^{n+1} = t^n + \Delta t$ to compute the new value for polarization and strain.

Equations 19-30 are only used when there are defects in the material. In the case of defects, the space charge is also integrated explicitly in parallel to the strain and the polarization. The gradient flow method is also described in much reference [14,15]. The convergence of the method is slow especially when the calculation becomes close to the final solution. The parameters chosen were slightly costly in terms of computation resource showever, the results were guaranteed to converge [17-20]. The numerical parameters used are given in Table 1.

Results and Discussion

180° domain wall in perfect Barium Titanate crystal

Figure 4 presents the variation of the normal polarization P_1 and the parallel polarization P_2 across the domain wall. Both polarizations are normalized with respect to the spontaneous polarization magnitude P_0. The Normal polarization P_1 is small compared to the polarization P_2 but it is not zero as it was always assumed. Similar results were found by Yudin et al. [8] and Gu et al. [21], where they found a polarization with a 2 order of magnitude smaller then P_2. The existence of the non-zero polarization induced an electric field around the domain wall with its potential shown in Figure 5. The effect of this polarization is clear in the Figure 6 where the induced electric field causes further distortion

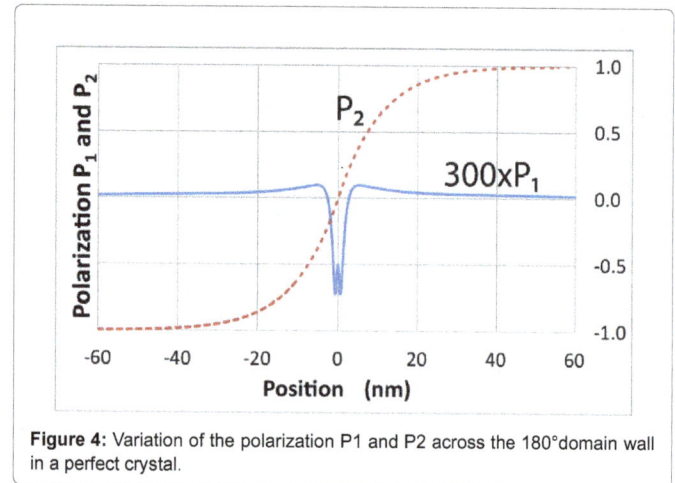

Figure 4: Variation of the polarization P1 and P2 across the 180°domain wall in a perfect crystal.

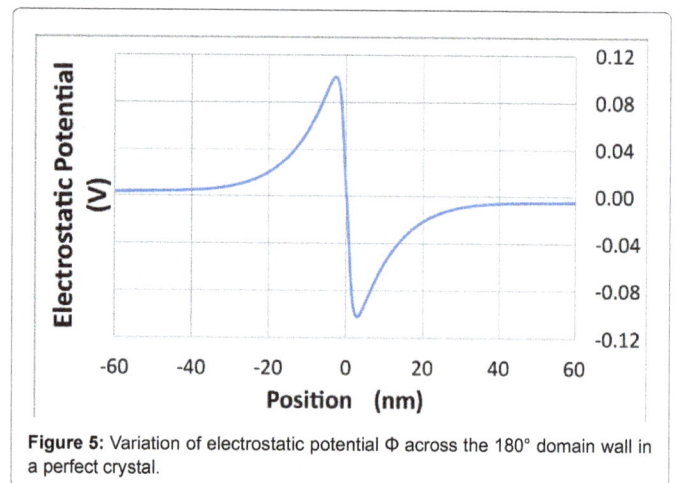

Figure 5: Variation of electrostatic potential Φ across the 180° domain wall in a perfect crystal.

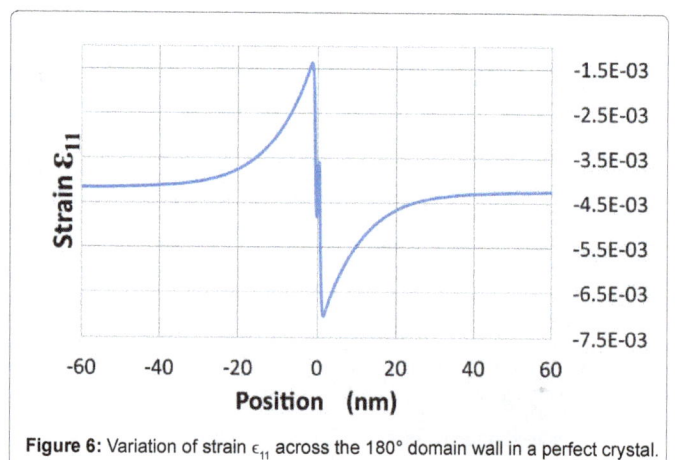

Figure 6: Variation of strain ϵ_{11} across the 180° domain wall in a perfect crystal.

of the structure. The strain variation shows the response of the material due to the force exerted by the non-zero polarization.

180° domain wall in oxygen vacancy doped Barium Titanate crystal

In this section, we check the effect of exoelectricity on the defects interaction with the domain wall. The typical values of doping in Barium Titanate ranges from 10 to 1000 ppm corresponding to value of

Parameters	Values	Ref
a_1 (Nm²/C²)	$6.6 \times 10^5(T-110)$	[17]
a_2 (Nm²/C²)	$1.44 \times 10^7(T-175)$	[17]
a_3 (Nm²/C²)	3.94×10^9	[17]
a_4 (Nm²/C²)	3.96×10^{10}	[17]
a_5 (Nm²/C²)	2.39×10^{14}	[17]
a_0 (Nm²/C²)	10^{-7}	
c_{11} (N/m²)	275×10^9	[18]
c_{12} (N/m²)	179×10^9	[18]
f_{11} (C/m)	0.35×10^{-9}	[19]
f_{12} (C/m)	5×10^{-6}	[20]
e_{12} (C/N)	-34.5	[22]
e_{22} (C/N)	85.6	[22]
e_{61} (C/N)	392	[22]
E_c (eV)	-3.6	[7]
E_v (eV)	-6.6	[7]
E_d (eV)	-4.0	[7]
$N_c;N_v$ (m⁻³)	10^{24}	[7]
K_1 (1=(eV ms))	10^{13}	[7]
K_2 (1=(eV ms))	10^{11}	[7]
k_1 (s⁻¹)	10^{11}	[7]
$k_2; k_3$ (s⁻¹)	10^8	[7]
k_4 (m⁻²s⁻¹)	10^{24}	[7]

Table 1: Material properties.

$Nd=10^{24}-10^{26}\,m^{-3}$. In this work, we take Nd equal to $10^{24}\,m^{-3}$. The platinum electrode has a work function $\Phi_e=5.3$ eV. Figure 6 shows the variation of polarization across the domain wall. The introduction of defects into the material has no apparent effect on the polarization. The polarization magnitude remains the same as well as the geometrical distortion of the structure. As illustrated in Figure 7, the electrostatic potential shows an average rise compared to the perfect crystal although the electric field remains the same. The total number of ionized donors varies with the electrostatic potential. Figure 8 presents the variation of the ionized donors distribution across the domain wall. At room temperature, the majority of the ionized donors are diffused to the right side of the domain wall. The defects distribution profile shown in this Figure 7 is normalized with respect to the total number of donors in the material. The diffusion of the ionized donors in the domain wall leads to the pinning of the domain wall. In fact, the electrons freed from the ionized donors are swept by the electrostatic potential and they get attracted to the high potential level. The mobile ionized donors are attracted to the potential with low level [22]. The energy required for domain wall motion becomes higher due to the required energy to counter the new electric field created by the distribution of the ionized donors near the domain wall. Xiao et al. [23] studied the interaction between domain walls and oxygen vacancies in tetragonal Barium Tianate. The authors found that the 180° domain has no effect on oxygen vacancy distribution however for the 90° domain wall, the result was similar to what we found in this paper. Yang et al. [24] presented an experimental study of domain wall pinning where they showed that a 180° domain wall can get pinned to defects and impurities in the material which can stop the motion of the domain wall. This behavior is analogous to the electron behavior in high injection of carriers in a PN junction of a semiconductor (such as diode); where it causes to violate one of the approximations made in the derivation of the ideal characteristics, namely that the majority carrier density equals the thermal equilibrium value. One can observe that the majority carrier (electron) density increases beyond the doping density and tracks the minority carrier (hole) density in an extended region away from the junction. High injection of carriers causes to violate one of the approximations made in the derivation of the ideal diode characteristics, namely that the majority carrier density equals the thermal equilibrium value. Excess carriers will dominate the electron and hole concentration. The carrier concentrations decay due to recombination as we move away from the depletion region. A similar behavior was observed in metal-semiconductor contacts for which a synthesis of the thermionic-emission and diffusion approaches has been proposed by Crowell et al. [25] that is derived from the boundary condition of a thermionic recombination velocity near the metal-semiconductor interface [26].

Conclusions

In this work, we presented a new theory for a ferroelectric material which takes into consideration the flexoelectric coupling between the polarization and the strain gradient. We showed that the 180° domain wall has mixed character of Ising and Bloch type. The magnitude of the normal polarization was found to be small compared to the spontaneous polarization. However the mixed character of the domain wall induced a new interaction with defects distribution. It was shown from this analysis that oxygen vacancies are attracted to the domain wall which can pin it and change the piezoelectric properties of the ferroelectric materials. Different parameters are used in this study and although the numerical results were carried only for tetragonal Barium Titanate, the results remain valid for all ferroelectric materials, though, the effect can be either negligible or it can be more important.

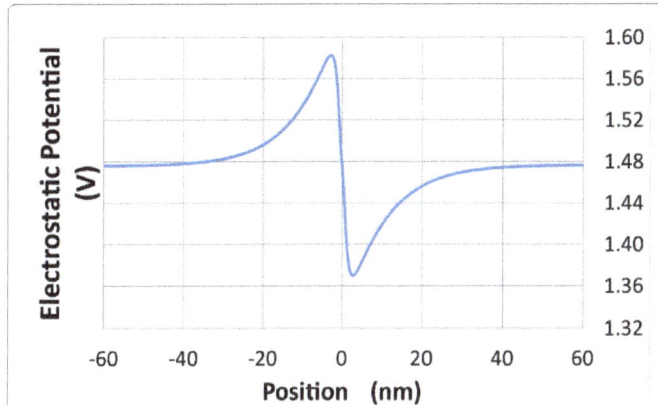

Figure 7: Variation of electrostatic potential Φ across the 180° domain wall in a doped crystal.

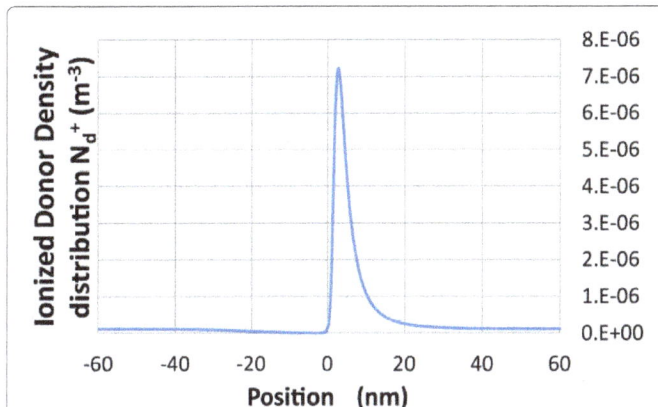

Figure 8: Variation of ionized donor's distribution N_d^+ across the 180° domain wall in a doped crystal.

References

1. Haun MJ (1988) The Pennsylvania State University.

2. Ghosh D, Sakata A, Carter J, Thomas PA, Han H (2013) Domain wall displacement is the origin of superior permittivity and piezoelectricity in batio$_3$ at intermediate grain sizes. Adv Funct Mater 24: 885-896.

3. Huang XR, Hu XB, Jiang SS, Feng D (1997) Theoretical model of 180° domain-wall structures and their transformation in ferroelectric perovskites. Phys Rev B 55: 5534.

4. Lee D, Behera RK, Wu P, Xu H, Li YL (2009) Mixed bloch-neel-ising character of 180° ferroelectric domain walls. Phys Rev B 80: 060102.

5. Dayal K, Bhattacharya K (2007) A real-space non-local phase field model of ferroelectric domain patterns in complex geometries. Acta Mater 55: 1907.

6. Padilla J, Zhong W, Vanderbilt D (1996) First-principles investigation of 180° domain walls in batio. Phys Rev B 53: 5969.

7. Suryanarayana P, Bhattacharya K (2012) Evolution of polarization and space charges in semiconducting ferroelectrics. J Appl Phys 111: 034109.

8. Yudin PV, Tagantsev AK, Eliseev EA, Morozovska AN, Setter N (2012) Bichiral structure of ferroelectric domain walls driven by flexoelectricity. Phys Rev B 86: 134102.

9. Cao W, Cross LE (1991) Theory of tetragonal twin structures in ferroelectric perovskites with a first- order phase transition. Phys Rev B 44: 5-12.

10. Majdoub MS, Sharma P, Cagin T (2008) Enhanced size-dependent piezoelectricity and elasticity in nanostructures due to the flexoelectric effect. Physical Review B 77: 125424.

11. Majdoub MS, Sharma P, Cagin T (2008) Dramatic enhancement in energy

harvesting for a narrow range of dimensions in piezoelectric nanostructures. Physical Review B 78: 121407.

12. Mindlin RD, Eshel NN (1968) On first strain-gradient theories in linear elasticity. Int J Solids Struct 4: 109-124.

13. Zhang W, Bhattacharya K(2005) A computational model of ferroelectric domains Part I: model formulation and domain switching. Acta Material 53: 185-198.

14. Behrman W (1998) An efficient gradient flow method for unconstrained optimization. Stanford University, Stanford, CA, USA.

15. Andrei N (2004) Gradient flow algorithm for unconstrained optimization. ICI Technical Report.

16. Burcsu E, Ravichandran G, Bhattacharya K (2004) Large electrostrictive actuation of barium titanate single crystals. J Mech Phys Solids 52: 823-846.

17. El-Naggar M, Dayal K, Goodwin DG, Bhattacharya K (2006) Graded ferroelectric capacitors with robust temperature characteristics. J App Phys 100: 114115.

18. Freire JD, Katiyar RS (1988) Lattice dynamics of crystals with tetragonal $batio_3$ structure. Phys Rev B 37: 2074-2085.

19. Hong J, Catalan G, Scott JF, Artacho E (2010) The flexoelectricity of barium and strontium titanates from first principles. J Phys Condens Matter 22: 112201.

20. Ma W, Cross LE (2006) Flexoelectricity of barium titanate. Journal of Appl Phys Lett 88: 232902.

21. Gu Y, Li M, Morozovska AN, Wang Y, Eliseev EA, et al. (2014) Flexo-electricity and ferroelectric domain wall structures: Phase-eld modeling and DFT calculations. Phys Rev B 89: 174111.

22. Berlincourt D, Jaffe H (1958) Elastic and piezoelectric coefficients of single-crystal barium titanate. Phys Rev 111: 143-148.

23. Xiao Y, Shenoy VB, Bhattacharya K (2005) Depletion layers and domain walls in semiconducting ferroelectric thin films. Phys Rev Lett 95: 247603.

24. Yang TJ, Gopalan V, Swart PJ, Mohideen U (1999) Direct observation of pinning and bowing of a single ferroelectric domain wall. Phys Rev Lett 82: 4106.

25. Crowell CR, Sze SM (1966) Current transport in metal-semiconductor barriers. Solid-State Electronics 9: 1035-1048.

26. Sze SM, Ng KK (2007) Physics of semiconductor devices. 3rd Edn. John Wiley and Sons, Inc, USA.

15

Polarization Dependent Reflectivity and Transmission for $Cd_{1-x}Zn_xte$/GaAs (001) Epifilms in the Far-Infrared and Near-Infrared to Ultraviolet Region

Talwar DN[1]* and Becla P[2]

[1]Department of Physics, Indiana University of Pennsylvania, Pennsylvania, USA
[2]Department of Materials Science and Engineering, Massachusetts Institute of Technology, Cambridge, Massachusetts, USA

Abstract

The results of a comprehensive experimental and theoretical study is reported to empathize the optical properties of binary GaAs, ZnTe, CdTe and ternary $Cd_{1-x}Zn_xTe$ (CZT) alloys in the two energy regions: (i) far-infrared (FIR), and (ii) near-infrared (NIR) to ultraviolet (UV). A high resolution Fourier transform infrared spectrometer is used to assess the FIR response of GaAs, ZnTe, CdTe and CZT alloys in the entire composition $1.0 \geq x \geq 0$ range. Accurate model dielectric functions are established appositely to extort the optical constants of the binary materials. The simulated dielectric functions $\tilde{\varepsilon}(\omega)$ and refractive indices $\tilde{n}(\omega)$ are meticulously appraised in the FIR $\rightarrow$ NIR $\rightarrow$ UV energy range by comparing them against the existing spectroscopic FTIR and ellipsometry data. These outcomes are expended eloquently for evaluating the polarization dependent reflectivity $R(\lambda)$ and transmission $T(\lambda)$ spectra of ultrathin CZT/GaAs (001) epifilms. A reasonably accurate assessment of the CZT film thickness by reflectivity study has offered a credible testimony for characterizing any semiconducting epitaxially grown nanostructured materials of technological importance.

Keywords: Fourier transform infrared spectroscopy; Reflectivity; Transmission; Ellipsometry; Dielectric functions; Epilayers

Introduction

The binary cadmium chalcogenides (CdX with X=S, Se, Te) and their ternary ($Cd_{1-x}Y_xX$ with Y=Be, Mg, Zn, and Hg) alloys belong to the group of a II-VI semiconductor family – exhibiting many intriguing properties with a wide-range of applications in photovoltaics, x-ray, γ-radiation sensors, electro-optical modulators including its usage as a substrate for HgCdTe based infrared (IR) detectors. As compared to the traditional Si and Ge detectors, requiring cryogenic cooling and consuming high power – the II-VI based devices [1-6] are compact, expend less power, operate at room-temperature and display unique features for processing more than one million photons/second/mm². In recent years, the growing interest for exploiting $Cd_{1-x}Zn_xTe$ (CZT) epifilms over group-IV semiconductors has been its ability to concoct alloys with accurate control of composition x and thickness d. The other advantages of utilizing these materials in device engineering include the accessibility of low-cost, large area, and electrically conductive substrates, such as GaAs. For the technological needs, the imperative qualities of CZT comprise of its higher atomic number Z, high mass density ρ and large bandgaps E_g [1.45 eV (CdTe), 2.26 eV (ZnTe)] for ensuring enhanced energy resolution and higher detection efficiency. While the key interest in $Cd_{1-x}Zn_xTe$ alloys, and $(CZT)_m$/$(ZnTe)_n$ superlattices (SLs) aspires assessing their electrical and optical characteristics by controlling x, m and n – the film thickness d plays an equally important role in regulating the efficiency of electro-optical devices, e.g., an accurate thickness of buffer and epifilm is required for fabricating sensors, detectors, and solar-cells [4-6].

The II-VI based electronics demand [7-10] high-quality crystalline materials with fewer defects. In the as grown CZT alloys, the constraints of phase diagram necessitating higher growth temperatures usually instigate intrinsic defects. In the electronic industry, as the veracity of using semiconductor materials intensifies – so does the compulsion of employing reliable and reproducible methods for appraising their distinctive qualities. In assessing the nature of defects and degree of crystallinity, many experimental techniques have been employed in the past, such as the Fourier transform infrared (FTIR) reflectivity and transmission [9,10], Raman scattering (RS) [8], photoluminescence

(PL) [11], synchrotron X-ray diffraction (S-XRD) topography [1-3], and deep level transient spectroscopy (DLTS) [12], etc. In addition, the spectroscopic ellipsometry (SE) is perceived as an equally valuable tool for appraising the optical constants of semiconductor materials and evaluating the epifilm thickness [13].

Despite the extensive technological needs, a limited number of experimental/theoretical studies are carried out on the fundamental properties of CZT alloys and SLs – especially the physics behind those attributes which ascertain their prominence at a practical level. Although, a significant amount of work exists dealing with the growth and electronic characteristics of II-VI materials – the structural, and optical properties of $Cd_{1-x}Zn_xTe$ are either scarcely known [6-8] or ambiguous. In the far-infrared (FIR) region 5 meV $\leq$ E $\leq$ 100 meV, while SE is recognized as an efficient method for exploring lattice dynamics and free carrier concentration in semiconductors – it has not yet been applied to study the phonons of CZT alloys. Again, no SE measurements are available for assessing the $Cd_{1-x}Zn_xTe$/GaAs epifilm thickness. Earlier, the vibrational properties of $Cd_{1-x}Zn_xTe$ were acquired by using FTIR [9,10] and RS [8] spectroscopy with a limited alloy composition, x. The results when analyzed by a classical method envisioned a "two-phonon-mode" behavior. Recently, we have performed extensive micro-Raman and extended X-ray absorption fine-structure (EXAFS) measurements on the Bridgman-grown $Cd_{1-x}Zn_xTe$ alloyed samples [14,15] in the entire composition range $1.0 \geq x \geq 0.0$ and comprehended their phonon and structural traits. A careful evaluation of the experimental data by an average-t-matrix

*Corresponding author: Talwar DN, Department of Physics, Indiana University of Pennsylvania, 975 Oakland Avenue, 56 Weyandt Hall, Indiana, Pennsylvania 15705-1087, USA, E-mail: talwar@iup.edu

Green's function (ATM-GF) approach has authenticated the "two-phonon mode" stance [14]. While the composition-dependent EXAFS data revealed a bimodal distribution of nearest-neighbor bond lengths–theoretical analysis by first-principles bond-orbital model permitted an accurate appraisal of the lattice relaxations around Zn/Cd atoms in CdTe/ZnTe materials. One must note that only limited efforts have been made by SE to uncover the optical properties of CZT alloys in the near-IR (NIR) to ultraviolet (UV) energy range [16-20]. From a theoretical stand point, it has now become possible [14,15] to expend pragmatic model dielectric functions (MDFs) of binary materials to elucidate structural and optical characteristics of semiconducting ternary alloys, and SLs grown on different substrates. Earlier, it was comprehended that only first principles methods could yield material characteristics with accuracies required of the experiments [21]. It is, therefore, quite intriguing to explore electrical and optical properties of novel CZT alloys [22-25] which are playing crucial roles in contriving IR detectors/sensors, photovoltaic-cells, and many other optoelectronic devices [1-6]. The purpose of this paper is to report the results of comprehensive experimental and theoretical investigations to apprehend the structural and optical characteristics of ultrathin (20 nm $\leq d \leq$ 1.0 μm) $Cd_{1-x}Zn_xTe$ epifilms prepared on GaAs substrate. Accurate model dielectric functions are established appositely to extort the optical constants of all the involved semiconductor materials. The simulated dielectric functions $\tilde{\varepsilon}(\omega)$ and refractive indices $\tilde{n}(\omega)$ are meticulously appraised in the FIR $\rightarrow$ NIR $\rightarrow$ UV energy range by comparing them against the existing spectroscopic FTIR [9] and ellipsometry [20] data. A traditional approach of multilayer optics is used to simulate polarization dependent reflectivity [$R(\lambda)$] and transmission [$T(\lambda)$] spectra for ultrathin $Cd_{1-x}Zn_xTe$/GaAs (001) epifilms of thickness ranging between 22 nm $\leq d \leq$ 1 μm. Theoretical results of $R(\lambda)$ and $T(\lambda)$ are compared, discussed and contrasted amongst the available experimental data [22-29] with concluding remarks presented in Section 7. An accurate assessment of film thickness by reflectivity study has offered a credible testimony for characterizing epitaxially grown nanostructured materials of diverse technological importance.

Experimental

$Cd_{1-x}Zn_xTe$ ternary alloys

The $Cd_xZn_{1-x}Te$ ($0 \leq x \leq 1$) samples used in the FIR (5 meV $\leq$ E $\leq$ 100 meV) reflectivity measurements [9] were grown at the Massachusetts Institute of Technology using the Bridgman technique. The CZT ternary alloys were prepared by reacting the 99.9999% pure elemental constituents in the evacuated sealed quartz tubes at ~ 1150°C. The Cd composition values x determined from the mass densities and the precast alloys were re-grown by directional solidification at the rates of 1.2 mm/h in the adiabatic zone of a Bridgman-Stockbarger type furnace with temperature gradient set at about 15°C/cm. The resultant boules cut into 1-2 mm thick slices perpendicular to the growth axis were annealed at 600°C in a Cd-saturated atmosphere for about 5 d to improve the crystalline perfection. Finally, the surfaces of the CZT sample were prepared by lapping, mechanical polishing, and etching in a bromine-methanol solution. The alloy composition x, set by the ratio of constituents before growth was confirmed by the x-ray diffraction and transmission measurements after preparation. All the samples were found to be single-crystal with the zinc-blende structure.

Far-infrared reflectivity

The room temperature FIR reflectance spectra on the $Cd_xZn_{1-x}Te$ ($0 \leq x \leq 1$) ternary alloy samples were measured at near normal incidence by using a Bruker IFS66 spectrometer with KBr beam-splitter and a

deuterated triglycine sulfate (DTGS) detector to achieve a good signal to noise ratio in the energy region 5 meV $\leq$ E $\leq$ 100 meV. The details of FIR measurements are described elsewhere [9] where we set the incident angle at about 9°C degrees–a negligible deviation from the near normal incidence. The experimental reflectance spectra was analyzed theoretically by using a classical "Drude-Lorentz" methodology – creating effective MDF's and including contributions from both the lattice phonons as well as free charge carriers.

Theoretical Background

The customary SE parameters Ψ and Δ are related [13] to the ratio ρ of the complex Fresnel reflection coefficients $\tilde{r}^p$ and $\tilde{r}^s$, respectively for the incident light polarized parallel (||) and perpendicular ($\perp$) to the plane of incidence:

$$\rho \equiv \frac{\tilde{r}^p}{\tilde{r}^s} = \tan \Psi \exp(i\Delta) \tag{1}$$

The SE method is considered quite methodical compared to the reflected intensity measurements – generally performed at a near normal incidence. Again, thickness measurement by in-line SE [13] has played an important role for monitoring the epitaxial film growth processes. For extracting the optical constants and thickness of layer structured materials in the energy range of 0.5 eV to 10 eV, one needs to establish reasonably accurate MDF's to simulate the complex dielectric functions, $\tilde{\varepsilon}(\omega)$ or refractive indices, $\tilde{n}(\omega)$ for both the epifilms and substrates.

In simulating the reflectance and transmission spectra of epifilms, a three phase (Figure 1) model (ambient (air)/film/substrate) [30] is considered to be convincingly adequate. For modeling the optical properties, we have pretended that both the epifilms and substrates form homogenous isotropic materials while the ambient is viewed as non-absorbing. In the framework of classical methodology of multilayer optics, the isotropic $Cd_{1-x}Zn_xTe$ film of thickness d (Figure 1) is described by a material of complex refractive index $\tilde{n}_1$ grown on a thick GaAs substrate of refractive index $\tilde{n}_2$. The incident light of wavelength λ from the ambient of refractive index $\tilde{n}_0$ ($\equiv$1) causes many reflections and transmissions at the substrate/film and film/air interface. Due to multiple reflections within the film, the reflected electric fields parallel and perpendicular to the plane of incidence adds up by a geometric series giving an Airy formula for the reflection $\tilde{r}^x$ coefficient ($x \equiv s$-, p- polarization) [14]:

$$\tilde{r}^x = \frac{\tilde{r}^x_{01} + \tilde{r}^x_{12}e^{2i\beta}}{1 + \tilde{r}^x_{01}\tilde{r}^x_{12}e^{2i\beta}} \tag{2}$$

where, $\tilde{r}^x_{01}$, $\tilde{r}^x_{12}$ are the Fresnel coefficients describing the reflection at the respective interfaces between media of refractive indices $\tilde{n}_0$, $\tilde{n}_1$ and among $\tilde{n}_1$, $\tilde{n}_2$. The film thickness d and the angle of incidence

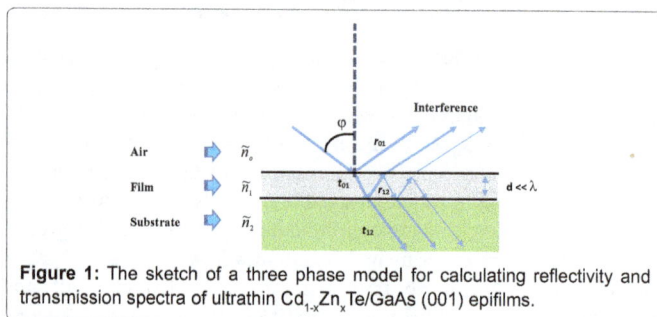

Figure 1: The sketch of a three phase model for calculating reflectivity and transmission spectra of ultrathin $Cd_{1-x}Zn_xTe$/GaAs (001) epifilms.

ϕ are restricted within the phase factor $\beta = 2\pi \dfrac{d}{\lambda} \sqrt{\tilde{n}_1^2 - \tilde{n}_0^2 \sin^2 \phi}$. By using Eq. (2) we have calculated the polarization dependent reflectivity $R^x = |\tilde{r}^x|^2$ spectra for ultrathin $Cd_{1-x}Zn_xTe$ epifilms prepared on GaAs (001) substrate. Similar calculations for the polarization dependent transmission $T^x = |\tilde{t}^x|^2$ are also performed. The results presented are compared and contrasted against the limited experimental data [22-29].

The model dielectric functions

Far-IR to mid-IR energy range: For $Cd_{1-x}Zn_xTe$ ternary alloys, the contribution of polar lattice phonons to the dielectric response $\{\tilde{\varepsilon}_L(\omega)\}$ is evaluated within the reststrahlen band region by exploiting a classical "Drude-Lorentz" model [9]:

$$\tilde{\varepsilon}_L(\omega) = \varepsilon_{\infty x}\left[1 - \frac{\omega_p^2}{\omega(\omega + i\gamma_p)}\right] + \sum_{j=1,2} \frac{S_{jx}\omega_{TOxj}^2}{\omega_{TOjx}^2 - \omega^2 - i\Gamma_{jx}\omega} \tag{3}$$

Here, the term $\varepsilon_{\infty x}$ represents the weighted high-frequency dielectric function; ω_{TOjx}- the CdTe-like and ZnTe-like TO-mode frequencies; S_{jx}- the oscillator strengths; Γ_{jx}-the broadening values of TO phonons; ω_p represents the plasma frequency and γ_p its damping constant. The plasma frequency $\omega_p\left(\equiv \sqrt{\dfrac{4\pi\eta e^2}{m^*\varepsilon_\infty}}\right)$ and $\gamma_p\left(\equiv \dfrac{e}{m^*\mu}\right)$ of the free carriers (electrons) are assessed from the effective mass m^*, the carrier concentration η, magnitude of the electron charge e and mobility μ. For simulating the reflectance and transmission spectra of $Cd_{1-x}Zn_xTe/GaAs$ (001) and/or $(CdTe)_m/(ZnTe)_n/GaAs$ (001) SLs, the required $\tilde{\varepsilon}_L(\omega)$ for the substrate (GaAs) is calculated independently. To attain the best-fit parameter values in Eq. (3), we followed an efficient Levenberg-Marquardt algorithm [31] and used the non-linear simulations to minimize the error function Ξ over n data points:

$$\hat{I} = \frac{1}{n}\sum_i^n |\mathfrak{R}_i^{\exp} - \mathfrak{R}_i^{cal}|^2 \tag{4}$$

where, $\mathfrak{R}_i^{\exp}$, $\mathfrak{R}_i^{cal}$ are the experimental and calculated values, respectively.

Near-IR to UV energy range: In the NIR to UV (0. 5 eV to > 7.0 eV) spectral range, the dielectric behaviors of crystalline materials are strongly allied to their energy-band structures [13]. It has been well established that in semiconductors both direct and indirect band gap transitions near the critical points (CPs) affect optical dispersion relations. In the indirect-band-gap semiconductors, while the transitions take place at energies below the onset of the lowest direct transitions–in the direct-band-gap materials the transitions take part at energies above the onset of the lowest direct transitions [32]. The electronic energy band structures of CdTe, ZnTe and GaAs have been extensively studied both theoretically [33] and experimentally [34-41]. In the experimental studies, several inter-band transitions related to CPs at different parts of the Brillouin zone (BZ) have been identified by exploiting reflectivity [34], SE [13], electro-reflectance [35], thermo-reflectance [36], and wavelength-modulated reflectivity [37] techniques.

In our simulations of the complex dielectric functions $\tilde{\varepsilon}(\omega)$ for the direct bandgap binary compounds CdTe, ZnTe and GaAs, we instigated Adachi's [16,20] optical dispersion mechanisms by exploiting the modified model dielectric functions. Based on the Kramers-Krönig (KK) transformation, this methodology [32] predicts very well the distinct optical features of the perfect materials near CPs in the BZ.

In this approach, one anticipates to have three fitting constraints for each CP transitions: the energy, strength and broadening parameter. For instance, the transition energies at CPs near Γ (i.e., at the center of the BZ); Λ or L (in the <111>); and X (in the <100>) points are labeled as E_0, $E_0+\Delta_0$; E_1, $E_1+\Delta_1$; and E_2, respectively [20]. While the E_0, $E_0+\Delta_0$, transitions at Γ point are of the three-dimensional (3D) M_0-type – the E_1, $E_1+\Delta_1$ transitions, take place in the <111> direction near Λ or L points in the BZ, are of 3D M_1-type. Since the M_1-CP longitudinal effective mass is much larger than its transverse counterparts, one can treat 3D-M_1 CPs as two dimensional (2D) minimum M_0 [20]. Again, a pronounced structure in the optical spectra of CdTe and ZnTe near X point in the <100> direction having energy higher than $E_1+\Delta_1$ is labeled as E_2. In general, the E_2 transition does not correspond to a single well-defined CP – it has been characterized by a damped harmonic oscillator.

For calculating the optical constants of ternary $Cd_{1-x}Zn_xTe$ alloys, the CP energies are represented by quadratic expressions involving binary energy values and composition, x while the strength and broadening parameters are assumed varying linearly with x. The relevant expressions of MDFs reported elsewhere [32] for each energy gaps are used in assessing the spectral dependence of various optical constants – linked to the dielectric function $\tilde{\varepsilon}(\omega)$. For instance, the complex refractive index $\tilde{n}(\omega)$ is allied to:

$$\tilde{n}(\omega) = n(\omega) + ik(\omega) = \sqrt{\tilde{\varepsilon}(\omega)} \tag{5}$$

where, the refractive index $n(\omega)$, extinction coefficient $k(\omega)$ and absorption coefficient $\alpha(\omega)$ are expressed as:

$$n(\omega) = \left[\frac{(\varepsilon_1^2 + \varepsilon_2^2)^{1/2} + \varepsilon_1}{2}\right]^{1/2} \tag{6a}$$

$$k(\omega) = \left[\frac{(\varepsilon_1^2 + \varepsilon_2^2)^{1/2} - \varepsilon_1}{2}\right]^{1/2} \tag{6b}$$

$$\alpha(\omega) = \frac{4\pi}{\lambda}k(\omega) \tag{6c}$$

with $\varepsilon_1(\omega) = \text{Re}\,\tilde{\varepsilon}(\omega)$ and $\varepsilon_2(\omega) = \text{Im}\,\tilde{\varepsilon}(\omega)$.

One must note that $\varepsilon_1(\omega)$ describes the refraction of photons at any energy range, while $\varepsilon_2(\omega)$ plays a crucial role near the characteristic resonances where the material absorbs electromagnetic radiation. For instance, in polar semiconductors, the distinct resonances in the FIR energy region arise from the transverse optical (ω_{TO}) vibrational modes. In bulk materials, the optical constants can be appraised by SE [13] and other experiments [34-41] – exploiting specific wavelengths ranging from FIR → NIR → UV. Theoretically, the dielectric functions $\tilde{\varepsilon}(\omega)$ or refractive indices $\tilde{n}(\omega)$ are extorted fitting SE data by expending the well-known KK analysis [13]. However, such a methodology cannot be offered to explicate the optical properties of epilayers prepared on a substrate. Again, from the traditional transmission/reflectance methods [27,28] adopted by others – it is equally impractical acquiring optical parameters of ultrathin films due to small and feeble interference patterns. Here, we have extorted the energy dependent dielectric constants by using accurate MDFs for both the binary GaAs and ternary $Cd_{1-x}Zn_xTe$ alloys and successfully evaluated epifilm thickness by assimilating a procedure outlined.

Numerical Computations and Results

Optical constants in the FIR region

In the photon energy range 100 meV $\geq$ E $\geq$ 5 meV, we have established the MDFs for CdTe, $Cd_{1-x}Zn_xTe$ and GaAs in terms of harmonic oscillators within the classical "Drude-Lorentz" methodology [9]– requiring contributions from the polar lattice phonons and free-charge carriers (Table 1a and 1b). In Figure 2a, we displayed our

experimental results of the FIR reflectivity at near normal incidence for GaAs and compared it with the best-fit model calculation (using Eq. (3)). The derived optical parameters n, k and ε_1, ε_2 are included in Figures 2b and 2c, respectively. The perusal of Figure 2b has revealed the long wavelength TO phonon energy near ~34 meV (~ 270 cm^{-1}) at the peak of $\varepsilon_2(\omega)$ while the LO phonon mode (Figure 2c) is perceived near ~37 meV (~ 295 cm^{-1}) at $n = k$ with $\varepsilon_1(\omega) = 0$. For $Cd_{1-x}Zn_xTe$ alloy [9] with $x = 0.2$, the analysis of our FIR reflectivity spectra (Figure 3a) at near normal incidence offered optical parameters (Figures 3b and 3c) in excellent affirmation to the RS results [14] revealing CdTe-like (TO$_1$, LO$_1$) and ZnTe-like (TO$_2$, LO$_2$) modes. Not only these observations provided strong corroboration to the polarization dependent results (Figure 3d) but are also found consistent with the recent elucidations of the two-phonon-mode behavior predicted by RS and modified random element iso-displacement (MREI) model [8-10]. In the absence of FIR-SE data for $Cd_{1-x}Zn_xTe$ alloys, our simulations of ε_1 and ε_2 agreed fairly well with the experimental data of $Cd_{0.925}Be_{0.075}Se$ [42].

Optical constants in the NIR-UV region

By exploiting Adachi's formalisms [13,16] and using the modified MDFs, we have numerically simulated ε_1, ε_2, n, k at the photon energy range of 10 eV ≥ E ≥ 0.5 eV for both the binary and ternary alloy semiconductors. Theoretical results of the optical constants shown in

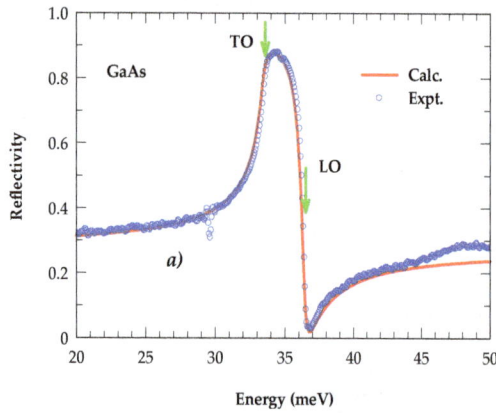

Figure 2a: Comparison of the experimental FIR reflectivity spectra for GaAs represented by blue color open circles (○) with the best fit simulated reflectivity results shown by red color solid line (—) using a classical Drude-Lorentz model (Eq. 3) with parameter values from Table 1 b) – green colored vertical arrows are used to represent TO and LO modes.

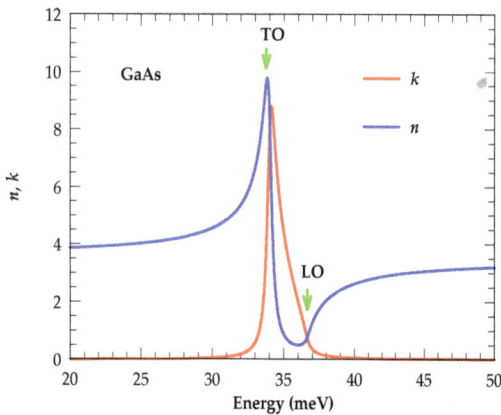

Figure 2b: The simulated results of index of refraction n(ω) and extinction coefficient k(ω).

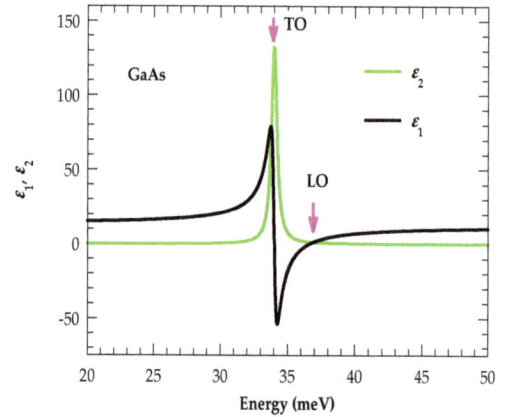

Figure 2c: The real $\varepsilon_1(\omega)$ and imaginary $\varepsilon_2(\omega)$ part of the complex dielectric function $\widetilde{\varepsilon}(\omega)$ for GaAs.

(a) $Cd_{1-x}Zn_xTe$[9]300K										
CdTe-like mode				**ZnTe-like mode**						
x	ε_∞	S_1	ω_{TO1} (cm^{-1})	Γ_1 (cm^{-1})	S_2	ω_{TO2} (cm^{-1})	Γ_2 (cm^{-1})	η 10^{14} cm^{-3}	m*/m	m cm^2/V-s
0.0	8.52	3.7	141.7	8.63				2.06	0.11	138.4
0.1	6.66	3.67	144.8	9.29				3.07	0.11	120.07
0.2	8.17	3.2	145.6	9.28	0.45	171.73	13.10	4.83	0.11	100.0
0.3	7.96	2.90	149.2	12.9	0.64	176.52	11.51	4.35	0.11	90.0
0.4	8.00	2.36	152.2	11.64	1.07	178.62	10.33	6.05	0.12	87.0
0.5	7.81	2.2	152.0	13.72	0.93	182.07	9.83	8.27	0.12	85.0
0.9	7.05	0.15	158.0	15.0	2.33	182.0	7.72	19.12	0.14	61.2
1.0	6.91				3.64	182.5	5.48	22.10	0.15	64.47

(b) GaAs[9] 300K			
ε_∞	S	ω_{TO} (cm^{-1})	Γ (cm^{-1})
11.1	1.95	268.2	2

[a] Ref. [9]

Table 1: Experimental FTIR (300K) data fitted exploiting the dielectric response model. The set of parameters are evaluated by least square fitting procedure a) for the ternary $Cd_{1-x}Zn_xTe$ alloys and b) for GaAs in the long wavelength limit.

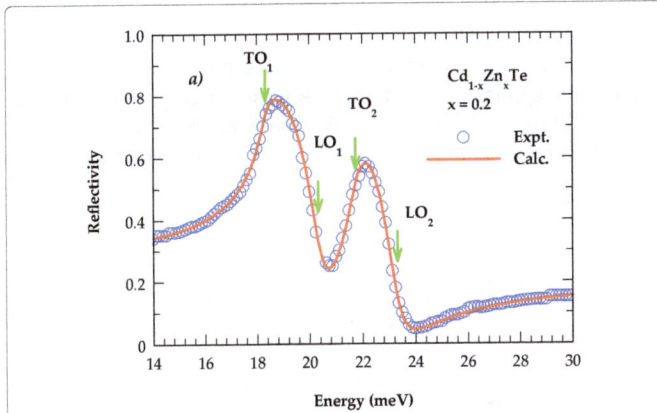

Figure 3a: Comparison of the experimental FIR reflectivity spectra of $Cd_{1-x}Zn_xTe$ (x=0.2) with the best fit calculation from the classical Drude-Lorentz model (Eq. 3) with parameter values from Table 1a.

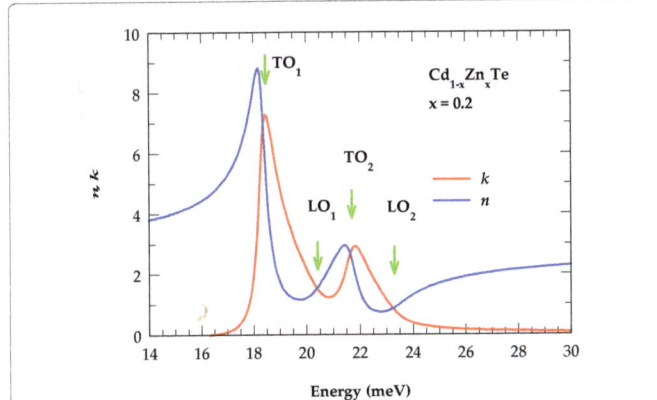

Figure 3b: The simulated results of index of refraction $n(\omega)$, extinction coefficient $k(\omega)$.

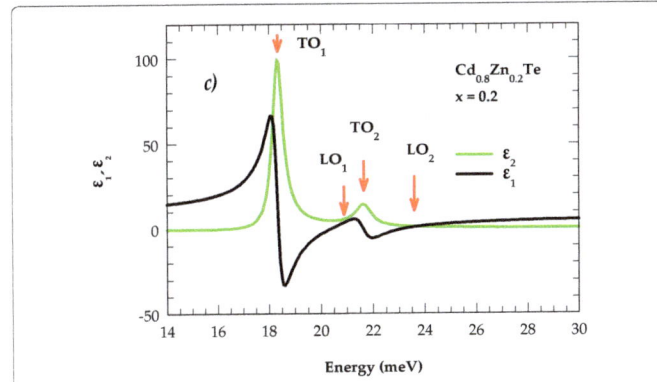

Figure 3c: The real $\varepsilon_1(\omega)$ and imaginary $\varepsilon_2(\omega)$ parts of the complex dielectric function $\tilde{\varepsilon}(\omega)$.

counterparts and used the quadratic expressions involving x, while the energy strength and broadening parameters are obtained deliberating linear dependence articulations on x. Once again, the results (Figures 5a and 5b) of the optical parameters for ternary $Cd_{1-x}Zn_xTe$ alloys are found not only consistent with the limited SE data [16-19] but has also provided clear revelations of the composition dependent CP energy shifts and widths.

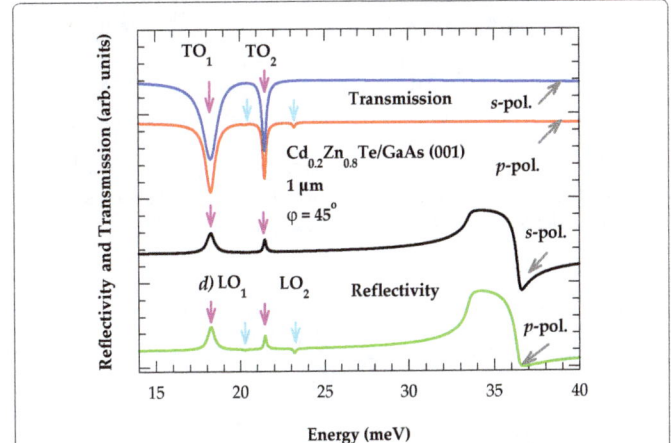

Figure 3d: The polarization dependent transmission (upper panel) and reflectivity spectra (lower panel) for $Cd_{1-x}Zn_xTe$ (x = 0.2) with an incident angle $\varphi = 45°$.

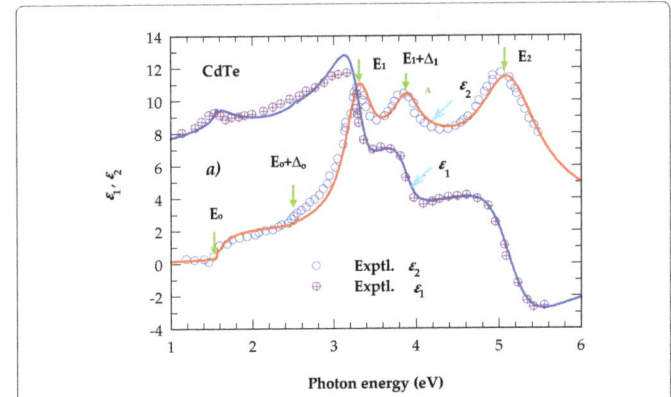

Figure 4a: Comparison of the spectroscopic ellipsometry data for the complex dielectric function $\tilde{\varepsilon}(\omega)$ with the best fit simulated spectra based on a modified MDF for CdTe.

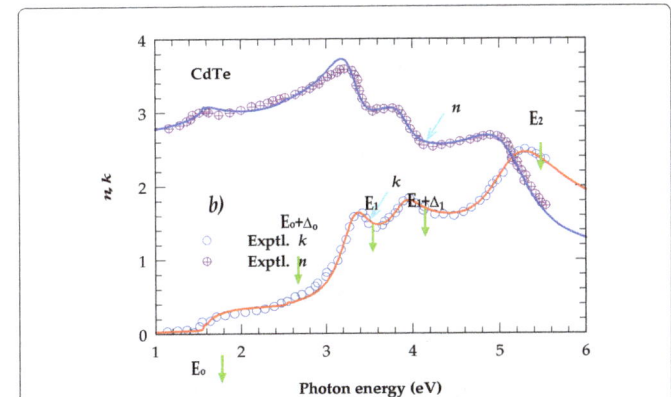

Figure 4b: The comparison of the experimental data with the best fit spectra for the complex refractive index $\tilde{n}(\omega)$ of CdTe.

Figures 4a-4e for CdTe, ZnTe and GaAs are compared with the existing SE data [38-40]. Not only, the dispersions of the pseudo-dielectric functions (Figures 4a-4e) concurred well with the experimental data for the binary materials–the results have clearly revealed distinct CP features from the band structures arising from inter-band transitions. In calculating the optical constants for the $Cd_{1-x}Zn_xTe$ ternary alloys with x=0.0, 0.11, 0.64, 0.86 and 1.0 (Figures 5a and 5b), we have evaluated the CP energy parameters from the values of its binary

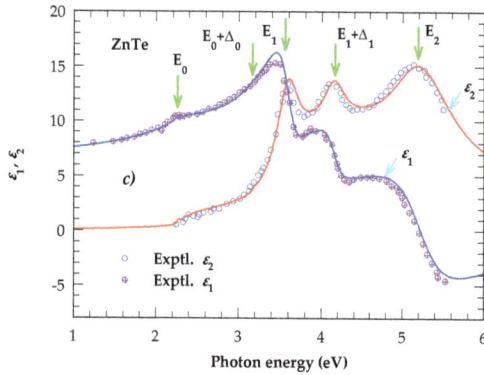

Figure 4c: Similar results [as of *a*)] showing comparison of the experimental complex dielectric function $\widetilde{\varepsilon}(\omega)$ with the best fit spectra for ZnTe.

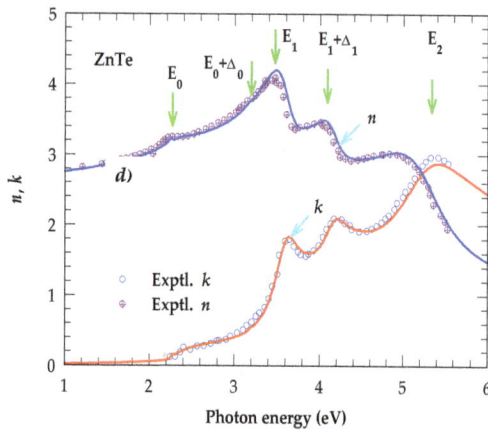

Figure 4d: Similar results [as of *b*)] showing comparison of the experimental complex refractive index $\widetilde{n}(\omega)$ with the best fit spectra for ZnTe.

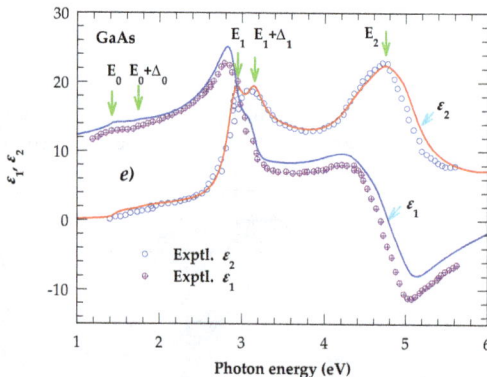

Figure 4e: Similar results [as of *a*)] showing comparison of the experimental complex dielectric function $\widetilde{\varepsilon}(\omega)$ with the best fit spectra for GaAs.

Reflectivity and transmission spectra

By incorporating the energy dependent optical constants and following the methodology outlined, we have calculated (Figures 6a-6d) the optical reflectivity [$R(\lambda)$] (blue line) and transmission [$T(\lambda)$] spectra (green line) as a function of photon wavelength $\lambda(\sim 150 - 1800$ nm) for several ultrathin $Cd_{1-x}Zn_xTe$ epifilms prepared on n-type GaAs

(001). Theoretical results of $R(\lambda)$ appraised at near normal incidence are displayed and compared against the existing experimental data (red squares) for various ultrathin ZnTe/GaAs (001) and $Cd_{0.89}Zn_{0.11}Te$/ GaAs (001) epifilms [26] having thicknesses ranging between 22 nm– 129 nm. The perusal of Figures 6a-6d clearly revealed that the simulated reflectivity spectra not only concurred well with the experiments – the

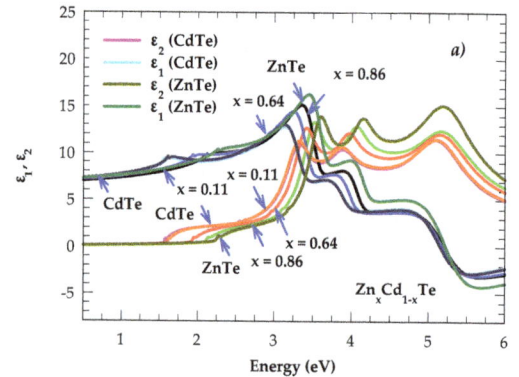

Figure 5a: Calculatedcomplex dielectric function $\widetilde{\varepsilon}(\omega)$ spectra based on a modified model dielectric functions for $Cd_{1-x}Zn_xTe$ ($x = 0.11$, 0.64 and 0.86).

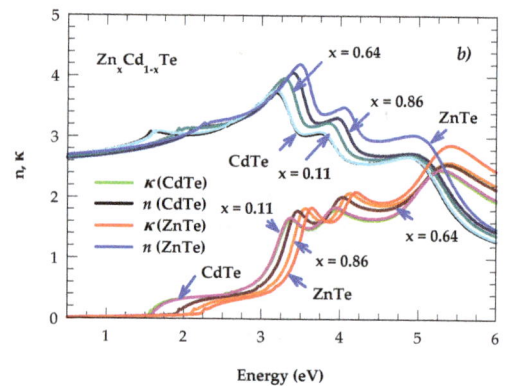

Figure 5b: The calculatedcomplex refractive index $\widetilde{n}(\omega)$ spectra based on a modified MDFs for $Cd_{1-x}Zn_xTe$ ($x = 0.11$, 0.64 and 0.86).

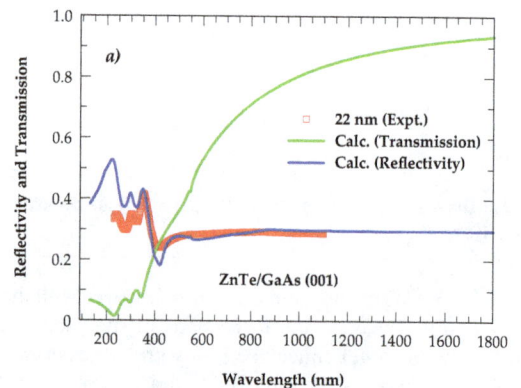

Figure 6a: Comparison of the experimental reflectivity (open red square Ref. [26]) spectra with the simulated reflectivity (solid blue lines) and transmission (solid green lines) spectra for different $Cd_{1-x}Zn_xTe$/GaAs (001) epifilms: for $x=1$ and $d=22$ nm.

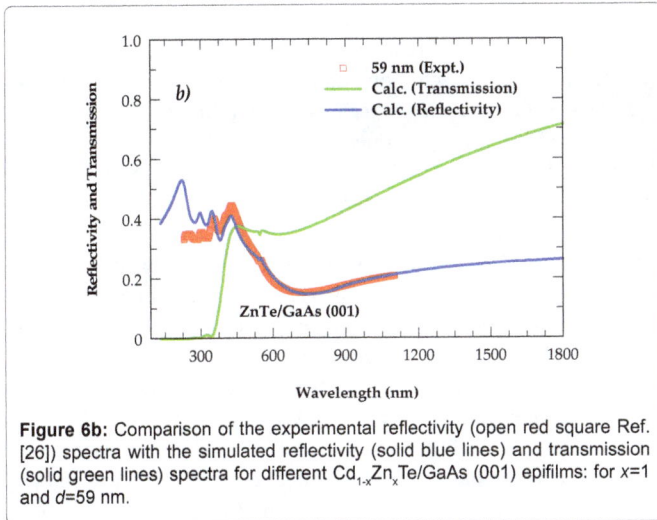

Figure 6b: Comparison of the experimental reflectivity (open red square Ref. [26]) spectra with the simulated reflectivity (solid blue lines) and transmission (solid green lines) spectra for different $Cd_{1-x}Zn_xTe/GaAs$ (001) epifilms: for $x=1$ and $d=59$ nm.

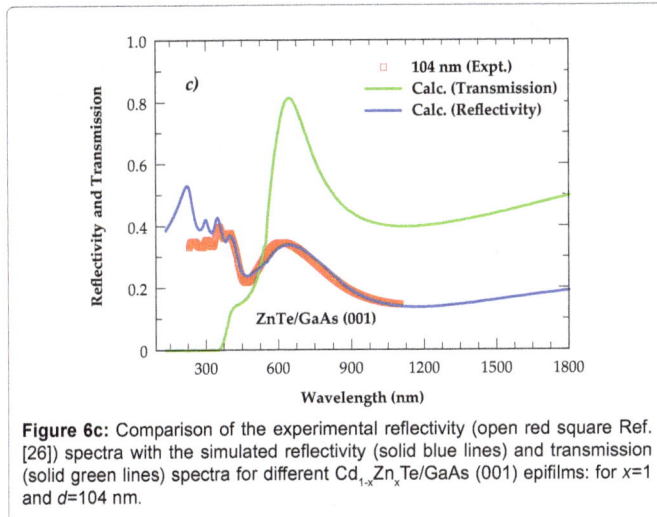

Figure 6c: Comparison of the experimental reflectivity (open red square Ref. [26]) spectra with the simulated reflectivity (solid blue lines) and transmission (solid green lines) spectra for different $Cd_{1-x}Zn_xTe/GaAs$ (001) epifilms: for $x=1$ and $d=104$ nm.

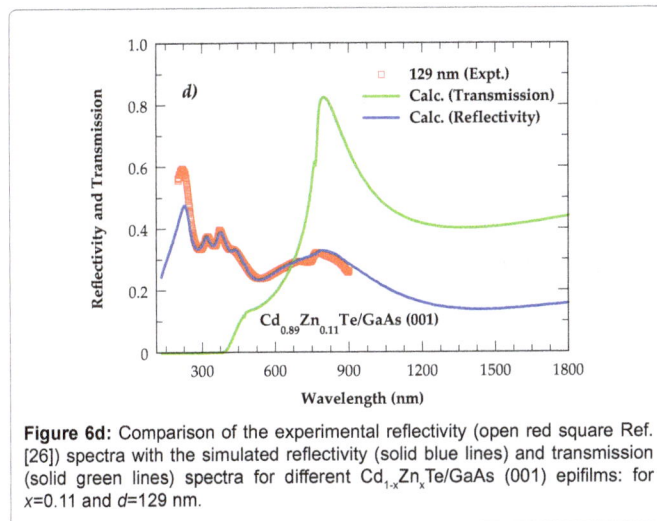

Figure 6d: Comparison of the experimental reflectivity (open red square Ref. [26]) spectra with the simulated reflectivity (solid blue lines) and transmission (solid green lines) spectra for different $Cd_{1-x}Zn_xTe/GaAs$ (001) epifilms: for $x=0.11$ and $d=129$ nm.

in the transparent photon energy region – except that they disclosed the broad intensity modulations. Moreover, for all the material samples studied here – the film thicknesses d assessed by reflectivity studies have concurred copiously with the apparent values appraised from the high resolution X-ray diffraction (HR-XRD) $\omega/2\theta$ scans and other measurements [41].

Thickness dependence: In Figures 7a and 7b, we have displayed the results of our model calculations for the optical reflectivity and transmission spectra of ZnTe/GaAs (001) as a function of λ (100 - 2500 nm) for film thicknesses d varied between 100 nm to 1000 nm. The reflectivity $R(\lambda)$ and transmission $T(\lambda)$ spectra have clearly revealed interference fringes in the highly transparent photon energy region ($\lambda>548$ nm). As expected for thicker films – we perceived emergence of fringes when the conditions of constructive and destructive interferences are met between the light waves reflected off the top and bottom of the film – causing maxima and minima, respectively. From the calculated results of transmission $T(\lambda)$ spectra for ZnTe/GaAs films, we also noticed that with the increase of film thickness from 100 nm to 1000 nm, the transmittance (Figure 7b) decreased – eliciting a sharp absorption edge near $\lambda \equiv 548$ nm. For ZnTe films with thickness $d>400$ nm, our assessed optical band gap E_g ($\equiv 2.26$ eV) from the absorption band edge concurred very well with the PL measurement of E_g for the bulk material. In thinner films ($d<400$ nm) we noticed the absorption band edge divulging a slight blue shift i.e., veering towards the lower λ – possibly prompting slightly larger optical band gaps. This

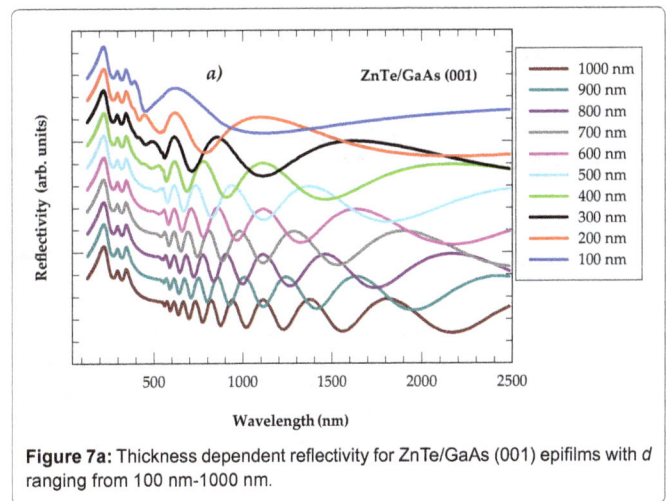

Figure 7a: Thickness dependent reflectivity for ZnTe/GaAs (001) epifilms with d ranging from 100 nm-1000 nm.

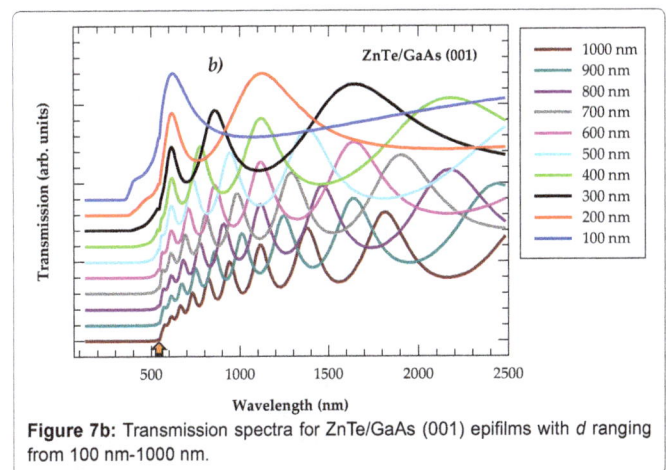

Figure 7b: Transmission spectra for ZnTe/GaAs (001) epifilms with d ranging from 100 nm-1000 nm.

theoretical results have certainly encapsulated all the major observed features. One must note that as the ZnTe ($Cd_{1-x}Zn_xTe$)/GaAs (001) epifilms are too lean, the simulated (Figures 6a-6d) reflectivity [$R(\lambda)$] and transmission [$T(\lambda)$] spectra have divulged no interference fringes

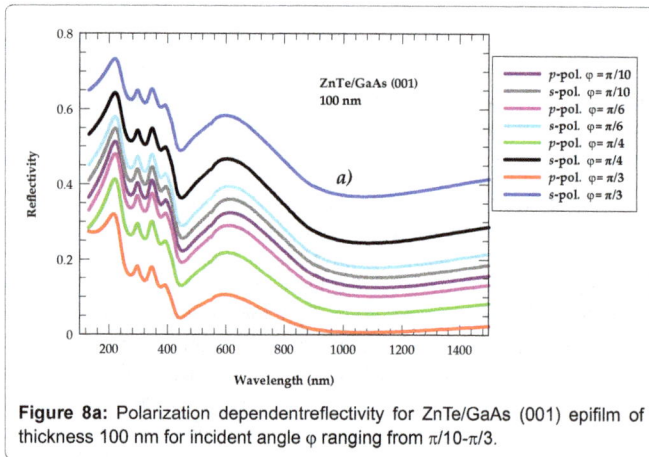

Figure 8a: Polarization dependentreflectivity for ZnTe/GaAs (001) epifilm of thickness 100 nm for incident angle φ ranging from π/10-π/3.

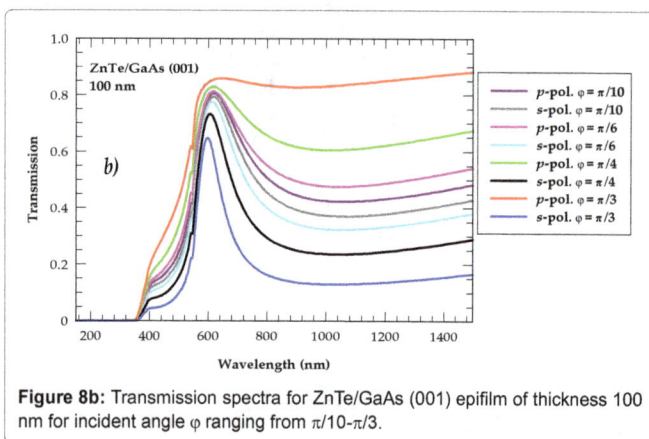

Figure 8b: Transmission spectra for ZnTe/GaAs (001) epifilm of thickness 100 nm for incident angle φ ranging from π/10-π/3.

manifestation in ultrathin ZnTe/GaAs films while not clearly implicit has been recognized, however, in a few recent experiments [26,27] and attributed to the probable misfit strain between the films and substrate.

Polarization Dependence: In Figures 8a and 8b, we have displayed our simulated results of the s- and p- polarized reflectance $[R^s(\lambda), R^p(\lambda)]$ and transmission $[T^s(\lambda), T^p(\lambda)]$ spectra for a 100 nm thick ZnTe/GaAs (001) epifilm by varying the angle of incidence ϕ from π/10 to π/3 (or from 18° to 60°). The reflectance and transmission spectra for λ between 100-1500 nm in the two polarization states have revealed the typical dispersion features. Our calculations (Figure 8a) have divulged that the s-polarized reflectivity $R^s(\lambda)$ increases as the angle of incidence increases from 18 to 60° while the p-polarized reflectivity $R^p(\lambda)$ decreases with the increase of incidence angle. On the other hand, the results of p-polarized transmission $T^p(\lambda)$ showed (Figure 8a) higher values at larger ϕ as compared to the s-polarized transmission $T^s(\lambda)$ spectra. Moreover, the $T^p(\lambda)$ results at 60° have indicated that the ZnTe film is more transparent (~85% in the visible to NIR region) than for the $T^s(\lambda)$ state. This outcome is further verified by the fact that the simulated s-polarized reflectivity $R^s(\lambda)$ is larger (Figure 8a) than the p-polarized reflectivity $R^p(\lambda)$.

Summary and Conclusions

In summary, we have reported the results of a comprehensive experimental and theoretical study for understanding the optical properties of ZnTe ($Cd_{1-x}Zn_xTe$) and GaAs in a wide spectral range covering the FIR (5 meV ≤ E ≤ 100 meV) and NIR to UV (0.75 eV ≤ E ≤ 10 eV) energy regions. By exploiting a Bruker IFS66 spectrometer, we

have measured the FIR response for GaAs, ZnTe, CdTe and $Cd_{1-x}Zn_xTe$ alloys (1.0 ≥ x ≥ 0). Different model dielectric functions are established for extracting the dielectric functions $\tilde{\varepsilon}(\omega)$ of the binary and ternary materials in the two energy regions. In the FIR spectral range, the necessary MDFs are ascertained within the classical Drude-Lorentz [9] methodology by assimilating contributions from both the lattice phonons and free-charge carriers. In the NIR to UV energy region, we have instigated Adachi's [16,20] optical dispersion mechanisms and extorted $\tilde{\varepsilon}(\omega)$ or $\tilde{n}(\omega)$ by exploiting the modified [32] model dielectric functions. The simulated energy dependent optical constants for the binary and ternary alloys compared favorably well with the existing FIR-SE and other experimental [22-29] data. These results are proven valuable for accurately assessing the film thickness d from the polarization dependent reflectivity $R(\lambda)$ and transmission $T(\lambda)$ spectra of ultrathin $Cd_{1-x}Zn_xTe/GaAs$ (001) epifilms. Clearly, the outcome of this methodology has offered a credible testimony for characterizing any semiconducting epitaxially grown nanostructured films of technological importance.

Acknowledgements

One of the authors (DNT) wishes to thank Dr. Deanne Snavely, Dean, College of Natural Science and Mathematics for the travel support and for an Innovation Grant that he received from School of Graduate Studies at Indiana University of Pennsylvania, Indiana, PA that made this collaborative research possible. The experimental results on the FIR measurements presented in details elsewhere were provided by Professor T. R. Yong (National Taiwan Normal University, Taipei).

References

1. Lohstroh A, Della Rocca I, Parsons S, Langley A, Shenton-Taylor C, et al. (2015) Cadmium zinc telluride based infrared interferometry for X-ray detection. Appl Phys Lett 106: 063507-063510.

2. Franc J, Dědič V, Rejhon M, Zázvorka J, Praus P, et al. (2015) Control of electric field in CdZnTe radiation detectors by above-bandgap light. J App Phys 117: 165702-165707.

3. Gan B, Wei T, Gao W, Zeng H, Hu Y (2013) Design of a low-noise front-end readout circuit for CdZnTe Detectors. Journal of Signal and Information Processing 4: 123-128.

4. Rodrigues BHG, Grindlay JE, Allen B, Hong J, Barthelmy S, et al. (2013) The high resolution X-Ray imaging detector planes for the MIRAX mission.

5. Bolotnikov AE, Babalola S, Camarda GS, Cui Y, Egarievwe SU, et al. (2010) Te inclusions in CZT detectors: New method for correcting their adverse effects. IEEE Trans Nucl Sci 57: 910-919.

6. Cai L, Meng LJ (2013) Hybrid pixel-waveform CdTe/CZT detector for use in an ultrahigh resolution MRI compatible SPECT system. Nucl Instr and Methods in Physics Research A 702: 101-103.

7. Hawkins SA, Aleman EV, Duff MC, Hunter DB, Burger A, et al. (2008) Light-induced tellurium enrichment on CdZnTe crystal surfaces detected by raman spectroscopy. J Electron Mater 37: 1438-1443.

8. Olego DJ, Raccah PM, Faurie JP (1986) Compositional dependence of the Raman frequencies and line shapes of Cd1-x ZnxTe determined with films grown by molecular-beam epitaxy. Phys Rev B 33: 3819-3824.

9. Talwar DN, Yang TR, Feng ZC, Becla P(2011) Infrared reflectance and transmission spectra in II-VI alloys and superlattices. Phys Rev B 84: 174203-174211.

10. Granger R, Marqueton Y, Triboulet R (1993) Optical phonons in bulk Cd1-XZnxTe mixed crystals in the whole composition range. J de Phys 3: 135-141.

11. Teng J, Sang W, Li G, Shi Z, Min J, et al. (2008) Influence of In dopant on PL Spectra of CdZnTe. Crystals J Korean Phys Soc 53: 146-149.

12. Gul R, Keeter K, Rodriguez R, Bolotnikov AE, Hossain A, et al. (2012) Point Defects in Pb-, Bi-, and In-Doped CdZnTe Detectors: Deep-Level Transient Spectroscopy (DLTS) Measurements. J Elect Mater 41: 488-493.

13. Fujiwara H (2007) Spectroscopic ellipsometry: principles and applications. John Wiley, USA.

Polarization Dependent Reflectivity and Transmission for Cd1-Xznxte/GaAs (001) Epifilms in the Far-Infrared...

83

14. Talwar DN, Feng ZC, Lee JF, Becla P (2013) Structural and dynamical properties of Bridgman-grown CdSexTe1−x (0<x _0.35) ternary alloys. Phys Rev B 87: 165208-165212.

15. Talwar DN, Becla P (2016) Infrared and Raman characteristics of bulk $Cd_{1-x}Mn_xTe$ and $(MnTe)m/(CdTe)n$ short periodsuperlattices. Mat Letts 175: 279-283.

16. Adachi S, Kimura T (1993) Synthesis of self-organized TiO_2 nanotube arrays: Microstructural, stereoscopic, and topographic studies. J Appl Phys 32: 3496-3501.

17. Yao HW, Erickson JC, Barber HB, James RB, Hermon H (1999) Optical Properties of Cd0.9Zn0.1Te studied by variable angle spectroscopic ellipsometry between 0.75 and 6.24 Ev. J Electron Mat 28: 760-765.

18. Daraselia M, Brill G, Garland JW, Nathan V, Sivananthan S (2000) In-situ control of temperature and alloy composition of Cd1-xZnxTe grown by molecular beam epitaxy. J Electron Mat 29: 742-745.

19. Sridharan M, Narayandass SAK, Mangalaraj D, Lee HC (2002) Optical constants of vacuum-evaporated Cd0.96Zn0.04Te thin films measured by spectroscopic ellipsometry. J Mat Sci: Mat Electron 13: 471-476.

20. Adachi S (2005) Properties of Group-IV, III-V and II-VI Semiconductors. John Wiley, USA.

21. Xu M, Li YF, Yao B, Ding Z, Yang G (2014) Structural, electronic and optical properties of $Cd_xZn_{1-x}S$ alloys from first-principles calculations. Phys Lett A 378: 3382-3388.

22. Polizzi E (2009) Density-matrix-based algorithm for solving eigenvalue problems. Phys Rev B 79: 115112.

23. Gygi F (2006) Large-scale first-principles molecular dynamics: moving from terascale to petascale computing. J Phys: Conf Ser 46: 268-277.

24. Windus TL, Bylaska EJ, Tsemekhman K, Andzelm J, Govind N (2009) Computational Nanoscience with NWChem. J Comp Theor Nanosci 6: 1297-1304.

25. Sridharan MG, Mekaladevi M, Rodriguez-Viejo J, Narayandass SK, Mangalraj D, et al. (2004) Spectroscopic ellipsometry studies on polycrystalline $Cd_{0.9}Zn_{0.1}Te$ thin films. Phys Stat Sol 201: 782-790.

26. Larramendi EM, Purón F, Melo O (2002) Thickness measurement and optical properties of Cd1-xZnxTe. Semicond Sci Technol 17: 8-12.

27. Salem AM, Dahy TM, El-Gendy YA (2008) Thickness dependence of optical parameters for ZnTe thin films deposited by electron beam gun evaporation technique. Physica B: Physics of Condensed Matter 403: 3027-3033.

28. Shaaban ER, Ahmad M, Abdel Wahab EA, Shokry Hassan H, Aboraia AM (2013) Structural and optical properties of varies thickness of ZnTe nanoparticle. Proc of Basic and Appl Sci 1: 244-257.

29. Khoshman JM (2005) Spectroscopic ellipsometry characterization of single and multilayer aluminum nitride/indium nitride thin film systems. Dissertation Abstracts International 66: 267.

30. Franta OD (2000) Ellipsometry of thin film systems. Progress in Optics 41: 181-282.

31. Levenberg K (1944) A method for the solution of certain non-linear problems in least square. The Quarterly of Applied Mathematics 2: 164-168.

32. Talwar DN (2009) Novel Dilute III-V-Ns from physics to Applications.

33. Chelikowsky JR, Cohen ML (1976) Nonlocal pseudopotential calculations for the electronic structure of eleven diamond and zinc-blende semiconductors. Phys Rev B 14: 556-566.

34. Guo Q, Ikejira M, Nishio M, Ogawa H (1996) Optical properties of zinc telluride in vacuum ultraviolet region. Solid Stat Commun 100: 813-815.

35. Yasuda K, Kojima K, Mori K, Kubota Y, Nimura T, et al. (1998) Electrical and optical properties of iodine doped CdZnTe layers grown by metalorganic vapor phase epitaxy. J Electron Mater 27: 527-531.

36. Cardona M, Shaklee KL, Pollak FH (1967) Electroreflectance at a semiconductor- electrolyte interface. Phys Rev154: 696-720.

37. Guizzetti G, Nosenzo L, Reguzzoni E, Samoggia G (1974) Thermoreflectance spectra of diamond and zinc-blende semiconductors in the vacuum-ultraviolet region. Phys Rev B 9: 640-647.

38. Aspnes DE, Studna AA (1983) Dielectric functions and optical parameters of Si, Ge, GaP, GaAs, GaSb, InP, InAs, and InSb from 1.5 to 6.0 Ev. Phys Rev B 27: 985-1009.

39. Viña L, Umbach C, Cardona M, Vodopyanov L (1984) Ellipsometric studies of electronic interband transitions in $Cd_xHg_{1-x}Te$. Phys Rev B 29: 6752-6760.

40. Palik ED (1991) Handbook of Optical Constants of Solids. Academic, New York.

41. Hernández LC (2003) Growth of ZnTe Semiconductor Thin films onto GaAs and Si Substrates by Isotherm closed space sublimation.

42. Wronkowska AA, Wronkowski A, Firszt A, Legowski S (2006) Investigation of II-VI alloy lattice dynamics by IR spectroscopic ellipsometry. Cryst Res Technol 41: 580-587.

16

Iriartea Deltoidea and Socratea Exhorriza: Sustainable Production Alternatives for Integrated Biosystems

Sánchez LM[1]* and Quiñonez MF[2]

[1]Environmental engineer and Research teacher, University of Meta, Villavicencio, 500001, Colombia
[2]Architect and Research teacher, University of Meta, Villavicencio, 500001, Colombia

Abstract

The objective of the present research is to analyze the structure, behavior and applicability of the integrated biosystems, through the study of the ancestral, current and potential uses of the sucker palms - Socratea Exhorriza and chonta - Iriartea Deltoidea; And how its ecosystem functioning provides us with environmental goods and services; Which until now have not been thoroughly reviewed and therefore many of its properties are unknown. It is for this reason that this research tries to demonstrate that there are tools that make it possible to formulate plans, programs or projects of sustainable use that allow the biosystem to maintain itself as the ecosystem.

It is understood by integrated biosystems such as those that link two or more biological systems to transform organic waste into value-added products, through the use of processes involving microorganisms, major organisms, animals and plants. One of the processes becomes the raw material for the start of the next or the following processes (Rodríguez, et al.). "Biosystems are semi-open chemical chemosystems (or chemical reactors), which take from their surroundings the matter and energy they employ, which synthesize all their other components and reproduce" (Bunge). They are made up of elements among which are living beings organized at different hierarchical levels. For example, genetic systems, organ systems, parasite systems, plant systems, among others, where intermediate levels can be observed between individual systems and population systems or communities (Jaramillo). These constitutive elements play a fundamental role in the functioning of systems and therefore of biosystems, such as: Inorganic substances, organic compounds, climatic elements, producers, consumers, disintegrators. The study will be carried out through seven stages: The first one refers to the bioconstruction and use of wood, the second to the use of biomass, the third, the fauna refuge, the fourth, the entomological analysis of species found in Cellulose; The fifth, to edible and medicinal uses, the sixth to the interaction of microorganisms in the soil; And the seventh and last to the proposal of normative tools and recommended uses.

Keywords: Chuapo; Chonta; Guadua; Integrated biosystems; Bioconstruction biomass; Entomological analysis; Orinoquia; Amazonia

Introduction

Iriartea deltoidea of common name CHONTA and the Socratea exorrhiza commonly recognized in the region of the Amazonia and Colombian Orinoquia like CHUAPO. Are tropical palm species [1,2] found in humid forest ecosystems, and have traditionally been used by the inhabitants of the Central American countries to Bolivia and Brazil, in these regions where they grow they are mainly used by the Indigenous and peasants for the elaboration of different products of daily use as furniture and handicrafts, that is to say that in Colombia it is common, traditional and widely used by communities of low economic resources, but its scientific study is scarce and surely with more investigation both Iriartea Like the Socratea could become an important source of income with diverse applications.

This study seeks to execute seven investigative stages and to execute it was divided into two parts beginning by taking one of the species, in this case the chonta, applying the respective phases of investigation and in parallel, to perform the same procedure with the chuapo.

Stages 1 and 2 of the process will be performed comparing the two specimens proposed with Guadua angustifolia, taking the latter as a reference, and providing qualitative and quantitative information on the research of each of the palms; Oriented to understand and consolidate the way in which it should be given the sustainable use or if possible a sustainable productive chain, based on the functioning of an integrated biosystem that allows to offer social, economic and environmental opportunities to the communities that currently work

With the chonta; And to those who could benefit from the correct use of the slur. As a result of the investigative procedure, environmental normative tools will be proposed to allow the correct use of these species; As well as the inclusion in the standard resistant earthquake that facilitates to dynamize the agroforestry structure of the region as already has been done in other places of the country with the guadua.

Experiment Materials and Methodology

The development of the investigation has begun with the palm chonta executing stages one and two in the following way:

Stage 1

A supplier of the certified material was located in the city of Mocoa in the department of Putumayo, which has been marketing "Chonta" for a long time (30 years approximately).

In parallel, mechanical tests were carried out in the laboratory,

*Corresponding author: Lina Mojica Sánchez, Research Center-CIAM José Antonio Candamo, and Faculty of environmental engineering, University of Meta, Colombia, E-mail: lina.mojica@unimeta.edu.co

corresponding to 15 test tubes for the penetration resistance tests (Pilodyn) and 15 test specimens for Perpendicular Compression tests.

The material was physically tested by installing a 12-square-meter weather deck with 7-cm-wide, 3-meter-long and 1.5-centimeter-thick boards with 1-cm slats dilating for the purpose to measure environmental degradation processes.

Stage 2

To explore the possible uses that can be given to biomass, the following parameters will be evaluated, preliminarily [3]:

Meteorological conditions: The experimentation will be carried out in close proximity to the avant-garde road very close to the weather station of the avant-garde airport of the city of Villavicencio, capital of the department of Meta (Colombia), taking into account the following atmospheric conditions: relative humidity, Altitude and geographical coordinates.

Conditioning of the material [4]: Milling of a certain amount of palm was carried out until passing through a sieve of a certain diameter.

Determination of apparent density of the sample [5]: Three different tests will be performed by the traditional method of gravimetry, using analytical balance and a pycnometer to calculate the weight of the dust sample from the difference of the filled and unoccupied pycnometer, and the chonta palace) Bibliographical source) and for the "chuapo" in the department of Meta in the Orinoquía region. In the cuts made it was possible to observe the differences and similarities between one species and the other in terms of fibers, texture, morphology, among other characteristics. The density is determined by the equation $d=m/V$

Determination of the hygroscopic moisture of the sample [6]: In the same way, 3 tests will be performed using the gravimetric method.

Determination of ashes [7]: Two tests will be carried out where the samples will be weighed, incinerated in the kiln, subsequently weighed and the difference in the masses calculated.

Determination of lignin [8]: Alkaline hydrolysis treatment at a temperature of 30°C, of 5 grams of sample, weighed in the analytical balance with an accuracy of 0.001g, the first experiment was one hour with sodium hydroxide and sodium sulfate.

Determination of hemicellulose [9]: Catalytic acid hydrolysis treatment of the sample previously treated by hydrolysis was performed.

Results

As is noted from the Figures 1 and 2, the different sites were searched to locate the specimens, for the Chonta in the department of Putumayo in the Amazon region (of which there is no photographic record, but was taken from a video of differences and similarities between one species and the other in terms of fibers, texture, morphology, among other characteristics.

For the process of construction of the prototype (DECK) the wood of the treated and legally marketed chonta was purchased, which will be exposed to the elements in the season of high and low rainfall. It will also be monitored for two years to see the evolution and behavior of the material and its probabilities of being classified as an element suitable for bioconstruction activities (Figure 3).

During the field work performed for the elaboration of the necessary cuts for laboratory tests, the presence of different Coleoptera

Figure 1: Material cut in the municipality of Restrepo, Meta, Orinoquía, Colombia. For the accomplishment of laboratory tests, wood of the plucking. Photography Francisco Quiñonez.

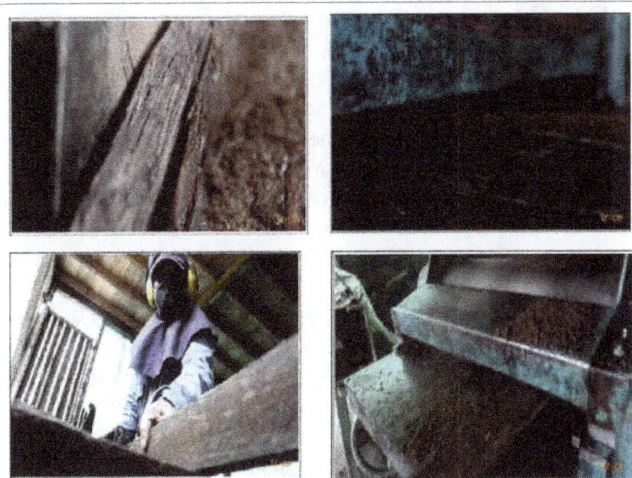

Figure 2: Material extracted and treated wood Iriartea deltoidea in the municipality of Mocoa, Putumayo, Amazonia, Colombia.

Figure 3: Wood of chonta, used in the construction of deck as experimental prototype, designed and constructed by the architect Francisco Quiñonez. Photography Francisco Quiñonez.

Figure 4 was found, which may be present in the specimens, possibly due to the existing symbiotic activity.

Through the secondary information consulted a very valuable revelation was found regarding the chuapo palm and its relation with the parrot orejiamarillo: With characteristics of life very different to the one that the country knew, was discovered in 2009 a population of Orejiamarillo parrot in the upper part of Cubarral (Meta), Orinoquía, which today reaches the 80 individuals in the reproduction stage.

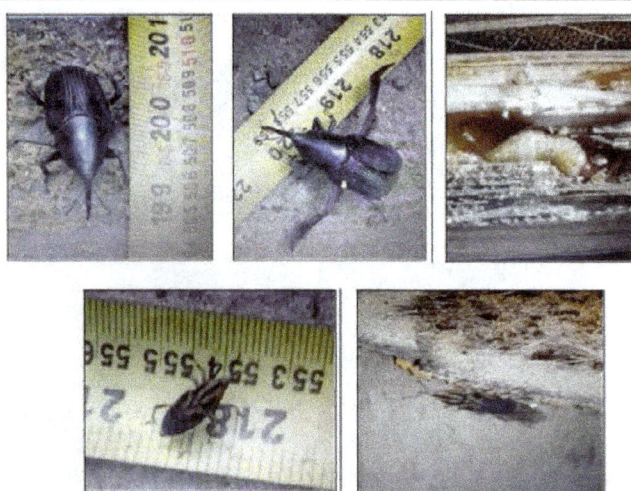

Figure 4: Finding coleoptera in the cellulose and trunk of the palms. Photography Francisco Quiñonez.

Since then, environmental authorities together with different entities are committed to its conservation with a project built on the basis of effort and dedication. But the struggle is not only to preserve the parrot but its 'home', that is why a study of the population dynamics of the Chuapo palm has been done since 2015, and in a recent phase the measurement of the rates of Mortality and growth in each. This impinges on the importance of analyzing the environmental services offered by the Socratea as a wildlife refuge.

Discussions

The use of biomass from both palms, such as roots, leaves, stems, fruits, leaf litter, will undergo different laboratory tests to experience the likelihood of obtaining liquid or gaseous fuels that contribute as energy to more efficient and cleaner mechanisms. That the gases produced in the combustion would have a much lower proportion of sulfur compounds than those coming from the processes of combustion of carbon.

The analysis will begin by making the parallel between the CHONTA and the GUADUA which is composed mainly of hemicellulose and lignin [10] 61-71% and 20-30% respectively, it is expected similarly a similar composition of the chonta due To that its morphology is similar, since the density and hardness of the tissue of the stem of the palms in general increase from the center to the periphery and are greater towards the base [10], the guadua instead presents a hollow structure in The center and that is populated and making it harder and denser towards the crust [10]. Due to its similarity in hardness and density in the stem, it is desired to verify the high lignin content in the chonta palm, since, if its chemical structure were similar, it could be used as the guadua in the construction of houses, as material Reinforcement and attractive color as landscaping element, it is known that the mechanical properties of lignin make it special, even today is the raw material of carbon fibers useful for the materials industry.

Conclusions

These palm varieties are extremely resistant and have been used ancestrally by the natives for their constructions, compared to the Guadua angustifolia, which is a very ductile fiber, easy to work and abundant in the tropical regions, but which in turn Is very delicate to the processes of deterioration. On the other hand, the chuapo and the chonta are today characterized by contemporary constructors as a "very hard material that wears and damages the cutting tools"

The behavior of the chonta wood during the construction of the prototype, was classified as excellent, since it is a material of great hardness and resistance that in turn counts on a degree of flexibility that allows to work it and to mold it according to the design. It is analyzed as a solid material, stable and beautiful finishes.

It is necessary to perform the entomological study, due to the coleopteran species found at the time of cutting the palms for laboratory tests, due to their variety and importance it is necessary to evaluate their biological and ecological function within the biosystem.

As the research stages are carried out, we confirm that the best way to propose a sustainable productive system or sustainable production project is to propose its exploration and evaluation under the theory of integrated biosystems, and it works understanding the hierarchy, interactions and interrelations of the System in general to understand ecosystems in an integral way, and in this way to interpret and use in a sustainable way the environmental goods and services of which we are kindly surrounded. The recognition, interpretation and encompassing of a strategic ecosystem as an integrated biosystem is the key to achieving sustainable development.

Acknowledgement

The authors thank to the University of Meta-Colombia and to Environmental.

References

1. Navarro J, Galeano G, Bernal R (2014) Manejo de la Palma Barrigona o Chonta (Iriartea Deltoidea) en el Piedemonte Amazónico Colombiano. Colombia Forestal 17: 5-24

2. Ruiz, Pavón, Árboles de Centroamérica. Descripción de Especies: Iriartea deltoidea.

3. Mauricio Andrés Beltrán Martínez. Chemical engineer. Industrial University of Santander T.P. 8721 del CPIQ.

4. NTC 1522: (1979) Suelos. Ensayo para determinar la granulometría por tamizado.

5. NTC 1974: (1984) Ingeniería Civil y Arquitectura. Suelos. Determinación de la densidad relativa de los sólidos.

6. NTC 1495: (1979) Suelos. Ensayo para determinar el contenido de humedad.

7. NTC 841. Papel. Pulpas. Determinación del contenido de cenizas. Research Center-CIAM José Antonio Candamo, and Faculty of environmental enginnering.

8. Srivastava AK, Agrawal P, Rahiman A (2014) Delignification of rice husk and production of bioethanol. International Journal of Innovative Research in Science, Engineering and Technology 2: 10187-10194.

9. Cuervo L, Folch J, Quiroz R (2009) Lignocelulosa como Fuente de Azúcares para la Producción de Etanol. Centro de Investigación de Tecnología UAEM. Instituto de Biotecnología UNAM.

10. Giraldo E, Sabogal O (2007) Una Alternativa Sostenible: La Guadua. Tercera Edición. ICBN 958-96754-0-9. Editor: Corporación Autónoma Regional del Quindío PRQ.

17

Raman Spectroscopy of Iron Oxide of Nanoparticles (Fe$_3$O$_4$)

Panta PC[1,2]* and Bergmann CP[1]

[1]Ceramic Materials Laboratory, Federal University of Rio Grande do Sul, Av. Osvaldo Aranha 99, sl. 705C, CEP 90035-190, Porto Alegre, RS, Brazil
[2]Department of Chemistry and Physics, University of Santa Cruz do Sul, Santa Cruz do Sul, Brazil, Av. Independência 2293, CEP 96815-900, Santa Cruz do Sul, Brazil

Abstract

Nanoparticles of iron oxide (Fe$_3$O$_4$) were obtained by Coprecipitation with synthesis time of 30, 60 and 90 min. The morphology of the samples was investigated by scanning electron microscopy (SEM) and structural characteristics were obtained by X-ray diffraction (XRD). The crystallite size was calculated from the spectrum X-ray diffraction with the application of the Scherrer equation and Winfit. The crystallite size varied from 4.6 to 14.4 nm when calculated by Scherrer equation and when calculated by the single line ranged from 7.5 to 22.3 nm Winfit. The degree of graphitization was studied by Raman spectroscopy where spectrums were analyzed with different lasers: 514 nm (0.75 mW power used) and 785 nm (1.2 mW power used). The dominant structures of the spectra are in 215, 276, 398, 487, 654 and 1300 cm^{-1} when using the laser 514 nm. The spectrum produced with laser 514 nm is characteristic peak of magnetite in 654 cm^{-1}. The spectrum produced by laser 785 nm has a peak at 670 cm^{-1}, shifted relative to the laser 514 nm. The spectrum generated by laser 785 nm peaks characteristic of maghemite encountered due to possible oxidation of the sample caused by the high power laser. The experimental results were satisfactory and are according to the survey.

Keywords: Nanoparticles; Coprecipitation; Raman Spectroscopy; Magnetite

Introduction

The nanoparticles of iron oxide are a major focus of research in physics, chemistry, engineering and materials science, among others. Research is because their main magnetic properties such as superparamagnetism, high coercivity, low Curie temperature, high magnetic susceptibility, among others, which occur due to the nanometric size. The Fe$_3$O$_4$ nanoparticles can have various applications in nanotechnology, such as pigment, drug delivery, targeting, magnetic resonance imaging for clinical diagnosis, recording material, hyperthermia, catalyst, etc., [1,2]. These nanoparticles have an ability to interact in different ways with different biological molecules due to their properties such as high specific area and wide variety of surface functionalization.

Many methods have been developed for preparing magnetic nanoparticles, such as polyols, microemulsions, laser pyrolysis, sonochemistry, coprecipitation, hydrothermal, etc., [3]. Coprecipitation method is suitable for low cost; this method has great potential due to direct production of nanoparticles Fe$_3$O$_4$ (water-based). Techniques for the synthesis of nanoparticles determine the particle size, which can be controlled by a surfactant. There are important conditions for the synthesis, such as pH of the solution, temperature and time of reaction, stirring speed, solute concentration and surfactant. All features are important in order to obtain particles of desired size and shape [1-3].

This work has focused on the preliminary results of the synthesis process of the nanoparticles of iron oxide by the coprecipitation method in which the particles can be well controlled. Subsequently characterizations were conducted by XRD, TEM and Raman spectroscopy.

Experimental

Materials

Ferrous chloride tetrahydrate (FeCl$_2$.4H$_2$O, 99%), ferric chloride hexahydrate (FeCl$_3$.6H$_2$O, 99%), were obtained from Sigma Aldrich (EUA), tetramethylammonium hydroxide (25-28%, w/w) was obtained from Vetec (Brazil). Distilled water was used for preparation of the solutions after deoxygenation with dry argon for 10 min. The divalent (FeCl$_2$.4H$_2$O), trivalent (FeCl$_3$.6H$_2$O) iron salts and aqueous ammonium hydroxide (25-28%, w/w) were also deoxygenated with dry argon before use. All the other chemicals used in this work were analytical reagent grade from commercial market without further purification.

Preparation of iron oxide magnetic nanoparticles

Method previously published by Kim was chosen for the iron oxide magnetic nanoparticles preparation. Nanoparticles were prepared by coprecipitation of Fe^{2+} and Fe^{3+} salts solution and NH$_4$OH solution using peristaltic pump. The principle of reaction is given by:

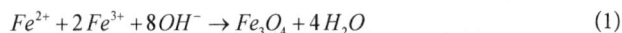

$$Fe^{2+} + 2Fe^{3+} + 8OH^- \rightarrow Fe_3O_4 + 4H_2O \qquad (1)$$

According to the results of thermodynamic model of this system, a complete precipitation of Fe$_3$O$_4$ is expected while maintaining a molar ratio of 1:2 to a non-oxidizing environment [4].

Salts, ferric chloride hexahydrate (FeCl$_3$.6H$_2$O, 99%) and ferrous chloride tetrahydrate FeCl$_2$.4H$_2$O (>99%) obtained from Aldrich were dissolved in deionized water (40 ml) previously degassed. During the reaction the pH must be controlled at the beginning and end. Besides these, the temperature and agitation rate should also be controlled in reaction [5,6]. The following is performed to pass the inert gas (argon) in the system under stirring and heating 75°C ± 15°C. Starts dripping of 100 ml of ammonium hydroxide (NH$_4$OH)(25%) obtained from Vetec. Just after mixing the solutions, the color of the solution changed from light brown to black, indicating the forming of Fe$_3$O$_4$

*Corresponding author: Panta PC, Ceramic Materials Laboratory, Federal University of Rio Grande do Sul, Av. Osvaldo Aranha 99, sl. 705C, CEP 90035-190, Porto Alegre, RS, Brazil, E-mail: pr.priscila@gmail.com

nanoparticles, which was allowed to crystallize completely for another 30, 60 and 90 min under rapid stirring. After this time period, the magnetic nanoparticles were washed three times with pure water with the help of a magnet [1].

Characterization of prepared magnetic nanoparticles

For the materials characterization were used the technique of X-ray diffraction using a Philips X 'equipment Pert MPD, 40 kV and 40 mA. The angular range was 5 to 75° and 0.05° and step of counting time 3 seconds for each step. The analysis of scanning electron microscopy (SEM) were performed on a JEOL microscope, model JSM 6060 with maximum operating voltage of 30 kV and nominal resolution of 3.5 nm. The voltage was 20 kV (see bar scale). Scanning Electron Microscopy (SEM) was used to determine an approximate particle size of prepared nanoparticles. The Raman spectra were recorded at room temperature by a Raman spectrometer (Renishaw brand - inVia Raman, with an Excitation laser line of 632.8 nm from a He-Ne laser.

Results and Discussion

The characterization of the resulting powders after heat treatment was performed by means of X-ray diffraction (XRD) to verify the presence of crystalline phases. The X-ray diffractometer, model Philips X'Pert MPD, was used with graphite monochromator and fixed anode operated at 40 kV and 40 mA, which uses a radiation wavelength (λ=0.154056 nm) of Cu-Kα. The angular range was 5 to 75° and 0.05° and step of counting time 3 seconds for each step.

The XRD peaks (Figure 1) showed the formation of a spinel type crystal structure of the inverse characteristic of magnetite. We also observed a mixture of $\alpha$and γ phases of Fe_2O_3 (hematite and maghemite), and magnetite (Fe_3O_4). The mean particle diameters were also calculated from the XRD pattern according to the linewidth of the (311) plane refraction peak using Scherrer equation (2)

$$D = \frac{K \cdot \lambda}{b \cdot \cos\theta} \tag{2}$$

The equation uses the reference peak width at angle θ, where λ is the X-ray wavelength (1.5418 Å), b is the width of the XRD peak at half height and K is a shape factor, about 0.9 for magnetite and maghemite [2]. The results were 9.5 nm for magnetite nanoparticles. The crystallite size was calculated from the X-ray diffraction spectrum, applying the Scherrer equation ranged from 4.6 to 14.4 nm. For calculation by single line Winfit ranged from 7.5 to 22.3 nm.

The micrographs show similar morphology as in Figure 2. The micrographs have the formation of large agglomerates of nanoscale particles, which can be attributed to the growth by coalescence of nuclei, resulting in particles that tend to aggregate toward a lower energy state free, by reduction of interfaces with the environment.

The Raman spectra were recorded at room temperature by a Renishaw Raman spectrometer mark - Raman inVia with a linear laser excitation of 514 nm and 785 nm (He-Ne). For each sample, exposed 10 sec, three distinct points were measured and displacement occurred between 100 and 1400 cm^{-1} (Figure 3).

Raman spectroscopy was employed to determine the nature of the iron oxide core (magnetite), where the Raman Effect is caused by the molecular effects produced from certain energy irradiated on the sample. However, it is known that magnetite has a weak Raman scattering, especially for lower laser powers that are low so that there are phase transformations induced by the laser. According to Li YS, the Raman spectrum peaks of magnetite were investigated, where

from certain vibrational modes, peak 670 cm^{-1} was identified as a band characteristic.

Shebanova conducted some experiments using various powers of the laser Raman spectrum for magnetite. And they concluded that above some critical value of the laser power, the Raman spectra passed to indicate the characteristics of maghemite bands due to the effects of oxidation of the material. Thus, in this work we used two different types of lasers 514 nm (power 0.75 mW) and 785 nm (power 1.2 mW) with different powers, which produced different Raman spectra from the same sample. The dominant structures of the spectra are in 215, 276, 398, 487, 654 and 1300 cm^{-1} when the laser used is 514 nm [7,8].

Figure 3 shows the spectrum produced by laser 514 nm, where the characteristic peak of magnetite is 670 cm^{-1}. The laser 785 nm has a peak at 654 cm^{-1} and shifted in relation to 670 cm^{-1}. The characteristics of the maghemite peaks occur due to high power laser, probably because the sample was oxidized.

To purchase these spectra we used the 785 nm laser with a power of 1.2 mW, which produced very similar spectra for all measured samples. The results confirm the formation of Fe_3O_4 since obtained Raman

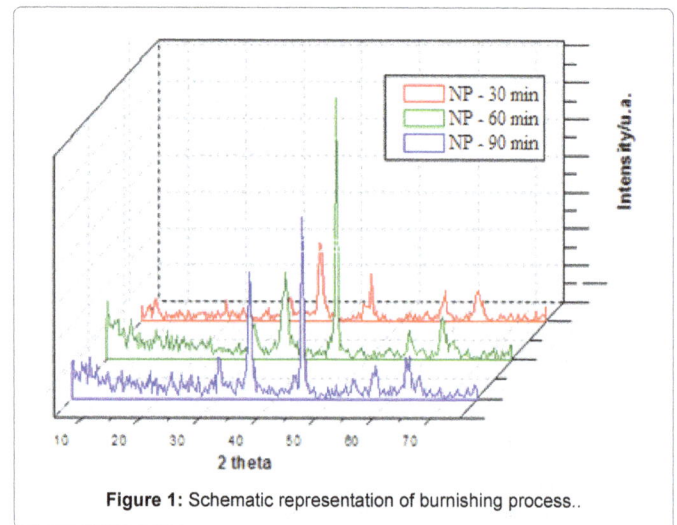

Figure 1: Schematic representation of burnishing process..

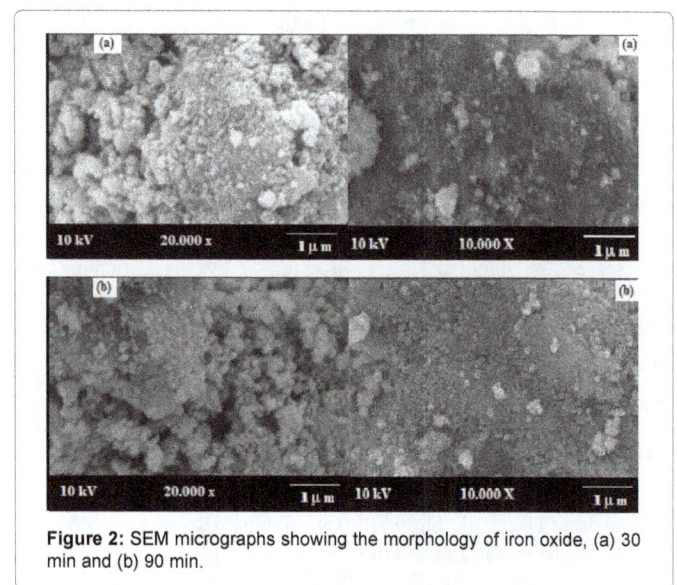

Figure 2: SEM micrographs showing the morphology of iron oxide, (a) 30 min and (b) 90 min.

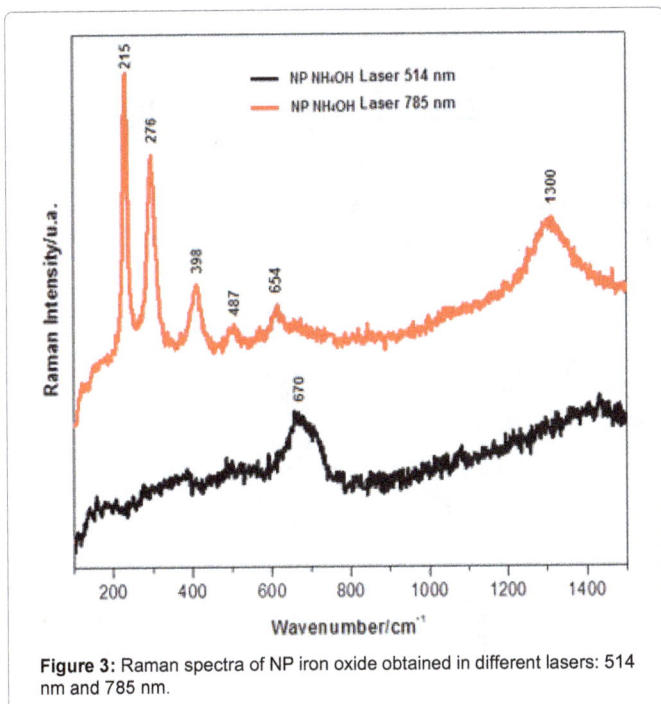

Figure 3: Raman spectra of NP iron oxide obtained in different lasers: 514 nm and 785 nm.

spectrum has identical features to the cited work by Si H and Slavov L. The spectra show magnetite and maghemite peaks because there is a partial oxidation by the incidence of high-power laser [9].

Conclusions

Magnetic nanoparticles of iron oxide were prepared by two-step process. In the first step, magnetite nanoparticles were produced by coprecipitation of Fe^{2+} and Fe^{3+} salts solution with NH_4OH solution formed magnetite nanoparticles. Resulting samples of nanoparticles were analyzed utilizing XDR, Raman and SEM. It was found that prepared nanoparticles revealed the magnetic properties in water suspension when external magnetic field was applied, that particles were generally spherical in shape and have particle size below 9.5 nm and that they tend to form agglomerates. When using Raman spectroscopy, the obtained one sample peaks vary with the potency and the laser 514 nm and 785 nm.

Acknowledgements

The authors acknowledge CNPq for financial support.

References

1. Sun J, Zhou S, Hou P, Yang Y, Weng J, et al. (2007) Synthesis and characterization of biocompatible Fe_3O_4 nanoparticles. Journal of Biomedical Materials Research 80: 333-341.

2. Wu W, Quanguo H, Changzhong J (2008) Magnetic Iron Oxide Nanoparticles: Synthesis and Surface Functionalization Strategies. Nanoscale Research Letters 3: 397-415.

3. Jiang W, Yanga HC, Yangb SY, Horngb HE, Hungd JC, et al. (2004) Preparation and properties of superparamagnetic nanoparticles with narrow size distribution and biocompatible. Journal of Magnetism and Magnetic Materials 283: 210-214.

4. Kim DK, Mikhaylova M, Zhang Y, Muhammed M (2003) Protective coating of superparamagnetic iron oxide nanoparticles. Chemistry of Materials 15: 1617-1627.

5. Gnanaprakash G, Mahadevan S, Jayakumar T, Kalyanasundaram P, Philip J, Raj B (2007) Effect of initial pH and temperature of iron salt solutions on formation of magnetite nanoparticles. Materials Chemistry and Physics 103: 168-175.

6. Si H, Zhou C, Wang H, Lou S, Li S, et al. (2008) Controlled synthesis of different types iron oxides nanocrystals in paraffin oil. Journal of Colloid and Interface Science 327: 466-471.

7. Shebanova ON, Lazor P (2003) Raman study of magnetite (Fe_3O_4): Laser-induced thermal effects and oxidation. Journal of Raman Spectroscopy 34: 845-852.

8. Rodenhausena KB, Davisa RS, Sekoraa D, Lianga D, Mocka A, et al. (2008) The retention of liquid by columnar nanostructured surfaces during quartz crystal microbalance measurements and the effects of adsorption thereon. Journal of Colloid and Interface Science 455: 226-235.

9. Slavov L, Abrashev MV, Merodiiska T, Gelev C, Vandenberghe RE, et al. (2010) Raman spectroscopy investigation of magnetite nanoparticles in ferro fluids. Journal of Magnetism and Magnetic Materials 322: 1904-1911.

18

Technology Geometric Decoration with Silver and/or Bronze Wires on Magnetite Patinas in the Armament of the Preroman Peoples of the Iberian Peninsula

Sánchez LG[1]*, Portal AJC[1], Valenzuela FP[2], Salazar Y Caso De Los Cobos JMG[1] and Martínez García JA[1]

[1]Departamento de Ciencia de Materiales e Ingeniería Metalúrgica, Facultad de Ciencias Químicas, Universidad Complutense de Madrid (U.C.M.), Madrid, España
[2]Museo del Cobre de Cerro Muriano, Córdoba. 14350 Obejo, Córdoba, España

Abstract

In this paper we studied weapons had found in archaeological finds in the Iberian Peninsula pre-Roman times, in which have been observed the presence of magnetite patinas. The novelty of the study is the presence of silver wires with these magnetite patinas and what technology was used to it.

Keywords: Patinas; Decoration; Armament; Archaeological; Pre-Roman

Introduction

The archaeological materials studied in the research belong to the prehistory of the Iberian Peninsula, more specifically, pre-Roman cultures of centuries V-II B.C. In these centuries the Peninsula was inhabited by the Iberians, Celts and Celtiberians [1-4]. The weapons of these villages are very characteristic and specific. Something striking is that the weaponry generally has a layer of artificial magnetite with a geometric decoration of silver or bronze wire inserted in it, and also, widely practiced the rite of incineration [5-11]. This incineration has made many of the pieces that have been studied show modifications of importance [12]. Research has been carried out with pieces from the Necropolis de la Hoya (Laguardia, Álava (Spain)) (Figure 1), cemetery of incineration mid-fourth century BC, with the presence of funerary pieces corresponding to tombs of warriors. Has been examined a large samples of such weapons; have been extracting samples for study of the most representative pieces [1].

The samples studied represent the two types of patinas of magnetite used in weapons studied in the Second Iron Age in the Iberian Peninsula: direct magnetite on steel and magnetite on an intermediate layer of bronze or silver deposited directly on steel [11] (Figure 2). The aim of study is to show the patina of magnetite can have, or not, an intermediate layer of silver or bronze patina and how to set magnetite geometric decoration with silver or bronze. As an initial image we

present reconstructions of Celtic weapons made by Encarnación Cabré [12] (Figures 3 and 4).

Experimental Technique

The samples studied are representative of these pre-Roman techniques of the Iberian Peninsula to get on weapons and other fixtures, a layer of matt black magnetite in which profuse geometric decorations were made with silver or bronze wires. The two varieties of patina, simple magnetite or bronze-magnetite, depended on the piece to be treated. If the piece was complex, was used bronze or silver to brazing its entirety and, on this layer, he appeared magnetite patina, for example, in handles or pods. If the piece was simple, the patina of magnetite was produced directly on steel; Examples of this are the steel sheets of daggers, swords and spearheads, etc. [10,11].

On the magnetite patina, in both cases, the decoration appears

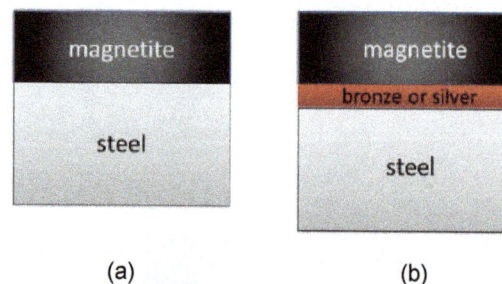

Figure 2: (a) The patina of magnetite is formed directly on steel. (b) The patina magnetite is formed on a film deposited on the steel bronze.

Figure 1: Location on the map of the Iberian Peninsula of the Necropolis de la Hoya (Laguardia, Álava).

*Corresponding author: Sánchez LG, Departamento de Ciencia de Materiales e Ingeniería Metalúrgica, Facultad de Ciencias Químicas, Universidad Complutense de Madrid (U.C.M.), Madrid, España
E-mail: gslaura@quim.ucm.es

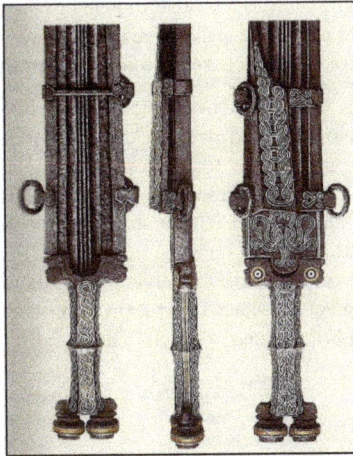

Figure 3: Celtic weapons from the Necropolis of the Osera (Chamartin de la Sierra, Ávila (Spain)) of the fourth century B.C. They are deposited in the National Archaeological Museum in Madrid (Spain).

Figure 4: Celtic weapons from the Necropolis of the Osera (Chamartin de la Sierra, Ávila (Spain)) of the fourth century B.C. They are deposited in the National Archaeological Museum in Madrid (Spain).

on silver and/or bronze. The event subsequent to incineration of the corpse with their weapons, as was the custom in pre-Roman Spain, deteriorates in some cases a lot, and others less, the magnetite layer, so that the pieces found in the necropolis have significant degree impairment. Counting on this fact, we have extracted samples of pieces that we think could provide us with the information we seek about the way to make the decoration with silver and/or bronze on the patina of magnetite.

There are two parts examined for this research:

A. The front half of one of the leaflets that make up the pommel of a sword (Necropolis de la Hoya. Archaeological Museum of Álava) (Figure 5).

B. The scabbard of sword Monte Bernorio type [12-14] (Necropolis de la Hoya. Archaeological Museum of Álava) (Figure 6). Samples

taken from the pieces were embedded in resin Mecaprex KM-U. Roughed by abrasive discs of Buehler grain 240, 320, 600 and 2000 in water; and subsequent polishing α alumina (0.3 microns) and alumina γ (0.03 microns) in Buehler polishing cloth. Chemical etching for metallographic observation by FEG should be very careful and free from residues of attack deposited on the target surface. The high quality of the microscopic observation of this instrument may be affected with a defective metallographic preparation and chemical attack.

The etching was performed with 4% Nital, and washed with distilled water in an ultrasonic bath. Samples were observed by conventional light microscopy (M.O.) and scanning electron microscopy (M.E.B.). For the observation of the samples in scanning electron microscopy, after preparation roughing, polishing and chemical etching, they were metalized with gold for 30 seconds with a current of 20 mA, and thickness of 3 nm gold. The scanning electron microscope with thermionic cathode of tungsten filament (FEG) used is JEOL model JSM 6400 that provides images and physic-chemical data of the sample surface. It has three sensors: secondary electron detector, the image resolution is 35 KV, detector to work at 8 mm distance with an image resolution of 3.5 nm and detector to work to 39 mm with an image resolution 10 nm. It provides backscattered electron images with an image resolution of 10 nm, an 8 mm working distance. In addition, you can perform qualitative elemental analysis (EDS) with a resolution of 133 eV.

Results and Discussion

The extracted sample of the part A, shows inlaid yarns silver artificial magnetite patina. It is an axial section; the images have been obtained by M.E.B., using backscattered electrons (Figures 7 and 8). We observed clearly the remains of the primitive patina of magnetite,

Figure 5: Image Part A: pommel of a sword presented magnetite black patina and geometric decoration with silver wires.

Figure 6: Image Part B: sword scabbard Monte Bernorio typology, with patina silver-magnetite and geometric decoration with silver wires.

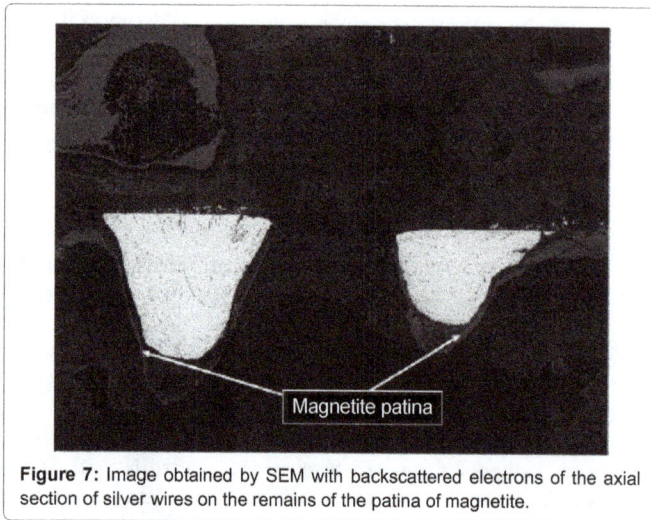

Figure 7: Image obtained by SEM with backscattered electrons of the axial section of silver wires on the remains of the patina of magnetite.

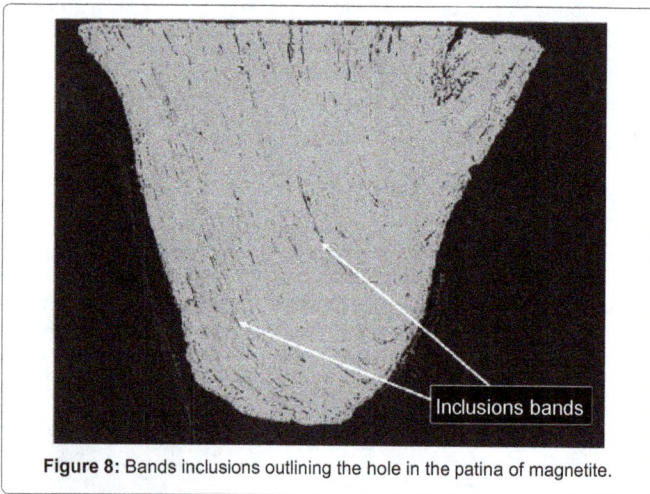

Figure 8: Bands inclusions outlining the hole in the patina of magnetite.

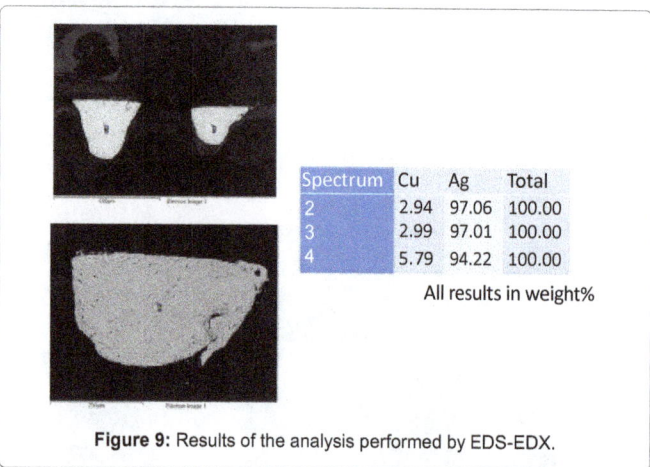

Spectrum	Cu	Ag	Total
2	2.94	97.06	100.00
3	2.99	97.01	100.00
4	5.79	94.22	100.00

All results in weight%

Figure 9: Results of the analysis performed by EDS-EDX.

It is evident that the silver wire decoration was applied after the patina of magnetite is created by cold drawing in the furrows of it. The grooves drawn in magnetite were carved on the base metal with a fine awl, before the creating process of it; following this order of operation (Figure 10).

A useful analogy of cold pressed, the silver wires in the furrows of the patina of magnetite, is that of a piece of metal by stamping. In Figures 11 and 12 a screw aluminum (DIN AlMgSi0.5, IN AW6063), cold stamping. Can be seen clearly bands or striations, inclusions adapted to the die (Figures 11 and 12). The extracted sample of part B, shows two silver wires on artificial silver magnetite patina (Figure 13). The sample has a very advanced corrosion state. The silver patina and the wires are visibly deformed by effects of incineration. In the Figure 14,

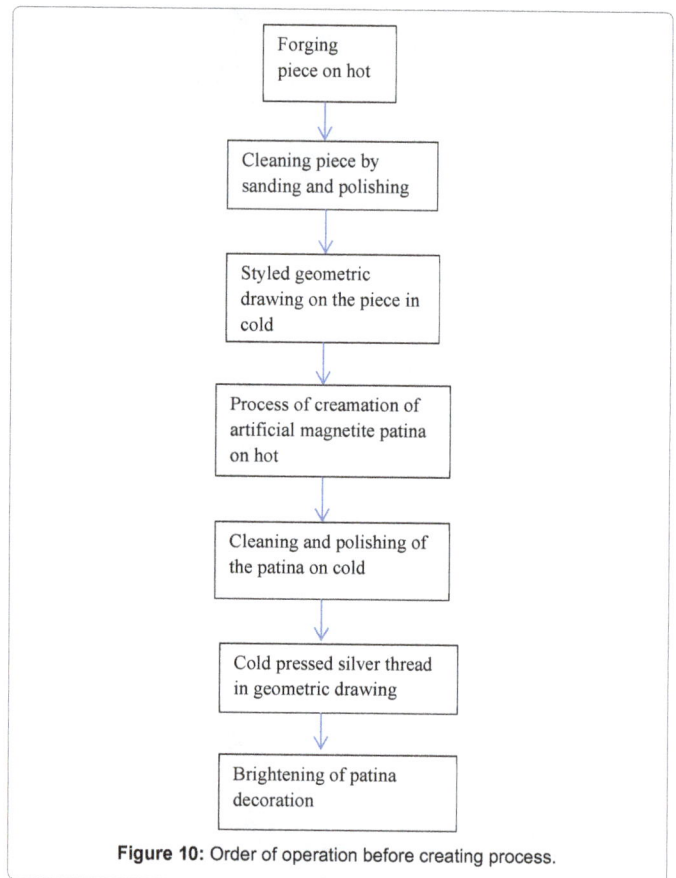

Figure 10: Order of operation before creating process.

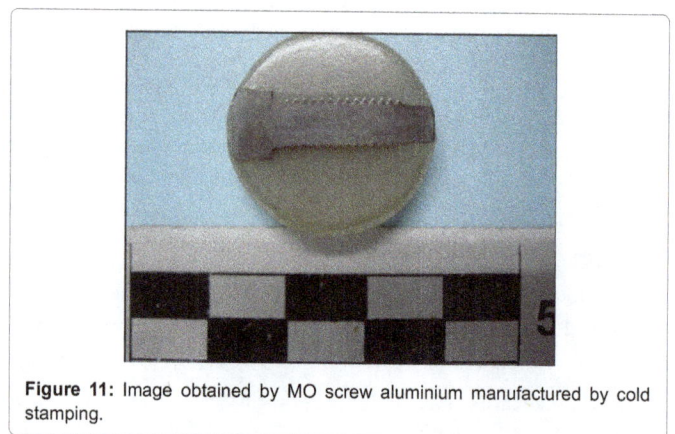

Figure 11: Image obtained by MO screw aluminium manufactured by cold stamping.

badly damaged by the incineration process produced (Figure 8). In detail, at higher magnification, of silver wires observed, in addition to the remains of magnetite patina, the internal structure of these silver wires (Figure 8). These bands of inclusions and impurities form the profile inlay yarn patina by cold stamping. Surely, they were beaten by a useful and hammer (Figure 8). The wires, analysis by EDS-EDX, confirm that the wires are of a very rich silver alloy (Figure 9).

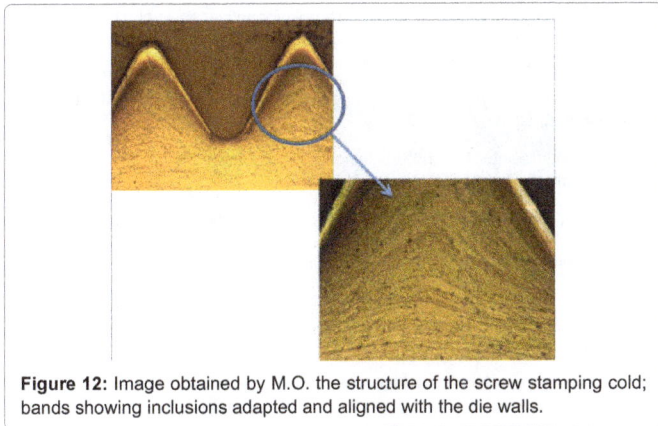

Figure 12: Image obtained by M.O. the structure of the screw stamping cold; bands showing inclusions adapted and aligned with the die walls.

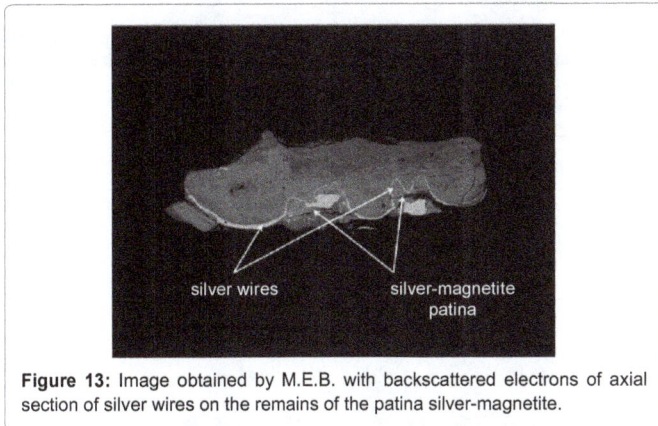

Figure 13: Image obtained by M.E.B. with backscattered electrons of axial section of silver wires on the remains of the patina silver-magnetite.

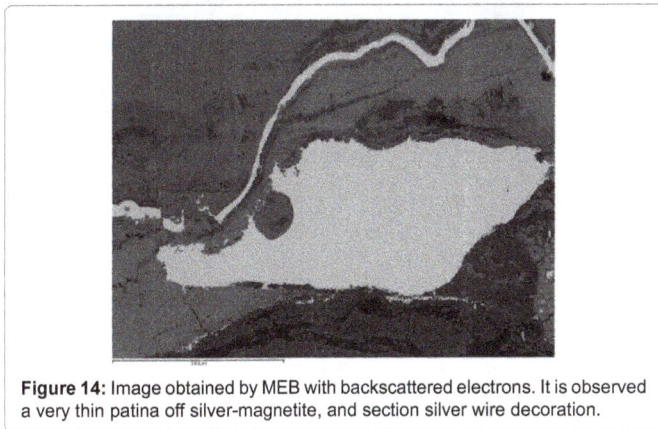

Figure 14: Image obtained by MEB with backscattered electrons. It is observed a very thin patina off silver-magnetite, and section silver wire decoration.

seen clearly, a fine patina of silver-magnetite, also the silver wire of the decoration, very deformed. What is very interesting about the piece B is that the patina is composed of bronzemagnetite, observed in numerous other pieces studied in this necropolis Hoya, which has been replaced by a silver-magnetite (Figure 15). We believe that it is a qualitative leap in the study of weapons of the Second Iron Age in the Iberian Peninsula, the existence of patinas silver-magnetite, which involves the use of brazing with silver, for the manufacture of parts composite, replacing the brazing. It is, at least, a curious detail, the use of silver, a more noble, scarce and expensive to manufacture bronzemagnetite metal patinas. Our proposal on the order of operations to perform geometric decoration with silver wires is as follows (Figure 16).

Conclusion

It has been demonstrated that silver wires of geometric decoration in pre-Roman weapons of the Second Iron Age (fifth to s.II BC) in the Iberian Peninsula were inlayed by pummel cold in the grooves made on the artificial matte black magnetite patina. The section of these yarns shows inclusions of bands aligned with the grooves in the magnetite, this is indicating they have been introduced by cold pressure. It has

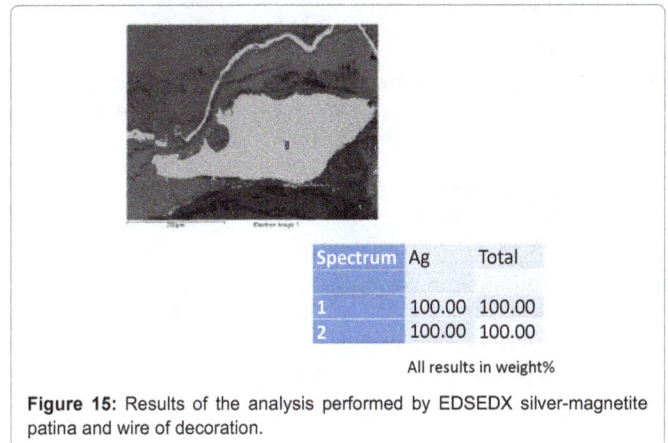

Spectrum	Ag	Total
1	100.00	100.00
2	100.00	100.00

All results in weight%

Figure 15: Results of the analysis performed by EDSEDX silver-magnetite patina and wire of decoration.

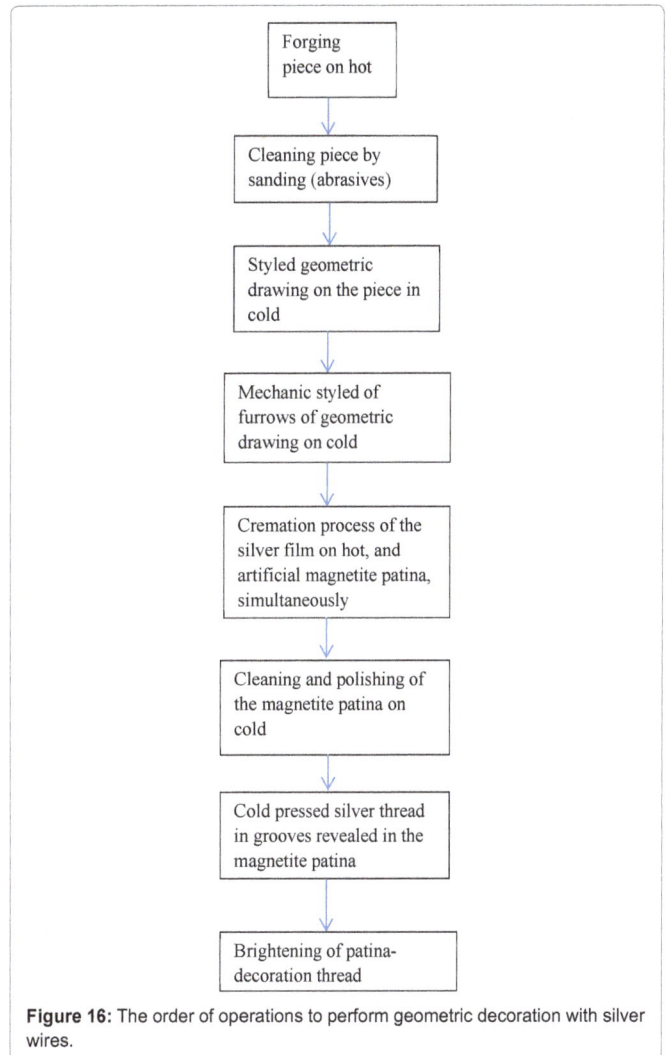

Figure 16: The order of operations to perform geometric decoration with silver wires.

also been shown that the decorative grooves in which silver wires are embedded are made in the clean metal part. The back and warm patina formation of magnetite or silver magnetite reproduces these grooves which is where the wires are inserted of the decoration.

A surprising fact of this research is to have shown that bronze-magnetite patina, in complex pieces of weaponry examined, can be silver magnetite. Silver is used instead of bronze as welding element in these composite parts.

References

1. Alonso J, Cerdán R, Nieva IF (1999) Nuevas técnicas metalúrgicas en armas de la II Edad del Hierro: arqueometalurgia y conservación analítica en la necrópolis de La Hoya (Laguardia, Álava). Diputación Foral de Álava, Vitoria-Gasteiz.

2. Bendala M (2000) Tartesios, y Celtic Iberians: pueblos, cultures y settlers de la antigua Hispania. Desc Physical 295: II.

3. Deamos MB, Chapa T (1997) Iron Age.

4. Sanz FQ, Mauné S (1997) El armamento ibérico. Estudio tipológico, geográfico, funcional, social y simbólico de las armas en la cultura ibérica (siglos VI-I a.C.). M Mergoil.

5. Alonso J (2010-2012) Magnetite coverings: state of the art and methodological Proposals fot ITS study and conservation. Institute of Prehistory and Archaeology Sautuola, Santander (Spain). pp: 389-483.

6. Alonso J (2009) Weapons of black color in the early history of the Iberian Peninsula. Conservation and macroscopic identification of finished magnetite. Congress of Conservation and Restoration of Heritage Metallic: Technology and Conservation of Archaeological Heritage III.

7. Mínguez CS (2008) A dagger vacceo relic found in Pintia (Padilla de Duero, Valladolid). Gladius 28: 177-194.

8. Sierra Montesinos M (2003) A lot of weapons from the Iberian necropolis of Torremorana (Baena, Córdoba). Gladius 23: 71-109.

9. Vega E (2007) Species transport in the corrosion products of ferrous archaeological analogues: contribution to the modelling of long term iron corrosion mechanisms. Corrosion of Metallic Heritage Artefacts.

10. García L, Criado AJ, Chamón J, Penco F, Alonso J, et al. (2011) Evidence for artificial magnetite coating on Iberian armoury. Revista de Metalurgia 47: 101-110.

11. García L (2010-2012) Scientific-technological contribution to an understanding of coating of magnetite and bronze-magnetite on Pre-Roman weapons. Sautuola/XVI-XVII. Instituto de Prehistoria y Arqueología Sautuola, Santander (Spain). pp: 435-456.

12. Criado AJ (2012) Archaeometry iron and fire. archaeometric techniques applied to the study of iron and steel proto-historic and Roman Iberian Peninsula subjected to incineration or fire. Doctoral thesis. TDX (Thesis Doctorals en Xarxa) National University of Distance Education, Madrid.

13. Cabré E (1997) War in antiquity: Celtic weapons in the Iron Age II. Edit. Fundación Caja Madrid, Madrid.

14. De Griñó B (1980) Los puñales del tipo Monte Bernorio-Miraveche. Zephyrus: Journal of prehistory and Archeology 39-40: 297-306.

19

Simulation to Study the Effect of Carrier Concentration on I-V Characteristics of Schottky Diode

Sharma R*

Department of Applied Sciences, Model Institute of Engineering and Technology, Jammu (J&K), India

Abstract

The current-voltage characteristics for Au/n-Si Schottky diode are generated by simulation. The simulation performed using Newton-Raphson iteration method yields current-voltage characteristics over wide temperature range. The data is analyzed using TDE-mechanism to study the temperature dependence of barrier height and ideality factor. Results obtained from simulation studies show the barrier height and ideality factor are independent of temperature for pure TE-mechanism. Thus the simulation of I-V characteristics is performed by incorporating ideality factor, obtained on the basis of carrier concentration from 1022-1024 atoms/m³.

The result of analysis yield barrier height is still independent of temperature but the ideality factor becomes temperature dependent and this dependence of ideality factor on temperature increases with increase in carrier concentration. Further, the temperature dependence of barrier height and ideality factor is discussed and simulation of current-voltage characteristics is performed.

Keywords: Schottky barrier diode; Carrier concentration; Electrical parameters of schottky diode

Introduction

Current transport across Metal-semiconductor (MS) contact is of great interest for device and material scientists. A number of attempts have been made so far to understand this mechanism, but a complete description of the conduction mechanism across the MS interface is still a challenging problem. For an ideal Schottky diode, the current flow is only due to thermionic emission (TE) mechanism with ideality factor equal to unity ($\eta=1$). However, due to various factors such as device temperature, dopant concentration, device area, density of interface states, structural properties of interface etc., the current-voltage characteristics of Schottky contact exhibit deviations from TE-mechanism with temperature dependent ideality factor [1-4]. Generally, the ideality factor increases with decrease in temperature, the phenomena is commonly known as "To-effect" and was first proposed by Padovani and Sumner [5], where as the barrier height for these diodes is found to decrease with fall in operating temperature.

Taking into account the effect of above mentioned anomalies of the I-V characteristics, different methods have been suggested. Padovani and Stratton proposed, in case of degenerate materials (i.e., $N_D \geq 10^{23}$ atoms/m³) the above anomalies can be explained on the basis of quantum mechanical tunneling [1,2]. Hackam and Harrop suggested, the ideality factor should be incorporated in saturation current expression [6]. Further, Zs. J. Horvath proposed, any mechanism which enhances electric field at the MS interface is responsible to enhance multistep tunneling at the interface [7]. Accordingly, the temperature dependence of ideality factor as well as barrier height (BH) of Schottky contacts can be explained on the basis of TFE-theory.

Another approach for explaining these anomalies is based on barrier inhomogeneity models. According to these models, barrier height of the contact will be affected by the non-uniformity of the interfacial layer and form a distribution over the contact. As a result, this fluctuation of the barrier height over Schottky contact may lead to non-ideal I-V characteristics [8,9].

Simulation is an emerging trend in the field of research and development and is widely used to study the I-V characteristics of

Schottky diode also. S. Chand and J. Kumar [10] and S. Chand and S. Bala [11,12] have carried out a detailed study on Schottky diode by using simulation. However, all these studies are based on barrier inhomogeneity model (Gaussian). In present paper a different approach has been adopted to explain the temperature dependence of barrier height through simulation of I-V characteristics of Schottky diode.

Method of Simulation

On the basis of Bathe's thermionic emission theory [13], the forward I - V characteristics of an ideal Schottky barrier diode can be expressed as

$$I = I_s \exp\left(\frac{qV}{kT}\right) \tag{1}$$

$$\text{where } I_s = A^* S T^2 \exp\left(\frac{-\varphi_{bo}}{kT}\right) \tag{2}$$

I_s is the diode saturation current, A^* is the Richardson constant, S is the diode area, φ_{bo} is the barrier height, T is the temperature and k is the Boltzman constant. The simulation of I-V characteristics using eq. (1) is performed through computer programming (GW-Basic) using Newton-Raphson iteration method. Parameters used for simulation are: diode area $S = 7.87 \times 10^{-7} \text{m}^2$ (corresponding to diameter 1mm), effective Richardson constant $A^* = 1.12 \times 10^6 \text{ Am}^{-2}\text{K}^{-2}$ (for n-type silicon), barrier height $\varphi_{bo} = 0.8$ eV (Au/n-Si device), and $Rs = 10\Omega$.

Results and Discussion

At any temperature, for $V >> 3kT/q$, eq (1) predicts, the $ln(I)$ vs V plot should be a straight line with unit slope and its intercept at zero-

***Corresponding author:** Sharma R, Department of Applied Sciences, Model Institute of Engineering and Technology, Jammu (J&K), India
E-mail: rbasotra@yahoo.com

bias give I_s. Whereas $ln(I_s/ST^2)$ *vs* $1/T$ plot (eq. 2,) should be another straight line with intercept at $1/T = 0$ and slope gives the value of A^* and φ_{bo} respectively. The $ln(I)$-V plots thus obtained at various temperatures are shown in Figure 1. These curves are linear over several orders of current. I-V data so obtained is analyzed using thermionic emission diffusion (TED) theory to obtain the BH and ideality factor.

Figure 2 shows the BH and ideality factor obtained from simulated data at various temperatures. The barrier height and ideality factor are found to be independent of temperature as expected for the case of pure thermionic emission. But in actual practice the barrier height and ideality factor is generally temperature dependent. Also, on the basis of pure TE – mechanism the simulation is possible upto 125K only. This is because the value of saturation current I_s below 125K is extremely low.

However, for diode from heavily doped semiconductor or with imperfection junction, the equation (1) is no longer valid. Podovani and Sumner [5] reported their non-ideal forward I-V characteristics to fit the expression.

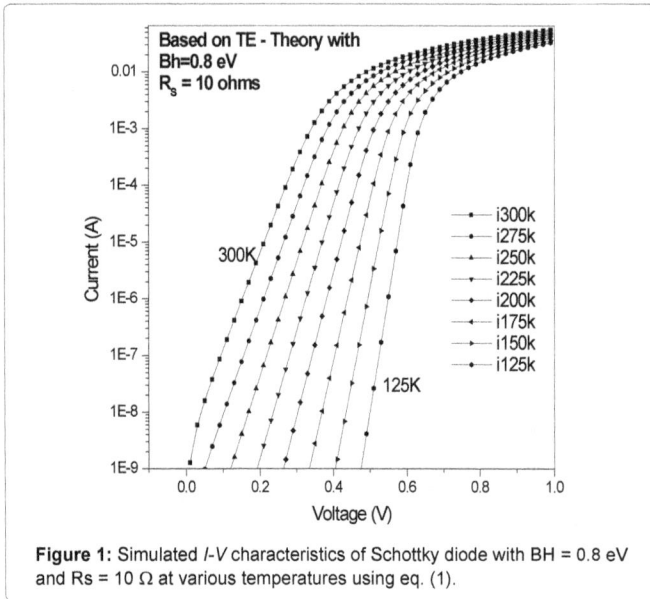

Figure 1: Simulated *I-V* characteristics of Schottky diode with BH = 0.8 eV and Rs = 10 Ω at various temperatures using eq. (1).

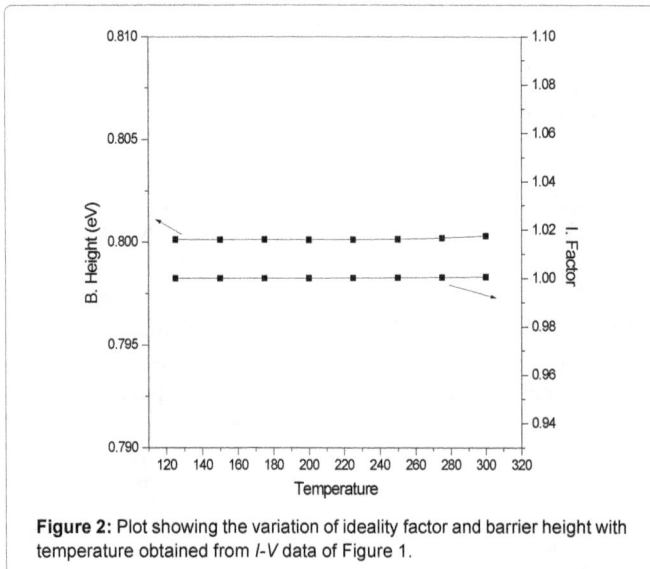

Figure 2: Plot showing the variation of ideality factor and barrier height with temperature obtained from *I-V* data of Figure 1.

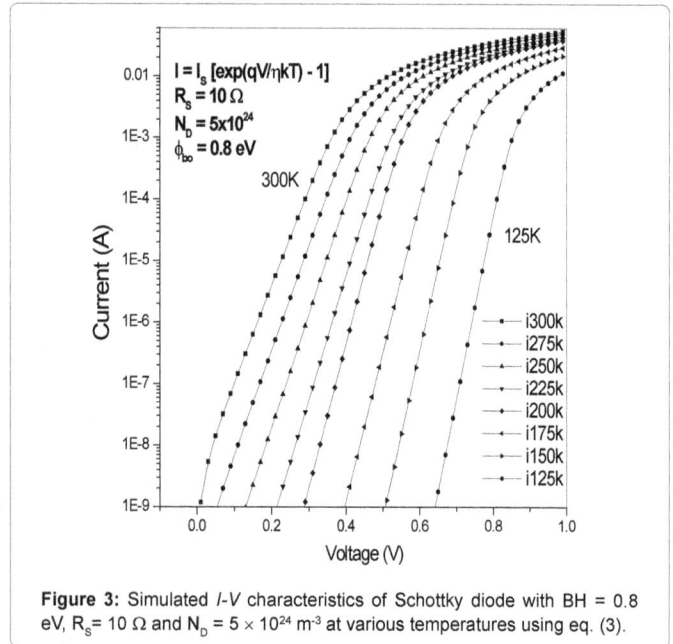

Figure 3: Simulated *I-V* characteristics of Schottky diode with BH = 0.8 eV, R_s= 10 Ω and N_D = 5 × 10^{24} m^{-3} at various temperatures using eq. (3).

$$I = A^* ST^2 \exp\left(\frac{-q\varphi_{bo}}{k(T+T_o)}\right)\exp\left(\frac{qV}{k(T+T_o)}\right)$$

where T_o is called excess temperature. Atalla and Soshea [14] fit their non-ideal I-V characteristics to the expression

$$I = A^* ST^2 \exp\left(\frac{-q\varphi_{bo}}{kT}\right)\exp\left(\frac{qV}{\eta kT}\right) \qquad (3)$$

where η is the ideality factor. Padovani and Stratton and later Rhoderick and Williams proposed, if current transport is controlled by the thermionic field emission theory, the relation between current and voltage can be expressed as [1]

$$I = I_s \exp\left(\frac{V}{E_o}\right) \qquad (4)$$

where $E_o = E_{oo} \coth\left(\frac{qE_{oo}}{kT}\right) = \frac{\eta kT}{q}$ and $E_{oo} = 18.5 \times 10^{-15}\left(\frac{N_D}{m_r\varepsilon_r}\right)^{1/2}$

The analysis of (4) shows, the ideality factor is temperature dependent through carrier concentration (N_D). Thus, I-V data at different temperatures for various values of N_D is simulated by incorporating ideality factor in eq. (1).

The I-V data so obtained, at $N_D = 5 \times 10^{24}$ atoms/m^3 only, is presented in Figure 3. Plot shows the I-V curves are still linear over several orders of current but these are not converging at higher bias voltage in spite of constant value of series resistance (R_s). Moreover they are shifting downward and this shift is more prominent at lower temperature. The simulation of I-V data is still possible upto 125K as there is no change in the saturation current. The ideality factor and barrier height obtained from the slope and intercept of $ln (I)$ *vs* V plot at various temperatures and for different carrier concentrations are presented in Figure 4. Analysis of plot shows, the ideality factor is temperature dependent as observed in most of the cases but the barrier height is still constant (i.e., independent of temperature).

One method to explain the temperature dependence of barrier height and ideality factor is barrier inhomogenities [8,9], which establishes a relation between barrier height and ideality factor also.

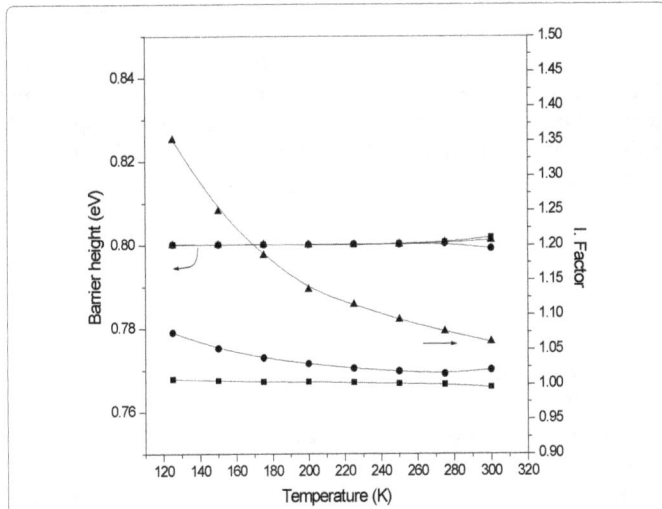

Figure 4: Plot showing the variation of ideality factor and barrier height as a function of temperature, obtained from *I-V* data shown in Figure 3.

But this method talks about TED - theory only. Hackam and Harrop [6] proposed diode equation, for non ideal *I-V* forward characteristics, as

$$I = A^{*}ST^{2} \exp\left(\frac{-q\varphi_{bo}}{\eta kT}\right) \exp\left(\frac{qV}{\eta kT}\right) \qquad (5)$$

Accordingly, the ideality factor should be included in the expression for saturation current. As the effects that cause deviation from $\eta = 1$ at higher bias voltage are present at zero-bias voltage also. Later, Bhuiyan [15] and more recently Rajinder [16,17], fitted their data by incorporating ideality factor into saturation current.

So, the inclusion of ideality factor in saturation current makes barrier height temperature dependent i.e.,

$$\varphi_{b} = \frac{\varphi_{bo}}{\eta(T)} \qquad (6)$$

Thus, the simulation of the *I-V* data of Schottky diode with different carrier concentration has been performed using eq (5) and (6). The *I-V* data thus obtained is further analyzed on the basis of TED - theory to study the effect of temperature as well as carrier concentration barrier height and ideality factor.

Figure 5a shows *I-V* curves from 50-300K at $N_{D} = 5 \times 10^{24}$ atoms/m³ only and Figure 5b shows *I-V* curves at 300, 200 and 125K for various values of N_{D}. A look at plot Figure 5a shows, the I-V characteristics are linear over several orders of current. These curves become more and more linear and linearity of curves shifts towards higher bias voltage with decrease in temperature. This is because of excess current at low temperature due to low barrier height. Also Figure 5b shows there is increase in current with increase in carrier concentration and variation becomes more prominent with decrease in temperature.

Figure 5a shows *I-V* curves from 50-300K at $N_{D} = 5 \times 10^{24}$ atoms/m³ only and Figure 5b shows *I-V* curves at 300, 200 and 125K for various values of N_{D}. A look at plot Figure 5a shows, the I-V characteristics are linear over several orders of current. These curves become more and more linear and linearity of curves shifts towards higher bias voltage with decrease in temperature. This is because of excess current at low temperature due to low barrier height. Also Figure 5b shows there is

increase in current with increase in carrier concentration and variation becomes more prominent with decrease in temperature.

The *I-V* data so obtained, for various values of N_{D}, is analyzed on the basis of thermionic emission theory. The barrier height and ideality factor as a function of temperature are shown in Figure 6. Plot shows the barrier height decreases and ideality factor increases with decrease in temperature in the same manner as most of the practical Schottky diodes exhibit and these changes become more prominent with increase in the donor concentration (N_{D}) and decrease in temperature.

The value of I_{s} at each temperature, obtained from Figure 5a, is used to plot $ln(Is/T^{2})$ vs $1/T$ as shown in Figure 7, the usual Richardson plot (curve-a). For an ideal diode this should be a straight line with intercept and slope yielding the value of A^{*} and BH respectively. But

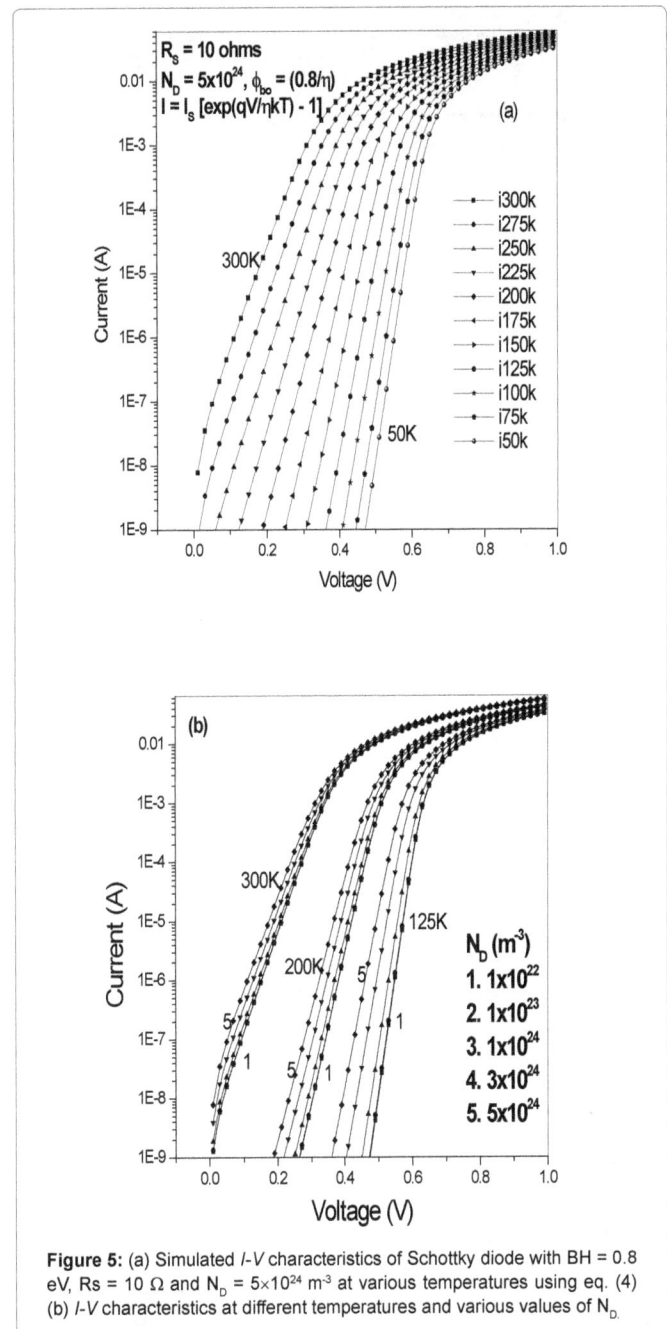

Figure 5: (a) Simulated *I-V* characteristics of Schottky diode with BH = 0.8 eV, Rs = 10 Ω and N_{D} = 5×10²⁴ m⁻³ at various temperatures using eq. (4) (b) *I-V* characteristics at different temperatures and various values of N_{D}.

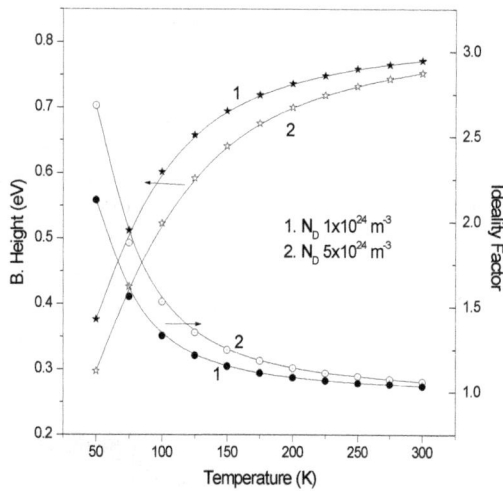

Figure 6: Variation of ideality factor and barrier height as a function of temperature at different values of N_D.

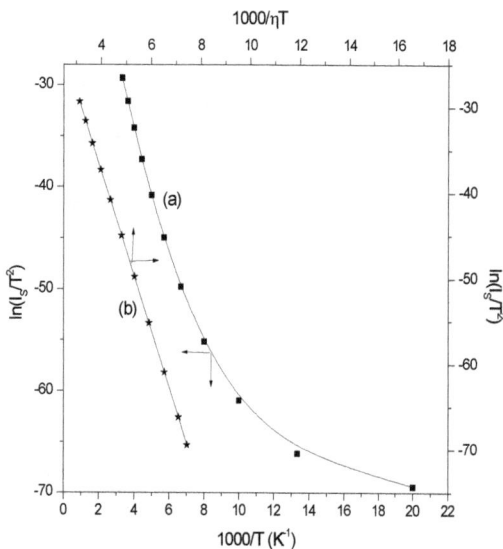

Figure 7: Plot of $ln(I_s/T^2)$ vs $1/T$ (curve-a) and $ln(I_s/T^2)$ vs $1/\eta T$ (curve-b).

the extracted data exhibit linearity upto 200K only with $A^* = 2.15 \times 10^3$ $Am^{-2}K^{-2}$ and BH = 0.60 eV. Clearly the values of A^* and BH, so obtained are very less as compared to the values used for simulation. This discrepancy is because of extra current at low temperature. Further, the data is analyzed by using modified Richardson plot i.e., $ln(Is/T^2)$ vs $1/\eta T$. Such a plot is shown in Figure 7 (curve-b), exhibit linearity over entire temperature range yielding the value of $A^* = 1.13 \times 10^6$ $Am^{-2}K^{-2}$ and BH = 0.8 eV and these values are in good agreement with those used for simulation.

Conclusion

The *I-V* characteristics of Schottky diodes are simulated over wide temperature range. The *I-V* data for Au/n-Si Schottky diode is generated with and without ideality factor as well as by incorporating ideality factor in saturation current equation. The data so obtained fitted into the pure TED-equation to see the temperature dependence

of ideality factor and barrier height. For *I-V* characteristics with $\eta=1$, the barrier height and ideality factor are independent of temperature. Whereas incorporation of ideality factor in saturation current makes barrier height temperature dependent. It is also observed that the temperature dependence of barrier height becomes stronger with increase in carrier concentration. Finally it is concluded that the temperature dependence of barrier height is due to the factors which are responsible for the deviation of *I-V* characteristics of Schottky diodes at higher bias voltage.

References

1. Rhoderick EH, Williams RH (1988) Metal Semiconductor Contacts. Clarendon Press, Oxford.

2. Sze SM, Lee MK (2011) Semiconductor Devices Physics and Technology. (3rd Eds.)Wiley New York, USA.

3. Saxena AN (1969) Forward current-voltage characteristics of Schottky barriers on n-type silicon. Surf Sci 13: 151-171.

4. Singh A, Rcinhardt KC, Anderson WA (1990) Temperature dependence of the electrical characteristics of Yb/p-InP tunnel metal-insulator-semiconductor junctions. J Appl Phys 68: 3475.

5. Padovani FA, Sumner GG (1965) Experimental Study of Gold-Gallium Arsenide Schottky Barriers. J Appl Phys 36: 3744.

6. Hackam R, Harrop P (1972) Electrical properties of nickel-low-doped n-type gallium arsenide Schottky-barrier diodes. IEEE Trans Electron Devices 19: 1231-1238.

7. Horvath ZJ (1996) On Advanced Semiconductor Devices and Microsystems. Worsaw, Poland.

8. Werner JH, Guttler HH (1991) Barrier inhomogeneities at Schottky contacts. J Appl Phys 69: 1522.

9. Chand S, Kumar J (1996) On the existence of a distribution of barrier heights in Pd2Si/Si Schottky diodes. J Appl Phys 80: 288.

10. Chand S, Kumar J (1997) Simulation and analysis of the I - V characteristics of a Schottky diode containing barrier inhomogeneities. Semicond Sci Technol 12: 899.

11. Chand S, Bala S (2007) Simulation studies of current transport in metal–insulator–semiconductor Schottky barrier diodes. Physica B: Condensed Matter 390: 179-184.

12. Chand S, Bala S (2005) A comparative study of numerical and analytical methods of simulating inhomogeneous Schottky diode characteristics. Semicond Sci Technol 20: 1143.

13. H. a. Bathe MTT (1942) Radiat Lab 43: 12.

14. Atalla MM, Soshea RN (1962) Hewlett-Packrd. Sci rep.

15. Bhuiyan AS, Martinez A, Esteve D (1988) A new Richardson plot for non-ideal schottky diodes. Thin Solid Films 161: 93-100.

16. Rajinder S (2010) Temperature Dependence of I-V Characteristics of Au/n-Si Schottky Barrier Diode. Journal of Electron Devices 8: 286-292.

17. Duman S (2008) Temperature dependence of current–voltage characteristics of an In/p-GaSe: Gd/Au–Sb Schottky barrier diode. Semicond Sci Technol 23: 075042.

20

Structural, Electronic and Magnetic Properties of Geometrically Frustrated Spinel CdCr$_2$O$_4$ from First-principles Based on Density Functional Theory

Bolandhemat N[1]*, Rahman M[1] and Shuaibu A[1,2]

[1]*Department of Physics, Faculty of Science, University Putra Malaysia, 43400 UPM Serdang, Selangor, Malaysia*
[2]*Department of Physics, Faculty of Science, Nigerian Defence Academy, P.M.B 2109, Kaduna, Nigeria*

Abstract

First-principles calculations are used to investigate the structural, electronic, and magnetic properties of CdCr$_2$O$_4$ with magnetic Cr cations, focusing on the changes through the magnetic phase transitions which shows relief of the geometric frustration of the ferromagnetic and antiferromagnetic orderings on the pyrochlore lattice. We computed the structural and electronic properties for the paramagnetic, ferromagnetic and antiferromagnetic orderings in cubic ($F\overline{d}3m$) and tetragonal (I4$_1$/amd) structures of CdCr$_2$O$_4$. We optimized the crystal structures with the PM, FM and AFM orderings using a pseudopotential plane wave (PP-PW) method within the generalized gradient approximation (GGA), and computed the electronic properties to investigate the magnetic properties in the geometrically frustrated ferromagnetic and antiferromagnetic spinel CdCr$_2$O$_4$ based on density functional theory and understanding of the principles of Quantum ESPRESSO in magnetic materials. On the other hand, the effect of magnetism were obtained and analyzed on the basis of density of states (DOS), projected density of states (PDOS), and charge density distribution.

Keywords: Spinels; Geometric frustration; Density functional theory; Phase transitions; Quantum Espresso; Electronic and magnetic properties; Ground states

Introduction

Spinels are a captivating class of materials that indicate rich complex behavior and novel ground states such as large magnetoresistance effects [1], non-collinear spin configurations [2], magnetodielectric coupling [3], and spin liquid states [4]. Spinel is the magnesium aluminum oxide member of this large group of materials with the Mg^{2+}Al^{3+}2O$_2^{-4}$ formula that gives its name to the family of compounds that are identified by two cation sites: an octahedral site and a tetrahedral site [5]. It is named as spinel to any material that have the general formula A^{2+}[B^{3+}]2[X^{2-}]$_4$ which crystallizes in the face-centered cubic crystal system and are described by the space group $Fd\overline{3}m$ (No. 227). In this structure, the X anions are located in a cubic close packed lattice, the cations A occupy tetrahedral (1/8, 1/8, 1/8) sites, and the cations B occupy octahedral (1/2, 1/2, 1/2) sites in the lattice. Tetrahedral and octahedral cations occupy the special Wyckoff positions 8a and 16d, respectively. The anions are located at general Wyckoff 32e positions that are assigned the parameter u, which takes on different values around the optimal position u=0.25 for various spinels. In particular, chromium spinels ACr$_2$O$_4$ (A=Cd, Zn) are an interesting class of frustrated antiferromagnets that are considering as the most frustrating lattice because of the direct overlap of the t$_{2g}$ orbitals of the neighboring Cr^{3+}(3d^3) ions with the dominant antiferromagnetic nearest neighbor interactions [6,7]. It also remains paramagnetic far below temperatures corresponding to the major exchange strength, i. e. the Curie-Weiss temperature |Θ$_{CW}$|= 88 K for Cd [8-10] and |Θ$_{CW}$|= 390 K [11]. Upon further cooling, however, phase transition occurs from a cubic paramagnet to a tetragonal Néel state at T$_N$=7.8 K for Cd [7,8] and T$_N$=12.5 K for Zn [12]. At high temperatures, ACr$_2$O$_4$ spinels have a cubic ($Fd\overline{3}m$) structure in which Cr^{3+} ions are enveloped by octahedral oxygen cages and form a pyrochlore lattice, while A^{2+} ions are in tetrahedral oxygen environment and form the diamond lattice (Figure 1). The octahedral crystal field splits the Cr 3d orbitals into a lower-lying t$_{2g}$ triplet and a higher-energy eg doublet. Cr^{3+} has three outer electrons that fill the majority t$_{2g}$ states which results in a net Cr spin S=3/2 [13]. In some magnetic materials, magnetic order does not appear even when the system is cooled down to the temperatures far below the characteristic strength of the interactions between the spins. We are particularly interested in the case where the crystal structure is responsible for the suppression of the magnetic phase transition; this is generally called geometrical frustration. The physics of frustrated magnetism is a subject of existing interest. Spinels with Cr^{3+} ions on the B sites are good examples to study the geometrical frustration [14-17]. Spinel oxides AB$_2$O$_4$ with magnetic B captions have received special attention because they are identified by three-dimensional geometrical frustration.

In this paper, we used first-principles calculations to investigate the effects of magnetic ordering on the minimum energy structure of geometrically frustrated spinel CdCr$_2$O$_4$. Obviously, with a nearest-neighbor antiferromagnetic (AFM) exchange interactions on threefold rings, CdCr$_2$O$_4$ is completely frustrated. CdCr$_2$O$_4$ has a Curie-Weiss temperature |Θ$_{CW}$|≈88 K, while its magnetic transition occurs at a temperature as low as T$_N$=7.8 K. The estimated frustration factor f≈31, demonstrating a large degree of geometrical frustration. This frustration is lifted from a cubic-to-tetragonal lattice distortion (c<a=b) that occurs with the magnetic transition, simultaneously. We begin by examining the crystal structure of spinel CdCr$_2$O$_4$ and then, analyzing the electronic and magnetic properties that are important in magnetic spinel oxides.

*Corresponding author: Bolandhemata N, Department of Physics, Faculty of Science, University Putra Malaysia, 43400 UPM Serdang, Selangor, Malaysia
E-mail: bolandhemat.n@gmail.com

Materials and Methods

Crystal structure

$CdCr_2O_4$ is a magnetic compound that crystallizes into a cubic spinel structure, and the magnetic properties stem from the Cr^{3+} magnetic ions, that are a three-dimensional network of corner-sharing tetrahedral. The crystal structure is consisting of Cd-centered tetrahedral and Cr-centered octahedral, in which Cd is tetrahedrally coordinated by oxygen and Cr is octahedrally coordinated by Oxygen. The cations occupy either the tetrahedral 8a site (Cd atoms) or the octahedral 16d site (Cr atoms). Figure 2 shows the primitive cell and the conventional lattice cell of the cubic spinel $CdCr_2O_4$.

We optimized the crystal structures with paramagnetic, ferromagnetic and antiferromagnetic orderings. The crystal structure of $CdCr_2O_4$ for the paramagnetic, ferromagnetic ordering are cubic (a=b=c=8.667 Å) with the space group $Fd\bar{3}m$ (No. 227), whereas for the antiferromagnetic ordering it is tetragonal (a=b=8.634 Å and c=8.694Å) (Figure 3). In order to visualize this system, we used XCrySDen graphic software which is a crystalline and molecular structure visualization program [18].

Calculation methods

We performed density functional theory calculations using a plane-wave basis set method with generalized gradient approximation (GGA) parameterized by Perdew-Burkew-Enzerhof (PBE) exchange correlation methods [19] as implemented in the QUANTUM ESPRESSO simulation package [20]. Ultrasoft pseudopotentials were used for both cubic and tetragonal structures. The pseudopotentials that we used are including twelve valence electrons for Cd ($4d^{10}$, $5s^2$), fourteen for Cr ($3s^2$, $3p^6$, $3d^5$, $4s^1$), and six for O ($3s^2$, $2p^4$). We obtained all pseudopotentials from the plane-wave self-consistent field (PWSCF) pseudopotentials online references [21].

In order to find the actual energy cut-off and k-point mesh, the scf convergence test was performed for all parameters and considered to be achieved with the minimum consecutive iterative steps with energy difference less than 1 meV (Figure 4). We applied the kinetic energy cut-off of 40 Ry for expanding the plain wave functions and a $8 \times 8 \times 8$ k-point mesh for the Brillouin zone (BZ) integration. All calculations are performed for collinear spins without spin-orbit coupling. The optimization of atomic positions was carried through the Broyden-Fletcher-Goldfarb-Shanno (BFGS) algorithm where the forces and energy minimization process are considered during structural relaxation.

Results

Spinel $CdCr_2O_4$ with the Cd non-magnetic and the Cr magnetic ions span a huge range of magnetic exchange strengths and different magnetic ground states. In this section, we discuss about the computational results on LSDA (or spin-polarized-GGA) calculations for both cubic and tetragonal structures of geometrically frustrated spinel $CdCr_2O_4$ in different magnetic configurations.

The Figure 5 shows the total density of states (DOS) for three different magnetic orderings of the spinel $CdCr_2O_4$. By looking at the charge density plot, we can find that in the paramagnetic configuration the two spin contributions are exactly the same due to its ground state is non-magnetic. In other word the valence bands (corresponding to bonding states) are all doubly occupied and the total magnetization is zero. Also, we presented the same density of states calculation

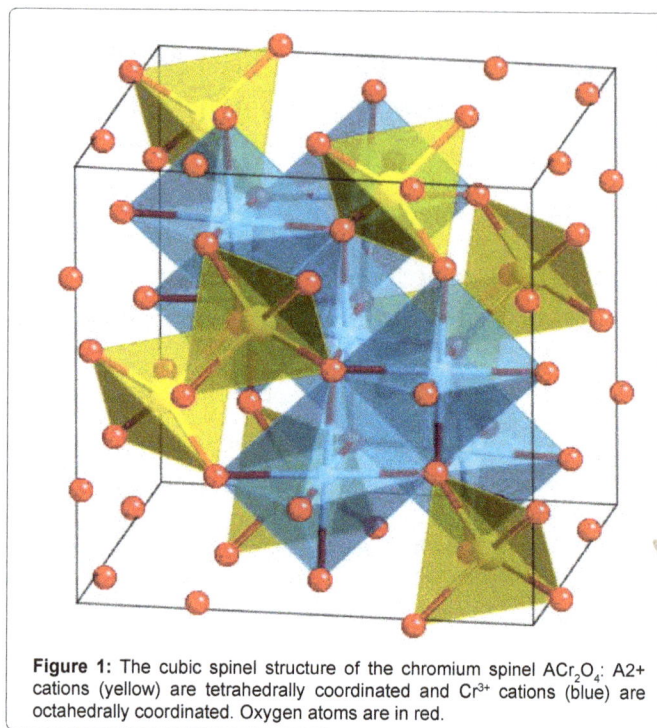

Figure 1: The cubic spinel structure of the chromium spinel ACr_2O_4: A2+ cations (yellow) are tetrahedrally coordinated and Cr^{3+} cations (blue) are octahedrally coordinated. Oxygen atoms are in red.

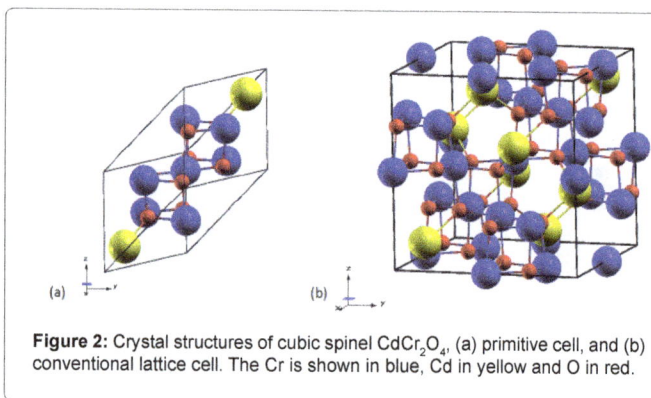

Figure 2: Crystal structures of cubic spinel $CdCr_2O_4$, (a) primitive cell, and (b) conventional lattice cell. The Cr is shown in blue, Cd in yellow and O in red.

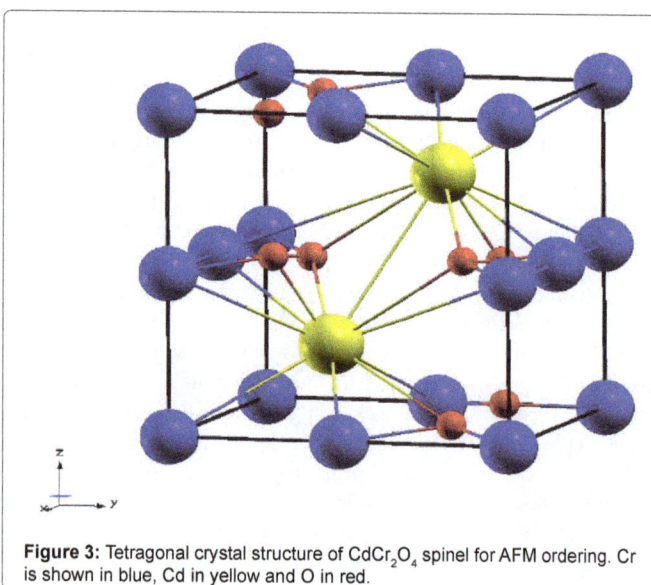

Figure 3: Tetragonal crystal structure of $CdCr_2O_4$ spinel for AFM ordering. Cr is shown in blue, Cd in yellow and O in red.

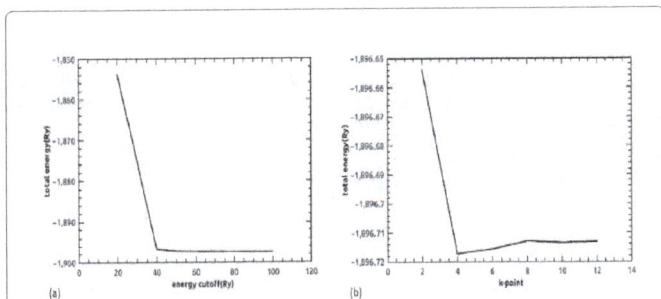

Figure 4: Convergence test with respect to (a) the energy cutoff, and (b) k-point.

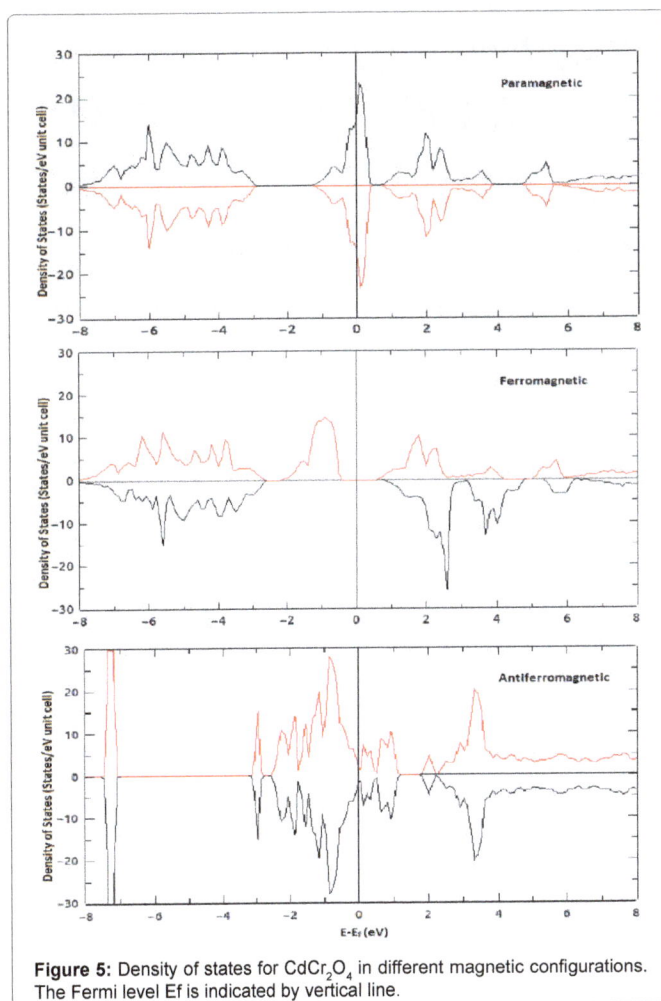

Figure 5: Density of states for CdCr$_2$O$_4$ in different magnetic configurations. The Fermi level Ef is indicated by vertical line.

for different magnetic configurations: ferromagnetic (FM) and antiferromagnetic (AFM) by applying the initial magnetization for Cr^{3+}(3d^3) magnetic atoms. In the FM configuration, we can observe that the density of states of the two spin are not aligned anymore and one spin population is larger than the other. Whereas, in the AFM configuration the two spin contributions are the same due to the total magnetization is zero as half of the atoms have a magnetization that is opposite to the magnetization of the other half.

In addition, the projected density of states (PDOS) is plotted for three different magnetic configurations, in order to observe the effect of magnetization by comparing the densities of Cd d, Cr d, and O p

states obtained from PDOS calculation for CdCr$_2$O$_4$ (Figure 6). As it is shown in Figure 6, in the PM and FM orderings of cubic CdCr$_2$O$_4$ the density of states of Cd and O atoms are remaining the same for both configurations, and only the density of Cr d states is changing from PM to FM. However, it is clear that by moving from cubic to tetragonal structure the density of states of the three elements have been changed, but the difference in densities is manly for Cr atoms.

In order to understand the distribution of the total electronic charge density of CdCr$_2$O$_4$ compound, we calculated the electronic charge density in the (110) crystallographic plane for both cubic and tetragonal structures (Figure 7). From Figure 7, we can conclude that in PM and FM orderings which both are in cubic structure the Cr-O makes the covalent bonding due to sharing of charge between Cr and O atoms while Cd and O atoms shows the ionic bonding. It is clear that in cubic structure, Cd atom shows a very week charge density but as we move to tetragonal structure, there is the stronger charge density, and also the Cd-O makes partial covalent bonding. As clear from the color charge density scale that the purple color (+1.0000) corresponds to the maximum charge accumulating site, so the chromium atoms have the greater charge density than the other atoms. By comparing the color of charge density scales in both structures, it is evident that the total electronic charge density in the tetragonal is stronger than in the cubic structure.

Summary

In the present work, we reported a comprehensive investigation of

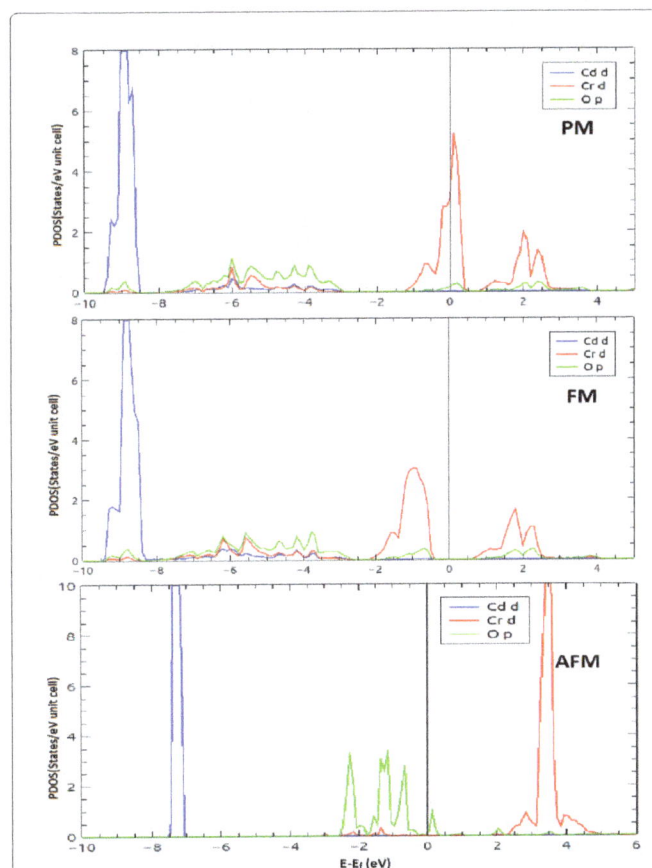

Figure 6: Projected density of states for cubic and tetragonal structure of CdCr$_2$O$_4$ in different magnetic configurations. The Fermi level Ef is indicated by vertical line.

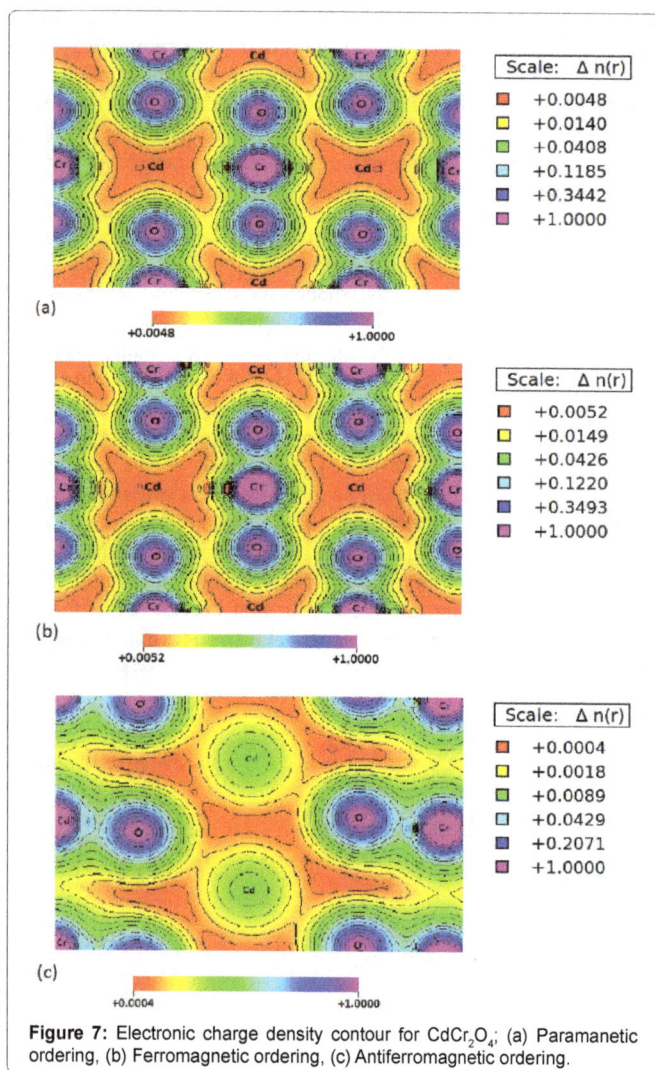

Figure 7: Electronic charge density contour for CdCr$_2$O$_4$; (a) Paramanetic ordering, (b) Ferromagnetic ordering, (c) Antiferromagnetic ordering.

the electronic structures, magnetic properties and chemical bonding properties of geometrically frustrated Spinel CdCr$_2$O$_4$ using the plane-wave Ultasoft pseudopotential technique within generalized gradient approximation (GGA) parameterized by Perdew-Burkew-Enzerhof (PBE) exchange correlation calculation. Density functional calculations are performed to observe the effects of magnetic ordering on the electronic and magnetic properties of CdCr$_2$O$_4$ with non-magnetic Cd cations and magnetic Cr cations from a pyrochlore lattice.

Acknowledgments

This work is supported by University Putra Malaysia and Ministry of Higher Education (MOHE) Malaysia, through Fundamental Research Grant Scheme (FRGS) (Project Code: FRGS/1/11/ST/UPM/02/11). Figures showing electronic configurations, charge density and charge density contour are generated using the XCRYSDEN program.

References

1. Ramirez AP, Cava RJ, Krajewski J (1997) Collossal magnetoresistance in Cr-based chalcogenide spinels. Nature 386: 156-159.

2. Yafet Y, Kittel C (1952) Antiferromagnetic arrangements in ferrites. Phys Rev 87: 290.

3. Lawes G, Melot BC, Page K, Ederer C, Hayward MA, et al. (2006) Dielectric anomalies and spiral magnetic ordering in CoCr$_2$O$_4$. Phys Rev B 74: 1-6.

4. Kemei MC, Barton BT, Moffitt SL, Gaultois MW, Kurzman JA, et al. (2013) Crystal structures of spinJahn-Teller-ordered MgCr$_2$O$_4$ and ZnCr$_2$O$_4$. J Phys Condens Matter 25: 326001.

5. Finger LW, Hazen RM, Hofmeister AM (1986) High-pressure crystal-chemistry of spinel (MgAl$_2$O$_4$) and Magnetite (Fe$_3$O$_4$)-Comparisons with silicate spinels. Phys Chem Minerals 13: 215-220.

6. Ueda H, Ueda Y (2008) Pressure-Enhanced Direct Exchange Couplings in Chromium Spinels. Physical Review B 77: 224411.

7. Tchernyshyov O, Chern GW (2005) Fractional vortices and composite domain walls in flat nanomagnets. Phys Rev Lett 95: 197204.

8. Rovers MT, Kyriakou PP, Dabkowska HA, Larkin MI, Savici AT (2002) Muon-spin-relaxation investigation of the spin dynamics of geometrically frustrated chromium spinels. Phys Rev B 66: 174434.

9. Ueda H, Katori HA, Mitamura H, Goto T, Takagi H (2005) Magnetic-Field Induced Transition to the 1/2 Magnetization Plateau State in the Geometrically Frustrated Magnet CdCr$_2$O$_4$. Phys Rev Lett 94: 047202.

10. Menyuk N, Dwight K, Arnott RJ, Wold A (1966) Ferromagnetism in CdCr$_2$Se$_4$ and CdCr$_2$S$_4$. J Appl Phys 37: 1387.

11. Yamashita Y, Ueda K (2000) Spin-Driven Jahn-Teller Distortion in a Pyrochlore System. Phys Rev Lett 85: 4960.

12. Lee SH, Gasparovic G, Broholm C, Matsuda M, Chung JH, et al. (2007) Crystal distortions in geometrically frustrated ACr$_2$O$_4$ (A = Zn, Cd). J Phys: Condens Matter 19: 145259.

13. Yaresko AN (2008) Electronic band structure and exchange coupling constants in ACr2X4 spinels (A=Zn, Cd, Hg; X=O, S, Se). Phys Rev B 77: 115106.

14. Lee SH, Broholm C, Kim TH, Ratcliff W, Cheong SW (2000) Local Spin Resonance and Spin-Peierls-like Phase Transition in a Geometrically Frustrated Antiferromagnet. Phys Rev Lett 84: 3718.

15. Lee SH, Broholm C, Ratcliff W, Gasparovic G, Huang Q, et al. (2002) Emergent excitations in a geometrically frustrated magnet. Nature 418: 856-858.

16. Tchernyshyov O, Moessner R, Sondhi SL (2002) Order by distortion and string modes in pyrochlore antiferromagnet. Phys Rev Lett 88: 067203.

17. Chung JH, Matsuda M, Lee SH, Kakurai K, Ueda H, et al. (2005) Statics and dynamics of an incommensurate spin order in a geometrically frustrated antiferromagnet CdCr$_2$O$_4$. Phys Rev Lett 95: 247204.

18. Kokalj A (2003) Computer graphics and graphical user interfaces as tools in simulations of matter at the atomic scale. Computer Mater Sci 28: 155-168.

19. Hohenberg P, Kohn W (1964) Inhomogeneous Electron Gas. Phys Rev 136: B864.

20. Giannozzi P, Baroni S, Bonini N, Calandra M, Car R, et al. (2009) QUANTUM ESPRESSO: a modular and open-source software project for quantum simulations of materials. J Phys: Condens Matter 21: 395502.

21. http://www.quantum-espresso.org/pseudopotentials/

Studying of Species Composition of Activators of Skin and Visceral Leishmaniasis in the South of Kazakhstan

Kozhabayevich KM*, Dossanuly SR and Zhanabayevich OA

M. Auezov South Kazakhstan State University, Shymkent City, Kazakhstan

Abstract

In this article is considered the data of species composition of pathogens of skin and visceral leishmaniasis in the South of Kazakhstan. The research results are shown that the studies within the stationary conditions landscapes differ by all indications, determining the intensity of epizootic: contamination of great gerbils, the nature of mosquito species complexes and their abundance on promastigots infected mosquitoes.

Considering the above, there is basis to assume existence on the studied by us territory of the independent natural centers or well shown morphological shares of the natural centers of zoonotic skin leishmaniasis which, most likely are dated for the certain studied landscape.

Keywords: Leishmania; Mosquitoes; Parasite; Skin; Viscerall; Zoonotic skin; Epizootic; Infectiousness

Introduction

There are two species of a parasite of Trypanosomatidae family gained distribution in Central Asia as well in the natural centers of the Republic of Kazakhstan. One of the – *Leishmania infantum*. This parasite is the activator of visceral leishmaniasis, other *Leishmania major* which proves as the activator of zoonotic skin leishmaniasis [1].

According to experiments in Kyzylorda region in 1997 by Professor Strelkova M.V. at white chickweed/sandwort - Rhombomysopimus - are allocated strains of *L. turanica*, *L. gerbiluu*, *L. major*, among which the first prevails. The European biobank of strains of leishmania of Central Asia has registered these data. From them only *L. major* has epidemiological value. However also the role of other activators in transfer of invasion is not excluded, for what additional researches are required [2].

Definitive or final owners of these parasites are domestic and wild animals, also as well as the person. The owner catches after stings of mosquitoes, infested by this parasite, besides he remains the constant carrier of the activator.

Historically leishmaniasis is very actual parasitic illness for Kazakhstan. In the Republic of Kazakhstan the active centers are located in the southern regions of the country: Kyzylorda, the South Kazakhstan and Zhambyl regions. Only for the last 9-10 years it is recorded skin leishmaniasis on the republic - 221 cases, and visceral leishmaniasis - 11 cases. The lethality has been registered at visceral leishmaniasis in 8 cases. Lethal outcomes from visceral leishmaniasis. Visceral leishmaniasis is lethal because of the late address of the population, serious associated diseases and that we consider important, absence at specialists of vigilance to this type of pathology. 2 lethal outcomes have been recorded In Kyzylorda region that are complication and weighting of epidemiological situation in region [3].

For the first time this parasite in the domestic and run wild dogs in Kyzylorda region was found in the middle of the XX century by the doctor F. Chun-Syun [4,5].

After that on studying of leishmaniasis diseases, and also by observation over ecological situation on distribution of mosquitoes the systematic clinic-epidemiological and other researches were not made. About 30 years ago the Institute's scientists of medical parasitology and tropical medicine named after E.I. Martsinovsky conducted applied researches on studying of the centers visceral in the territory of Central Asia, including Kazakhstan and Transcaucasia. In Kazakhstan visceral leishmaniasis cases generally were registered in the floodplain of the Syr-Darya River in Kyzylorda region where the natural source of the activator – Canisaureus L and Vulpescorsac, and carrier - Phlebotomussmirnovi has been established. The synanthropic centers of visceral leishmaniasis which were registered earlier in the cities of Zhambyl (nowadays Taraz) and Shymkent where dogs were probable sources of infection, and *P. longiductus* carrier, were not active. In the present time the researches on this subject in Kazakhstan are not conducted, human cases skin and visceral and cases of lethal outcome (generally children) from the visceral are only registered.

Materials and Methods

Work was carried out in 2015 at the laboratories of parasitology of "Pathology of Animals" Department of the South Kazakhstan State University named after M. Auezov. In the course of researches the general technique of the organization of such experiences has been fulfilled, where some specifications on collecting materials and their processing, and also methodical character have been made.

Researches were conducted only in stationary conditions in Kyzylorda, the South Kazakhstan and Zhambyl regions. Leishmaniasis infectiousness of mammals was established on visually confirmed defeat of integuments with rare indumentums on different sites of the head (it is established that auricles and very exceptional cases of lip, nose or eyelids are struck, most often). Dabs were made of these sites. Dabs for researches of the big sandwort, both sick, and healthy, were made of extreme (regional) site of auricles.

*Corresponding author: Kozhabayevich KM, M. Auezov South Kazakhstan State University, Shymkent City, Republic of Kazakhstan, Shymkent city, Tauke-khan avenue 5, Kazakhstan, E-mail: ortaev@mail.ru

Researches in Kyzylorda region were conducted in the Zhanakurgan district in the neighborhood of the settlements of Tugusken and Birlik (about 50 sq.km). This site is on average watercourse Syr-Darya. This area is known long since as one of the most unsuccessful in the epidemiological plan for zoonotic skin leishmaniasis. Because lately it has not been registered cases of this disease, the decision on special inspection of the population of this area has been made by us.

Under supervision there were only 27 holes in different natural boundaries and sand of the studied territories. During the whole season of activity of mosquitoes in each hole weekly caught by means of tin funnels with the cages from capron thread on the narrow end established in 20 openings of hole (exposure - 2,5 hours after calling of the sun), an also on the sheets of special sticky paper inserted into 20 openings of hole (flypapers kept in holes from 18 to 6 o'clock morning). The distributions of mosquitoes given about the general patterns have been as a result received.

Results and Discussion

According to the planned work of researchers certain work on studying of species composition of activators of skin and visceral leishmaniasis in Kyzylorda, the South Kazakhstan and Zhambyl regions is carried out now. In the village of Tugusken of the Zhanakurgan district of Kyzylorda region, test was selected from point of marrow from 50 dogs of whom did dabs for microreproduction. Aflagellar forms of leishmaniasis (amastigota) was not revealed at research under microscope and also sample from suspects on clinical sign is taken from 2 dogs.

From the study it was found that the small firms knots were found in one dog on the head, and at another on the head which has two cankers on 5 mm everyone are revealed. From the first dog sample of syringe was taken from small knots; which have cleared canker of the second dog sterile tampon. After the preparation of granulation fabrics, dab was made. Final step was coloring the agent according to Romanovsky-Gimz and it was then micro-reproductioned. The activator of leishmaniasis was not revealed from the first dog, and from the second dog the sample was taken from canker which had shown positive results.

The sample was then taken for aflagellar forms-Leishmania major- the activator of zoonotic skin leishmaniasis was found. When coloring agent dab was added the kernel of parasite becomes red color, and cytoplasm becomes blue. In the meanwhile, sampling was carried out from mouse-like rodents. In particular at white sandwort - 40 tests, from them in 2 tests leishmaniasis infantum - the activator of visceral leishmaniasis was revealed. Similarly, work was carried out in the Merke district of Zhambyl region and in the Otrar district of the South Kazakhstan region where activators of leishmaniasis were found of both types. The study was carried out collecting mosquitoes of 1000 copies. During research, it was revealed from 15 copies, that aflagellar forms of leishmaniasis (promastigota) were found. Specific accessory identification of activators was carried out *in vitro*.

Dominant and subdominant species of mosquitoes had been defined like their number and seasonality of flying. Usually mosquitoes flights begin at the beginning of May and ultimately climate change influences flight of mosquitoes on the settlement. Mass flight begins at the end of May. From 1000 mosquitoes of 700 mosquitoes were found to be dominant. *Phlebotomus papatasi*, and subdominant 300 mosquitoes of *Sergentomyia graecovi* and *Ph. caucasicus*. The researcher on study reveals that district the following species of

mosquitoes meet: *Ph. andrejevi*, *Sergentomyia graecovi*, *Ph. sergenti*, *Phlebotomus papatasii*, *S. Clydei*, *Ph. Caucasicus* and *Ph. smirnovi*. From the study, it was revealed that by quantity the special leader *Ph. papatasii* was dominants, subdominants were *Sergentomyia graecovi* and *Ph. caucasicus* were not considered because of their small number. While considering the climatic conditions of the studied regions was necessary to expand coverage zone of catching of mosquitoes.

Difficulties in implementation of actions for protection of the population living in these areas from diseases of the studied pathologies which in big degree was caused by insufficient study of features of epizootic and epidemic process in areas with different environment. Infectiousness of mosquitoes were determined by aflagellar form of leishmaniasis (promastigota) through various research of intestinal contents of females (more than 1000 researches) and also, by allocation of live leptomonadny cultures (about 600 dabs). From the study of dabs it was shown that average infectiousness of *Ph. papatasii* makes 7.4%, and that time as *S. Graecovi* – 4.4%. *Ph. caucasicus* was defined only in 2.6% of cases 203 *Ph. papatasii* were investigated by method of inoculations of medium. From them 20 cultures of promastigote were allocated; *Sergentomyia graecovi* (N = 295) 7 of cultures promastigote *Ph. caucasicus* (N = 80). In all cases results were negative. On all studied sites of the valley of the Syr-Darya River mosquitoes have been found.

It is obvious that in holes of big sandwort there passes the most part of life of mosquitoes, and there is transfer of the activator of leishmaniasis by them from one little wild beast to another. For explanation of differences of infectiousness of big sandwort and mosquitoes, the attention has been paid on to holes of animals. Studying of these holes has shown that the structure of these dwellings of animals depends on their location in landscape and soil features of the soil, such as density, mechanical structure, humidity and temperature of soil.

The studied reveal is provided by three types of landscape:

1. Landscape of ridge and hilly sand;

2. Valley landscape of average watercourse Syr-Darya;

3. Landscape of the hollow - wavy sand.

They objectively differ through character of different forms of relief, on soil vegetable cover and on the origin. In the studied territory, it was possible to meet practically all mammals registered as epizootic of visceral and skin leishmaniasis. In spite, of the fact that the number of long-clawed ground squirrel and big sandworts was at the high level in all desert landscapes, total quantity of individuals of small sandwort during our researches was low.

Except big sandworts, in epizootic cases of skin leishmaniasis, retailed and midday sandworts, big-eared hedgehog and long-clawed ground squirrel was also the participant. Patients of individuals have been found only among big sandworts (N = 1860 individuals). Among other rodents (redtailed sandwort (N = 87), midday sandwort (N = 128), gophers (N = 121)) noticeable damages of skin was noted.

From the studied 22 individuals of hedgehogs at 3 the defeats visually similar to leishmaniasis have been found, however parasites in dab preparations there was not revealed. Researchers want to note that despite big infectiousness of big sandworts to 15% in the intensive epizooty, among individuals which is conditionally called by minor types of carriers of sick animals it was not revealed.

Considering the above study, assumptions can be made on the existence of the study by the territory of the independent natural centers or well shown morphological shares of the natural centers of zoonotic skin leishmaniasis which most of them are dated for the certain studied landscape. Owing to insufficient development of complex questions of structural features of the natural centers, more specific definition of taxonomical situation was not represented at the possible moment that demands further researches.

It allows study of those structural features of the natural centers of everyone where studying of the epizootic occurring in smaller area of dwelling of animals-carriers can be seen. There is high probability of skin leishmaniasis other patterns of existence of the natural centers will be revealed. However we hope that the experience of carrying out stationary works stated by us can be considered when developing the general techniques.

Conclusion

As have shown results of our experiences, in the landscapes investigated in the conditions of hospital distinctions on many signs which are defined by intensity of epizootic are noted. It is infectiousness of sandworts, mosquito infectiousness of promastigote, character of specific mosquito complexes and on their abundance. We highlight that for standardization of data on different types of the natural centers it is necessary to bring together and to compare techniques of researches in this field.

References

1. Pronin AI, Kudryavtsev NA, Ivanov AA, Dikov VA, Rybkin GS (1996) Apparatuses having batteries of hydrocyclones made of plastics. Chemical and Petroleum Engineering 32: 375-379.

2. Lipton M, Longhurst R (1977) New Seeds and Poor People. Taylor & Francis, UK.

3. Sukhomlinov VN, Manzhurina OA, Romashov BV, Skogoreva AM (2014) Epizootic situation on cryptosporidiosis in cattle in the cattle farms of the belgorod region. Theory and practice of parasitic diseases of animals 15: 298-301.

4. Chun HF, Lee L, Chang FR (1994) Robust control analysis and design for discrete-time singular systems. Automatica 30: 1741-1750.

5. Huang CY, Wang DY, Wang CH, Chen YT, Wang YT (2010) Efficient light harvesting by photon down conversion and light trapping in hybrid ZnS Nanoparticles/Si Nanotips solar cells. ACS Nano 4: 5849-5854.

22

Rheology Properties of Castor Oil: Temperature and Shear Rate-dependence of Castor Oil Shear Stress

Abdelraziq IR* and Nierat TH

Physics Department, An-Najah National University, Nablus, Palestine

Abstract

The dynamic viscosity of castor oil was measured as a function of shear rate at different temperature ranged from 1.7°C to 62°C. In this study, shear stress and the dynamic viscosity as a function of temperature of castor oil decrease with increasing temperatures. Three and multi-constant formulas were proposed to obtain more suitable prediction of temperature dependence of shear stress and dynamic viscosity of castor oil. The best AAD% was calculated using our proposed formulas to be 0.03%. This work shows that the behavior of castor oil at the temperature ranged from 1.7°C to 62°C is Newtonian behavior by fitting the model of power law.

Keywords: Castor oil; Viscosity; Shear stress; Temperature; Formulas

Introduction

The rheology is the science of deformation and the study of the manner in which materials respond to applied stress or strain [1,2]. There are several studies that proposed alternative equations to describe the dynamic viscosity of liquids as a function of temperature. Two constant, three constant and multi-constant forms were expressed to describe the viscosity of vegetable oil as a function of temperature. Stanciu in his study, proposed four relationships of dynamic viscosity temperature dependence for vegetable oils. The purpose of his study was to find a polynomial or exponential dependence between temperature and dynamic viscosity of vegetable oil [3]. Giap derived an equation to replace the well-known Arrhenius-type relationship, his model was tested by using six vegetable oils [4]. Multi-constant forms to represent the liquid viscosity as a function of temperature was proposed by Thorpe and another by Daubert [5,6].Three-constant forms also were proposed by De Guzman and by Vogel [7,8]. The modified versions of the Andrade equation were used by Abramovic to describe the effect of temperature on dynamic viscosities for a number of vegetable oils. In addition, new forms to describe the effect of temperature on viscosity were suggested by Abramovic [9].

The flow behavior of liquids (Newtonian or non-Newtonian) was explained by Zhou. In his paper, the shear rate and temperature dependencies of viscosity of alumina nano-fluids were investigated experimentally [10]. The dynamic viscosity and shear stress, as a function of shear rate of chitosan dissolved in weakly acid solutions were studied by Esam. His results showed that shear thinning behavior is as pseudoplastic (non- Newtonian behavior) at temperature range from 20°C to 50°C, but it is showed more remarkable at lower temperature [11]. In Akhtar's study, Newtonian or Non-Newtonian behaviors of different oil samples (olive, almond, coconut, castor, sesame, cotton seed, sunflower and paraffin) were investigated. All the samples of oil investigated were found to possess Newtonian behavior with little deviation in olive and coconut oils [12]. Giap in his study, few food grade vegetable oils were subjected to viscometer measurements of viscosity at shear rate and temperature ranged from 3 to 100 rpm, and 40°C to 100°C, respectively. Results have shown that vegetable oils behaved as pseudoplastic [13]. Rheological properties of different samples of olive oils were obtained in a wide range of temperature by Ashrafi [14]. The dynamic viscosity of three commercial coconut fats was measured as a function of shear rate. The flow behavior of coconut fats was explained according to Newton's equation. The dynamic viscosity of coconut fats at different temperatures was used to calculate the flow

activation energy [15]. Fasina in his study, the viscosities and specific heat capacities of twelve vegetable oils were experimentally determined as a function of temperature (35°C to 180°C) [16]. The temperature-dependent rheological behavior of un-used (soybean, sunflower, olive, rapeseed, corn, rice and the mixtures soybean + olive and sunflower + olive) and used vegetable cooking oils was evaluated by Santos [17]. The viscosities of the three oil types (soybean oil, Sunflower oil and Canola oil) were investigated as a function of the shear rate and also shear stress as a function of shear rate at temperatures ranging from 10°C to 80°C [18]. In Stefanescu's paper the variation of the dynamic viscosity with the shear stress for to non-additive vegetable oils rape seed and soya bean [19]. The viscosity as a function of temperature from 24°C to 110°C was measured by Noureddini for a number of vegetable oils (crambe, rapeseed, corn, soybean, milk- weed coconut, and lesquerella) [20].

The aim of this study is to evaluate the rheological properties of castor oil. The dependence of dynamic viscosity of castor oil on temperature, shear rate and shear stress will be studied. The relationship between the dynamic viscosity of castor oil with temperature shear rate and shear stress will be found by fitting equations. The flow behavior of castor oil will be explained whether it is Newtonian or non-Newtonian.

Theory

Viscosity is a measure of the resistance to flow or shear. There are two types of viscosity:

Dynamic viscosity (η)

Dynamic viscosity which is defined as the ratio of shear stress (force over cross section area) to the rate of deformation (the difference of velocity over a sheared distance) and it is presented as:

*Corresponding author: Abdelraziq IR, Physics Department, An-Najah National University, Nablus, Palestine, E-mail: ashqer@najah.edu

$$\eta = \frac{\tau}{\frac{\partial u}{\partial x}} \qquad (1)$$

Where, η is the dynamic viscosity in Pascal-second (Pa.s); τ is shear stress (N/m²); and, $\frac{\partial u}{\partial x} = \gamma$ is rate of deformation or velocity gradient or better known as shear rate (1/s).

Kinematic viscosity (v)

The Kinematic viscosity requires knowledge of mass density of the liquid (ρ) at that temperature and pressure. It is defined as:

$$v = \frac{\eta}{\rho} \qquad (2)$$

Where, v is kinematic viscosity in centistokes (cSt), ρ is in g/cm³ [21]. Fluids such as water and benzene are Newtonian. This means that a plot of shear stress versus shear rate at a given temperature is a straight line with a constant slope that is independent of the shear rate. This slope is the viscosity of the fluid. Also, the plot passes through the origin, that is, the shear rate is zero when the shear stress is zero [22].

Any fluid that does not obey the Newtonian relationship between the shear stress and shear rate is called non-Newtonian. In this case, the slope of the shear stress versus shear rate curve will not be constant as one changes the shear rate. When the viscosity decreases with increasing shear rate, one calls the fluid shear-thinning. In the opposite case where the viscosity increases as the fluid is subjected to a higher shear rate, the fluid is called shear-thickening. Shear-thinning fluids also are called pseudoplastic fluids. The behavior of liquids is usually evaluated by applying the power law model of the form:

$$\tau = k\gamma^n \qquad (3)$$

Where τ is shear stress (N/m²), γ is shear rate (1/s). K is the consistency coefficient (N/m².sⁿ) and n is the flow behavior index [2].

The flow index of Eq.(3) will be calculated. If the flow index (n) of a liquid is 1, then it is Newtonian and if the value of flow index deviated from 1 then it shows the Non-Newtonian behavior. Several studies have been carried out on the effect of temperature on dynamic viscosity of vegetable oil. Many empirical relations have been proposed to describe this temperature dependence. Simple form of Andrade equation generally used to describe the effect of temperature on dynamic viscosity given by:

$$Ln\eta = A + \frac{B}{T} \qquad (4)$$

Nierat used the Andrade equations of three-constant formula are represented in the following equation [23,24]. $Ln\eta = A + \frac{B}{T} + CT \qquad (5)$

Neelamegam and Abramovic also used Eq.5 in their study [9,25]. Another formula of three-constant of Andrade equation was used by Neelamegam and Abramovic, which is given by [9,25]:

$$Ln\eta = A + \frac{B}{T} + \frac{C}{T^2} \qquad (6)$$

Where η is the dynamic viscosity in Pa.s, T is the temperature in Kelvin. A, B and C are constants. The constants of Eq.(5) of castor oil and other oils are presented by Neelamegam [25].

A polynomial dependence between temperature and dynamic viscosity of vegetable oil was found by Stanciu, which are:

$$\eta = A + BT + CT^2 \qquad (7)$$

$$\eta = A + BT + CT^2 + DT^3 \qquad (8)$$

Where η is the dynamic viscosity in Pa,s and T is the temperature in Kelvin. A, B, C and D are constants [3].

Methodology

The viscosity of grad castor oil samples of crop 2012 from AL-Zahra factory was measured as a function of shear rate at different temperature. Shear rate and temperature ranged from 6 to 60 rpm, and 2°C to 60°C, respectively. The experimental data were fitted and the correlation constants of the best fits were estimated.

Experimental apparatus

The viscosity of castor oil samples were measured using the NDJ-1 Rotational Viscometer with accuracy ± 5%. The spindles SP-2, SP-3 and SP-4 were operated at different speeds (6, 12, 30 and 60 rpm). Temperature was measured using Digital Prima Long Thermometer with accuracy ± 1% which measures temperature ranges from –20°C to +100°C. The Refrigerated and Heating Circulators F25-HD was used to increase and decrease the temperature of the castor oil samples to a specific temperature.

Statistical analysis

Some empirical relations were found to describe the dependence of dynamic viscosity and shear stress on temperature. The statistical analysis of the data was done by using Microsoft Excel Program. The correlation constants for the best fit were estimated. The best fit equation was chosen based on the percentage of average absolute deviation (%AAD) of the data [21].

Results and Analysis

The dependence of shear stress on shear rate

Newtonian or Non-Newtonian behavior of castor oil was investigated at different temperature ranged from 1.7°C to 62°C. Castor oil shows Newtonian behavior at the temperature ranged from 1.7°C to 62°C. Our experimental data of dynamic viscosity of castor oil was fitted using the model of power $\tau = k\gamma^n$ law. The values of flow index (n) and the consistency coefficient (K) are tabulated in Table 1. According to power law model, if the flow index (n) of a liquid is 1, then it is Newtonian and if the value of flow index deviates from 1 then it shows the Non-Newtonian behavior. Table 1 shows that castor oil has flow index value very close to 1 at the temperature ranged from 1.7°C to 62°C.

Figures 1a-2b show our experimental data and our fitting curves using power law model of shear stress of castor oil as a function of shear rate.

The dependence of dynamic viscosity on temperature

The dynamic viscosity of castor oil as a function of temperature is plotted in Figure 3. A comparison was made between the measured experimental data of dynamic viscosity (η_{exp}) and the previously calculated values (η_{cal}). The calculated values found by three-constant formula of Andrade's $Ln\eta = A + \frac{B}{T} + CT$. Where A, B and C are constants for castor oil (Table 2). It was found that the literature values didn't fit our experimental data. Using Andrade's formulas, the AAD% values were found by this work to be 99.96%. As a result, Andrade's formula was not the best fit for our experimental data of dynamic viscosity of castor oil. Accordingly, a modification was introduced to Andrade's formula in order to obtain a suitable description of our experimental data of dynamic viscosity as a function of temperature. The constants of Andrade's formula were determined using the modification. Our experimental values (η_{exp}) and calculated values (η_{cal}), using the modified form of Andrade's formulas of dynamic viscosities at different temperatures is given. AAD% value of the modified form of Andrade's formulas was found to be 0.48%. The results indicate that the modified form of Andrade's formula didn't fit exactly our experimental data. The values of the constants of the modified form of Andrade's formula are given in Table 3. These values are in disagreement with Neelamegam's values (Table 2) [25].

Multi-constant formulas were proposed by this work to obtain more suitable prediction of temperature dependence of dynamic viscosity of castor oil. The η_{exp} and η_{cal} were used to propose the formulas that fit our experimental data. That is, AAD% value is chosen to select the suitable prediction. It is found that our proposed formula of multi-constant to be $\eta = A + \frac{B}{t} + CLn(t) + Dt^E$ which fits our experimental data of dynamic viscosity. Our calculated values of A, B, C, D, E and AAD% of the data, are given in Table 4 shows that AAD% = 0.03%; therefore, our proposed multi-constant formulas are more suitable to describe the temperature dependence of dynamic viscosity of castor oil. Figure 4 shows our experimental data and our fitting curves using our proposed multi-constant formula of dynamic viscosity of castor oil as a function of temperature.

The dependence of shear stress on temperature

The shear stress of castor oil as a function of temperature at 60 rpm is plotted in Figure 5. Three and multi-constant formulas were

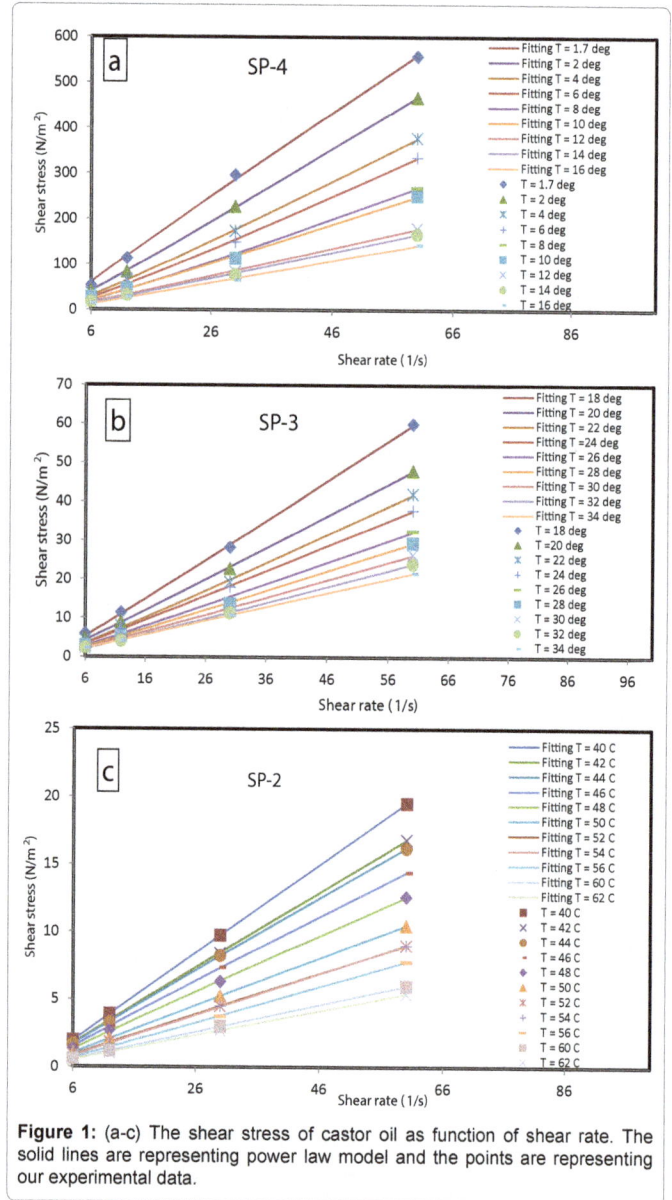

Figure 1: (a-c) The shear stress of castor oil as function of shear rate. The solid lines are representing power law model and the points are representing our experimental data.

proposed by this work to obtain suitable prediction of the dependence of shear stress of castor oil on temperature. The τ_{exp} and τ_{cal} were used to propose the formulas that fit our experimental data. That is, AAD% values are chosen to select the suitable prediction.

The experimental values of the shear stress of castor oil were fitted by using our three and multi-constant formulas. Our three and multi-constant formulas are proposed to be:

$$\tau = A + Bt^C \qquad (9)$$

$$Log\,\tau = A + \frac{B}{(C+T)^D} \qquad (10)$$

Where τ is shear stress (N/m²), T is the temperature in Kelvin, t is the temperature in degrees Celsius. Our calculated values of A, B, C, D, AAD% are given in Table 5. Table 5 shows that AAD% values of three and multi-constant are- 0.03% and -0.08% respectively; therefore, our proposed three and multi-constant formula are suitable to describe the dependence of shear stress of castor oil as a function of temperature.

T (°C)	K	n	SP
1.7	10.81	0.96	
2	06.34	1.05	
4	04.40	1.09	
6	03.60	1.11	
8	02.84	1.11	4
10	03.21	1.06	
12	02.54	1.04	
14	02.36	1.04	
16	02.07	1.03	
18	00.82	1.05	
20	00.66	1.05	
22	00.52	1.07	
24	00.52	1.04	
26	00.43	1.05	3
28	00.39	1.06	
30	00.35	1.06	
32	00.31	1.06	
34	00.36	1.00	
40	00.33	1.00	
42	00.30	0.99	
44	00.29	0.98	
46	00.27	0.97	
48	00.23	0.98	
50	00.18	1.00	
52	00.15	1.00	2
54	00.14	1.02	
56	00.11	1.04	
58	00.09	1.08	
60	00.10	1.00	
62	00.09	1.00	

Table 1: Our values of flow index (n) and the consistency coefficient (K) using power law model.

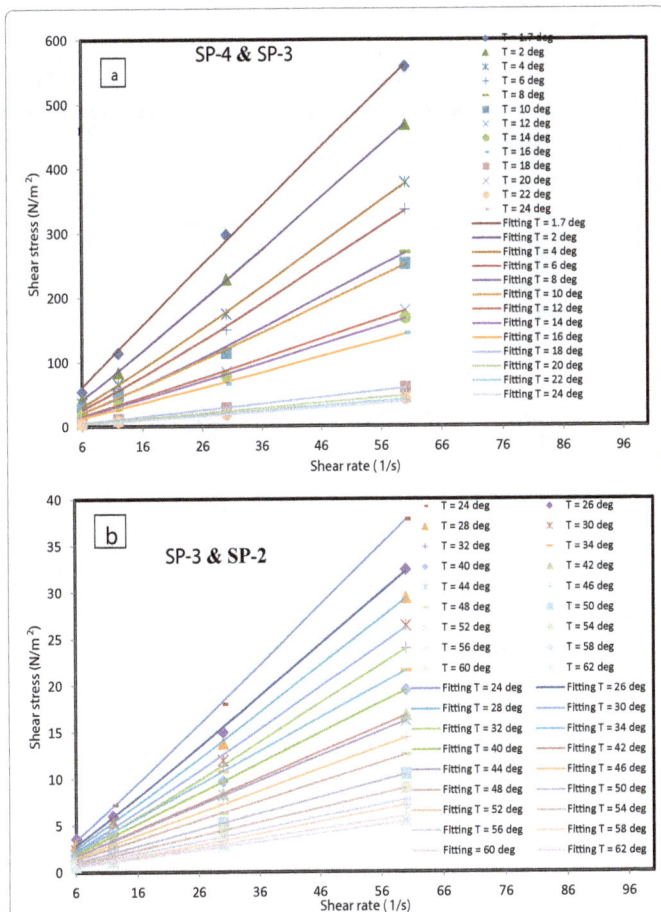

Figure 2: (a and b) The shear stress of castor oil as function of shear rate. The solid lines are representing power law model and the points are representing our experimental data.

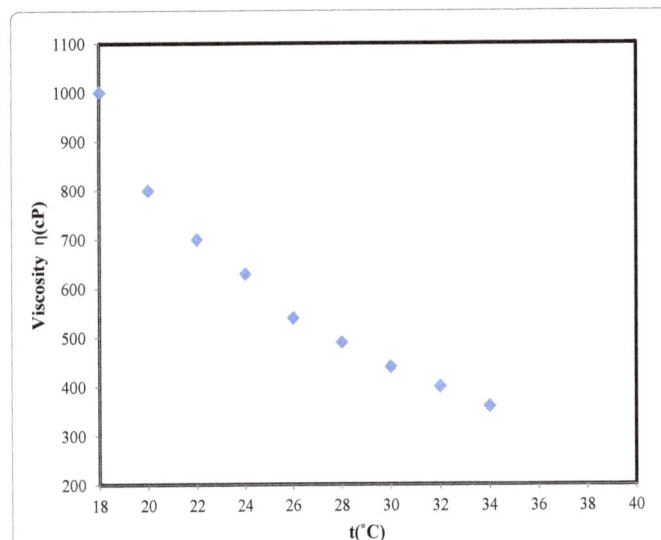

Figure 3: The measured values of the dynamic viscosity of castor oil as a function of temperature.

A	B	C	Temperature range (K)
3.1924	24.8450	-0.0389	303 - 363

Table 2: The constants given by Neelamegam using Andrade's formula.

A	B (K)	C (K²)	Temp. Range (K)	AAD%
-10.5381	4393.397	-0.01584	291 - 307	0.48

Table 3: Our values of A, B and AAD% using the modified Andrade's formula of castor oil.

A (cP)	B (cP.°C)	C (cP)	D(cP/°C^E)	E	Temp Range (°C)	AAD%
-2238.88	21547.23	500.5564	60258.99	-1.60718	18 - 34	0.03

Table 4: Our values of A, B, C, D, E and AAD% using our proposed multi-constants formula.

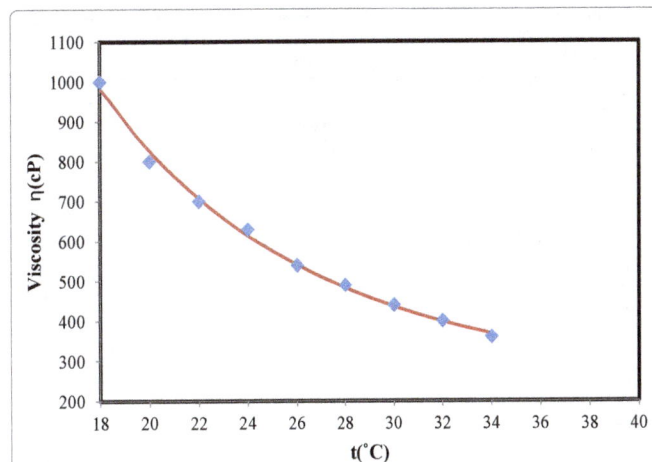

Figure 4: The dynamic viscosity castor oil as function of temperature. The solid lines are representing our proposed multi-constant formula and the points are representing our experimental data.

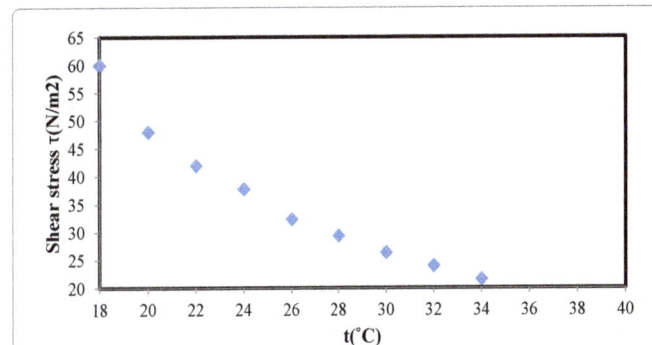

Figure 5: The measured values of shear stress of castor oil as a function of temperature at 60 rpm.

The formula	A	B	C	D	Temp Range	AAD%
Three-constant	7.214N/m²	15297.458	-1.967((N/m²)/°C^C)		18 – 34°C	-0.03
Multi constant	0.638	5.343K^D	-264.468K	5.638	291 – 307 K	-0.08

Table 5: Our values of A, B, C, D and AAD% using our proposed three &multi-constants formula.

Figures 6 and 7 show our experimental data and our fitting curves using our proposed three and multi-constant formulas of shear stress of castor oil as function of temperature.

Discussion

Our measured flow index (n) values of castor oil ranged between 0.96 - 1.11 (which are very close to 1) at the temperature that ranged from 1.7°C to 62°C. Our calculated values of the coefficient (K) of

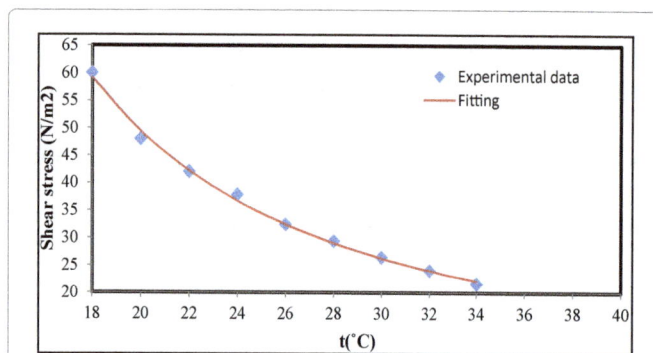

Figure 6: The shear stress of castor oil as a function of temperature. The solid lines are representing our proposed three-constant formula and the points are representing our experimental data.

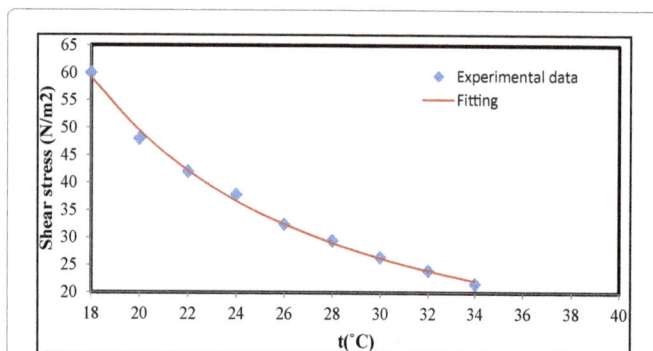

Figure 7: The shear stress of castor oil as a function of temperature. The solid lines are representing our proposed multi-constant formula and the points are representing our experimental data.

castor oil are found to be from 0.09 to 10.81. This indicates that castor oil shows Newtonian behavior at the temperature ranged from 1.7°C to 62°C. Akhtarin his study also shows the behavior of castor oil to be Newtonianin nature. Akhtar obtained the flow index of castor oil to be 0.98°C at 25°C [12].

Our dynamic viscosity of castor oil was measured to be 800.00 cP at 20°C, 440.00 cP at 30°C, and 325.00 cP at 40°C. Robert obtained the dynamic viscosity of castor oil to be 986.00 cP at 20°C [26]. Robert and Neelamegam obtained the dynamic viscosity of castor oil to be 451.00 cP and 329.37 cP, at 30°C, respectively. Robert and Neelamegam obtained also 231.00 cP and 271.35 cP, at 40°C, respectively [25,26].

Our values of dynamic viscosity of castor oil at high temperatures (only) are in good agreement with Neelamegam's and Robert's values. The measured experimental results of dynamic viscosity of castor oil are compared against the previously calculated values found by Andrade's formula of three-constants which is given by $Ln = A + - + CT$. For instance, the calculated value of dynamic viscosity at 30°C was found to be 0.20 cP. Where, our measured value at 30°C is 440.00 cP. This shows significant difference between our result and the literature value. This indicates that Andrade's formulas are not the best fit to be used for dynamic viscosity of castor oil. So, Andrade's formula is modified to fit our experimental values. As a result of this modification, the constants A, B and C were determined using Andrade's formula. The calculated dynamic viscosity using the modified form of Andrade's formulas at 30°C is found to be 432.84 cP, which indicates that Andrade's modified formula does not fit exactly our experimental data. The multi-constant equation given by $(\eta = A + \frac{B}{t} + CLn(t) + Dt^E)$ is proposed to obtain more

suitable prediction of temperature dependence of dynamic viscosity of castor oil. The constants of our proposed formula were estimated to give the best fit. The results of shear stress of castor oil in this study indicate that the shear stress of castor oil decreases as a function of temperature increases.

References

1. Barnes HA, Hutton JF, Walters K (1989) An introduction to rheology. Elsevier, USA.

2. Steffe JF (1996) Rheological methods in food process engineering. Freeman Press, USA.

3. Stanciu I (2012) A new viscosity-temperature relationship for vegetable oil. J Pet Technol Altern Fuels 3: 19-23.

4. Giap SGE (2010) The hidden property of arrhenius-type relationship: Viscosity as a function of temperature. Journal of Physical Science 21: 29-39.

5. Thorpe TE, Rodger JW, Barnett RE (1897) On the relations between the viscosity (internal friction) of liquids and their chemical nature. Part II, Phil Trans 189: 71-107.

6. Daubert TE, Danner RP (1994) Physical and thermodynamic properties of pure chemicals - data compilation design institute for Physical properties data. AIChE, Taylor and Francis, Washington DC.

7. De Guzman J (1913) Relation between fluidity and heat of fusion. Anales Soc Espan Fis Y Quim 11: 353-362.

8. Vogel H (1921) Das temperature abhängigkeitsgesetz der viskosität von flüssigkeiten. Physics 22: 645-646.

9. Abramovic H, Klofutar C (1998) The temperature dependence of dynamic viscosity for some vegetables oils. Acta Chim Slov 45: 69-77.

10. Zhou SQ, Ni R, Funfschilling D (2010) Effects of shear rate and temperature on viscosity of alumina polyalphaolefins nano fluids. J Appl Phys 107: 1-6.

11. Esam A, Elham ES, Mainal A, Yahaya AH (2010) Characterization of chitosan in acetic acid: Rheological and thermal studies. Turk J Chem 34: 47-56.

12. Akhtar N, Adnan Q, Ahmad M, Mehmood A, Farzana K (2009) Rheological studies and characterization of different oils. J Chem Soc Pak 31: 201-206.

13. Giap SGE, Nik WMNW, Ahmad MF, Amran A (2009) The assessment of rheological model reliability in lubricating behavior of vegetable oils. Engineering e-Transaction 4: 81-89.

14. Ashrafi N (2009) Effects of temperature on rheology of olive oils. ASME 2009 International Mechanical Engineering Congress and Exposition 9: 97-102.

15. Tipvarakarnkoon T, Blochwitz R, Senge B (2008) Rheological properties and phase change behaviors of coconut fats and oils. Ann Trans Nordic Rheol Soc 16.

16. Fasinaa OO, Colleya Z (2008) Viscosity and specific heat of vegetable oils as a function of temperature: 35°C to 180°C. International Journal of Food Properties 11: 738-746.

17. Santos JCO, Santos IMG, Souza AG (2005) Effect of heating and cooling on rheological parameters of edible vegetable oils. Journal of Food Engineering 67: 401-405.

18. Hojjatoleslamya M, Dehghannejada N, Zahedia A, Gharachorloob M (2011) The effect of heating on rheological behavior of vegetable edible oil.

19. Stefanescu I, Calomir C, Spanu C (2003) Studies concerning the vegetable oils viscosity used as biodegradable lubricant. National Tribology Conference 237-242.

20. Noureddini H, Teoh BC, Davis Clements L (1992) Viscosities of vegetable oils and fatty acids. JAOCS 69: 1189-1191.

21. Viswanath S, Ghosh K, Prasad H, Dutt NVK, Rani KY (2007) Viscosity of Liquids Theory, Estimation, Experiment and Data. Springer.

22. Kim S (2002) A Study of Non-Newtonian Viscosity and Yield Stress of Blood in a Scanning Capillary-Tube Rheometer.

23. Nierat TH, Mohammad S, Abdel-Raziq IR (2014) Temperature and Storage Age (Weakly Basis)-Dependence of Olive Oil Viscosity. MSAIJ 9: 445-451.

24. Nierat TH, Mohammad SM, Abdel-Raziq IR (2014) Temperature and Storage Age (Yearly Basis)-Dependence of Olive Oil Viscosity in Different Locations in Palestine. J Material Env Sci 5: 245-254.

25. Neelamegam P, Krishnaraj S (2011) Estimation of liquid viscosities of oils using associative neural networks. IJCT 18: 463-468.

26. Robert C (1979) CRC Handbook of Chemistry and Physics (59th eds.).

23

The Free Energy of Solvation for N-Decane in Ethanol-Water Solutions

Mongelli GF*

Department of Chemical Engineering, Case Western Reserve University, USA

Abstract

Within this work GROMACS computational molecular dynamics simulations were performed to determine the free energy of solvation of an n-decane molecule. The pull-code was utilized to draw the molecule across z-space to span from an initial gas state to a final solubilized state. The free energy was computed at each location of the molecule and this quantity is presented as a function of space for multiple alcohol contents. A single simulation of n-dodecane is considered for comparison in 100% water.

Keywords: Solutions; Molecule; Surface; Dynamics; Polymeric

Introduction

Recent molecular simulations and force field development have indicated that computational methods can offer significant insight into the behavior of surfaces [1], polyethoxy [2] and polysilicone [3] systems, hydrogen bonding [4], solvent-water interactions [5], and alkane properties [6]. Of particular interest are several studies which treat mixing thermodynamically [7], as well as those discussing the source of the hydrophobic effect [8]. According to market evaluation firm Transparency Market Research, the "specialty surfactants industry is expected to reach USD 29.2 Billion Globally in 2017 [9]". Furthermore, Transparency indicates that until at least that time non-ionic surfactants, which are solubilized in alcohol co-solvated systems, are expected to be the fastest growing of all surfactant classes. Such surfactants require solvation with water and alcohol for surface tension tuning. Although alcohol solvated systems are very apparently significant, it is surprising that there is still a lack of understanding regarding solvation free energy relationships in these systems. In order to ascertain what gives these polymers their surface presence in mixed solvent solutions, a critical examination of the free energy of solvation relationships is first necessary. That is what this work seeks to study. It is expected that this information might be useful in predicting solvophilicity of other materials. If the free energy of solvation of a new target molecule in a future study is simulated in any of the solutions studied in this paper and its free energy of solvation is determined, then the relative free energy of solvation for all of the other alcohol contents in this study is readily determinable. This is the concept of a free energy landscape. The free energy landscape as an abstraction can be a function of space, conformation or mass content. Model systems of interest for this study are composed of ethanol. Ethanol is a very common solvent in commercial products such as shampoos and shaving creams.

Methods

GROMACS molecular dynamics (MD) software was utilized to simulate the macroscopic properties of polymeric materials from classical principles in simple binary mixtures. This molecular dynamics package creates an ensemble of interacting particles and simulates molecular motions by solving Newton's equation of motion for incrementally small time steps on the order of a femtosecond. GROMACS accounts for bond stretching, hybridization changes, 1.4-diaxial, Lennard-Jones, and Coulombic interactions by parameterizing the relative strength of these forces for a molecular structure. Then it determines how the atoms in the system move under the action of those forces.

The water model used in this work is the well-established Simple Point Charge Extended (SPCE) model. Each individual mass fraction composition directly corresponds to a specific number of molecules incorporated into the simulation box with a constrained total number of molecules implying the number of water and non-water molecules.

Within this manuscript, NVT simulations of solvents with polymer are analyzed to determine key properties and trends. The pull code creates a harmonic potential along the z-axis that interacts only with the N-decane molecule. The spring constant of the system is tuned to be stronger than the intermolecular repulsions of the solvent molecules and N-decane. For a molecule which is insoluble in a specific solvent system, there are increased repulsions the deeper into the bulk the molecule travels. In order to determine the free energy of solvation, the solute material must travel at least as far into the solvent as the cut-off radius for non-bonded interactions. The temperature of these simulations is 293.15 K. For the pull code constrained ensemble the equation which governs the key relationship between n-decane density and Gibb's free energy is as follows:

$$\frac{\Delta G(z)}{k_B T} \propto Log_e(\rho(z)) \qquad (1)$$

$$Log_e(\rho(z)) = Log(Prob(z)) \qquad (2)$$

The simulation time average N-decane density as a function of z is supplied by the GROMACS density utility and is non-symmetrized. The density is convoluted into the free energy as a function of z by equation 1 given that equation 2 relates the density to the probability of finding the target molecule at position z. These results could be viewed as an indirect route to the same information that would come from the partition function of the system. The software does not readily report the ensemble properties of individual or sub-groups of molecules with the energy command, so this method was necessary. It is possible that looking at the free energy of the system as the molecule was pulled from the vacuum state to the solvated state would not be helpful since the free energy of solvation may be on the order of magnitude of the fluctuations in the ensemble properties of the system, though this

*Corresponding author:** Mongelli GF, Department of Chemical Engineering, Case Western Reserve University, 10900 Euclid Ave., A.W. Smith 116, Cleveland, OH 44120, USA, E-mail: Gfm12@case.edu

was not exclusively checked. Simulation equilibration was observed by increasing the simulation time until the density profile decreased by less than 5% with increased simulation time at 1 ns intervals. The simulation time was greater than 15 ns in all cases (Figures 1-7).

Results

The free energy of the simulation is plotted versus the center of mass of the molecule of interest as it is pulled throughout the system. The solvents occupy approximately 2 to 8 nm. The solvent thickness varies in each alcohol content as is fixed by the constrained total number of molecules. The negative free energy arbitrarily corresponds to system instability. The free energy scale is relative and can be shifted to be completely positive.

Discussion

The free energy of solvation is determined by taking the maximum

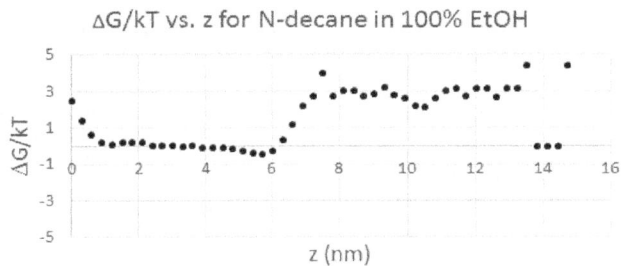

Figure 1: This figure depicts the free energy increasing as the polymer enters into the bulk. The polymer prefers to be in the surface state. The free energy of solvation is 4.87*kT or 2.00˙10-20 J.

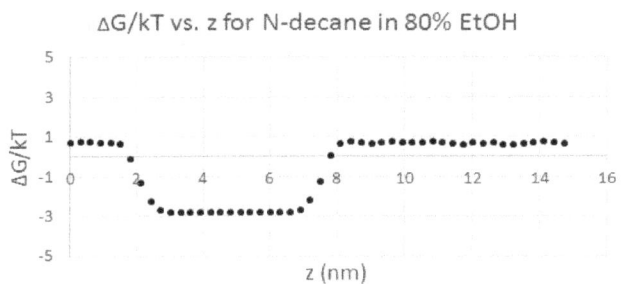

Figure 2: This figure depicts the free energy decreasing as the polymer enters the bulk. The solubilized state is preferred over the surface state. The free energy of solvation is 3.87*kT or 1.88*10-20 J.

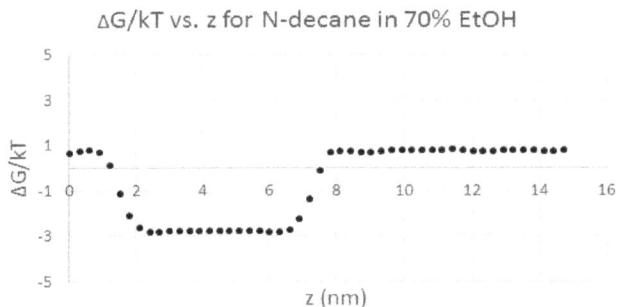

Figure 3: This figure depicts the free energy as a function of z for a ten carbon alkane material in 70% by mass ethanol. The solubilized state is preferred over the surface state. The free energy of solvation is 3.63*kT or 1.76*10-20 J.

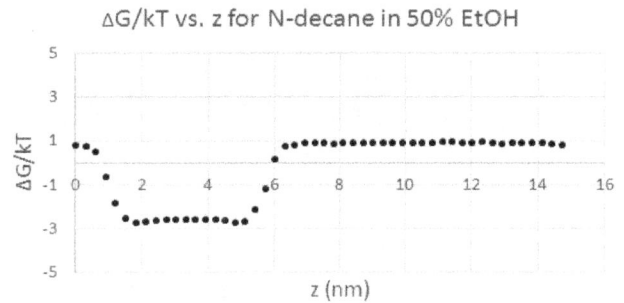

Figure 4: This figure depicts the free energy as a function of z for a ten carbon alkane material in 50% by mass ethanol. The solubilized state is preferred over the surface state. The free energy of solvation is 3.69*kT or 1.80810-20 J.

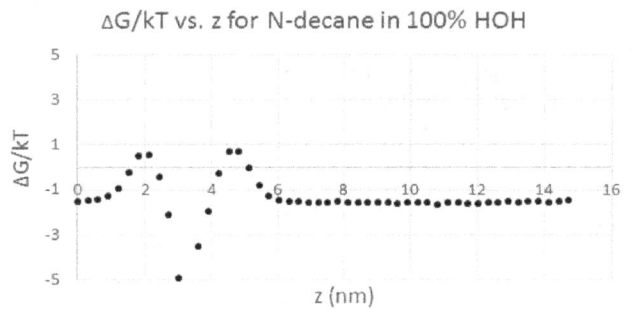

Figure 5: This figure depicts the free energy as a function of z for a ten carbon alkane in 100% by mass ethanol. The interface state is preferred over the solubilized state. The free energy of solvation is 5.64*kT or 2.74*10-20 J.

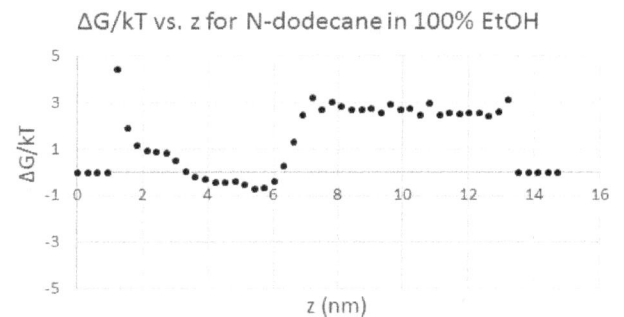

Figure 6: This figure depicts the free energy as a function of z for a twelve carbon alkane in 100% by mass ethanol. The interface state is preferred over the solubilized state. The free energy of solvation is 5.1˙kT or 2.48˙10-20 J.

free energy and subtracting the minimum value. Each of those values corresponds to either the gas state or the solubilized state. The free energy of solvation decreases with increasing alcohol content then decreases after a certain point. That point represents the content at which solvent-solvent attraction interactions overtake the solvent A-solvent B repulsion interactions modified by the solvent A or solvent B attraction to the alkane.an N-decane molecule was simulated in 100% ethanol and the free energy of solvation of this system is greater than n-decane in the same content. The trend is expected to continue at higher and lower contents. Meaning that the free energy of solvation of n-novae in 100% EtOH is expected to be less than that of N-decane

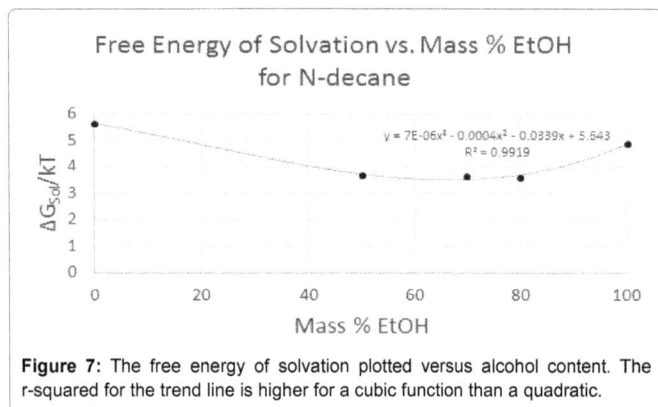

Figure 7: The free energy of solvation plotted versus alcohol content. The r-squared for the trend line is higher for a cubic function than a quadratic.

and that the free energy of solvation of simple n-$C_{13}H_{28}$ in 100% Et OH is expected to be larger than that of N-dodecane in the same solvent.

Conclusions

This manuscript details the free energy as a function of z-space for a small alkane in several alcohol contents. It is determined from an inversion of the time-averaged alkane density profile and allows for the calculation of the free energy of solvation. A longer alkane free energy of solvation was presented for single alcohol content and a trend that is expected to hold as a function of molecular weight is asserted. Though there are entropic considerations associated with the configurations of the backbone carbons rotating in solubilized versus bulk states. In small molecule alkanes, the end carbons LJ and Coulombics could impact free energy of solvation more than in high molecular weight Polk alkanes where such forces would have a much smaller contribution to the free energy of solvation. Future work should seek to determine the alcohol content at which free energy of solvation begins to increase. Furthermore, to determine the order of the best fit polynomial for the free energy of solvation as a function of alcohol content for different alkanes. Additionally to determine to what extent dihedral interactions contribute to the entropy of solvation and whether LJ or Coulombics contribute more to the mass content dependent, molecular weight dependent free energy of solvation relationships for various target molecules.

Acknowledgements

I would like to acknowledge the National Science Foundation Award Abstract #1159327.

References

1. Kinoshita M (2003) Interaction between surfaces with solvophobicity or solvophilicity immersed in a solvent. J Chem Phys 118: 8969.

2. Fischer J, Paschek D, Geiger A, Sadowski G (2008) Modeling of aqueous poly(oxyethylene) solutions, 1 Atomistic simulations. J Phys Chem B 112: 2388-2398.

3. Frischknecht AL, Curro JG (2003) Improved united atom force field for poly(dimethylsiloxane). Macromolecules 36: 2122-2129.

4. Smith G, Bedrov D, Borodin O (2000) Molecular dynamics simulation study of hydrogen bonding in aqueous poly(ethylene oxide) solutions. Phys Rev Lett 85: 5583.

5. Bedrov D, Smith GD (2002) A molecular dynamics simulation study of the influence of hydrogen bonding and polar interactions on hydration and conformations of a poly (ethylene oxide) oligomer in dilute aqueous solution. Macromolecules 35: 5712-5719.

6. Makrodimitri ZA, Dohrn R, Economou IG (2007) Atomistic simulation of poly (dimethylsiloxane): Force field development, structure, and thermodynamic properties of polymer melt and solubility of n-alkanes, n-perfluoroalkanes, and noble and light gases. Macromolecules 40: 1720-1729.

7. Shah PP, Roberts CJ (2008) Solvation in mixed aqueous solvents from a thermodynamic cycle approach. J Phys Chem Lett 112: 1049-1052.

8. Dalvi VH, Rossky PJ (2010) Molecular origins of fluorocarbon hydrophobicity Proc Natl Acad Sci USA 107: 13603-13607.

9. Specialty Surfactants Market -Global Scenario, Raw Material and Consumption Trends, Industry Analysis, Size, Share & Forecast 2011-2017.

24

Study of the Adsorption of Bright Green by a Natural Clay and Modified

Ltifi I*, Ayari F, Hassen Chehimi DB and Ayadi MT

Laboratory of Applications of Chemistry to Resources and Natural Substances and the Environment (LACReSNE), Department of Chemistry, Faculty of Science of Bizerte, University of Carthage, Tunisia

Abstract

The adsorption of Bright Green (BG), a cationic dye, was studied by clay treatment experiments by modification with an aqueous solution of a cationic surfactant. Hexadecyltrimethylammonium bromide (HDTMA) and Cetylpyridinium chloride (CPC) were used for the modification of the clay. Clay modified HDTMA showed the greatest adsorption capacity compared to the other adsorbents studied. The adsorption of HDTMA on BG depended on the adsorbent dose, the pH of the solution, the contact time and the initial dye concentration studied.

The adsorption data to correspond to the HDTMA experiments have been better described by the Langmuir isotherm model. The isothermal adsorption capacity of BG on HDTMA modified clay was found to be 45.5 mg/g (for an initial BG concentration of 50 mg/L), which is significantly higher than that of other adsorbents. The kinetics of adsorption of BG on clay modified by HDTMA has been described more precisely by the pseudo-second order kinetics model. The adsorbent was characterized by analysis of the Brunauer-Emmett-Teller surface (BET), Fourier transform infrared spectroscopy (FTIR) and X-ray diffraction (XRD). The BG adsorption mechanism on the surfactant-modified clay may comprise a hydrophobic interaction or van der Waals interaction or a combination of the two.

Keywords: Adsorption; Cationic dye; Surfactant; HDTMA; CPC; Modified clay

Introduction

Nowadays, wastewater effluents from different industries have become a major environmental concern. The treatment of water contaminated with textile dyes has been the subject of several studies aimed at reducing the intensity of the colors and the quantity of organic matter [1]. There are many methods for removing dyes from wastewater such as flocculation, chemical coagulation, oxidation, precipitation and filtration [2-6]. Among these methods, adsorption is the most effective technique for the treatment of wastewater [5-7]. Many adsorbents have been tested to reduce dye concentrations from aqueous solutions such as activated carbon [8], adsorbents including agricultural waste [9,10], natural phosphate [11], chitosan [12], kaolinite [13], montmorillonite [14]. However, the use of natural materials is a promising alternative because of their relative abundance and low commercial value. The surface properties of the natural clays can be substantially modified with large organic surfactants such as long chain quaternary alkylammonium salts such as HDTMA by ion exchange reaction. The intercalation of the cationic surfactants modifies only the surface properties, from hydrophilic to hydrophobic, but also greatly increases the basal spacing of the layers. The organo-clay becomes a more efficient adsorbent. In particular, the hydrophobic nature of the organo-layer suggests that the material can be used as a filter material to leach water from organic pollutants [15], transport of non-ionic contaminants into groundwater [16].

This work deals with the study of the potentiality of Tunisian natural clay with surfactants, Hexadecyltrimethylammonium bromide (HDTMA) and Cetylpyridinium chloride (CPC) as a low cost adsorbent for the removal of organic textile dye.

Experimental

Materials

Bright Green (BG) was purchased from Sigma–Aldrich. The formula weight of BG is 482.62 and its chemical formula is $C_{27}H_{34}N_2O_4S$. The maximum wavelength (λmax) of BG is 625 nm. The molecular structure of BG is illustrated in Figure 1. Organic surfactants used were

Hexadecyltrimethylammonium bromide (HDTMA, formula weight: 364.45, and chemical formula: $C_{19}H_{42}BrN$) and Cetylpyridinium chloride (CPC, formula weight: 339.9, and chemical formula: $C_{21}H_{38}ClN$) were obtained from Sigma–Aldrich. The molecular structure of CPC and

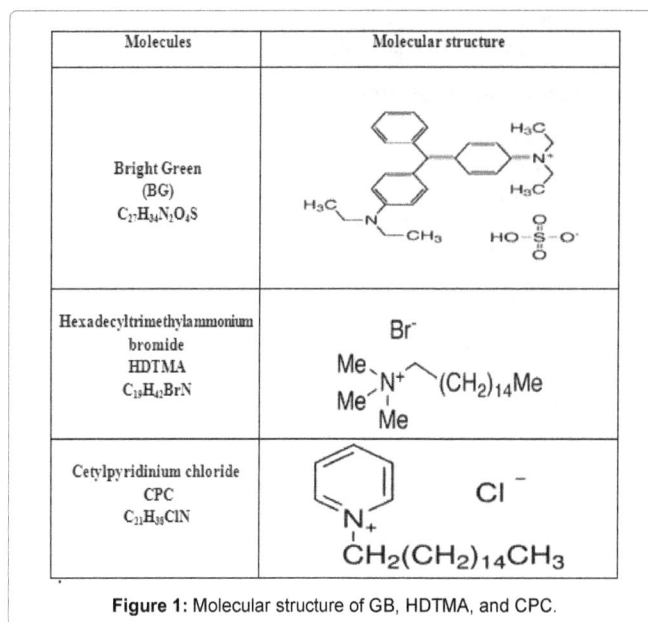

Figure 1: Molecular structure of GB, HDTMA, and CPC.

*Corresponding author: Ltifi I, Department of Chemistry, Faculty of Science of Bizerte, University of Carthage 7021, Tunisia
E-mail: ltifiismail@gmail.com

HDTMA are illustrated in Figure 1. Other chemical reagents, such as NaOH, HCl, and KCl were of analytical grade.

Preparation of the adsorbent

The absorbent clay used in this work which is named Gafsa clay located in the south of Tunisia. The mineralogical composition was determined from the fraction <2 μm with XRD. Clay Gafsa is mainly composed of smectite as shown in Table 1 the techniques of surface area (Ss) and the cation exchange capacity (CEC) and the BET were measured respectively. The point of zero charge (PZNPC) clay of Gafsa was conducted by the potentiometric titration method acid-base (Table 1).

Synthesis of the surfactant modified adsorbent

The exchange capacity of the outer cation (CEC) of the clay, determined by the MANTIN method, is 91 meq/100 g of purified clay [17]. It can be seen that: the value of the CEC decreased due to the application of an organophilic treatment adsorbent. Theoretically, CEC is defined as the number of monovalent cations that can replace compensating cations to compensate for the 100 g mineral electrical charge. Adsorbents modified by the following procedure [18] were prepared: on the one hand, 20 g of the adsorbent (Gafsa Clay) was dispersed in about 500 ml of water in distilled water. Then, a desired amount of surfactants (HDTMA or CPC) was stirred in 100 mL of distilled water until completely dissolved and then added drop wise to the clay solutions. The amounts of each surfactant were calculated on the basis of the CEC of the adsorbent. The reaction mixtures were mechanically stirred at room temperature for 48 hours. The resulting modified adsorbent surfactant was then filtered by filter papers and washed with distilled water until complete disruption of Br⁻ and Cl⁻ (AgNO₃ test). The products were dried at 80°C. for 12 h. Finally, the adsorbents were ground in an agate mortar and stored in a sealed glass container to be vented and labeled.

Characterization of the modified adsorbents

The prepared organoclays were characterised by X-ray diffraction (XRD), surface area measurement (BET), Fourier transform infrared spectroscopy (FT-IR). XRD for obtaining basal spacing d(001) values was operated and the method was described in the paper by Park et al [19]. The BET specific surface area, pore structure parameters were characterize from N_2 adsorption–desorption isotherms using a Micromeritics Tristar 3000 instrument.

Determination of the point of zero charge

The point of zero charge of the clay adsorbent in aqueous phase was analyzed using the solid addition method [20]. For this purpose, 0.1M KNO₃ solutions were applied and its pH was adjusted in the range of 2-12 by adding either 0.1 N HCl or NaOH and measured by a pH meter (Selecta Lab, PHW 100 Model, China). And then 0.2 g of the clay adsorbent was taken to each solution. The solutions were agitated for 48 h and the final pH values of the solution were measured.

Adsorption experiments

To study the adsorption isotherms of dyes by the raw and modified clays, volumes of 0.05 L of different concentrations of dye from (10 to 500 mg/L) are brought into contact with a mass of 0.1 g of the adsorbent. The experimental conditions are analogous to those of adsorption kinetics.

Modelling of the adsorption isotherm: The last stage of the study is to model isothermal curve, or more specifically, to report by a mathematical equation of the entire curve. Conventional models of Langmuir and Freundlich characterizing the formation of a monolayer are used for their simplicity of artwork. The model Langmuir [21] is based on the following hypotheses. Forming a single layer of adsorbate on the surface of the adsorbent, the existence of adsorption sites defined, the surface is uniform with no interaction between the adsorbed molecules.

The Langmuir equation is as follows:

$$q_e = \frac{q_m b C_e}{1 + b C_e}$$

With:

q_m(mg/g): Adsorptive capacity at saturation (characteristic of the formation of the monolayer of adsorbed molecules), and b (L/mg): Constant characteristic of adsorbent equilibrium temperature dependent and experimental conditions.

The model Freundlich [22] is based on an empirical equation reflects a change in energy with the amount adsorbed. This distribution of energy interaction is explained by heterogeneity of the adsorption sites. Unlike the model of Langmuir, Freundlich equation does not plan to limit higher than adsorption which restricts its application to dilute media. However, this model admits the existence of interactions between the adsorbed molecules [23]. It is of the following form:

$$Q_e = K_{f\times} C_e^{1/n}$$

Where K: adsorbent's capacity (L/g) and n: heterogeneity factor.

Results and Discussions

The study of the point of zero charge

The PZNPC or pH zero corresponds to the pH value for which the net charge of the adsorbing surface is zero [24]. This parameter is very important in the adsorption phenomena, especially when electrostatic forces are involved in the mechanisms. A quick and easy way to determine the PZNPC is to place 50 mL of distilled water in closed bottles and adjust the pH of each (values between 2 and 12) by addition of NaOH solution or HCl (0.1M). Then added to each flask, 50 mg of sample material to be characterized. The suspensions should be kept in agitation at room temperature for 24 h, and the final pH is then determined. It relates to a graph pH=f (pH$_i$) where pH=(pH$_f$-pH$_i$), the intersection of the curve with the axis that passes through the zero gives the isoelectric point (Figure 2).

Initial considerations

To determine the best adsorbent for the removal of Bright Green, several adsorption experiments were carried out using 0.1 g of

Parameter	Unit	Gafsa clay < 2 μm fraction
Specific surface	m²/g	83
CEC	meq/100 g	91
PZNPC		9.9
SiO₂	%(weight)	54.3
Al₂O₃	%(weight)	16.4
Fe₂O₃	%(weight)	8.2
MgO	%(weight)	4.7
CaO	%(weight)	4.6
SO₃	%(weight)	2.2
K₂O	%(weight)	1.3

Table 1: Physico-chemical characteristics of Gafsa clay.

adsorbent prepared. Each adsorbent was added to 50 mL of 100 mg/L BG dye to 30°C and the solution was stirred at a speed of 200 rpm for 12 h. The results are illustrated in Figure 3. It can be seen that the modified surfactants adsorbents have a capacity significantly higher for the adsorption of BG with respect to the clay. Indeed, adsorbents produced by intercalation by Cationic surfactants improved the adsorption capacity of the adsorbents. This is consistent with the results of other researchers [25-28]. According Figure 3, the surfactant is a modified adsorption capacity slightly higher compared to the surfactant that can be attributed to the higher value of CEC clay. The CEC is a value characterizing parameter representing the adsorbent with a higher amount of the CEC is more likely to be able to exchange cations with the cationic surfactants [29]. We continue the rest of experiences with the organophilic clay with HDTMA due to its high adsorption capacity compared to other adsorbents.

The adsorption mechanism

Technically, the adsorption of cationic surfactant onto the surface of adsorbent may follow two approaches. The first approach: the surfactant molecules interact with clay through their non-polar (alkyl) groups; hence the positive head of the surfactants points toward the bulk of the solution. The second approach [26]: in this approach, the adsorption of the cationic surfactant onto the negatively charged surface of the adsorbent can be considered to be controlled by two steps; (1) the formation of surfactant monolayer through the ion exchange and electrostatic attraction and (2) the formation of surfactant bilayer via hydrophobic interactions [30-32]. As a matter of fact, firstly, the positive head of the surfactants are exchanged with the interlayer exchangeable cations within the clay, thereby forming a surfactant monolayer with outward pointing head groups. Secondly, the bilayer is organized by the attachment of the surfactant alkyl chains to the outer surface of the monolayer by means of the hydrophobic–hydrophobic interactions. Therefore, the external surface of the modified adsorbent has become positive and accordingly, more appropriate for the adsorption of the cationic adsorbantes like the BG molecules. The first stage is more probable to occur at low surfactant concentrations (at about 100% CEC or below) and the second stage takes place at higher concentrations hemimicelles or micelles (more than 100% CEC) [33,34]. In the present research, the amount of the surfactant is provided at about 200% of CEC; hence it can be assured that the bilayer is formed. Figure 4 best schematizes the modification procedure of clay using cationic surfactant. Owing to the different configurations of the CPC and HDTMA surfactants on clay, various interactions may be involved in the adsorption of the BG from aqueous solution. On the one hand, the positive head of the surfactants covering the exterior surface of the adsorbent may be the main responsible for the increase of BG sorption in the case of organically modified adsorbents. In fact, the electrostatic attraction between the anionic SO_4^-H group of BG molecules and positively charged adsorbent is the dominant phenomenon for the adsorption of BG. On the other hand, as reported in the literature, the hydrophobic portion of the adsorbent surface has more preference for dissociated species of BG in aqueous solution. Furthermore, the van der Waals interaction between the phenyl ring of BG and CH_2 group of the modified adsorbent can be considered as one of the driving forces through the adsorption process [28]. It should be pointed out that in the case of the CPC modified adsorbent, the phenyl ring of BG can be bound to the pyridine ring of CPC molecules via the π stacking interaction. But then, it can be questioned that why the HDTMA modified adsorbent is superior to the CPC modified adsorbent in the adsorption of BG if the latter adsorbent take advantage of the π stacking interaction. This can be justified by the spatial hindrance arising from the pyridine ring around the CPC head (Figure 4).

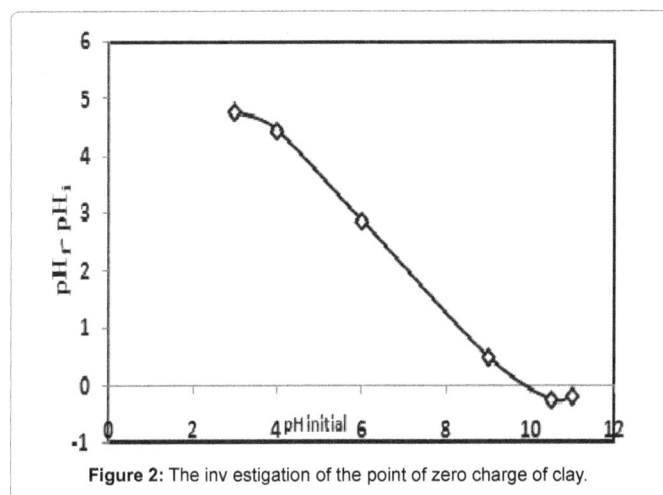

Figure 2: The inv estigation of the point of zero charge of clay.

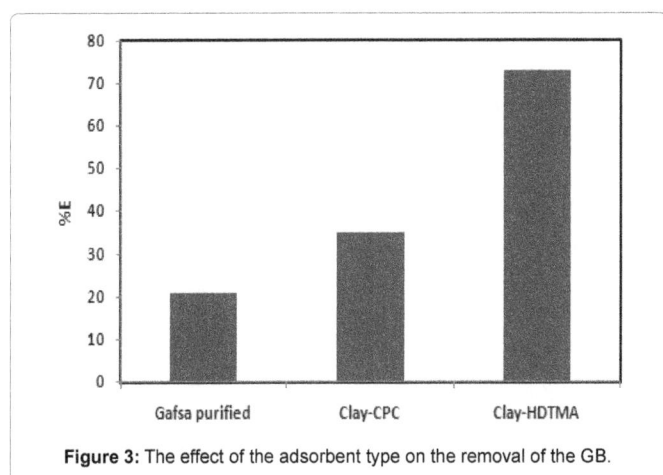

Figure 3: The effect of the adsorbent type on the removal of the GB.

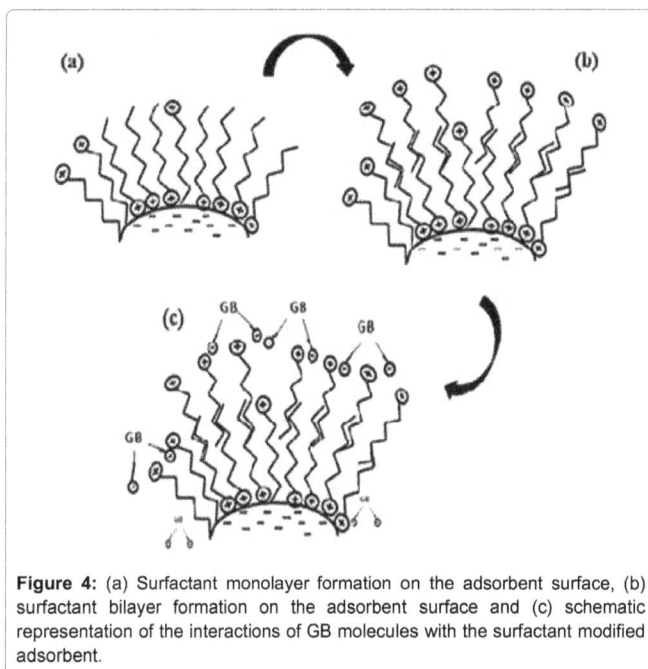

Figure 4: (a) Surfactant monolayer formation on the adsorbent surface, (b) surfactant bilayer formation on the adsorbent surface and (c) schematic representation of the interactions of GB molecules with the surfactant modified adsorbent.

The study of the adsorbent dosage on the sorption of BG

Figure 5 shows the effect of the adsorbent dose on the removal of BG from aqueous solution. It can be seen that the adsorption of BG has rapidly increased with the increase of the adsorbent dose. The optimum amount of the adsorbent is 0.7, 0.4 and 0.2 for clay, clay-CPC, and clay-HDTMA, respectively, and after the optimum amount of each adsorbent, the increase of the adsorbent do not effect on the removal of BG and the adsorption is nearly constant. The sharper adsorption curve was observed in the case of the modified adsorbent with the cationic surfactants. It is revealed that the implementation of the surfactant has an influential effect on the adsorption of BG and the optimum amount of the adsorbent has decreased.

The study of the initial pH of the solution on the sorption of GB

pH is an important factor in any adsorption study, because it can influence the adsorbent and adsorbate structure as well as the adsorption mechanism. In this article, we studied the adsorption efficiency of a bright green dye by varying the pH from 4 to 11 using a solution of hydrochloric acid HCl (0.1M) or soda NaOH (0.1 M) according to the desired pH. Under these pH conditions, a mass of 100 mg of the adsorbent was stirred in 100 mL of the colored solution at 20 mg/L. The results obtained in these tests are shown in Figure 6.

The results obtained show that the variation of the residual dye

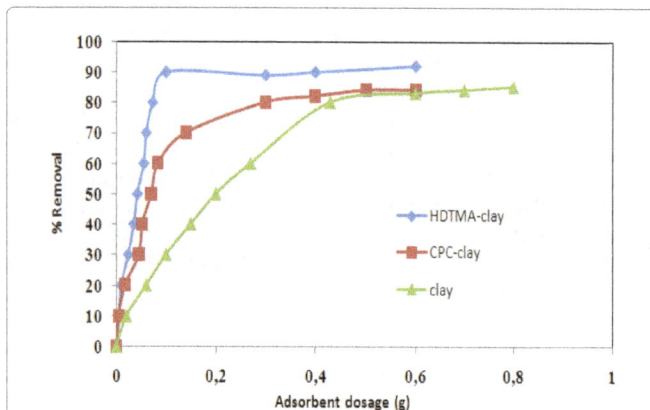

Figure 5: The effect of the adsorbent dosage on the adsorption of GB on three different adsorbent.

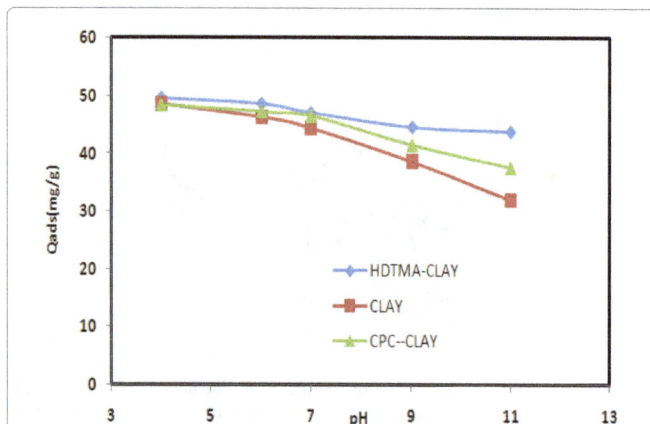

Figure 6: The effect of pH on the adsorption of GB on three different adsorbent.

concentrations is relatively low. Discoloration is therefore little influenced by the variation in pH. In the light of these results, all the discoloration tests on the fly ash and bottom ash were carried out at the natural pH of the colored solution (between 6 and 7) for the bright green.

Kinetics of adsorption of dyes by different clays (natural and modified by HDTMA)

Figure 7 shows the evolution of the adsorbed amount as a function of time. The kinetics of dye adsorption on the clays used shows a strong adsorption from the first minutes of dye-clay contact, followed by a slow increase until reaching an equilibrium state.

The kinetics of adsorption rapid during the first minutes of reaction can be interpreted by the fact that at the beginning of the adsorption the number of active sites available on the surface of the adsorbent material is much higher than that of the sites remaining after a certain time

Adsorption isotherms

The adsorption isotherm is an important technique providing precious information to predict the adsorbent efficiency for the removal of a specific adsorbate. It is reported in several studies that non-linear analysis should be considered as a better approach to obtain the isotherm parameters as sometimes linearization of non-linear experimental data may distort the error distribution structure of isotherm [36-38]. Hence, the non-linear procedure was carried out for describing the adsorption isotherms and predicting the overall sorption behavior for the removal of BG using clay-HDTMA adsorbent. In this section two-parameter isotherm models (Freundlich and Langmuir) were examined to find out the best fit for the experimental data. The experimental adsorption isotherm of BG on clay and clay-HDTMA are illustrated in Figure 8. It is apparent from Figure 8 that clay-HDTMA has higher adsorption capacity compared to clay for the removal of BG from aqueous solution. It is important to be noted that the shape of the isotherm can be used for the interpretation of the adsorption process. According to classification of Giles [39], the adsorption isotherms are classified into four groups: L, S, H, and C. According to the aforementioned classification, the adsorption of BG onto clay and clay-HDTMA followed the L curve pattern which indicates that there is no strong competition between solvent and the adsorbate to occupy the adsorbent surface sites.

The isotherm model parameters and the statistical results are presented in Table 2. Figure 8 shows the experimental data as well

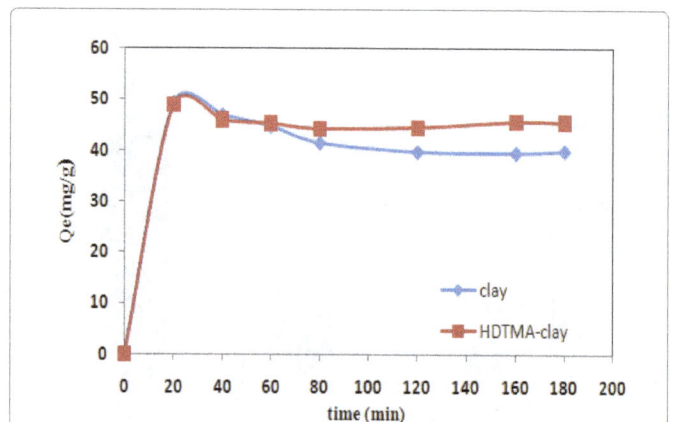

Figure 7: Kinetics of adsorption of Brilliant Green on the clays used.

as the isothermal models for the adsorption of BG on the HDTMA-clay. According to Table 1, all isothermal models correctly describe the adsorption of BG on the adsorbent; However, the R^2 statistical parameter amounts are related to the Langmuir model ($R^2=0.9997$) higher than those of the other isothermal models that are closer to the unit.

The quantity of the dye increases more or less rapidly for low concentrations in solution and then equilibrates to reach a plateau, corresponding to saturation of the adsorption sites and reflecting an adsorption in monolayer. The isotherm obtained is of type L according to the classification of Giles [39].

It is clear that the maximum adsorption capacity of the surfactant-modified adsorbent (clay-HDTMA; 45.5 mg/g) was significantly higher than that of the clay (30.26 mg/g). Furthermore, it is found that the BG adsorption capacity of the HDTMA-clay is relatively high compared to the other adsorbents reported in the literature which indicates that the HDTMA clay is a promising adsorbent for the removal of BG.

Adsorption kinetics

The modeling of the adsorption kinetics of the removal of BG on clay-HDTMA was studied by the two most common models, namely pseudo-first-order model and pseudo-second-order model. The pseudo-first-order model is represented by the following equation:

$$\text{Log}\left(q_e - q_t\right) = \log\left(q_e\right) - \left(\frac{k1}{2,303}\right)t$$

Where k_1 (1/min) is the rate constant of pseudo-first-order adsorption. A linear plot of log $(q_e - q_t)$ versus t was used to determine

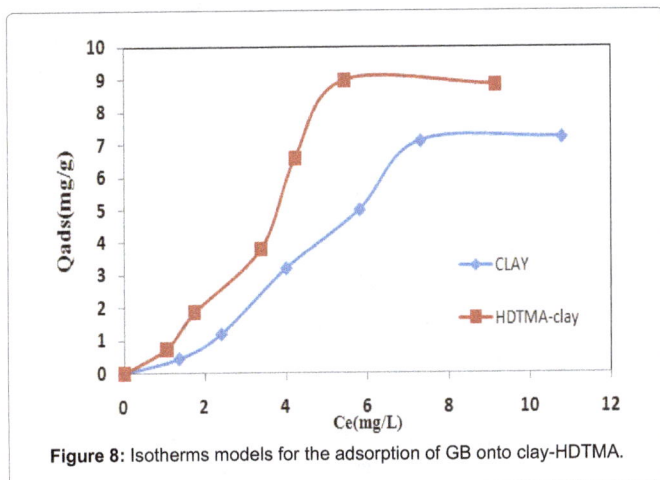

Figure 8: Isotherms models for the adsorption of GB onto clay-HDTMA.

Model	Model parameter	R^2
Frendlich	$K_F=5.11$, n=1.498	0.998
Langmuir	$q_m=45.5$, b=0.029	0.9997

Table 2: The isotherm constants and coefficients for the adsorption of GB on HDTMA-clay.

the values of k_1 and the equilibrium adsorption capacity (q_e). The model parameters used to evaluate the experimental data and the corresponding correlation coefficient are presented in Table 3. A comparison between the resulted correlation coefficients (Table 3) implies that the pseudo-first-order model cannot provide a suitable description for the adsorption of GB on clay-HDTMA adsorbent. Besides, the calculated q_e resulted from the pseudo-first-order model were obviously different from that of the experimental values. The pseudo-second-order model can be expressed as:

$$\frac{t}{q_t} = \frac{1}{k_2 q_e^2} + \frac{t}{q_e}$$

Where k_2 (g/mg min) relates to the constant of pseudo-second-order adsorption. The straight line plots of t/qt versus t permits the calculation of k_2 and q_e (Figure 9). The results are shown in Table 3. According to Figure 9, the pseudo-second-order model was found to be most appropriate and accommodate with the experimental results. It can also be observed from the correlation coefficient values from Table 3 that R^2 values were tending to ward unity. In addition, the calculated q_e values from the pseudo-second-order model fairly agree the experimental ones better than the pseudo-first-order. All in all, this can be concluded that the adsorption of BG on clay-HDTMA obeys the pseudo-second-order kinetic model.

Structural characteristics

BET analysis of clay and surfactant modified HDTMA and CPC: Table 4 shows the BET surface area (m²/g), the total pore volume (cm³/g) and the average pore diameter (nm) and the basal distance (d_{001}) of the clay, clay-CPC adsorbent and Clay-HDTMA. The BET surface area decreased from 54 to 1.28 m²/g for CPC-modified clay and to 2.19 m²/g for HDTMA clay modification, which can be attributed to blockage and pore screening of clay by surfactant alkyl chains [40]. It should be noted that, despite our hypothesis, the average pore diameter decreases slightly from 4.0 to 3.3 nm from clay to HDTMA clay, which is likely due to the complete removal of Micro-pores of the adsorbent structure [40].

S_{BET}: specific surface area.

VP: pore volume determined by BJH method from N2 desorption isotherm.

APD: average pore diameter determined by the curve of BJH desorption dV/dD pore volume. d_{001}: Basal distance

The results of XRD of the untreated natural clays and those modified with the cationic surfactant HDTMA are given in Figure 10. The purified natural clay (from GAFSA) shows a position of the basal distance (001) of reflection (d = 12.72 Å) characteristic of a smectite exchanged with sodium. The modified clay model (HDTMA-GAFSA clay) compared to the others shows a spacing of (d=26 Å), indicating the intercalation of the HDTMA molecules. The spacing d is in agreement with that obtained by Slade and Gates [41] for various smectites.

Initial concentration (mg/L)	q_e, exp (mg/g)	Pseudo-first-order kinetic model			Pseudo-second-order kinetic model		
		K_1(g/mg min) 10^3	q_e, cal (mg/g)	R^2	K_2(g/mg min) 10^3	q_e, cal (mg/g)	R^2
50	4.20	11.52	0.69	0.91	53.99	11.63	0.999
100	9.08	0.02	0.88	0.071	16.13	22.73	0.999
150	16.19	13.82	25.27	0.994	1.2	40.00	0.999
200	24.04	29.94	570.15	0.563	0.68	47.62	0.997
250	35.09	9.21	54.2	0.826	0.35	58.58	0.999

Table 3: Kinetic parameters of pseudo-first and pseudo-second order models for the adsorption of GB on clay.

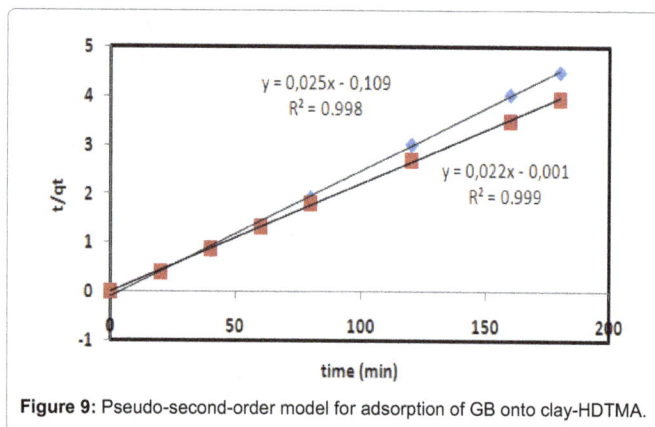

Figure 9: Pseudo-second-order model for adsorption of GB onto clay-HDTMA.

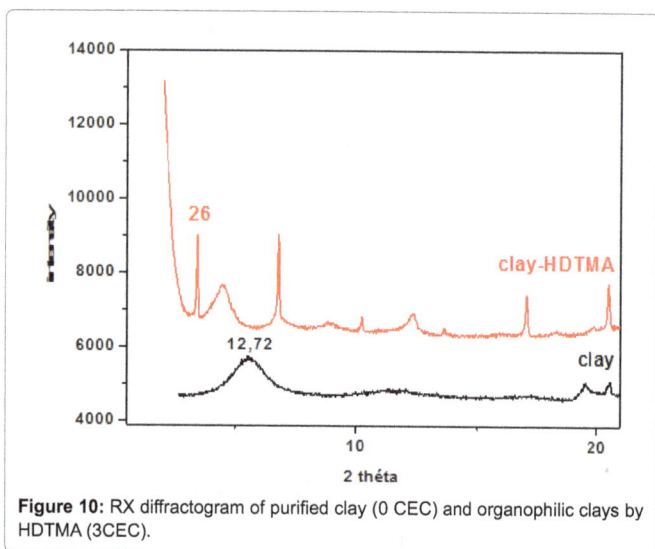

Figure 10: RX diffractogram of purified clay (0 CEC) and organophilic clays by HDTMA (3CEC).

Adsorbent samples	S_{BET} (m²/g)	VP (cm²/g)	APD (nm)	d_{001} (nm)
Clay	54	0.107	4.0	1.27
HDTMA-clay	2.19	0.011	3.3	2.6
CPC-clay	1.28	0.007	2.7	2.1

Table 4: The BET surface area, total pore volume, and average pore diameter for clay, and HDTMA-clay.

X-ray diffraction analysis of clay and surfactant modified clay-HDTMA: The results of XRD of the untreated natural clays and those modified by the cationic surfactant HDTMA are given in Figure 11. The potted Na-stah gafsa clay shows a position of the basal distance (001) of reflection (d = 12,72 Å) characteristic of a smectite exchanged with sodium. The modified clay model (HDTMA-stah gafsa) with respect to the others shows a spacing of (d = 26 Å), indicating the intercalation of the HDTMA molecules. The value of the spacing d is in agreement with those obtained by the slats and the grids for various smectites.

FTIR analysis of clay and surfactant modified clay: Based on the results of XRD, it is possible to confirm that the exchange of HDTMA polycations with Na⁺ alkalin cations in the interlayer space is successful (Figure 11).

The main peaks are:

- A peak at 1480 cm⁻¹ indicating the presence of the functional group N-C, corresponding to the tertiary amine.

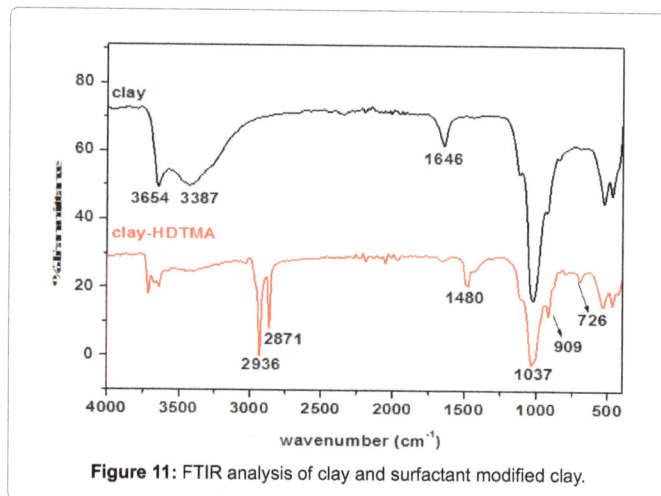

Figure 11: FTIR analysis of clay and surfactant modified clay.

- Two absorption peaks at 726-780 cm⁻¹ corresponding to the deformation vibration mode outside the plane of the CH_2 group.

- Two strong peaks at 2871 and 2936 cm⁻¹, corresponding to the symmetrical stretch vibration mode and incline the methyl amine.

Conclusions

The equilibrium and dynamics of Bright Green (BG) adsorption on the surfactant-modified adsorbent were studied in this study. It has been found that the value of the zero charge point of the clay is about 9.9 and above that the surface of the adsorbent is negative. Several adsorbents such as clay, HDTMA clay and clay-CPC have been used for the adsorption of BG from aqueous solutions. The modified HDTMA clay remarkably has the highest adsorption capacity compared to the other adsorbents prepared. It was observed that the BG adsorption capacity for the clay increased with the contact time and the initial dye concentration.

The equilibrium data have been well described by the Langmuir model. Kinetic studies of BG adsorption on clay indicate that the pseudo-second order kinetic model agrees very well with experimental adsorption data (Qe=45.5 mg/g).

The clay studied in the GAFSA region (south of Tunisia) has a very high adsorption capacity because of their specific properties (CEC=91 meq/100 g...Surfac Specific=54 m²/g). Several characterizations analyzes, including BET, FTIR and XRD, confirmed the modification of the surfactants of the adsorbent. The results revealed that the modified HDTMA clay could be applied as a low cost material for the adsorption of the BG dye from aqueous solutions.

References

1. Nejib A, Joëlle D, Abdellah E, Amane J, Malika TA (2014) Textile dye adsorption onto raw clay: influence of clay surface properties and dyeing additives. Journal of Colloid Science and Biotechnology 3: 98-110.

2. Gupta VK (2009) Application of low-cost adsorbents for dye removal–A review. J Environ Manage 90: 2313-2342.

3. Rafatullah M, Sulaiman O, Hashim R, Ahmad A (2010) Adsorption of methylene blue on low-cost adsorbents: a review. J Hazard Mater 177: 70-80.

4. Preethi S, Sivasamy A, Sivanesan S, Ramamurthi V, Swaminathan G (2006) Removal of safranin basic dye from aqueous solutions by adsorption onto corncob activated carbon. Ind Eng Chem Res 45: 7627-7632.

5. Mohan D, Singh KP, Singh G, Kumar K (2002) Removal of dyes from wastewater using flyash, a low-cost adsorbent. Ind Eng Chem Res 41: 3688-3695.

6. Acemioglu B (2004) Adsorption of Congo red from aqueous solution onto calcium-rich fly ash, J Colloid Interface Sci 274l: 371-379.

7. Abid N (2015) Traitement d'effluents contenant des colorants par de l'argile naturelle. J Cleaner Production 86 : 432-440.

8. Barka N, Assabbane A, Ichou YA, Nounah A (2006) Decantamination of textile wastewater by powdered activated carbon. J Appl Sci 6: 692-695.

9. Ahmad MA, Puad NAA, Bello OS (2014) Kinetic, equilibrium and thermodynamic studies of synthetic dye removal using pomegranate peel activated carbon prepared by microwave-induced KOH activation. Water Resources and Industry 6: 18-35.

10. Dogan M, Abak H, Alkan M (2009) Adsorption of methylene blue onto hazelnut shell: kinetics, mechanism and activation parameters. J Hazard Mater 164: 172-181.

11. Barka N, Assabbane A, Nounah A, Laanab L, Ichou YA (2009) Removal textile dyes from aqueous solution by Natural Phosphate as new adsorbent. Desalination 235: 264-275.

12. Tsai FC, Ma N, Chiang TC, Tsai LC, Shi JJ, et al. (2014) Adsorptive removal of methyl orange from aqueous solution with crosslinking chitosan microspheres. Journal of Water Process Engineering 1: 2-7.

13. Doğan M, Karaoğlu MH, Alkan M (2009) Adsorption kinetics of maxilon yellow 4GL and maxilon red GRL dyes on kaolinite. J Hazard Mater 165: 1142-1151.

14. Gemeay AH, El-Sherbiny AS, Zaki AB (2002) Adsorption and kinetic studies of the intercalation of some organic compounds onto Na+-montmorillonite. J Colloid Interface Sci 245: 116-125.

15. Zhu LZ, Su YH (2002) Cationic carbonyl complexes of manganese (I). Clays and Clay Miner 49: 421-427.

16. Yurdakoc M, Akc May Y, Tonbul OK, Yurdakoc KF (2008) The mechanism of epoxide reactions. Part VII. The reactions of 1,2-epoxybutane, 3,4-epoxybut-1-ene, 1,2-epoxy-3-chloropropane, and 1,2-epoxy-3-methoxypropane with chloride ion in water under neutral and acidic conditions. Microporous Mesoporous Mater 111: 211

17. Mantin I (1969) Mesure de la capacité d'échange cationique des minéraux argileux par l'éthylène diamine et les ions complexes de l'éthylène diamine. CR Sci Paris 269: 815-818.

18. Khenifi A, Zohra B, Kahina B, Houari H, Zoubir D (2009) Removal of 2, 4-DCP from wastewater by CTAB/bentonite using one-step and two-step methods: a comparative study. Chem Eng J 146: 345-354.

19. Özcan A, Ömeroğlu Ç, Erdoğan Y, Özcan AS (2007) Modification of bentonite with a cationic surfactant: an adsorption study of textile dye Reactive Blue 19. J Hazard Mater 140: 173-179.

20. Wibowo N, Setyadhi L, Wibowo D, Setiawan J, Ismadji S (2007) Adsorption of benzene and toluene from aqueous solutions onto activated carbon and its acid and heat treated forms: influence of surface chemistry on adsorption. J Hazard Mater 146: 237-242.

21. Langmuir L (1918) 1,1'-Diaminobicyclohexyl and the stability of its metal complexes. Journal of American Chemical Society 40: 1361.

22. Freundlich H (1926) Heterocyclic chemistry. Part II. Nuclear magnetic resonance studies of purines and pteridines. Colloid and Capillary Chemistry, Metheum, London.

23. Mc Kay G (1996) Use of Adsorbents for the Removal of Pollutants From Wastewaters. CRC Press, USA.

24. Wibowo N, Setyadhi L, Wibowo D, Setiawan J, Ismadji S (2007) Adsorption of benzene and toluene from aqueous solution onto activated carbon and its acid heat treated forms: Influence of surface chemistry on adsorption. J Hazard Mater 146: 237-242.

25. Chatterjee S, Lee DS, Lee MW, Woo SH (2009) Congo red adsorption from aqueous solutions by using chitosan hydrogel beads impregnated with nonionic or anionic surfactant. Bioresour Technol 100: 3862-3868.

26. Chatterjee S, Lee DS, Lee MW, Woo SH (2009b) Enhanced adsorption of cationic dye from aqueous solutions by chitosan hydrogel beads impregnated with cetyl trimethyl ammonium bromide. Bioresour Technol 100: 2803-2809.

27. Taffarel SR, Rubio J (2010) Adsorption of sodium dodecyl benzene sulfonate from aqueous solution using a modified natural zeolite with CTAB. Miner Eng 23: 771-779.

28. Xia C, Jing Y, Jia Y, Yue D, Ma J, Yin X (2011) Adsorption properties of Congo red from aqueous solution on modified hectorite: kinetic and thermodynamic studies. Desalination 265: 81-87.

29. Pouya ES, Abolghasemi H, Assar M, Hashemi SJ, Salehpour A, et al. (2015) Theoretical and experimental studies of benzoic acid batch adsorption dynamics using vermiculite-based adsorbent. Chem Eng Res Des 93: 800-811.

30. Ghadiri SK, Mahvi RNAH, Nasseri S, Kazemian H, Mesdaghinia AR, et al. (2010) Methyl tert-butyl ether adsorption on surfactant modified natural zeolites. Iran J Environ Health Sci Eng 7: 241-252.

31. Seifi L, Torabian A, Kazemian H, Bidhendi GN, Azimi AA, et al. (2011) Adsorption of petroleum monoaromatics from aqueous solutions using granulated surface modified natural nanozeolites: systematic study of equilibrium isotherms. Water, Air, Soil Pollut 217: 611-625.

32. Torabian A, Kazemian H, Seifi L, Bidhendi GN, Azimi AA, et al. (2010) Removal of Petroleum Aromatic Hydroc arbons by Surfactant-modified Natural Zeolite: The Effect of Surfactant. Clean–Soil, Air, Water 38: 77-83.

33. Guan H, Bestland E, Zhu C, Zhu H, Albertsdottir D, et al. (2010) Variation in performance of surfactant loading and resulting nitrate removal among four selected natural zeolites. J Hazard Mater 183: 616-621.

34. Lin J, Zhan Y, Zhu Z, Xing Y (2011) Adsorption of tannic acid from aqueous solution onto surfactant-modified zeolite. J Hazard Mater 193: 102-111.

35. Foroughi-Dahr M, Abolghasemi H, Esmaili M, Shojamoradi A, Fatoorehchi H (2015) Adsorption characteristics of Congo red from aqueous solution onto tea waste. Chem Eng Commun 202: 181-193.

36. Ho YS (2004) Selection of optimum sorption isotherm. Carbon 42: 2115-2116.

37. Kumar KV, Porkodi K (2006) Relation between some two- and three-parameter isotherm models for the sorption of methylene blue onto lemon peel. J Hazard Mater 138: 633-635.

38. Kumar KV, Porkodi K, Rocha F (2008) Comparison of various error functions in predicting the optimum isotherm by linear and non-linear regression analysis for the sorption of basic red 9 by activated carbon. J Hazard Mater 150: 158-165.

39. Giles CH, MacEwan T, Nakhwa S, Smith D (1960) 786 Studies in adsorption. Part XI. A system of classification of solution adsorption isotherms, and its use in diagnosis of adsorption mechanisms and in measurement of specific surface areas of solids. J Chem Soc 1960: 3973-3993.

40. Gładysz-Płaska A, Majdan M, Pikus S, Sternik D (2012) Simultaneous adsorption of chromium (VI) and phenol on natural red clay modified by HDTMA. Chem Eng J 179: 140-150.

41. Slade PG, Gates WP (2004) The swelling of HDTMA smectites as influenced by their preparation and layer charges. Appl Clay Sci 25: 93-101.

25

The Benefit of 3D Printing in Medical Field: Example Frontal Defect Reconstruction

Singare S*, Shenggui C and Nan Li

School of Mechanical Engineering, Dongguan University of Technology, Dongguan, Dongguan 523808, China

Abstract

This study describes a methodology to design a custom-made cranial prosthesis for a patient who suffered injuries from road traffic accident. Computer based cranial defect reconstruction techniques is developed. The design approach was based on the 3D reconstruction of the skull of the patient, obtained by a CT scan. Then a reverse engineering (RE) method is used to reconstruct the defect prosthesis computer-aided design (CAD) model. Once the prosthesis CAD design was completed, the 3D models the skull and the prosthesis were transported into Rapid Prototyping (RP) machine to fabricate the physical model. Finally, the RP model is directly used to produce the biomaterial calcium phosphate cement (CPC) prosthesis. The prosthesis was successfully implanted and a satisfactory result was obtained by using this design method.

Keywords: Tomography; Designed prosthesis; Computer-aided reconstruction; 3D printing

Introduction

The medical imaging such as Computerized Tomography (CT) is an important tool to diagnose the defect and advances in computer software algorithms has allowed the 3D reconstruction of anatomical structures for several medical applications, including the design of custom-made prosthesis. Several studies have reported the use of CAD and advance manufacturing platforms such as computer aided manufacturing/computer numerical control (CAM/CNC), 3D printing and Rapid Tooling (RT) in the production of customized prosthesis and surgical resection template [1-16].

This paper presents a clinical cases study of frontal reconstruction using CT/RE/3D printing, with skull template to design the prosthesis geometry. The results demonstrate that the use of 3D printing to produce the custom made prostheses reduces the possibility of errors during surgery, and perfect fit of the prosthesis was obtained, as result the surgical time was reduced.

Case Study: Cranial Defect

3D image reconstructions

A patient with frontal injury from traffic accident was admitted to hospital for defect reconstruction. It was decided to use calcium phosphate cement (CPC) as the cranioplasty prosthesis through a rapid prototyping stereo lithographic technique. A CT scan was performed using standard craniofacial CT Scanning Protocol; the CT raw data in the form of DICOM files was transferred into Mimics software to convert a set of 2D CT images into a 3D volumetric image, at this time, the craniofacial osseous structures and the defect area were clearly demonstrated (Figure 1). After the 3D volumetric reconstruction using medical imaging software, a STL file of the entire skull was generated and exported into stereo lithography machine to produce a life-size physical skull model (Figure 2a), with this stereo lithography skull model, the cranial defect was clearly shown and evaluated. For prosthesis geometry modeling, a point cloud data of the 3D volumetric image was generated and transferred into Geomagics Studio 6.0 (Raindrop Geomagic, Inc., Research Triangle Park, NC) to design the prosthesis CAD model (Figure 2b).

Image based prosthesis design

Once a 3D reconstruction of the skull was obtained, a point cloud data of the skull 3D volumetric image is imported into reverse engineering software (Geomagic Studio 6-Raindrop Geomagic, Inc., Research Triangle Park, NC) to design the prosthesis CAD model. In reverse engineering environment, the points cloud data are denoised and wrapped as polygonal surfaces as shown in Figure 2c.

Using the information acquired from imaging diagnostics and 3D printing skull models, the extent of the necessary resection, including a margin of safety, is determined before surgery, and the implant custom made prosthesis is designed to cover the resulting bone after defect resection This is done as follows:

First, the approximate area of defect is identified on the 3D volumetric image in reverse engineering environment, then cut out to remove the defect area feature (image data). Next, the design of prosthesis geometry for this case is based on another intact cranial data. A sound Individuals skull use as reference skull template is chosen from the CT database, then the reference skull template 3D image is positioned such that it matches the orientation of the target skull (patient skull), then superimposed on the target skull image. Next, the individual sound skull 3D image data is scaled to better fit with the target skull (patient skull), when the reference template is well fitted to the target skull, the shape of the reference template that match the defect area is used to cover the defect on the target skull as well as to build the prosthesis for final craniofacial reconstruction.

All the data surrounding defect are removed from the reference skull template image leaving only the defect area feature which will be

---

*Corresponding author: Sekou Singare, Associate Professor, College of Mechanical Engineering, Dongguan University of Technology, Dongguan, Dongguan 523808, China, E-mail: sekou2d@yahoo.com

used to generate the prosthesis geometry, and the part of the prosthesis which is used to close the defect area is derived from the reference skull template. Because the precise and individual fit results from determining the implant margins by the borders of the defect, the prosthesis margins that contact the defect surrounding bone is derived from the patient skull nonaffected neighboring contours. Thus, the defect area surface and margin area surfaces of the prosthesis are connected, and a three-dimensional geometry results (Figure 2d). The prosthesis was fabricated in a Rapid Prototyping (RP) machine using stereolithography. Finally, the prosthesis SLA pattern is directly used to produce the biomaterial calcium phosphate cement (CPC) prosthesis (Figure 2e).

The patient underwent frontal bone resection and reconstruction using the customized biomaterial CPC prosthesis. The customized CPC prosthesis for frontal bone defect repair was then successfully implanted into the bone defect area at the correct position during surgery, and the surgery time was significantly reduced by using the 3D printing technique in the fabrication of the prosthesis and surgical template. Figure 3 show the patient photos before and after surgery operation.

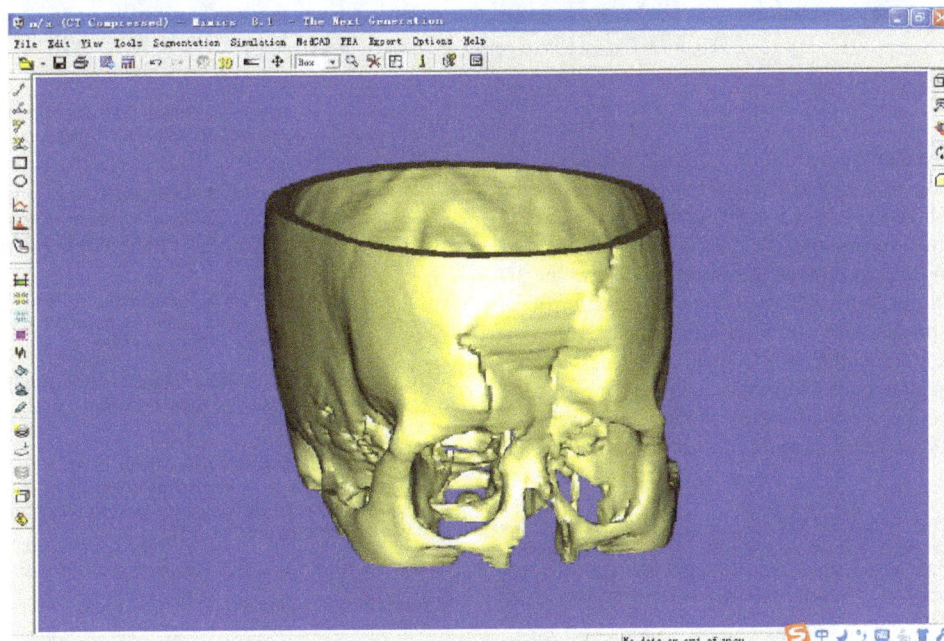

Figure 1: 3D reconstruction of the skull from DICOM data.

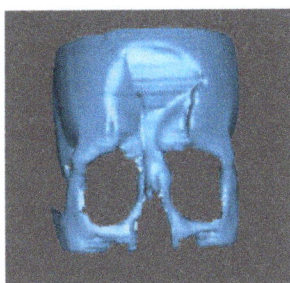

a) 3D volumetric image transferred into Geomagics software

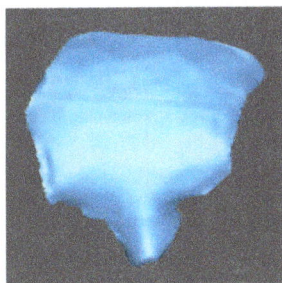

b) CAD prosthesis

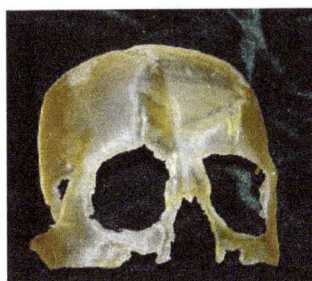

c) SLA skull model

d) SLA prosthesis model

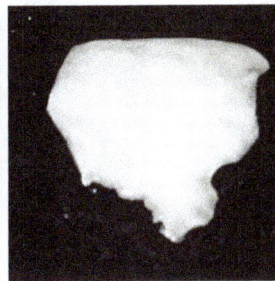

e) CPC prosthesis

Figure 2: Customized prosthesis modelling.

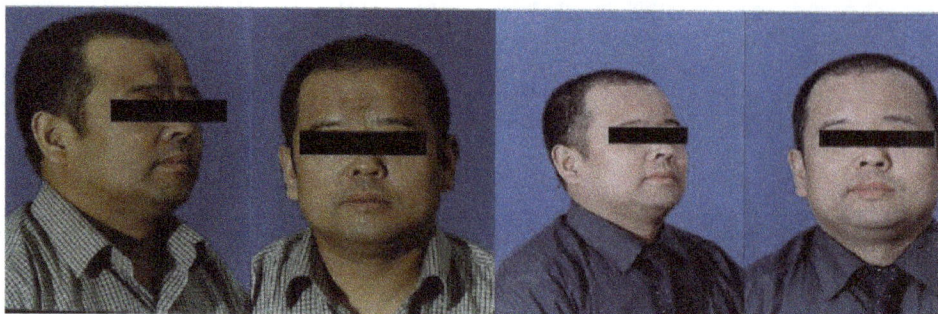

Figure 3: Patient photos before and after surgery operation.

Conclusion

This paper presents a clinical cases study of frontal reconstruction using CT/RE/3D printing, with a reference skull template to design the prosthesis geometry. Three-dimensional reformatted images and 3D printing were used in the evaluation of the defects, custom prosthesis design, surgery planning and reconstruction of cranial defect. The 3D printing skull model of the patient has allowed a clear visualization of the defect area and enable to better assess the localization of bone resection contour. Moreover, the combination of 3-D imaging, physical models and reference skull template have allows the design and production of precise fit prosthesis; the operation time was reduced as well as a satisfactory result was obtained by using this design method.

Acknowledgement

This project is supported by National Natural Science Foundation of China (Grant No. 51445008), Science and Technology Planning Project of Guangdong Province (Project No. 2013B090500130, 2015A010101305, 2013B090600047), Project supported by Guangdong Provincial Key Laboratory construction project of China (2011A060901026).

References

1. Morrison DA, Guy DT, Day RE, Lee GY (2011) Simultaneous repair of two large cranial defects using rapid prototyping and custom computer-designed titanium plates: a case report. Proc Inst Mech Eng H 225: 1108-1112.

2. Lee SC, Wu CT, Lee ST, Chen PJ (2009) Cranioplasty using polymethyl methacrylate prostheses. Journal of clinical neuroscience 16: 56-63.

3. Singare S, Dichen L, Bingheng L, Zhenyu G, Yaxiong L (2005) Customized design and manufacturing of chin implant based on rapid prototyping. Rapid Prototyping Journal 11: 113-118.

4. Singare S, Lian Q, Ping Wang W, Wang J, Liu Y, et al. (2009) Rapid prototyping assisted surgery planning and custom implant design. Rapid Prototyping Journal 15: 19-23.

5. Winder J, Bibb R (2005) Medical rapid prototyping technologies: state of the art and current limitations for application in oral and maxillofacial surgery. Journal of oral and maxillofacial surgery 63: 1006-1015.

6. Müller A, Krishnan KG, Uhl E, Mast G (2003) The application of rapid prototyping techniques in cranial reconstruction and preoperative planning in neurosurgery. Journal of Craniofacial Surgery 14: 899-914.

7. Eppley BL, Sadove AM (1998) Computer-generated patient models for reconstruction of cranial and facial deformities. Journal of Craniofacial Surgery 9003A: 548-556.

8. Zhou LB, Shang HT, He LS, Bo B, Liu GC, et al. (2010) Accurate reconstruction of discontinuous mandible using a reverse engineering/computer-aided design/rapid prototyping technique: A preliminary clinical study. J Oral Maxillofac Surg 68: 2115-2121.

9. Rotaru H, Stan H, Florian IS, Schumacher R, Park YT, et al. (2012) Cranioplasty with custom-made implants: analyzing the cases of 10 patients. Journal of Oral and Maxillofacial Surgery 70: e169-e176.

10. Aakash Arora MDS, Datarkar AN, Borle RM (2013) Custom-made implant for maxillofacial defects using rapid prototype models. J Oral Maxillofac Surg 71: e104-e110.

11. Wang G, Li J, Khadka A, Hsu Y, Li W, et al. (2012) CAD/CAM and rapid prototyped titanium for reconstruction of ramus defect and condylar fracture caused by mandibular reduction. Oral surgery, oral medicine, oral pathology and oral radiology 113: 356-361.

12. Oral Medicine (2012) Oral Pathology and Oral Radiology 113: 356-361.

13. Mustafa SF, Evans PL, Bocca A, Patton DW, Sugar AW, et al. (2011) Customized titanium reconstruction of post-traumatic orbital wall defects: a review of 22 cases. International journal of oral and maxillofacial surgery 40: 1357-1362.

14. Taft RM, Kondor S, Grant GT (2011) Accuracy of rapid prototype models for head and neck reconstruction. The Journal of prosthetic dentistry 106: 399-408.

15. Markiewicz MR, Bell RB (2011) The use of 3D imaging tools in facial plastic surgery. Facial plastic surgery clinics of North America 19: 655-682.

16. Lai JB, Sittitavornwong S, Waite PD (2011) Computer-assisted designed and computer-assisted manufactured polyetheretherketone prosthesis for complex fronto-orbito-temporal defect. Journal of Oral and Maxillofacial Surgery 69: 1175-1180.

26

Size-Exclusion Chromatography and Its Optimization for Material Science

Netopilík M* and Trhlíková O

Institute of Macromolecular Chemistry, Academy of Sciences of the Czech Republic, Heyrovský Sq. 2, 162 06 Prague 6, Czech Republic

Abstract

A theoretical analysis of improving the separation power of size exclusion chromatography by decreasing flow-rate is conveyed. The variance of the elution curves is larger than expected form estimated dispersity in molecular weight, M_w/M_n. The shape of the elution curves depends strongly on experimental conditions. When the experimental conditions are approaching those to the ideal separation, judged by statistical properties of the curves, both skew and excess kurtosis of the elution curves increase near the exclusion limit in accord with theoretical prediction. In analyses of polystyrene standards by size exclusion chromatography for polymers of molecular weight about thirty thousand and higher, the longitudinal diffusion is not important even at low flow-rates.

Keywords: Separation mechanism; Size-exclusion chromatography; Band broadening function; Skew; Excess kurtosis

Introduction

The material science has ever been the driving force of the progress which has always advanced with the progress in analytical sciences. For polymer materials, the distribution of chain lengths and, therefore, molecular weight distribution (MWD) is of primary importance. MWD is characterized by its averages, M_n, M_w, M_z... and dispersities, M_w/M_n, M_z/M_w. Material properties depend on molecular weight of polymers in different ways and they are correlated with different averages. Thus M_n may be correlative with polymer colligative properties, e.g. freezing point depression, M_w with melt viscosity and M_z with toughness [1]. Several methods for the determination of the molecular weight averages were developed, e.g., membrane osmometry, cryoscopy and vapor pressure osmometry for the determination of M_n [2], static light scattering for the determination of M_w [2], sedimentation equilibrium for the determination of M_z [3] etc. However, all methods for the determination of individual molecular-weight averages have been dwarfed by size-exclusion chromatography (SEC) [4] which is a method widely used for the determination of MWD as well as of all averages and dispersities, M_w/M_n and M_z/M_w.

SEC with single concentration detector and results evaluated according to a calibration dependence, constructed from elution curves (maxima) of reference standards, frequently polystyrene or poly(methyl methacrylate), is a relative method and for polymers of different chemical composition the error may reach up to 200% [5]. Even if the analyte is chemically identical with the calibration reference standards, the correct value of M_w/M_n is not obtained because of the effect of the band-broadening [6]. This phenomenon follows from the principle of the method [7,8] and increases the value of M_w/M_n, influencing both M_w and M_n [6]. According to the theory, MWD of polystyrene prepared by anion polymerization (reference standards) is Poissonian type with dispersity decreasing with M [9]. This leads us to the conclusion that the dispersities of the reference standards given by the producers are systematically overestimated [8,10]. However, there is some evidence that efficiency of separation increases with decreasing flow rate, r_f [11,12]. and there is evidence that the values of M_w/M_n of the reference standards are much lower than commonly believed [8,10,13,14]. The knowledge of the dispersity of the analyzed samples is extremely important for our understanding of the SEC separation of polymers, because of the above discussed dependences of the material properties on MWD and its averages. The experimental elution curve is related with broadening function (BBF) and theoretical elution curve, characterized by the sample dispersity, M_w/M_n, by the Tung equation [4,6].

In the theory of chromatographic separation and band broadening, BBF is called the elution curve of an analyte uniform in molecular weight and chemical composition [15]. For its description, several mathematical models of separation were developed. The kinetic model by Giddings and Eyring assumes that the capture of molecules of the analyte (analyzed substance) is described by two kinetic constants of ingress and egress [15].

The equilibrium model is based on the concept of theoretical plate on which the equilibrium is formed between molecules of the analyte moving together with MP and those anchored on the surface by enthalpic attractive forces or penetrated into the pores by entropic process basically of Brownian diffusion into pores of the stationary phase (SP). The plate model was first proposed by Martin and Synge [16]. The spacial distribution of the analyte with respect to the longitudinal axis of the separation system, developing in time, was expressed by the binomial distribution. However, further treatments of this physical situation were approximative. The exact solution to the problem gives the probability that an experiment with given probabilities of success and failure is successful, i.e., the molecule is eluted from the column in a volume of a plate after a given number of failures [17].

The obtaining of the separation parameters and the description of elution curves by the Giddings-Eyring equation was shown for low-molecular weight substances separated in the adsorption mode by liquid chromatography (LC). This was possible because the LC separation is effective, the retention (capacity) factor [18] high and the baseline-separation of the sample-components is in many cases achieved. For polymers, separated by SEC such phenomena were not observed. In SEC compared to LC, the situation concerning BBF is much less favourable. The capacity factor [18] in SEC is low and the analyzed polymer samples are almost always disperse in molecular weight. The elution curve of such sample is then the convolution of BBF with theoretical elution curve corresponding with the sample MWD

*Corresponding author: Milos Netopilík, Institute of Macromolecular Chemistry, Academy of Sciences of the Czech Republic, Heyrovský Sq. 2, 162 06 Prague 6, Czech Republic, E-mail: netopilik@imc.cas.cz

[6]. On the other hand, the elution curves of narrow-MWD are much broader than expected from their MWD [19] and therefore we expect that their shape reflects properties of BBF rather than those of narrow-MWD sample MWD. It has been known soon that the concentration detector signal gives unacceptably broad MWD and the evidence suggests that MWD of polymer standards are much narrower than believed [13,14], which supports the idea of the peak-shape formation by the BBF mechanism. According to the theory, its skew and excess kurtosis (referred to sometimes as `excess' only [7,20]) increase and variance decreases near the exclusion limit of the separation system [7].

From the plate-height mechanism [7,16] it follows strictly that the polymer occurring in MP only is not separated at all and is eluted in the volume approaching that of one plate. In this case, the variance of its BBF approaches zero. For the separation of low-molecular weight substances, characterized by high capacity factor [18], the expectation of the decrease in variance has been verified [7]. With high-molecular wight polymers, the situation is more complicated. Just below the exclusion limit the capture of macromolecules in the pores causes hindered diffusion and the egress process is expected to be slow and the BBF variance to increase. This effect has been frequently observed in times when the injected concentration was usually independent of molecular weight [21,22] and it may have been the consequence of concentration effects as discussed [7]. With concentration decreasing with molecular weight the BBF variance does not change considerably [7]. Tung and Runyon [23] report increasing variance with decreasing elution volume. This is followed by a steep decrease in the vicinity of the exclusion limit. An ample decrease throughout the whole span of elution volume cannot be expected. Such effects do not follow any consistent pattern and can hardly form a part of a theory describing the separation. Polymers eluted far after the exclusion limit show almost entirely Gaussian peaks (and therefore BBF). This can be expected from the random character.

The polymers which elute near the exclusion limit in a given separation system are those of the highest molecular weight. It is therefore necessary to increase the separation power of the system. As recognized by Giddings, two kinetic processes, viz., tortuous and obstructed flow of eluent through and around the particles of packing and the flow profile in the mobile phase whose dispersive effect is checked by transversal diffusion, may combine to reduce and control band broadening. To avoid band broadening, the individual molecule has to sample the complete range of linear flow velocities in a random way as it moves along the column; this is achieved by transverse diffusion [24]. Compared to low-molecular weight substances studied by Giddings by the gas chromatography, the rate of diffusion of polymers in solutions separated by SEC is low [25]. Therefore, by decreasing the flow-rate, an increase in separation power of polymers elution close to the elution limit is expected just by letting time enough to molecules just to compensate for the low rate of transversal diffusion as well as to reach the equilibrium of the polymer between volumes inside and outside the pores of SP in accord with experimental observation [11,12].

The aim of this paper is the demonstration of the improvement of efficiency of separation in SEC aimed at the material research with decreasing flow-rate on the basis of expected statistical parameters of BBF, derived on theoretical model [26].

Materials and Methods

The SEC analyzes were performed using a Pump Deltachrom (Watrex Ltd.) with computer-controlled piston movement, autosampler MIDAS (Spark Ltd.), one column PL gel mixed D

(Polymer Laboratories), separating according to the producer in the range approximately $10^2 \leq M \leq 4$–5×10^5, particle size 5 µm. UV/VIS DeltaChrom UVD 200 detector (Watrex) with flow-cell volume of 8 µl, operating at wavelength λ=264 nm, the light-scattering photometer DAWN Heleos II, measuring at 18 angles of observation and Optilab T-rEX differential refractometer (both Wyatt Technology Corp.) were the detectors in the order of flow. The data were collected into the Astra 6.1 (Wyatt Technology Corp.) and Clarity (DataApex Ltd.) softwares, communicating with detectors using U-PAD2 USB acquisition device.

Results and Discussion

The dependence of a physical property P on the particular average of molecular weight M_x can often be described by

$$P=P_\infty(1+k/M_x)^{-1} \qquad (1)$$

where P_∞ is the value of the property at infinite molecular weight and k is a constant. As an example, Figure 1 shows the dependence of the melting point on the chain length expressed as the number of monomer units published by Flory and Vrij [27]. The dependence (Figure 1) can be described by eqn. (1) with $M_x=M_n$, P_∞=418.8 and k=7.33. Hence, the determination of the correct chain-length and therefore molecular weight is of highest importance in macromolecular science.

The improvement of the separation can be shown on elution curves for the standard M_p=4 × 10^5. With decreasing r_f the curves are becoming narrower and in the near the exclusion limit they become moreover non-symmetric (Figure 2). The Figure suggests that flow-rate r_f=0.5 ml·min^{-1} as a very good compromise between separation power and the time of analysis. On the other hand, in the region 0.01 ≤ r_f/ml·min^{-1} ≤ 0.035 the shape of the curves changes showing the expected increase in skew, indicating thus increasingly efficient separation. The change in variance of elution curves with r_f can be seen in the van Deemter plot [28]

$$\sigma^2 = a+b/r_f+cr_f \qquad (2)$$

where the constants a, b and c are associated, in the first approximation, with eddy diffusion (the contribution of MWD is in this case neglected), longitudinal diffusion and mass transfer [4], respectively, in logarithmic scale on the r_f axis (Figure 3). Only the curve for toluene rises continuously which is apparently due to the effect of longitudinal diffusion of low molecular weight substance, expressed by the term b/r_f in eqn. (2). For the standards of higher M minima appear at rf=0.02 ml min^{-1}.

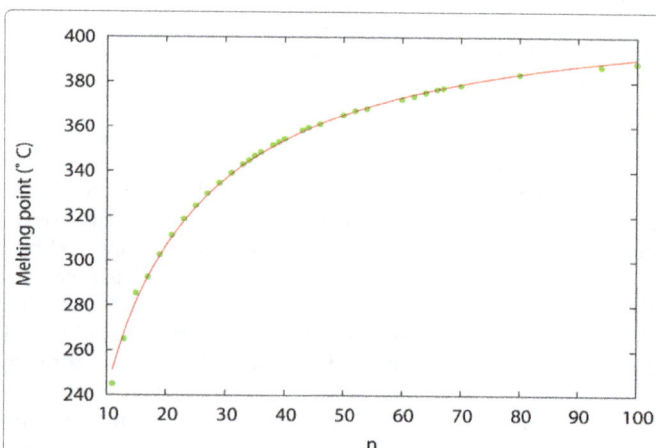

Figure 1: The dependence of the melting point temperature on the number of monomeric units in the molecule [27].

The decrease in broadness of the elution curves can be seen also in values of the apparent dispersity in molecular weight, M_w/M_n, obtained according to calibration dependence constructed at each r_f (Table 1). The values of M_w/M_n for the majority of them significantly lower than those given by the producer were obtained at very low r_f. This is in contrast with relatively high values obtained at widely used $r_f \approx 1$ ml·min^{-1} (Table 1).

The variability of the shape with the change of experimental conditions (r_f) and low dispersity justify the approximation of BBF for the studies of its statistical properties by elution curve. The values of the skew are highly scattered (Figure 4), but the lowest values are always those for the lowest flow-rate r_f=0.01 ml·min^{-1} and are close to the theoretical line [8] calculated for parameters of the standard of $M_p=4 \times 10^5$ (see below). The order of points from the left to the right is the same as the order of standards in Table 1. Near the elution limit, estimated as 4.64 ml, the curve as well as the experimental value rises abruptly together with the theoretical line and for the standard of $M_p=4 \times 10^5$ at r_f=0.01 ml·min^{-1} the value of σ^2 is the highest and is close to the rising part of the theoretical curve.

The large scatter of points at lower r_f suggests that operational conditions and changes of the character of the flow, characterized by irregular change in shape of elution curves with r_f with r_p, discussed Figure 2, rather than instrumental defects (e.g., voids in the separation system) are the cause of the skew.

The theoretical line in Figure 4 was calculated from parameters of the sample of $M_p=4 \times 10^5$, which is eluted close to the exclusion limit,

shows a curvature. The points at r_f=0.01 ml·min^{-1} are for polymers of $M_p < 4 \times 10^5$ the lowest and close to the theoretical dependence which was calculated for data of sample $M_p=4 \times 10^5$. However, for this sample the value of γ_s is the highest of all, in accord with the theoretical curve. The points for the standard of $M_p=9 \times 10^5$, which are in front of V_0, are scattered according no rule, indicating imperfect separation, the point at r_f=0.01 ml·min^{-1} being just in the middle.

The excess kurtosis, γ_E calculated shows also a strong increase of values obtained at the lowest flow-rate, r_f=0.01 ml·min^{-1}, near the exclusion limit (Figure 5). Theoretical curve obtained by the same procedure [8] as that of γ_s, shows again a strong increase near the exclusion limit and the curve calculated for the standard of $M_p=4 \times 10^5$ fits the experimental data for this standard best. The points at r_f=0.01 ml·min^{-1} for polymers of $M_p < 4 \times 10^5$ are the lowest and close to the theoretical dependence. For data of sample $M_p=4 \times 10^5$ the value of γ_E is the highest of all in accord with the theoretical curve. The points for the standard of $M_p=9 \times 10^5$ in front of V° do not also obey any rule.

Conclusions

Due to the transversal diffusion and increases the time necessary for a molecule to sample the complete range of flow velocities which is a condition for a good separation. This is achieved in the region of high M at extreme low r_f in accord with the idea of transversal diffusion as a factor decreasing band broadening, as recognized by Giddings [24].

For polymers with approximately $M \geq 3 \times 10^4$ the longitudinal

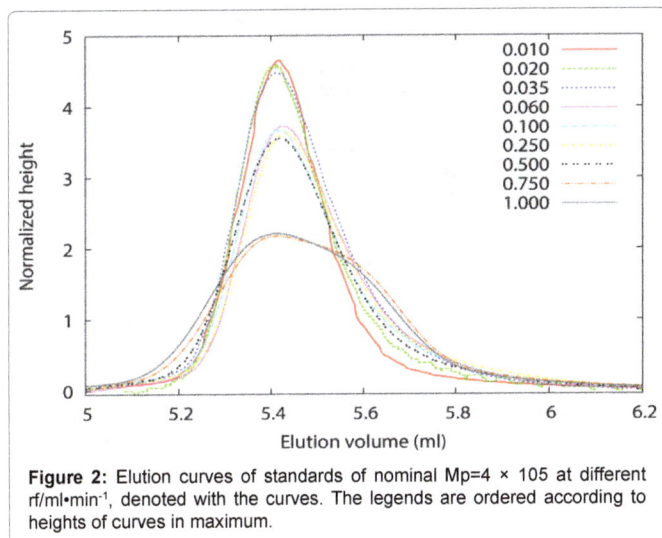

Figure 2: Elution curves of standards of nominal Mp=4 × 105 at different rf/ml·min^{-1}, denoted with the curves. The legends are ordered according to heights of curves in maximum.

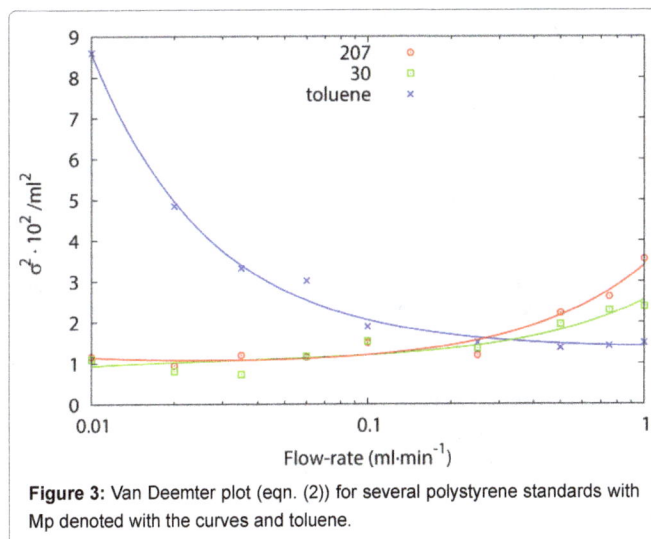

Figure 3: Van Deemter plot (eqn. (2)) for several polystyrene standards with Mp denoted with the curves and toluene.

$M \times 10^{-3}$			M_w/M_n				
				r_f/ml min^{-1}			SEC with LS
Producer	LS	Producer	0.01	0.5	1	
900	1084	1.1	1.027	1.087	1.196	1.026
400	381	1.06	1.029	1.092	1.019	1.028
207	206	1.05	1.033	1.066	1.068	1.021
90	84	1.04	1.024	1.068	1.064	1.014
30	31	1.06	1.023	1.072	1.071	1.012
4	4.5	1.06	1.106	1.113	1.116	1.026
Toluene						

Table 1: The molecular weights M of polystyrene standards given by the producer and obtained by the on-line light scattering (LS), dispersity M_w/M_n given by the producer and obtained by the calibration dependence constructed at a given flow-rate F and obtained from the dual light-scattering concentration detection corrected for band broadening ('SEC with LS).

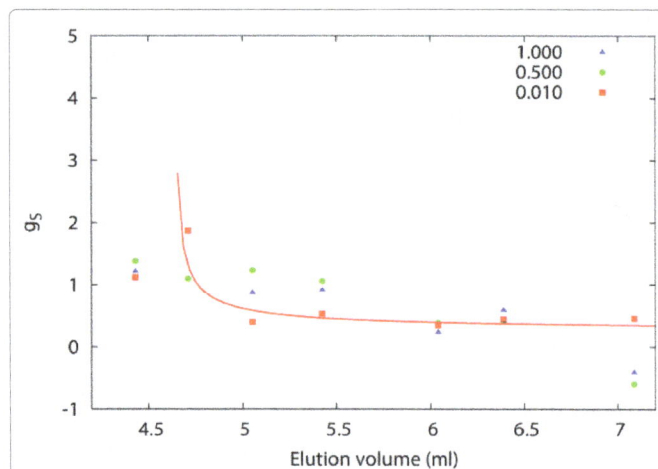

Figure 4: Skew of elution curves calculated as a function of elution volume V at rf=0.2 ml.min⁻¹. The flow-rate, rf/ml•min⁻¹, denoted with the points and M_p in thousands with curves. The order of analytes from left to right is the same as in Table 1 from top to bottom.

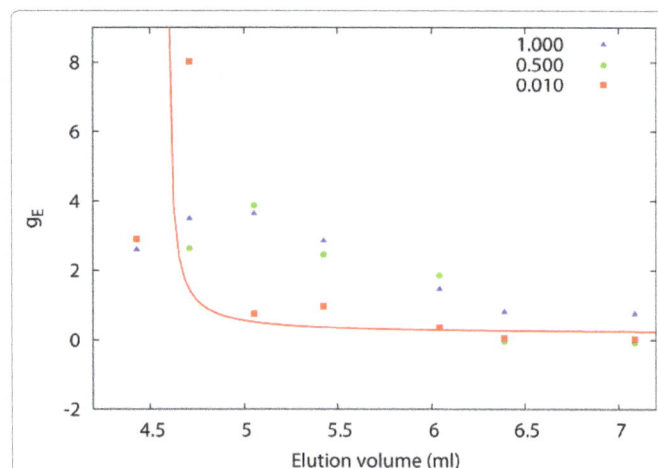

Figure 5: Excess kurtosis of elution curves as a function of elution volume. For the details of the points and lines calculated see legend to Figure 4 and text.

diffusion ceases to play an important role in the SEC separation process. With decreasing r_f the efficiency of separation increases and the separation mechanism appears, using γ_s and γ_E as a criterion, to approach the mechanism of equilibrium on theoretical plate combined with longitudinal shift, proposed by Martin and Synge [16] and solved exactly by Netopilík [26].

For polystyrene in THF for $9 \times 10^5 \leq M \leq 4 \times 10^3$, the flow-rate r_f=0.5 ml·min⁻¹ appears, rather than usual r_f=1 ml·min⁻¹, a good compromise between efficiency of separation and the time of analysis. Further decrease in r_f has no observable effect down to $r_f \approx 0.06$ ml·min⁻¹, then a new region of increase in separation power is for some standards of higher M observed.

Acknowledgements

The financial support of the Grant Agency of the Czech Republic (project 17-04258J) is gratefully acknowledged. The research was also supported by European Research Development Fund, project OPPK CZ.2.16/3.1.00/24504 and Egyptian Academy of Sci. & Technology for funding the joint mobility project 2016/2018 (ASRT-16-2).

References

1. Carraher CE (2003) Seymour/Carraher's Polymer Chemistry: Sixth Edition, MarcelDekker.
2. Flory PJ (1953) Principles of polymer chemistry, Cornell University Press, Ithaca, New York.
3. Tanford C (1965) Physical Chemistry of Macromolecules. (3rdedn), New York: Interscience.
4. Yau WW, Kirkland JJ, Bly DD (1979) Modern size-exclusion chromatography, J Wiley & Sons, Inc.
5. Netopilík M, Kratochvíl P (2003) Polystyrene equivalent molecular weight versus true molecular weight in size-exclusion chromatography. Polymer 44: 3431-3436.
6. Tung LH (1966) Method of calculating molecular weight distribution function from gel permeation chromatogram. J Appl Polym Sci 10: 375-385.
7. Netopilík M (2006) Statistical properties of the band-broadening function. J Chromatogr A 1133: 95-103.
8. Netopilík M (2017) Towards ideal separation by size-exclusion chromatography. J Chromatogr A 1487: 139-146.
9. Flory PJ (1940) Molecular size distribution in ethylene oxide polymers. J Am Chem Soc 62: 1501-1505.
10. Netopilík M, Podzimek S, Kratochvíl P (2001) Estimation of width of narrow molecular-weight distributions by size-exclusion chromatography with concentration and light scattering detectors. J Chromatogr A 922: 25-36.
11. Hong P, Koza S, Bouvier ESP (2012) Size-exclusion chromatography for the analysis of protein biotherapeutics and their aggregates. J Liquid Chromatogr Relat Tech 35: 2923-2950.
12. Boborodea A, O'Donohue S (2016) Low solvent consumption gel permeation chromatography method. International Journal of Polymer Analysis and Characterization 21: 657-662.
13. Vander Heyden Y, Popovici ST, Staal BBP, Schoenmakers PJ (2003) Contribution of the polymer standards' polydispersity to the observed band broadening in size-exclusion chromatography. Journal of Chromatography A 986: 1-15.
14. Shortt DW (1993) Differential molecular weight distributions in high performance size exclusion chromatography. Journal of Liquid Chromatography & Related Technologies 16: 3371-3391.
15. Giddings JC, Byring H (1955) A molecular dynamic theory of chromatography. The Journal of Physical Chemistry 59: 416-421.
16. Martin AJP, Synge RM (1941) A new form of chromatogram employing two liquid phases: A theory of chromatography. 2. Application to the micro-determination of the higher monoamino-acids in proteins. Biochemical Journal 35: 1358.
17. Spiegel MR (1992) Theory and Problems of Probability and Statistics. New York: McGraw-Hill.
18. IUPAC (1997) Compendium of Chemical Terminology, (2ndedn), (the "Gold Book"). Online corrected version: (2006) "Retention factor, k in column chromatography"
19. Netopilík M, Podzimek Š, Kratochvíl P (2004) Determination of the interdetector volume by s-detection in size-exclusion chromatography of polymers with on-line multiangle light-scattering detection. Journal of Chromatography A 1045: 37-41.
20. Foley JP, Dorsey JG (1983) Equations for calculation of chromatographic figures of merit for ideal and skewed peaks. Analytical Chemistry 55: 730-737.
21. Balke ST, Hamielec AE (1969) Polymer reactors and molecular weight distribution. VIII. A method of interpreting skewed GPC chromatograms. Journal of Applied Polymer Science 13: 1381-1419.
22. Vozka S, Kubín M, Samay G (1980) Calibration of spreading function in GPC using statistical moments. Journal of Polymer Science: Polymer Symposia 68: 199-208.
23. Tung LH, Runyon JR (1969) Calibration of instrumental spreading for GPC. Journal of Applied Polymer Science 13: 2397-2409.
24. Giddings JC (1959) Eddy diffusion in chromatography. Nature 184: 357-358.

25. McDonnell ME, Jamieson AM (1977) Quasielastic light-scattering measurements of diffusion coefficients in polystyrene solutions. Journal of Macromolecular Science, Part B: Physics 13: 67-88.

26. Netopilík M (2002) Relations between the separation coefficient, longitudinal displacement and peak broadening in size exclusion chromatography of macromolecules. Journal of Chromatography A 978: 109-117.

27. Flory PJ, Vrij A (1963) Melting points of linear-chain homologs. The normal paraffin hydrocarbons. Journal of the American Chemical Society 85: 3548-3553.

28. Van Deemter JJ, Zuiderweg FJ, Klinkenberg AV (1956) Longitudinal diffusion and resistance to mass transfer as causes of nonideality in chromatography. Chemical Engineering Science 5: 271-289.

Removal of Chromium from Industrial Wastewater by Adsorption Using Coffee Husk

Dessalew Berihun*

Department of Urban Environmental Management, Kotebe Metropolitan University, Addis Ababa, Ethiopia

Abstract

Fresh water is vital to human life and economic well-being, and societies extract vast quantities of water from rivers, lakes, wetlands, and underground aquifers but most of these freshwater sources are polluted by different chemicals discharged from industries. Our need for fresh water has long caused us to overlook equally vital benefits of water that remains in streams to sustain healthy freshwater habitats.

Heavy metals are discharged from different industries into freshwaters and are easily absorbed by fish and other aquatic organisms. Small concentrations can be toxic because heavy metals undergo bio concentration. Chromium is an essential element that is required in small amounts for carbohydrate metabolism, but becomes toxic at higher concentrations. The most bioavailable and therefore most toxic form of chromium is the hexavalent Cr (VI) ion. It is well recognized as an element of environmental and public health concern. The objective of this study was to examine the potential of coffee husk in removing chromium from polluted water.

In this study, the adsorption potential of activated carbon for the removal of Cr (VI) ions from industrial wastewater has been investigated. The adsorption of hexavalent chromium from aqueous solution by coffee husk activated carbon prepared by chemical method and its application to real wastewater was studied. The extent of adsorption was studied as a function of pH, contact time, adsorbent dose, and initial adsorbed concentration. Optimum results were found to be 60 min, 80 mg/l, 2 g/l, 3 g/l and 200 rpm for time contact, initial concentration, pH, adsorbent dose and stirring speed respectively at the optimal condition the adsorption of hexavalent chromium was found to be 98.19%.

Keywords: Activated carbon; Adsorbent; Adsorption isotherms; Adsorption kinetics; Coffee husk; Hexavalent chromium

Introduction

Fresh water is vital to human life and economic well-being, and societies extract vast quantities of water from rivers, lakes, wetlands, and underground aquifers but most of these freshwater sources are polluted by different chemicals discharged from industries. Our need for fresh water has long caused us to overlook equally vital benefits of water that remains in streams to sustain healthy freshwater habitats.

Advances in science and technology have brought remarkable improvement in many areas of development, but in the process, also contributed to degradation of environment all over the globe which increased demand for new technologies for proper treatment facilities before discharge to the environment [1,2]. Discharging of heavy metals which are major pollutants in marine, ground and surface waters into the environment by human activity has enormously increased since industrialization there by impacting geochemical cycling and food chain [3-5]. Chromium is a transition metal which occurs in nine different forms of oxidation states ranging from Cr (-II) up to Cr (+VI), but the two common valence states are trivalent and hexavalent chromium forms [6]; however, concerns regarding the presence of Chromium in the environment focus on the potential adverse health effects of Cr (VI)-contaminated soils, groundwater, and drinking water supplies. Hexavalent chromium is more hazardous, carcinogenic, and mutagenic and the most water soluble which easily enters to living cells [7]. The contamination of environment by Chromium is a critical problem because of adverse effects on aquatic life and human health [8-11].

Hexavalent chromium is well thought out to be a group "A" human carcinogen because of its mutagenic and carcinogenic properties .It is included in the priority list of hazardous substances since it affects both; human and aquatic life. It has been also reported that excessive intake of hexavalent chromium by plants severely affects the mitotic process and reduce seed germination in extensively cultivated pulse crops (Altaf et al., 2008, Mina et al., 2011).

Majority of industries use chromium compounds in attempt to improve human living standards, but the discharge of those chemicals into the environment without proper treatment reverse the intended living standard. Leather Tanning industries are ranked as the highest contributors of chromium pollution.

Materials and Methods

Preparation of the adsorbent

The coffee husk was used for activated carbon preparation due to the large amount generated and burned as a waste. This husk was obtained from coffee processing unit at, Jimma zone, Gomma 2 Limmu coffee farm. The coffee husk was washed with tap water then it was sun dried for 2 days. The dried husk was crushed and ground to a sieve size of 1,000 μm. Soon after, it was impregnated with 40% H_3PO_4 for 2 hours. Then after, the impregnated powder was soaked with the same solution overnight followed by decantation through 212 μm sieve, air drying for 12 hours at room temperature and activation overnight at 105°C in a

***Corresponding author:** Dessalew Berihun, Faculty of Urban Development Studies, Department of Urban Environmental Management, Kotebe Metropolitan University, Addis Ababa, Ethiopia, E-mail: dessalewb@gmail.com

tabular oven. Then the activated coffee husk powder was subjected to carbonization at 500°C with a constant heating rate of 5°C/min in a tubular furnace for 1 hour. After completing the carbonization process it was cooled overnight in a tubular furnace. The produced activated coffee husk carbon was washed with distilled water and neutralized by NaOH. Then after, it was dried overnight at 105°C, ground and sieved with 150 μm sieve size (Figure 1).

Preparation of adsorbate solution

The stock solution of Cr (VI) containing concentration of 1000 mg/l was prepared by dissolving 2.829 g of potassium dichromate, $K_2Cr_2O_7$, analytical grade, in 1000 ml of deionized water. The stock solution was further diluted with distilled water to desired concentration of test solution. The required pH of the solution was adjusted by drop wise addition of 0.1 N HCl and NaOH depending on the acidity or basicity of the sample.

Applicability to industrial wastewater

The adoptability of the technique enlarged with the activated coffee husk carbon for chromium removal was undertaken with some actual effluent samples. Chrome tanning effluent was collected from Batu Tannery PLC, which is a private leather tanning industry in Addis Ababa, at discharge point. The chrome tanning liquor had pH of 3.85 and 1220.2 mg/l chromium concentration which was measured before application with adsorbent. In order to study the efficiency of the activated coffee husk carbon for the actual sample, the effluent was digested with concentrated Nitric acid and then filtered through Whatmann No.41 filter paper. The resulting solution was dissolved at different time intervals, pH values including initial pH of the wastewater and at optimum conditions of activated coffee husk carbon dose, stirring speed and with actual adsorbate concentration. Finally, the solutions were subjected to atomic absorption spectroscopy for extraction of total chromium as Cr (VI).

Experimental procedures for digestion

The sample digestion and extraction procedures necessary to give values for dissolved (free) Cr (VI) or total chromium as Cr(VI).The chrome waste of 100 ml was transferred to a beaker and 5 ml HNO_3 (conc.) was added with a few boiling chips. Then this was subjected to a slow boiling and evaporated in a hood to the lowest possible volume of 20 ml before precipitation occurs. Since digestion is not complete after final volume of 20 ml, 5 ml of HNO_3 (conc.) was added until a clear solution was observed. This was then cooled and filtered through 100 ml cylinder using Whatmann No.41 filter paper. To oxidize Cr (III), a portion of digested filtrate sample was taken into 125 ml conical flask by adding several drops of methyl orange indicator, then NH_4OH (conc.) was added until solution just begins to turn yellow. Then after, 1+1 H_2SO_4 was added drop wise until it becomes acidic and the volume was adjusted to 40 ml followed by adding boiling chips and subjected to heat for boiling. When two drops of $KMnO_4$ solution was added, the solution became faded then an excess of two drops of $KMnO_4$ solution was added and this gave a dark blue color. This was boiled for 5 min, and then 1 ml sodium acid solution was added and boiled for a minute gently. Since red color didn't fade completely after boiling for a minute, another 1 ml sodium acid solution was added and continued boiling for 1 min until color had faded completely and allowed to cool. The final solution was subjected to atomic absorption spectroscopy for total chromium concentration as Cr (VI) ions (Figure 2).

Instrumentation

The determination of amount of chromium in the effluent solutions before and after adsorption takes place was done using Flame Atomic

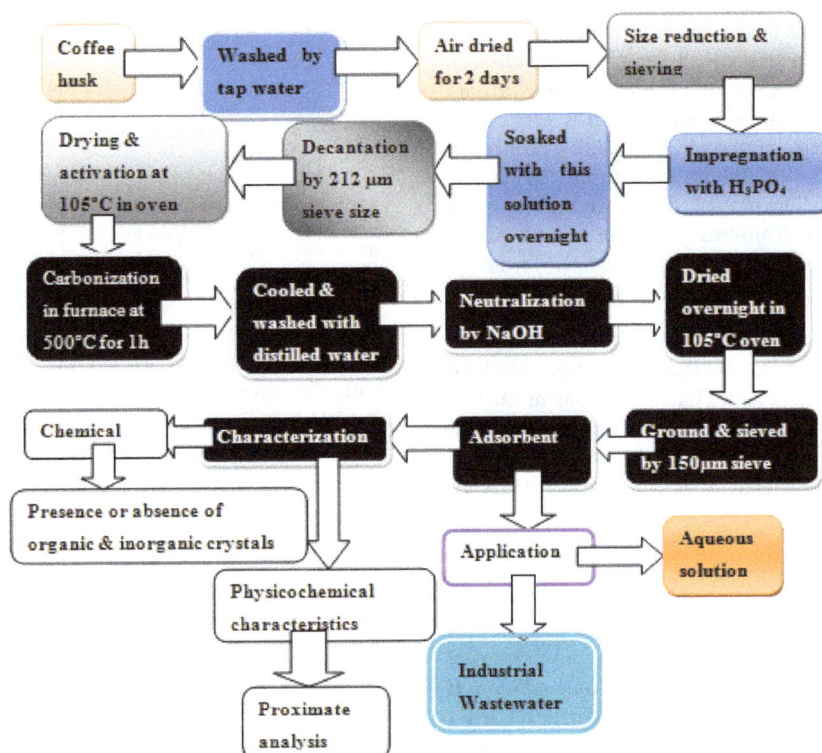

Figure 1: Flow chart for preparation of adsorbent from coffee husk.

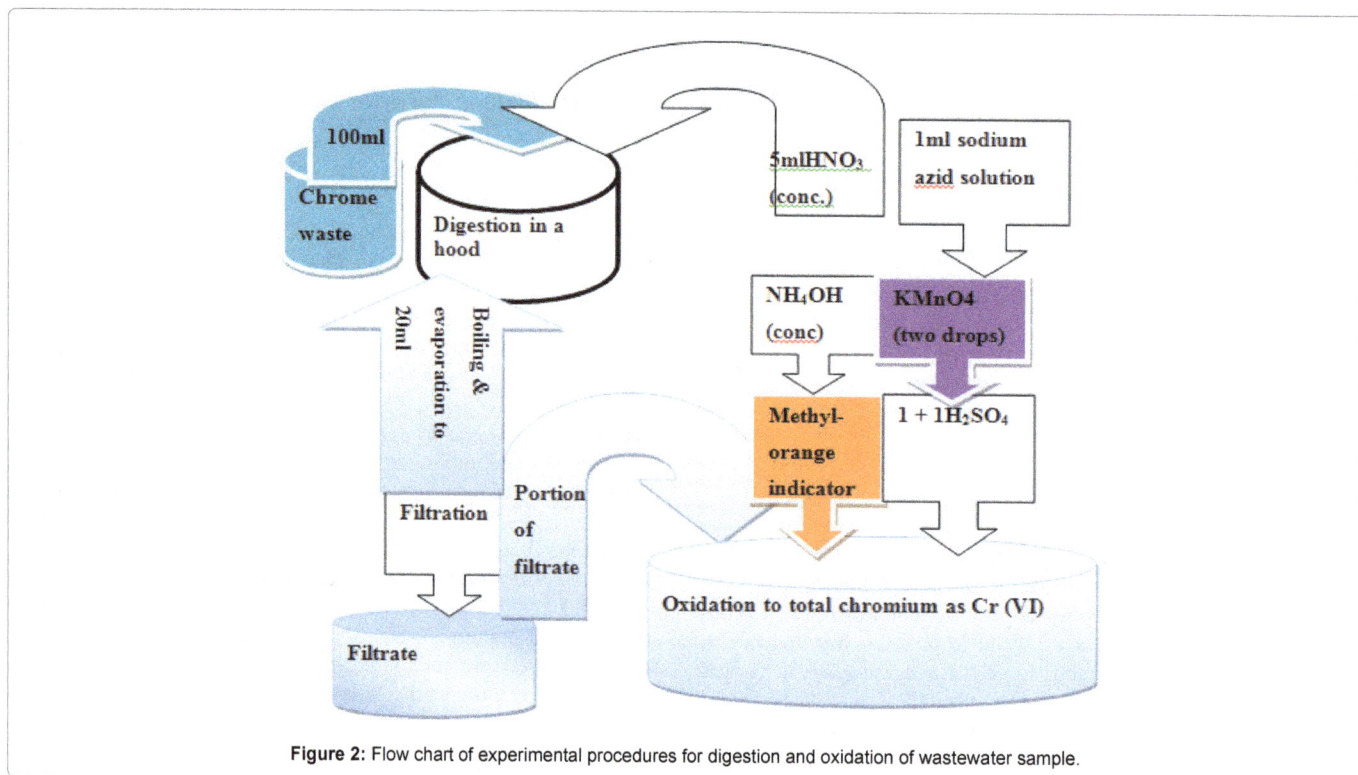

Figure 2: Flow chart of experimental procedures for digestion and oxidation of wastewater sample.

Absorption Spectroscopy (FAAS) BUCK SCIENTIFIC MODEL 210 VGP, East Norwalk, USA) equipped with deuterium ark background corrector, nebulizer and hallow cathode lamp corresponding to metal of interest, in this case chromium, using air-acetylene flame. The operating conditions of AAS employed for each analyte.

To determine concentration of chromium in the filtrates, four series of standard chromium solutions in the range of 0.05-8 mg/L were prepared by diluting the stock solution of chromium with de-ionized water. A blank (de-ionized water) and standards were run in flame atomic absorption spectrometer and four point's calibration curves were established. Then, sample solutions were aspirated in to the AAS instrument and direct readings of total chromium concentrations were recorded. Three replicate determinations were carried out on each sample. The amount of Cr adsorbed was then calculated from the difference between the amount before and after adsorption.

Mass of adsorbents and mass of different chemicals whenever the stock solutions were prepared from solid chemicals was measured using analytical balance of 0.01 g accuracy (Adam Equipment Co. Ltd, Mil Ton Kenyes, U.K.; Model No. WL3000). pH of different solutions was measured using pH meter (pH 301 GLP Bench pH/mv/Ion/C meter microprocessor, PC compatible, Serial No. 511919, HANNA instruments, Portugal).

Results and Discussions

Effect of contact time

Figure 3 depicts the effect of contact time on the adsorption of chromium on activated coffee husk carbon from aqueous solution. It is clear that increasing contact time increases the removal efficiency but at a certain time the percentage removal becomes almost constant. As shown in Figure 3 the percentage removal at a contact time of 5 min was 70.7% this was increased to 97.8% as time passes to 150 min.

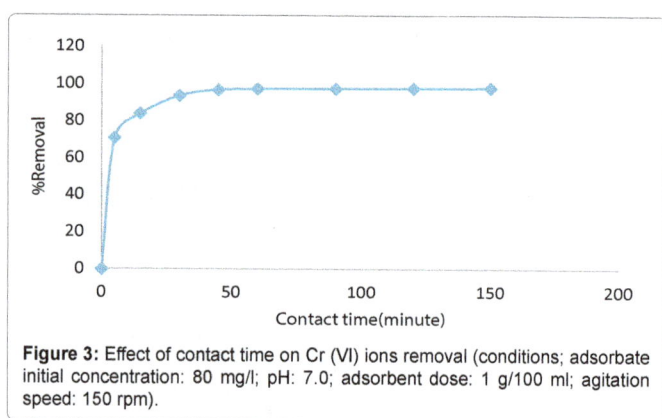

Figure 3: Effect of contact time on Cr (VI) ions removal (conditions; adsorbate initial concentration: 80 mg/l; pH: 7.0; adsorbent dose: 1 g/100 ml; agitation speed: 150 rpm).

At initial stages the rate of adsorption was high then after the rate of adsorption decreases but the removal efficiency still increased in some extent. The percentage removal at different time intervals was 93.6%, 96.7%, 97.2%, 97.4%, 97.6% as contact passes from 30, 45, 60, 90 and 120 min respectively. This result clearly shows that increasing contact time after 60 min for this specific adsorbent is simply wastage of time because the increase in adsorption efficiency after contact time of 60 min is insignificant. Therefore; the equilibrium time for the removal of hexavalent chromium from aqueous solution by activated coffee husk is taken as 60 min.

Effect of pH

The influence of pH on the adsorption of Cr (VI) ions onto activated coffee husk carbon was examined in the pH range of 2-8. Figure 4 shows the effect of pH on the adsorption of chromium on activated coffee husk carbon. The adsorption of chromium was highly dependent on the pH of the adsorbate solution prepared in the laboratory. The adsorption efficiency rapidly decreased as the pH of the solution increased. The percentage removal of chromium by activated coffee husk carbon was greater than 90% up to solution pH of 5 but the maximum percentage removal was found be between solution pH of 2 and 3 which is 98% and 97.2% respectively. The chromium removal efficiency of activated coffee husk carbon was less than 90% when the solution pH was increased above 5. The decrease in chromium adsorption efficiency of activated coffee husk carbon was seen from 98.2%-70% when the solution pH increased from 2-8 respectively. This shows that the activated coffee husk carbon is applicable for the removal of chromium from aqueous solution under acidic environment up to pH 5.

Effect of stirring speed

The effect of the sorbent/sorbate system on adsorption of chromium from aqueous solution was studied at optimum conditions of contact time and pH of the solution by dissolving 1 g of activated coffee husk carbon 100 ml chromium solution containing 100 mg/l of chromium concentration. The effect of stirring sorbent/sorbate system was monitored at low, medium and high-agitation speeds (90, 100, 120, 150, 200 rpm). As depicted in Figure 3 the chromium adsorption of activated coffee husk carbon rapidly increased from 65.67% to 98.6% when the stirring speed increased from low to medium agitation speed. When the stirring speed is increased beyond the medium agitation, the chromium removal efficiency of the activated coffee husk decreased to 98.2% at stirring speed of 200 rpm. Therefore, the adsorption of chromium on activated coffee husk carbon depends of agitation speed (Figure 5).

Effect of initial concentration

The percentage removal of Cr (VI) was studied by varying Cr (VI)

ions concentration from 80 to 300 mg/l stirring with 1.0 g of activated coffee husk carbon keeping other parameters at optimum conditions. It is clear from Figure 6 that as the concentration of chromium ions in the solution increases, the percent adsorption of chromium on activated coffee husk decrease rapidly from 98.91% to 57.11% when initial chromium concentration increased from 80 mg/l to 300 mg/l respectively. But the actual amount of chromium ions adsorbed per unit mass of the activated coffee husk carbon was increased with increasing in chromium ions concentration in the aqueous solution as represented below. As a result, the maximum and the minimum amount of chromium adsorbed were found to be 5.7113 mg/g and 2.5698 mg/g at initial chromium concentration of 300 mg/l and 80 mg/l respectively.

Effect of adsorbent dose

The different activated coffee husk carbon doses (0.5, 1.0, 1.5, 2.5 and 3.0 g/100 ml) were studied to see the effect on chromium adsorption keeping other parameters at optimum conditions (Figure 7). The results showed that with increase in activated coffee husk carbon dose, the percentage adsorption of chromium was increased; however, unit adsorption of chromium was decreased with increasing in activated coffee husk carbon dose as showed in Figure 8. Moreover, the maximum chromium removal efficiency was observed at 3 g/100 ml activated coffee husk carbon dose of the aqueous solution containing 100 mg/l chromium concentration.

Applicability to industrial wastewater

Figure 9 shows that percentage adsorption of chromium on activated coffee husk carbon increased with time at optimized parameters in aqueous solution. After 90 min of contact time increase in percentage removal was negligible. The maximum percentage

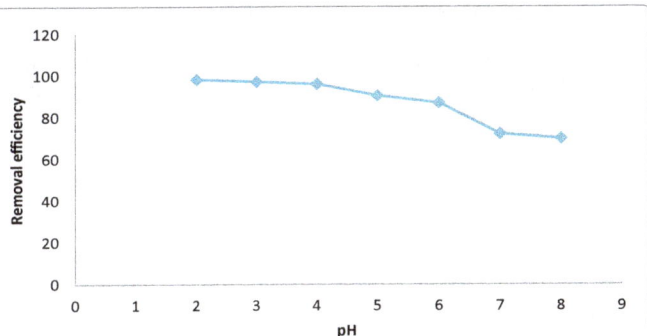

Figure 4: Effect of pH for the adsorption of Cr (VI) ions onto activated coffee husk carbon (conditions; adsorbate initial concentration: 100 mg/l; adsorbent dose: 1 g/100 ml; agitation speed: 150 rpm; contact time: 60 min).

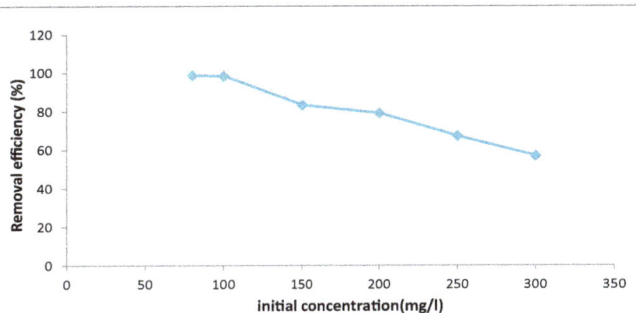

Figure 6: Effect of initial Cr (VI) ions concentration (conditions; pH: 2.0; adsorbent dose: 1 g/100 ml; agitation speed: 150 rpm; contact time: 60 min).

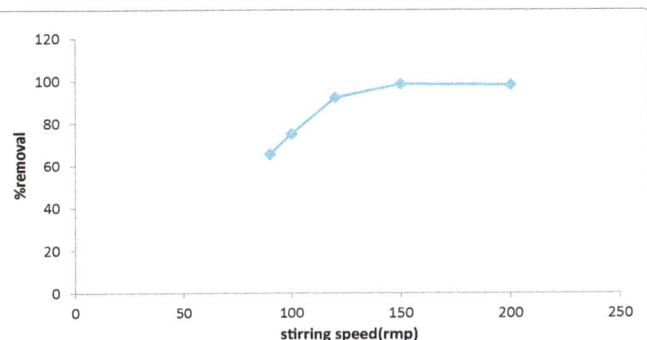

Figure 5: Effect of stirring speed on Cr (VI) ions adsorption (conditions; adsorbate initial concentration: 100 mg/l; pH: 2.0; contact time: 60 min; adsorbent dose: 1 g/100 ml)

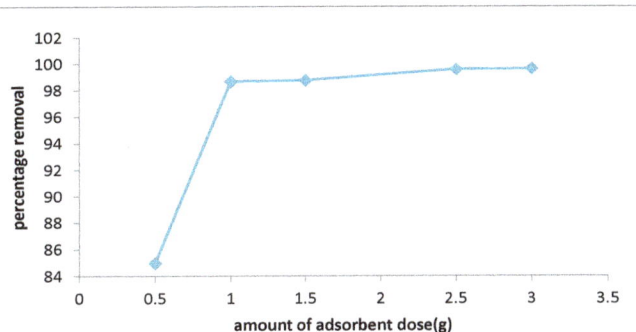

Figure 7: Effect of adsorbent dosage on adsorption of Cr (VI) ions; adsorbate initial concentration: 100 mg/l; contact time: 60 min; pH 2, agitation speed: 150 rpm).

adsorption was 83.7% at contact time of 150 min. The rate of adsorption of chromium on activated coffee husk carbon at initial stages as well as after optimum contact time is almost similar with the results obtained in aqueous solution.

Adsorption experiments were also carried out at two different pH values; pH 3.85, which is the pH of the waste effluent and at pH 2 which is the optimum pH for removal of Cr (VI) ions from aqueous solution. The percentage adsorption was 79% at pH 3.85 and 89% at pH 2 (Figure 10).

Adsorption equilibrium

Adsorption isotherm is required to show the adsorption process and to limit the adsorption efficiency of an activated coffee husk

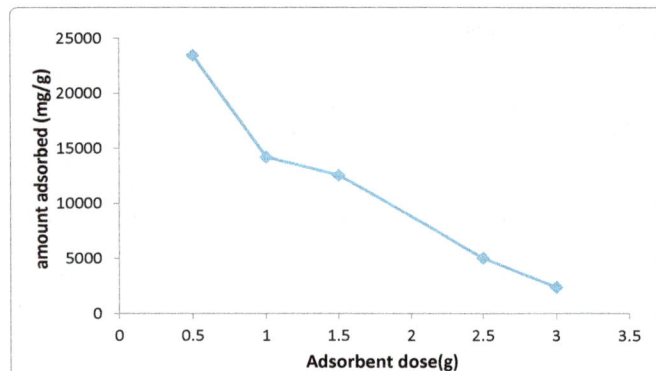

Figure 8: Effect of adsorbent dosage on adsorption of Cr (VI) ions; adsorbate initial concentration: 100 mg/l; contact time: 60 min; pH 2; agitation speed: 150 rpm).

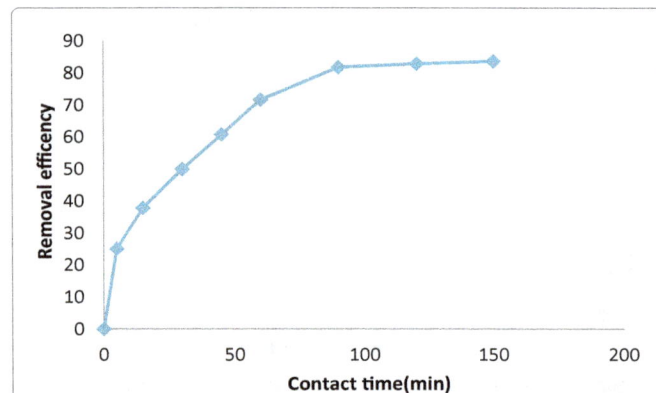

Figure 9: Effect of contact time on Cr (VI) ions adsorption from Batu Tannery wastewater (Optimum conditions; pH: 2.0; adsorbent dose: 3 g/100 ml; agitation speed: 150 rpm).

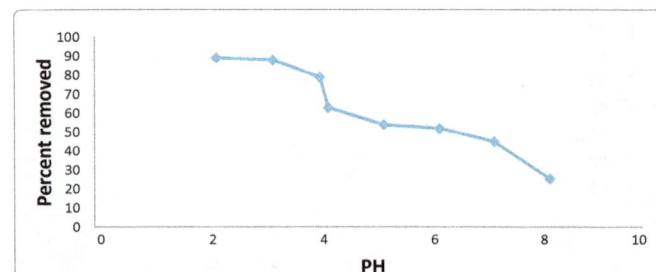

Figure 10: Effect of pH on Cr (VI) ions removal from Batu Tannery wastewater (Optimum conditions; adsorbent dose: 3 g/100 ml; agitation speed: 150 rpm; contact time 60 min).

carbon. Out of the different adsorption isotherm models, Langmuir and Freundlich models have been preferred by most of the studies concerning about adsorption sciences. The present investigation also used Langmuir and Freundlich equations to explain the adsorption ways of the activated coffee husk carbon.

Adsorption isotherms: The study was conducted by using 0.5 g, 2.5 g, 4.5 g, 6.5 g, 8.5 g, and 12.5 g dose of activated coffee husk carbon. Figure 11 shows the relationship between the amount of Cr (VI) ions removed per gram of activated coffee husk and the equilibrium liquid phase concentration at room temperature in one hour contact time. The graph obtained by Langmuir equation showed that the relation between Ce/qe in the y axis and Ce (mg/l) in the x axis.

Langmuir Equation

$$Ce/qe = 1/axb + Ce/a$$

Where qe = the amount of solute adsorbed per unit weight of adsorbent at equilibrium

b = The Langmuir constant related to the heat of adsorption

a = amount of solute adsorbed per unit weight of adsorbent required for monolayer capacity when Ce/qe is plotted vs. Ce a straight line Figure 11 show the applicability of Langmuir isotherm. The values of (a) and (b) have been determined from the slop (1/a) and intercept (1/(b * a)) for this line (Ismaeel et al., 2012).

According to Figure 11:

Slop (1/a) = 0.019, a = 52.6

Intercept Y (1/axb) = 1.5273, b = 0.029

Langmuir's equation, $qe = Ce + 0.029/52.6Ce$

Using Langmuir's equation, the maximum adsorption capacity by activated coffee husk carbon was obtained as 52.6 mg/g. That is one gram of the coffee husk can absorb 52.6 mg Chromium.

Freundlich Equation

$$lnqe = lnKf + 1/n \, lnCe$$

Where: the Freundlich constants Kf and n, which respectively indicate the adsorption capacity and the adsorption intensity, were calculated from the intercept and slope of the plot of $lnqe$ versus $lnCe$ as shown in Figure 10.

For values in the range $0.1 < 1/n < 1$, adsorption is favorable. The greater the values of Kf better is the favorability of adsorption.

The linear plots of $lnqe$ vs. $lnCe$ show that the adsorption of metal ions onto the adsorbent follows the Freundlich isotherm model.

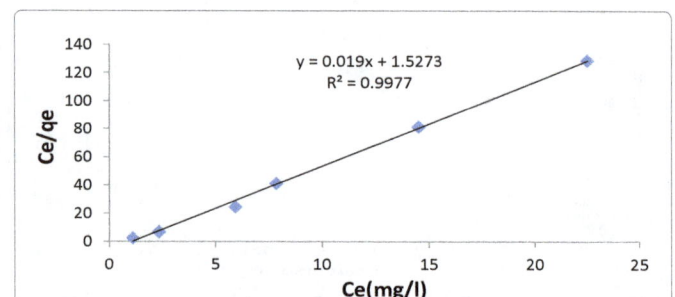

Figure 11: Langmuir adsorption isotherm for Cr (VI) ions adsorption by activated coffee husk carbon at room temperature and contact time of 1 hour.

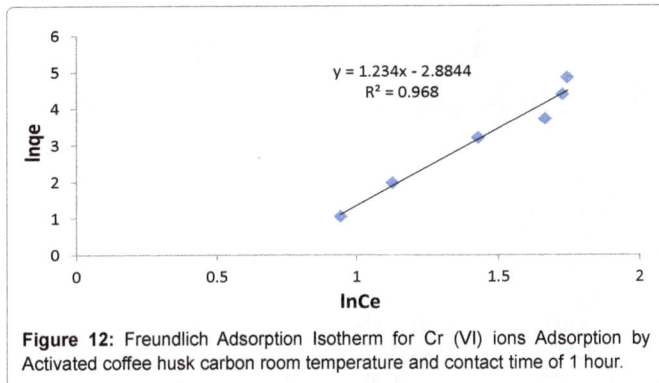

Figure 12: Freundlich Adsorption Isotherm for Cr (VI) ions Adsorption by Activated coffee husk carbon room temperature and contact time of 1 hour.

According to Figure 12:

Intercept Y (lnkf)=-2.8844; constant, Kf=1.06

Slop (1/n)=1.234

Freundlich equation, lnqe=-2.8844+1.234 lnCe.

The values of 1/n and n obtained from Freundlich equation were 1.23 and 0.81 respectively. It means that the surface of activated coffee husk carbon becomes less heterogeneous which leads to less adsorption intensity. The Chromium adsorption on activated coffee husk carbon best fitted to Langmuir model since its correlation coefficient, R^2=0.9977.

Conclusion

This paper has evaluated the chromium adsorption capacity of activated coffee husk carbon. Adsorption of Cr (VI) was highly pH dependent and the results showed that the removal efficiency increased, as decreasing pH, increasing the adsorbent dose and contact time. The optimum adsorption percentage of Cr (VI) ions using activated coffee husk carbon as an adsorbent was achieved within 60 min. The maximum removal of Cr (VI) ions by this adsorbent was carried out at pH 2).

The efficiency of adsorbent, coffee husk activated carbon, towards the removal of chromium was also tested using Batu tannery waste. The mode of purifying heavy metal contamination of surface waters must be given to strategies designed to high comprehensive methods while keeping cost at minimum. The activated coffee husk carbon can be successfully used for removing of chromium (VI) ions from wastewaters.

Acknowledgements

I gratefully acknowledge the constructive comments and suggestions of anonymous reviewers.

References

1. Chandra R, Gupta M, Pandey A (2011) Monitoring of river Ram Ganga: Physico-chemical characteristic at Bareilly. Recent research in science and technology 3.

2. Rane NM, Sapkal RS, Sapkal VS, Patil MB, Shewale SP (2010) Use of naturally available low cost adsorbents for removal of Cr (VI) from waste water. International Journal of Chemical Sciences and Applications 1: 65-69.

3. Dhanakumar S, Solaraj G, Mohanraj R, Pattabhi S (2007) Removal of Cr (VI) from aqueous solution by adsorption using cooked tea dust. Indian Journal of Science and Technology 1: 1-6.

4. Ghaderi AA, Abduli MA, Karbassi AR, Nasrabadi T, Khajeh M (2012) Evaluating the effects of fertilizers on bioavailable metallic pollution of soils, case study of Sistan farms, Iran. International Journal of Environmental Research 6: 565-570.

5. Rajendran A, Mansiya C (2011) Extraction of Chromium from Tannery Effluents Using Waste Egg Shell Material as an Adsorbent.

6. Occupational Safety and Health Administration (2003) Safety and health regulations for construction. 29 Code of federal regulation Part 1926.

7. Gholipour M, Hashemipour H, Mollashahi M (2011) Hexavalent chromium removal from aqueous solution via adsorption on granular activated carbon: adsorption, desorption, modelling and simulation studies. J Eng Appl Sci 6: 10-18.

8. Nasrabadi T, Bidhendi GN, Karbassi A, Mehrdadi N (2010) Evaluating the efficiency of sediment metal pollution indices in interpreting the pollution of Haraz River sediments, southern Caspian Sea basin. Environmental monitoring and assessment 171: 395-410.

9. Haruna A, Uzairu A, Harrison GFS (2011) Chemical fractionation of trace metals in sewage water–irrigated soils. International Journal of Environmental Research 5: 733-744.

10. Serbaji MM, Azri C, Medhioub K (2012) Anthropogenic Contributions to Heavy Metal Distributions in the Surface and Sub-surface Sediments of the Northern Coast of Sfax, Tunisia. Int J Environ Res 6: 613-626.

11. Manfe MM, Attar SJ, Parande M, Topare NS (2012) Treatment Of Cr (Vi) Contaminated waste water Using Biosorbent Prunus Amygdalus (Almond) Nut Shell Carbon. International Journal of Chemical Sciences 10: 609-618.

28

The Effect on Extracting Solvents using Natural Dye Extracts from *Hyphaene thebaica* for Dye-sensitized Solar Cells

Mohammed IK[1]*, Kasim Uthman ISAH[1], Yabagi JA[2] and Taufiq S[3]

[1]*Department of Physics, Federal University of Technology, Minna, Nigeria*
[2]*Department of Physics, Ibrahim Badamasi Babangida University Lapai, Niger State, Nigeria*
[3]*Department of Preliminary Studies, Umaru Waziri Federal Polytechnics, Birnin Kebbi, Kebbi State, Nigeria*

Abstract

This study covers the fabrication and characterization of dye sensitized solar cell using *Hyphaene thebaica* as the natural dye sensitizer for DSSCs. Ethanol and water in separate container was used as the extracting solvent for the natural dyes. Titanium dioxide (TiO_2) was deposited on fluorine doped tin oxide (FTO) conductive glass forming a TiO_2 thin film, underwent sintering at 400ºC for 40 minutes. The photo electrochemical performance of the dye sensitized solar cell (DSSC) based on the doum palm pericarp shows open circuit voltage (V_{oc}) of 0.37 V and 0.50 V, and short circuit current density (J_{sc}) of 0.005 mA/cm² and 0.010 mA/cm² for ethanol and water extracts respectively. This study further inspected the fill factor as 0.63 and 0.66 for the ethanol and water extract respectively. The conversion efficiency for the ethanol extract was 0.012% and water extract is up to 0.033% under light intensity of 1000 m/Wm² (AM 1.5).

Keywords: Dye-sensitized solar cell; *Hyphaene thebaica*; Doum pericarp; Titanium dioxide

Introduction

Energy technology is one of the most important technologies in the 21st century that dominated people's life and people's consumption. Moreover, environmental pollution has increasingly become a worldwide concern in the past few decades. Thus, how to enhance the efficiency of natural energy use and to recycle regenerated energy has become an important research field for developed countries [1].

Dye-sensitized solar cells (DSSCs) have been widely investigated as one of the next-generation solar cell because of their simple structure and low manufacturing cost [2]. Generally, DSSC comprises of a nanocrystalline titanium dioxide (TiO_2) electrode modified with a dye fabricated on a transparent conducting oxide (TCO), a platinum (Pt) counter electrode, and an electrolyte solution with a dissolved iodine/triiodide ion redox couple between the electrodes [3]. Although certified conversion efficiency using black dye has been reported to be 10.4% by the Swiss Federal Institute of Technology in Lausanne (EPFL) [4]. It is well known that the conversion efficiency (η) of solar cells can be represented as follows Kimpa [5].

$$\eta = FF \times I_{sc} \times V_{oc} / P_{in} \qquad (1)$$

where FF, I_{SC}, V_{OC}, and P_{in} are fill factor, short circuit current, open circuit voltage, and incident power, respectively.

Organic dye have higher absorption efficient used for DSSCs with efficiencies of up to 9% have been reported [6]. Organic dyes with high absorption coefficient could translate into thinner nanostructure metal oxide film. This advantageous of transporting charge both in the metal oxide and in the permeating phase, allowing for the use of higher viscosity materials such as ionic liquids, solid electrolytes or holes conductors.

In nature, some fruits, flowers, leaves and so on show various colors and contain several pigments that can be easily extracted and then employed in DSSCs. The leaves of most green plants are rich in chlorophyll and the application of this kind of natural dye has been frequently investigated in many related studies. Anthocyanins are natural compounds that give color to fruits and plants and are also largely responsible for the purple–red color of autumn leaves and for the red color of flower buds [7].

Kimpa used the extract of flame tree flower and pawpaw leaf as photosensitizer and the open-circuit voltage (V_{OC}) of fabricated DSSCs is 0.51 V and 0.50 V [5]. Zhu investigated the extract of frozen blackberries to serve as photosensitizer and the open-circuit voltage (V_{OC}) of fabricated DSSCs is 0.33 V [8]. Polo extracted the blue violet anthocyanin of Jaboticaba and Calafate respectively to serve as photosensitizer and the (V_{OC}) of prepared DSSCs is 0.59 V and 0.4 V respectively [9]. Furukawa investigated the extract of red-cabbage, curcumin and red-perilla to serve as natural dye sensitizers for DSSC and the (V_{OC}) is 0.52 V, 0.53 V and 0.49 V respectively. Patrocinio adopted the extract of blueberries and Jaboticaba's skin to serve as photosensitizer and the (V_{OC}) of prepared DSSCs is 0.59 V and 0.45 V respectively [10]. In this paper, extracts of doum pericap was used as the natural dyes as dye-sensitizers for the preparation of DSSCs. Doum palm fruit (*Hyphaene thebaia*) is a desert palm tree with edible oval fruit, originally native to the Nile valley. It also grows very well in the northern part of Nigeria. It is a member of the palm family, Arecaceae. The trunk of this small palm commonly branches into two like Y and often each branch divides again in a Y form, giving the tree a very distinctive appearance; it is dichotomous and arborescent in nature. It is listed as one of the useful plants of the world. It is represented by the genus Hyphaene, the fruit of interest in the current study. Its fibre and leaflets are used by people along the Nile to weave baskets. Doum palm fruit is also a source of potent antioxidants. The fruit has a brown outer fibrous flesh which is normally chewed and spewed out. Doum palm

---

*Corresponding author: Mohammed IK, Department of Physics, Federal University of Technology, Minna, Nigeria, E-mail: kimpa@futminna.edu.ng

kernel is edible when it is unripe but hard when it is ripe. The fruit is depicted in Figure 1.

Experimental

Doum fruit was peeled and the pericarp was used as the natural plant. The doum pericarp was crushed with a porcelain mortar and pestle, the crushed sample were kept on two different conical flask. The samples were mixed separately with 50 cm³ of ethanol (99% absolute) at room temperature and 50 cm³ of distilled water in a dark room. The solution was filtered separately using filter paper to acquire a pure and natural dye solution. The TiO$_2$ film was prepared by blending 0.2 g of commercial TiO$_2$ powder (Degussa, P25), 0.4 cm³ of nitric acid (0.1 M), 0.08 g of polyethylene glycol (MW 10000) and one drop of a Triton x-100 (a non-ionic surfactant). The mixture was well mixed using an ultrasonic bath for 1 h and the resulting paste was spread over an FTO conductive glass plate (SOLARONIX) having 15 Ω/cm². TiO$_2$ pastes were deposited on the FTO conductive glass by rigid squeegee and screen printing procedure (polyester mesh of 90) in order to obtain a TiO$_2$ film with a thickness of 9 μm. The active area of DSSC was 1.04 cm² (1.3 cm × 0.8 cm). The TiO$_2$ thin film was sintered at 450°C for 45 minutes to increase compact- ness of the thin film. The TiO$_2$ film was consolidated through heat treatment, increasing the internal voids of film organization and thus enhancing its absorption performance. Then the sintered TiO$_2$ thin film was immersed for 24 h in the natural dyes prepared, allowing the natural dye molecules to be adsorbed on the surface of TiO$_2$ nanoparticles. Anhydrous alcohol was used to remove any natural dye that had not been adsorbed on the surface of TiO$_2$ nanoparticles. DSSCs were assembled following the procedure described in the literature [5], the catalyst-coated counter electrode was placed on the top so that the conductive side of the counter electrode faces the TiO$_2$ film. The iodide electrolyte solution (0.5M potassium iodide mixed with 0.05M iodine in water-free ethylene glycol) was placed at the edges of the plates. The liquid was drawn into the space between the electrodes by capillary action. Two binder clips were used to hold the electrodes together. In the performance test of the prepared DSSC, xenon (Xe) light of 1000 W was selected to simulate sunlight (AM 1.5), and an I-V curve analyzer (Model 4200 SC) was employed to measure the photoelectric conversion efficiency of the prepared DSSC. The measured results were plotted on I-V curve.

Results and Discussion

Figure 2 shows the absorption spectra of doum pericarp dye extracts and doum pericarp dye extracts adsorped on TiO$_2$ surface using ethanol and distilled water as extracting solvent. Absorption spectra provide necessary information on the absorption transition between the dye ground state and excited states and the solar energy range absorbed by the dye. The absorption range of doum pericarp dye extracts adsorbed on TiO$_2$ was found within the range of 300 – 450 nm. Doum water extract has two absorption peaks at 350 nm and 400 nm, while the absorption peak of the doum ethanol extract adsorbed on TiO$_2$ was only at one absorption peak of 353 nm. It can be seen that after TiO$_2$ nanoparticles was added to doum pericarp extract, its absorption intensity decreases from 440 nm to 350 nm. This property reduces the charge transfer ability of the fabricated DSSCs under normal sunlight thereby reducing the efficiency. The dye pigments belongs to the existence of chromophores and it represents the chemical group that is responsible for the colour of the molecule, that is its ability to absorbed photon. Conjugation of chromophores makes them absorb light of different wavelength and energy. Peak wavelength tend to be shifted towards long wavelength as the size of conjugated system or

chromophores increases. The double or two absorption peaks of the doum pericarp dye extracts with distilled water as extracting solvent could be due to the presence of complex chromophores conjugated with other group of chromophores. Figure 3 compares the absorption spectrum of doum pericarp ethanol extract and doum pericarp distilled water extract.

The dye sensitization effect was demonstrated with a DSSC using natural dye from doum pericarp and characterized by current – voltage measurements in Figure 4. Energy conversion efficiency of 0.012% at about 0.1 mW/cm² solar illumination was obtained from the dye extracts using ethanol as the extracting solvent while 0.033% efficiency was obtained for the dye extract using water as the extracting solvent. Table 1 shows the data acquired from measuring the photoelectric conversion efficiency of the DSSCs. From the result obtained, It was observed that using ethanol as our extracting solvent has better absorption capability than water. In this research work the efficiencies of the cells are both very low.

Conclusion

The absorption range of the dye and the dye adsorbed on TiO$_2$ using ethanol and water as the extracting solvent are within the wavelength of 300 nm, 440 nm. The absorption range for pure dye is between 300 nm – 600 nm as shown in Figure 3. The power conversion efficiency η of the DSSC using doum pericap ethanol extract is 0.012% while that of distilled water as the extracting solvent is up to 0.033%. DSSC efficiency from doum pericarp extract has a very low efficiencies compared to that obtained by Zhu [8], Kimpa [5], Polo [9] and Furukawa [11].

Figure 1: Doum Palm Fruit.

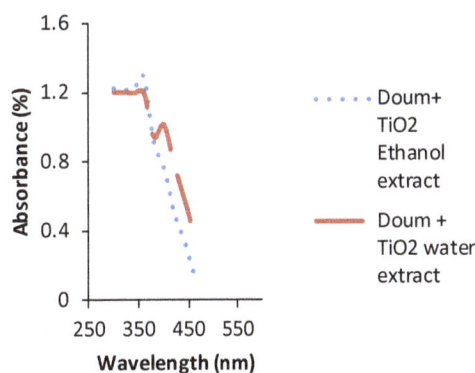

Figure 2: Absorption spectra of doum pericarp with TiO$_2$ dye extract using ethanol and water as extracting solvent

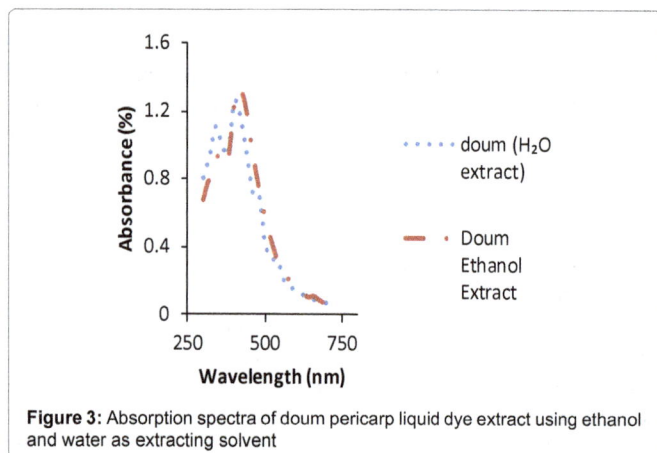

Figure 3: Absorption spectra of doum pericarp liquid dye extract using ethanol and water as extracting solvent

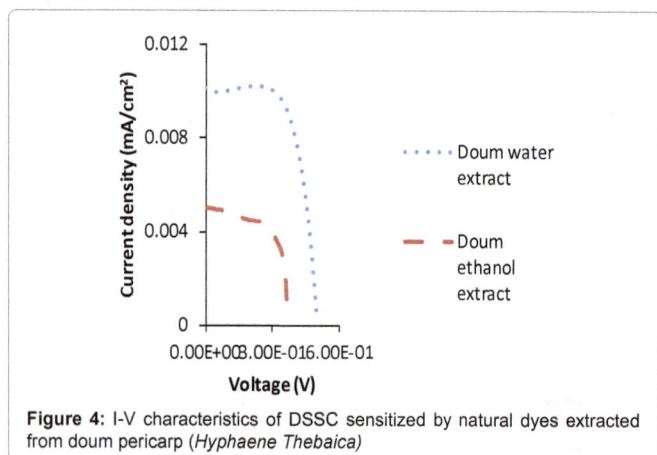

Figure 4: I-V characteristics of DSSC sensitized by natural dyes extracted from doum pericarp (*Hyphaene Thebaica*)

Dye	Jsc (mAcm^{-2})	Voc (mV)	FF	η %
Doum pericarp (ethanol extract)	0.005	0.37	0.63	0.012
Doum pericarp (distilled water extract)	0.010	0.50	0.66	0.033

Table 1: Photoelectrochemical parameters of the DSSCs sensitized by natural dyes extracted with water in the UV-vis light range

The poor performance of the cell could be linked with various factors, one of which could be the dye used and the ability of the electron transportation to the conduction surface of the TiO_2.

Acknowledgment

The authors would like to thank the entire staff of the Physics Advanced Lab, Sheda Science and Technology Complex (SHESTCO) Abuja for allowing us to used their lab and Federal University of Technology Minna University Board of Research (UBR) for their sponsorship.

Reference

1. Chang H, Kao MJ, Chen TL, Kuo HG, Choand KC, et al. (2011) Natural sensitizer for dye-sensitized solar cell using three layers of photoelectrode thin films with a schottky barrier. American Journal of Engineering and Applied Sciences 4: 214-222.

2. O'Regan B, Grätzel M (1991) A low-cost, high-efficiency solar cell based on dye- sensitized colloidal TiO_2 films. Nature 353: 737-740.

3. Chiba Y, Islam A, Watanabe Y, Komiya R, Koide A, et al. (2006) Dye-sensitized solar cells with conversion efficiency of 11.1%. Japanese Journal of Applied Physics 45: L638-L640.

4. Nazeeruddin MK, Pechy P, Renouard T, Zakeeruddin SM, Humphry-Baker R, et al. (2001) Engineering of Efficient Panchromatic Sensitizers for Nanocrystalline TiO_2-Based Solar Cells. Journal of American Chemical Society 123: 1613-1624.

5. Kimpa MI, Momoh M, Isah KU, Yahya HN, Ndamitso MM (2012) Photoelectric Characterization of Dye Sensitized Solar Cells Using Natural Dye From Pawpaw Leaf and Flame Tree Flower as Sensitizers. Materials Sciences and Applications 3: 281-286.

6. Hara K, Dan-Oh Y, Kasada C, Arakawa HN (2004) Effects of additives on the photovoltaic performance of coumarin-dye-sensitized nanocrystalline TiO_2 solar cell. Langmuir 20: 4205-4210.

7. Chang H, Lo YJ (2010) Pomergranate leaves and mulberry fruit as natural sensitizers for dye-sensitized solar cells. Solar energy 84: 1833-1837.

8. Zhu H, Zeng H, Subramanian V, Masarapu C, Hung KH, et al. (2008) Anthocyanin-Sensitized Solar Cells Using Carbon Nanotube Films as Counter Electrodes. Nanotechnology 19: 465204.

9. Polo AS, Iha NY (2006) Blue sensitizers for solar cells: natural dyes from calafate and jaboticaba. Solar Energy Material and Solar Cells 90: 1936-1944.

10. Patrocinio AOT, Mizoguchi SK, Paterno LG, Garcia CG, Iha NYM (2009) Efficient and Low Cost Devices for Solar Energy Conversion: Efficiency and Stability of some Natural Dye-Sensitized Solar Cells. Synthesis Metals 159: 2342-2344.

11. Furukawa S, Iino H, Iwamoto T, Kukita K, Yamauchi S (2009) Characteristics of Dye-Sensitized Solar Cells Using Natural Dye. Thin Solid Films 518: 526 529.

29

Stress-Strain Characterization in Fiber-Reinforced Composites by Digital Image Correlation

Heredia AS[1], Aguilar PAM[2*] and Ocampo AM[2]

[1] *Research Center on Engineering and Applied Sciences - (IICBA)*
[2] *Department of Engineering, Autonomous Univiersity of Morelos State, Mexico*

Abstract

Composite materials are widely used in structural mechanics as they can withstand high loads; although after a while, they can present relative strain due to these loadings. In the present work, it is implemented digital image correlation using one laser-beam and the speckles created by its reflection to describe the mechanical behavior of fiber-reinforced composites submitted under compression test. Composites were tested in two set of arrays: the first was done with fibers orientated parallel to the load and the second was done with fibers randomly orientated, as it is known that the stress-strain evolution change according to the orientation of the fibers. Our method allows us to evaluate the heterogeneous strain evolution observed during the tests. Validity of optical strain-measurements is assessed against the results of an universal testing machine, a good correlation was found by comparing the results.

Keywords: Laser speckle; In-plane strain measurement; Digital image correlation; Compression test; Composite materials

Introduction

It is important to measure local strain and stress distributions for understanding the mechanical properties of structural materials. The stress-strain diagram helps to understand materials under loads [1]. There are two methods to obtain those diagrams: contact and non-contact; in the former case, mechanics excels doing physical tests such as compression test, in which a material is placed in the universal machine, a continuous load is applied to it and the resulting deformation is measured [2]. Optical methods are also used as an invasive way to determine residual stress, in-field displacements and strain, in which hole-drilling is one of the most used techniques, developed in 1930 by Mathar [3]. Nowadays this technique is standardized by ASTM [4]. When a laser light felt on a matt surface such as paper or rough-like material, a speckle pattern is formed and a high-contrast grainy pattern will be reflected. This effect was called â€œgranularityâ€ by Rigden and Gordon [5]. The speckle pattern is formed when coherent light is reflected from a rough surface or when light is propagated through a medium with random refractive index fluctuations [6]. The simplest image matching procedure is cross-correlation (CC) that can be performed either in physical space [7,8]; in Fourier space by using the Fast Fourier Transform (FFT), one can evaluate the CC function very quickly [9]. Anuta took advantage of the high speed of FFT algorithm doing digital multispectral and multi-temporal statistical pattern-recognition [10]. Kuglin and Hines observed that information about the displacement of one image with respect to another is included in the phase component of the cross-power spectrum of the images [11].

Composite materials are widely used in structures as they can reach high loads and strain measurements had been studied through different techniques: one of them is the strain gauges used to analyze the strain [12], the electron MoirÃ© method used to measure the strain distribution [13] and fiber optics sensors for strain measurements [14]. The development of non-destructive testing methods is the main challenge for the assessment of structural elements in existing constructions. This paper presents an alternative method for measuring deformation in composite materials, in which we use a laser beam focusing on the cross-section of our sample which is under a compression test; the studied material is formed by ARMEX plus concrete. As the reflection of the laser is a speckle pattern, we need to study it by the means of Digital Image Correlation (DIC) and the Infinitesimal Strain Tensor in order to get in-field displacements measurements. We calculate the accuracy, error and sensitivity of the method, as well as a theoretical demonstration of our phenomenological process; assessing that the present work would be a cheap technique as it only uses one laser beam for stress-strain measurements.

In section 4 we put forward concepts that are used in the present work. Section 4.1 explains the correlation technique and the equations we are following, in section 4.2 we put forward the equations involving the composite materials study, including the well-known mixture law, section 4.3 shows the stress-strain relation with the Youngâ€™s modulus and how the stress is calculated in a composite material and in section 4.4 we explain a theoretical analysis of the physical phenomena. Experimental set-up is presented in section 5, which includes the dimensions of our samples, how compression tests are done and how DIC is taking place. Finally the results and discussion are shown in section 6, in which there are two plots of interest: the stress-strain diagram which is obtained from the universal machine and the DIC plot which is programmed using a code written in Matlab.

Developments

Correlation technique

Digital image correlation is an optical method that examines image data taken while samples are during mechanical tests. This technique consists on capture consecutive images with a digital camera in order to evaluate the change in surface characteristics and understand the behavior of the specimen while it is subjected to a continous axial load.

*Corresponding author: Aguilar PAM, Department of Engineering, Autonomous Univiersity of Morelos State, Mexico
E-mail: pmarquez@uaem.mx

It is well known that for correlation is needed to obtain the spectrum between two images; let's call U and V a pair of 2-D images, where U represents the reference image and V the displaced image [15], therefore is taken the FFT form both and cross-spectrum is defined by:

$$C_s(u,v) = F(u,v) \cdot G^*(u,v) \qquad (1)$$

where F is the Fourier transform of the first image and G^* is the complex conjugate Fourier transform of the second; once obtained the spectrum, it is taken the cross-correlation between those images, defined by:

$$C_c = F^{-1}[F(U) \cdot G^*(V)] \qquad (2)$$

The use of FFT requires that images U and V are the same size and have dimensions that are powers of 2. In the present work, we analyze 2^{10} square images in order to map a bigger area of the whole image and the shift $\delta x = (\delta y)$ between two consecutive images is 128 pixels in order to get a better result. These two parameters define the mesh formed by the images used to describe the displacement field [16].

Stress-strain analysis in composite materials

When unidirectional continuous-fiber laminate is loaded in a direction parallel to its fibers, the longitudinal modulus E_{11} can be estimated from its constituent properties by using the well-known rule of mixtures:

$$E_{11} = E_f V_f + E_m V_m \qquad (3)$$

where E_f is the fiber modulus, V_f is the fiber volume percentage, E_m is the matrix modulus and V_m is the matrix volume percentage [17]. We treat the composite material as acting in a purely elastic manner and neglect the viscoelastic effects of the matrix; Hooke's Law then can be applied:

$$\sigma_f = E_f \varepsilon_f \qquad (4)$$

$$\sigma_m = E_m \varepsilon_m \qquad (5)$$

Therefore, the longitudinal tensile strength σ_{11} also can be estimated by:

$$\sigma_{11} = \sigma_f V_f + \sigma_m V_m \qquad (6)$$

Where σ_f and σ_m are the ultimate fiber and matrix strengths respectively. As properties of fibers dominate for all practical volume percentage, the values of the matrix can be ignored and Eq. (6) is reduced:

$$\sigma_{11} \approx \sigma_f V_f \qquad (7)$$

This type of composite materials is called fiber-reinforced composites (FRC), which can be classified as either continuous or discontinuous; generally, the highest strength is obtained with continuous reinforcement. When FRC are used with continuous and aligned fibers orientated parallel to the load, the efficiency of the reinforcement is 98 % as Eq. (7) shows, but when fibers are randomly orientated through the 3-D space, the efficiency decreases to 1/5 of its value [18].

Compression tests

Physical tests are used in order to know mechanical properties of materials and compression test is one of these tests which enable the user to understand the behavior of a material under a continuous axial load; from this test we obtain the stress-strain diagram [2]. In this work we made a mixture of sand-cement with a ratio of 3x1 respectively with dimensions of 5×5×4.5 cm. and they undergo into compression test according to ASTM E-9 [19]. We did two set of tests, one for fibers

parallel orientated and the other for randomly orientated; both of them were performed with a speed ratio of 0.5 mm/s, the first up to 140 GPa and the second up to 7 GPa approx. Since we are working on the elastic part of the diagram, therefore we can apply Hooke's law:

$$\sigma = E \cdot \varepsilon \qquad (8)$$

where σ is the stress, E is the Young's modulus of the material: 53 GPa when fibers are aligned parallel to the load and 2.2 GPa when they are randomly aligned, both experimentally obtained by the universal machine and ε is the dimensionless strain.

In DIC procedure, when the same reference picture is used, it is not possible to measure large displacements in a sequence of pictures, but when they remain small enough it is possible to assume the first image as the reference for the whole analysis. In the present work, we work with deformations less than two millimeters. It is well-known that the infinitesimal strain tensor ε_c is well adapted to small displacements and it can be evaluated as:

$$\varepsilon_c = \frac{1}{2}(R + R^T) - 1 \qquad (9)$$

where ε_c is the strain, R is an orthogonal second rank tensor and R^T implies transpose. Taking this outset, we show the theoretical results of strains: ε obtained by the universal machine and ε_c obtained by DIC technique.

Theoretical analysis

In the present work, we are working with a laser beam with an output tye of TEM_{00} which corresponds to a Gaussian beam, this kind of beam has an intensity distribution of:

$$I(r) = I_0 exp[\frac{-2r^2}{w^2}] \qquad (10)$$

where $r = (x^2 + y^2)^{1/2}$ and w is the spot size and depends on the z-coordinate [20]. The equation of such beam is deduced from Helmholtz equation and is represented by [21]:

$$E(r,z) = A\frac{w_0}{w(z)}exp[\frac{-r^2}{[w(z)]^2} + \frac{kr^2}{2R(z)} + kz - \eta(z)] \qquad (11)$$

where w_0 is the beam waist, $w(z)$ is the beam spot size, $R(z)$ is the curvature radius of the spherical waves and $\eta(z)$ is the beam phase angle [22]. And the intensity distribution of a Gaussian beam according to Eq. (11) is:

$$I(r,z) = I_0 \frac{w_0^2}{[w(z)]^2} exp[\frac{-2r^2}{[w(z)]^2}] \qquad (12)$$

Gaussian beams are able to pass through different media; the light reflection occurs when it arrives to the boundary separating two media of different optical densities and some of the energy is reflected back into the first medium [23], taking this outset, our laser-beam strikes a rough material and the reflection can be studied as a speckle pattern. The ratio between the intensity of the reflected beam and the incident beam is called reflectivity R and is expressed by:

$$R = \frac{I_r}{I_i} \qquad (13)$$

where I_r and I_i are the reflected and incident beams respectively; when a beam pass through the media, there exist transmissivity T and according to the conservation law of energy [24]:

$$T + R = 1 \qquad (14)$$

In the present work, as we are working with solid materials, the transmissivity is zero, thus we can assume that:

$$I_i \cdot R = I_r \qquad (15)$$

for R ≤ 1, invoking Fresnel diffraction equation and according to cross-spectrum analysis Eq. (1), taking U as a speckle pattern with an intensity distribution [25], expanding formally and changing coordinates we obtain

$$F(u,v) = \iint_{-\infty} \frac{1}{4\pi\sigma^2} exp[\frac{-(x^2+y^2)}{2\sigma^2}]exp^{-i2\pi(ux+vy)}dxdy \quad (16)$$

$$U(\rho:t) = \frac{1}{4\sigma^2} exp[\frac{\pi^2\sigma^2}{2}\rho^2] \quad (17)$$

Being $U(\rho:t)$ the first image Fourier transform at an initial time t_0; the same process is done for $V(\rho:\Delta t)$, where V is the second image Fourier transform at a time t_1 and cross correlation is defined by:

$$CC = [\frac{1}{4\sigma^2}]^2 exp^{i2\pi(\Delta x+\Delta y)} \quad (18)$$

Finally it is taken the correlation phase from the exponential and infinitesimal strain tensor takes place obtaining in-field measurements. From Eq. (18) is possible to see that the relation keep a Gaussian form, thus this case can be studied as a linear behaviour.

Experimental Set-Up and Data Processing

Six probes were made by adding 3×1 sand-cement mixture and each with four fibers of ARMEX, the fibers of three of them were randomly orientated in order to compare and study if the technique would be accurate in the measurement of a possible human error. The samples were placed in the universal testing machine in order to begin the compression tests. A diagram of this method is shown schematically in Figure 1a. The output of a Diode-Pumped Solid-State laser (L) with a wavelength λ=532 nm and power of 220 mW [26], is propagated through a positive lens (l) is placed in front of the sample (Sa) in order to irradiate the cross-section face and the scattering reflection impacts a screen (Sc) which is placed aside the laser beam; the material is first completely flat and it is compressed by the machine (M). During the compression tests, the speed of the compression load was 0.5 mm/s with duration of 20 minutes approximately and while they were taking place the speckle reflection was recorded with a high resolution video camera (Vc).

Once the video is recorded, Video to JPG free software is used to divide it into frames in order to load each image and process it, this software enable us to turn our videos with an average of 54 frames per second. Therefore a program is written in Matlab for Digital Image Correlation (DIC). The cross-correlation calculation is described below: in Figure 2a it is shown how the code loads the image and it is converted into a gray-scale image as it is shown in (b); the image is crop into a 1024 × 1024 in order to begin the FFT analysis of all the images, taken the first as the reference; from the second image and forward, they are considered as deformed images.

Once it is done the cross correlation technique we obtain the phase component as Kuglin and Hine. Therefore we apply a shift between each interval δx=δy 128 pixels as is shown in Figure 3.

The first strain measured by the universal machine is 0.002 while cross correlation technique allows us to measure 0.002162; the shear component is equal to or less than 1.6×10⁻⁴. It has been reported that for pure rotation measurements they obtained 2×10⁻⁴ [16].

Results and Discussion

In the section above, we mentioned that six samples were prepared for compression test, the first three probes that are shown below are done with continuous and aligned fibers orientated parallel to the load, for each one has been obtained two graphs of interest, the strain diagram is shown in Figure 4 and the DIC plot which is shown in Figure 5, which correspond to the first sample.

In Figure 4 we plot the strain diagram obtained from the machine and in Figure 5 we plot the stress-strain diagram from the experimental results obtained by Eq.(9). For both graphs we see a linear behavior. In the next Figure 6 we compare the real strain and the experimental data obtained, also some statistical results are shown in Table 1.

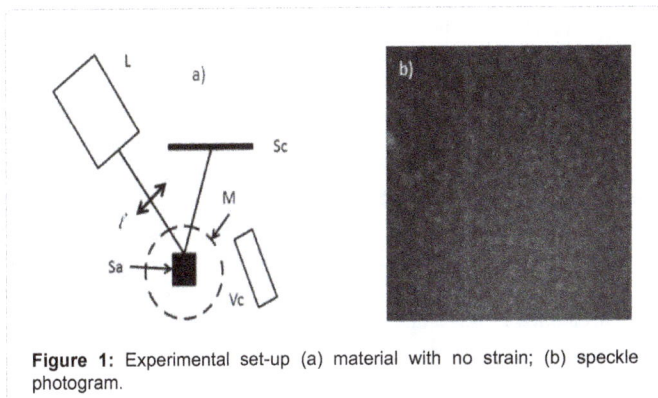

Figure 2: Cross correlation technique: (a) loaded image; (b) grey-scale image.

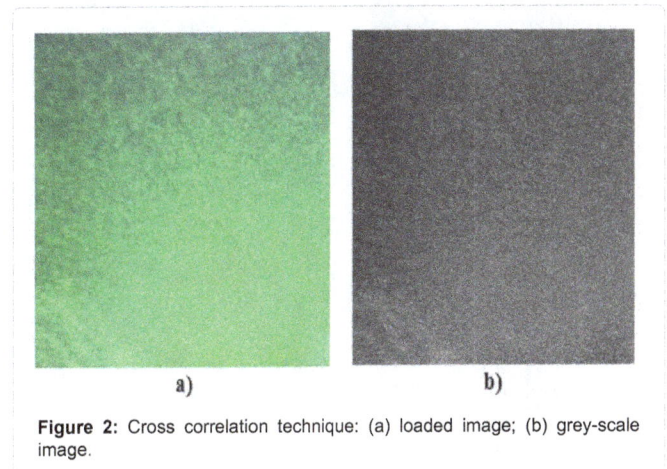

Figure 1: Experimental set-up (a) material with no strain; (b) speckle photogram.

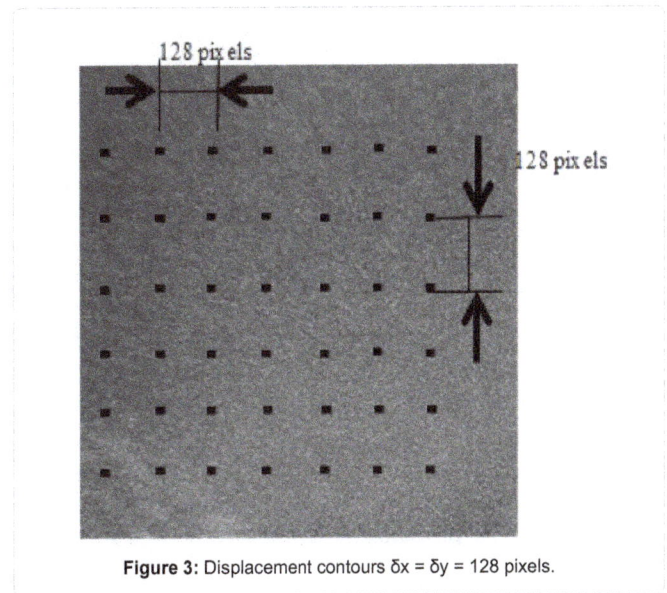

Figure 3: Displacement contours δx = δy = 128 pixels.

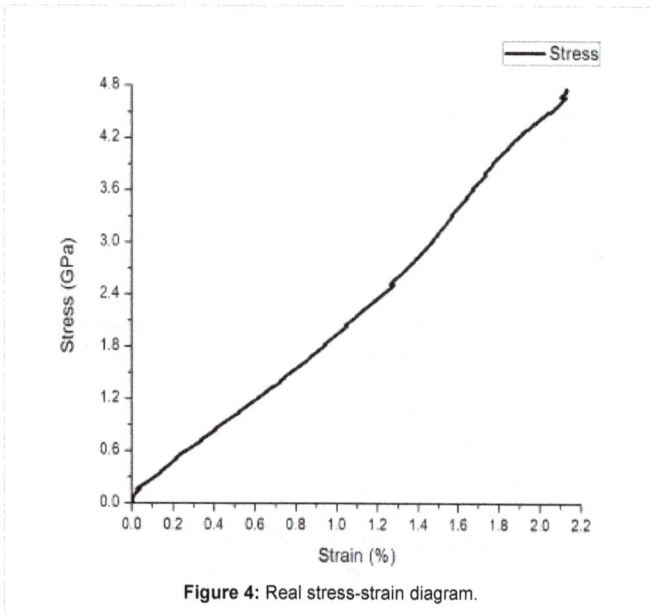

Figure 4: Real stress-strain diagram.

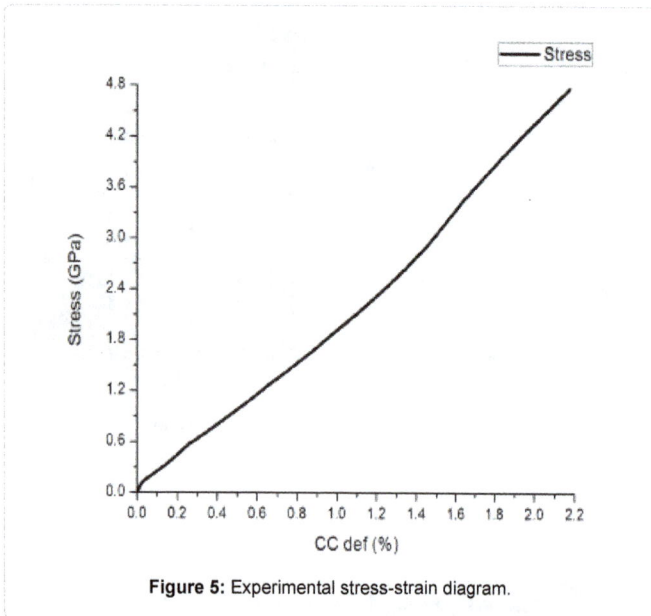

Figure 5: Experimental stress-strain diagram.

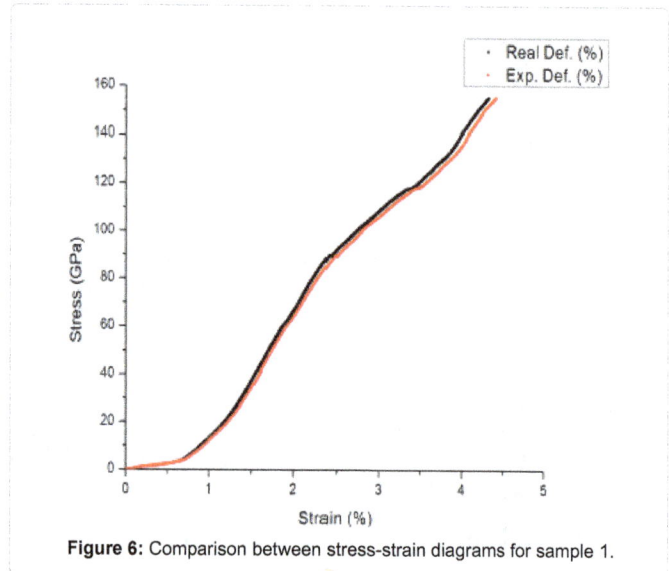

Figure 6: Comparison between stress-strain diagrams for sample 1.

Young's Modulus E	50 +/- 1.8 GPa
% difference from 52 GPa	5.4%
Standard deviation	2.45

Table 1: Statistical results for sample 1 during the compression test and cross-correlation analysis; fibers aligned and parallel orientated to the load.

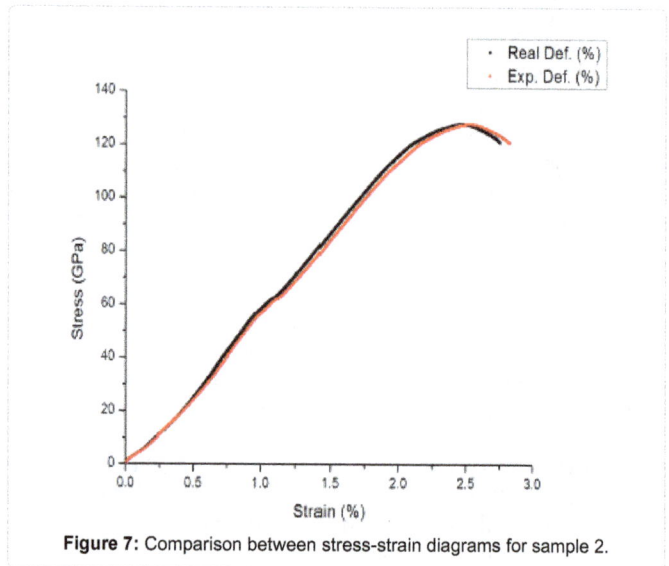

Figure 7: Comparison between stress-strain diagrams for sample 2.

Young's Modulus E	51 +/- 1 GPa
% difference from 52 GPa	4%
Standard deviation	1.5

Table 2: Statistical results for sample 2 during the compression test and cross-correlation analysis; fibers aligned and parallel orientated to the load.

In Figure 6 we plot both graphs of interest, stress-strain from the machine and cross-correlation strains from Eq.(8) (black) and from Eq.(9) (red) respectively. These results are shown for each sample; in Table 1 is shown the statistics of this test, it was observed that there is a good correlation obtaining 5.4% error from the real strain measurement.

In Figure 7 is represented the second test and is shown both plots of interest; Table 2 summarizes it showing the statistics of this test, it was observed a better correlation than sample 1 obtaining 4% error from the real strain measurement.

The test for the third sample is shown in Figure 8, in which is shown both plots of interest and Table 3 summarizes showing the statistics of this test, it was observed a simmilar behaviour than sample 2 obtaining 4.6% error from the real strain measurement.

Table 4 summarizes these tests performed with fibers aligned and

parallel orientated to the load, we can assume that the mean accuracy is 95.3%. Now we present the results for the three composites with fibers randomly orientated to the load. In Figure 9 we plot both plots of interest, stress-strain from the machine and cross-correlation strain measurements. These results are shown for each sample in Table 5 shows the statistics of the first test, it is appreciate a better correlation than parallel ones, obtaining 3% error from the real strain measurement. By

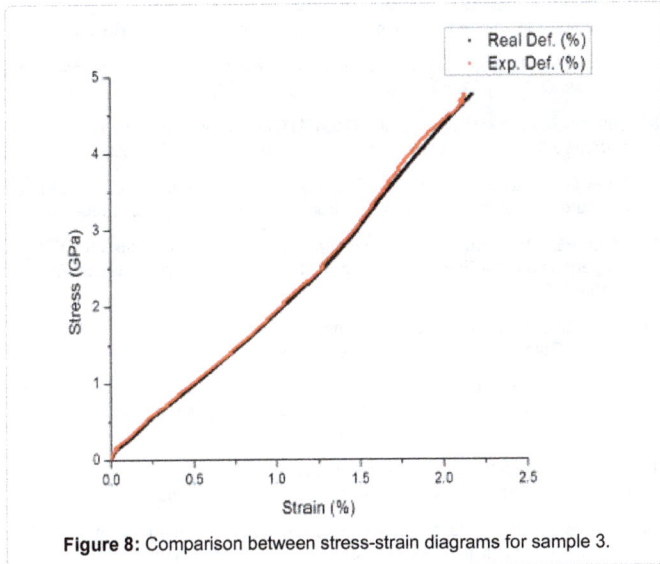

Figure 8: Comparison between stress-strain diagrams for sample 3.

Young's Modulus E	50.5 +/- 1.5 GPa
%difference from 52 GPa	4.6%
Standard deviation	1.95

Table 3: Statistical results for sample 3 during the compression test and cross-correlation analysis; fibers aligned and parallel orientated to the load.

Sample	Accuracy	Mean error
	94.6%	5.4%
	96%	4%
	95.4%	4.6%

Table 4: Accuracy and error of the three tests; fibers aligned parallel to the load.

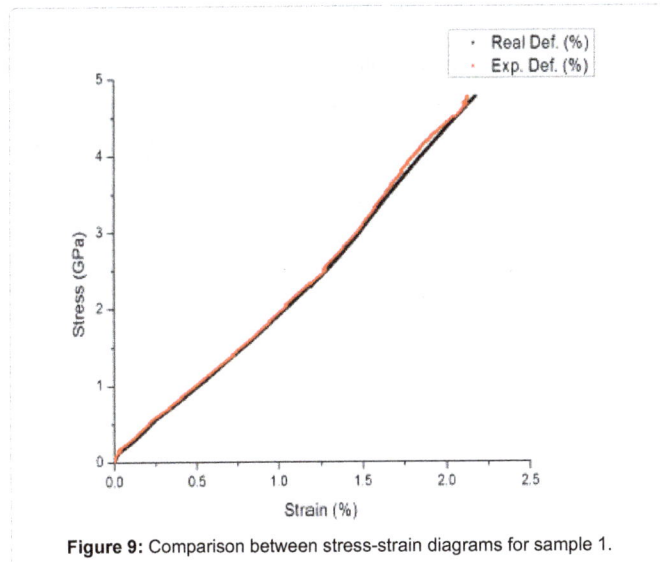

Figure 9: Comparison between stress-strain diagrams for sample 1.

Young's Modulus E	1.9 +/- 0.04 GPa
%difference from 2 GPa	3%
Standard deviation	1.2

Table 5: Statistical results for sample 1 during the compression test and cross-correlation analysis; fibers randomly orientated to the load.

doing a second test with fibers randomly orientated, it was observed not as good correlation than sample 1 obtaining 5% error from the real

strain measurement. In Figure 10 we can see both plots of interest and Table 6 summarizes showing the statistics of this test.

A third test was done and Figure 11 shows both plots of interest, Table 7 summarizes showing the statistics of this test, it was observed a similar correlation to sample 1 obtaining 3% error from the real strain measurement. Table 8 summarizes these tests performed with fibers randomly orientated to the load, we can assume that the mean

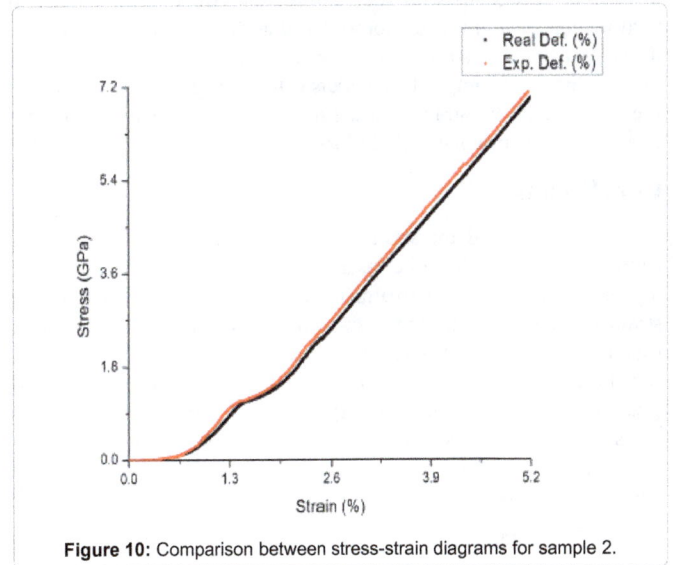

Figure 10: Comparison between stress-strain diagrams for sample 2.

Young's Modulus E	1.7 +/- 0.4 GPa
%difference from 2 GPa	5%
Standard deviation	1.9

Table 6: Statistical results for sample 2 during the compression test and cross-correlation analysis; fibers randomly orientated to the load.

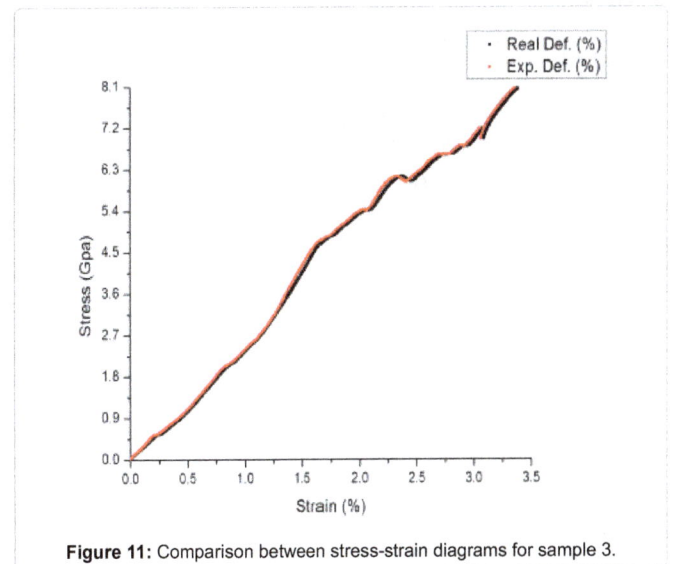

Figure 11: Comparison between stress-strain diagrams for sample 3.

Young's Modulus E	1.89 +/- 0.05 GPa
%difference from 2 GPa	3%
Standard deviation	1.92

Table 7: Statistical results for sample 3 during the compression test and cross-correlation analysis; fibers randomly orientated to the load.

Sample	Accuracy	Mean error
	97%	3%
	95%	5%
	97%	3%

Table 8: Accuracy and error of the three tests; fibers randomly orientated to the load.

accuracy is 96%. By comparing the average strain determined for 6 different specimens with the compression stress values measured with the load cell, it was demonstrated that the present technique can measure the relative strain with an average uncertainty of 5% for strain measurements of composite materials with fibers parallel orientated to the load and 4% for strain measurements of composite materials with fibers randomly orientated to the load.

Conclusions

It was measured the strain of two composite materials under a compression test using one laser-beam, its speckles reflection and digital image correlation treatment, we obtain 95% of accuracy for strain measurements in composites with fibres parallel orientated to the load and 96% of accuracy for strain measurements in composites with fibres randomly orientated to the load. Also we obtained 1.05×10^{-4} sensitivity, which is less than the reported 2×10^{-4}. We demonstrate that cross-correlation plus speckle metrology could be a reliable technique for measuring strain in composite materials under standardized compression tests.

Acknowledgements

Alonso Salda Ã ± a Heredia wants to thank CONACYT for the grant No. 360140.

References

1. Beasley F, Beasley DE (2011) Theory and design for mechanical measurements (5th edn.) John Wiley & Sons, USA.

2. Gere J, Goodno B (2009) Materials mechanics (7th edn.) Cengage Learning, Mexico.

3. Mathar J (1934) Determination of initial stresses by measuring the deformation around drilled holes. Transactions ASME 56: 249-254.

4. ASTM (2008) Determining residual stresses by the hole-drilling strain gage method. ASTM Standard Test Method E837-08. American Society for Testing and Materials, West Conshohocken, PA.

5. Rigden JD, Gordon EI (1962) The granularity of scattered optical laser light. Proceedings of the Institute of Radio Engineers 50: 2367-2368.

6. Dainty JC, Ennos AE, Francon M, Goodman JW, McKechnie TS, et al. (1975) Laser speckle and related phenomena. Springer-Verlag Berlin Heidelberg.

7. Peters WH, Ranson WF (1982) Digital image techniques in experimental stress analysis. Opt Eng 21: 427-431.

8. Sutton MA, Mc Neill SR, Helm JP, Chao YJ (2000) Advances in two dimensional and three dimensional computer vision. Photomechanics 77: 323-372.

9. Chen DJ, Chiang FP, Tan YS, Don HS (1993) Digital speckle-displacement measurement using a complex spectrum method. Appl Opt 32: 1839-1849.

10. Anuta PE (1970) Spatial registration of multispectral and multitemporal digital imagery using fast Fourier transform techniques. IEEE Trans Geosci Electron 8: 353-368.

11. Kuglin CD, Hines DC (1975) The phase correlation image alignment method. Proc Int Conf Cybernetics and Society, Japan.

12. Stanciu A, Cotoros D, Baritz M, Rogozea L (2010) Analysis by strain gauges of the strains in a composite material. Recent advances in signal processing, robotics and automation. pp: 254-257.

13. Satoshi K, Yoshihisa T, Kimiyoshi N, Yutaka K (2011) Measurement of strain distribution of composite materials by electron moire method.

14. Domanski AW, Bieda M, Lesiak P, Makowski P, Szelag M, et al. (2013) Polarimetric optical fiber sensors for dynamic strain measurement in composite materials. Acta Phys Pol A 124: 399- 401.

15. Pradille C, Bellet M, Chastel Y (2010) A laser speckle method for measuring displacement field. Application to resistance heating tensile test on steel. Applied Mechanics and Materials 24: 135-140.

16. Chevalier L, Calloch S, Hild F, Marco Y (2001) Digital image correlation used to analyze the multiaxial behavior of rubber-like materials. Eur J mech A/solids 20: 169-187.

17. Campbell FC (2010) Structural composite materials. ASM International.

18. Mitchell BS (2004) An introduction to materials engineering and science. John Wiley & Sons, USA.

19. ASTM E-9, ICS Number Code 77.040.10, Mechanical testing of metals.

20. Sirohi RS (2009) Optical methods of measurement wholefield techniques (2nd edn.) Taylor and Francis Group, USA.

21. Alda J (2003) Laser and gaussian beam propagation and transformation. Encyclopaedia of optical Engineering. pp: 999-1013.

22. Yariv A (1990) Quantum Electronics (3rd edn.) John Wiley & Sons, USA.

23. Wood R (1988) Physical Optics (3rd edn.) Optical Science of America, Washington DC.

24. Born M, Wolf E (1970) Principle of optics (4th edn.) Pergamon Press, U.K.

25. Goodman JW (1976) Some fundamental properties of speckle. J Opt Soc Am 66: 1145-1150.

26. (2015) Laser Product Datasheet generated on.

30

Scratch Wear Resistance of TiALN And AlCrN Coated EN-353 Steel

Chandrashekhar A[1]*, Kabadi VR[2] and Bhide R[3]

[1]Department of Mechanical Engineering, Basaveshwar Engineering College, Bagalkot, India
[2]Department of Mechanical Engineering, Nitte Meenakshi Institute of Technology, Bangaluru, India
[3]Application Support Centre, Oerlikon Balzers Coating India Limited, Bhosari, Pune, Maharashtra, India

Abstract

TiAlN and AlCrN coatings were deposited on mild steel (EN 353 steel) by cathodic arc evaporation technique. Coefficients of friction, critical load, adhesive and cohesive properties of these coatings were studied using scratch tester. Failure mode for coatings with thickness of 2 and 4 μm were studied using scratch channels and acoustic activities as basis. Studies presented relationship between progressive and in situ emission signals. AlCrN coating exhibited higher critical load as determined by the acoustic emission signal. Additionally wear mechanisms were analyzed. Analysis of the experimental results showed that load carrying capacity of AlCrN coating is better than TiAlN coating. Scratch wear test of AlCrN coated substrates showed reduction in cracking and spalling of coated layers.

Keywords: TiAlN and AlCrN coatings; Magnetron sputtering; Scratch testing; Scratch resistance

Introduction

For variety of applications, improved tribological properties has become an important aspect in the development of high performance materials [1]. In this direction, coatings are widely used to reduce the coefficient of friction as well as wear loss in all types of sliding contacts. Much excellent information on coated components in automobile applications has been achieved. Majority of sliding components have shown reduction in the coefficient of friction by two orders of magnitude. For example, by coating a steel substrate with molybdenum disulphide or diamond, this slides against a steel counterface, the wear loss has been reduced several orders of magnitudes compared with the uncoated pairs in contact [2].

Thermal spraying, electro deposition, chemical vapour deposition (CVD), and physical vapour deposition (PVD) etc. are some of the coating methods commercialized recently [3-5]. Improvement in hardness, lower friction and corrosion resistance of surfaces in tool like drilling etc., is achieved by coating them with TiN, CrN, TiAlN, AlTiN, TiCN, WC, AlCrN, multi layers of TiN and TiAlN [6]. The tribological properties of steels are generally improved with deposition of PVD coatings; the durability of the coating depends not only on the properties of coating but also on the adhesion of the coating and the substrate as reported by Totik [7]. High hardness, low coefficient of friction, high wear resistance and good oxidation resistance makes titanium nitride (TiN) an industrially popular coating. These properties make this an ideal candidate material for various tribological applications. TiN coatings have been most commonly used in the applications of forging tools, molds, cutting tools, bearing spindles and many mechanical components to decorative items because of its resistance to wear, corrosion and temperature [8-12]. The tribological behaviour of TiN, AlTiN and AlCrN films were investigated that vary with substrate roughness, thickness of the film coating, hardness of the substrate, deposition method, type of wear, stoichiometry and type of heat treatment. For hard substrates are relatively brittle like ceramics, the hard coating materials are TiN, AlTiN and AlCrN films. They have a great tendency to fracture and spall from the substrate during wear. Such tendency increases with increasing the applied load, sliding speed and coating thickness [13]. In recent years, studies on aluminium titanium nitride (AlTiN) and aluminium chromium nitride (AlCrN) coatings have been increased [14-16].

In most of the mechanical industries, EN-353 low carbon steel is widely used because of its good ductility and weldability. This material has poor tribological properties such as high coefficient of friction, low wear resistance, and low hardness [17]. Two methods are commonly used to improve the tribological behaviour of steel. Firstly, adding alloy elements to the steel during smelting of the integral alloy and secondly, by surface modification. The former is impractical because number of interstitial elements doped into the steel is harmful to ductility and increase the cost of products. Thus, surface modification is an effective and economical method that has become one of the most popular research fields in recent years. A wide variety of surface modifications of steels are viable routes to improve tribological properties viz., PVD and ion plating. An advantage of the technology is that the coatings are metallurgically bonded to the substrate with good interface. It also exhibits a high deposition rate, good coating uniformity, and controllability of coating thickness to complex shaped substrates.

The scratch test is one of the most convenient methods to use and no special dimension specimen or preparation is required compared to other methods. Therefore, to investigate the adhesion of thin coating-substrate scratch tester has increasingly been used [18]. Adhesion strength of thin films is measured using ramp progressive load scratch tester where in a normal load applied to the coating surface is increased and the load, at which coating fails adhesively, is taken as the critical load for coating failure. This test can be used to determine the wear behaviour of a material under different stresses. The strength of adhesion is influenced by many factors: the substrate hardness, the coating thickness, the surface quality, the coating hardness, the loading rate, interface bonding, indenter dimensions, and the friction between the indenter and a coating [19].

Although a great deal of research work has been carried out on friction and wear behaviour of various coatings, very limited work has

---

*Corresponding author: Chandrashekhar A, Department of Mechanical Engineering, Basaveshwar Engineering College, Bagalkot, India
E-mail: chandrashekhar_dev@yahoo.co.in

been done on scratch behaviour of TiAlN and AlCrN coatings. Against this background, the present work, we have studied the relation between the wear resistance and the scratch behaviour for comparing different hard coatings with different soft and hard substrates. The critical load values depend upon the hardness of the substrate, when scratching a specimen with a thin coating. Increasing the thickness of coating the critical load values causes to rise.

Materials and Methods

Materials

The EN-353 steel substrate of the actual chemical composition were analysed with the help of Optical Emission spectrometer (Thermo Electron S. A. En Vallaire Quest 1024, Ecublens, Switzerland make). The nominal and actual chemical composition of the substrates used in the present investigation are listed in Table 1.

Deposition method

The substrate used was mild steel (EN 353 steel), consisting of 0.217% C. The substrate surface was ground with SiC paper to remove the oxides and other contamination. Prior to deposition, substrates were initially degreased in ultrasonic bath to remove excessive oils and greases. This is followed by cleaning with alkaline solutions and rinsing with Demineralized Water (DM). Final stage of cleaning includes drying with hot air blower. Cleaning of these parts was carried out using Oerlikon Balzers proprietary (standard) cleaning procedure.

TiAlN and AlCrN coatings were carried out with arc evaporation method using Oerlikon Blazers commercial coating process. Customized TiAl and AlCr target in reactive nitrogen atmosphere were used to obtain stoichiometric TiAlN and AlCrN coating. The thicknesses of the AlCrN coatings were approximately 2 ± 0.3 and 4 ± 0.2 μm. Coating was carried out at temperature of about 500 ± 10°C with nitrogen as reactive gas. A DC-substrate bias voltage was maintained in the range of -50 to -150V during coating.

Microstructure

A Zeiss Axiovert 200 MAT inverted optical microscope, fitted with image software Zeiss Axiovision Release 4.1, was used for optical microscopy. The porosity measurements were made with image analyser, having software of Dewinter Materials Plus 1.01 based on ASTM B276. A PMP3 inverted metallurgical microscope was used to obtain the images.

Hardness

Hardness measurements for the compound system of substrate + coating were carried out as per the IS 1501-2002 procedures by using Vickers hardness tester (MH6). The hardness of the coatings on steel was measured using a VH-1 (METATECH) hardness tester, applying a load of 10, 25 and 50 g and a dwell time of 15 seconds. To measure the hardness of the coating a nano-indenter was used. The hardness was measured in three different locations and the readings reported are the average of the three readings.

X-ray diffraction

The coated specimens were subjected to XRD pattern was recorded using computer controlled XRD-system, JEOL, and Model: JPX-8030 with CuK radiation. The scan rate used was 2°/min and the scan range was from 10° to 120°. The grain size of the thin films was estimated from Scherrer formula, as given in Eq. (1). In this expression, the grain size D is along the surface normal direction, which is also the direction of the XRD diffraction vector.

$$D = \frac{0.9\,\tau}{B\cos\theta} \tag{1}$$

where B is the corrected full-width at half maximum (FWHM) of a Bragg peak, λ is the X-ray wavelength, and θ is the Bragg angle. B is obtained from the equation $B^2 = B_r^2 - B_{strain}^2 - C^2$, where B_r is the FWHM of a measured Bragg peak, $B_{strain} = \varepsilon\,\tan\theta$ is the lattice broadening from the residual strain ε measured by XRD using the $\cos^2\alpha\,\sin^2\psi$ method, and C is the instrumental line broadening.

Surface roughness

The surface morphology (2D and 3D) of the thin films was characterized by atomic force microscopy (AFM) to calculate the surface roughness and particle size. Innova SPM atomic force microscope works in both contact and tapping mode. Contact Mode is the most straightforward, basic topography imaging mode of the AFM. In contact mode, the AFM tip has a direct contact with the sample. While the tip is scanned along the surface, the sample topography induces a vertical deflection of the cantilever. This deflection is measured by a fiber- optical interferometer. On the other hand, the tapping mode maps topography by lightly tapping the surface with an oscillating probe tip. The cantilever's oscillation amplitude changes with sample surface topography, and the topography image is obtained by monitoring these changes and closing the z feedback loop to minimize them. It overcomes some of the limitations of both contact and non-contact AFM.

Scratch testing

Wear behaviour tests were carried out on the DUCOM scratch tester TR101 (Figure 1) under dry conditions, in ambient air at room temperature (≈ 25°C). Micro scratch tester have a Rockwell shaped C spherical indenter with a normal radius of 200 μm and an angle of 120° was used. A schematic diagram of scratch tester is presented in Figure 2. The sample for scratch test experiments substrates were machined with dimensions of 60 mm × 25 mm × 10 mm.

The indenter velocity was used over a wear tracks of 14 mm, with different normal loads applied. PLST (Progressive Loading

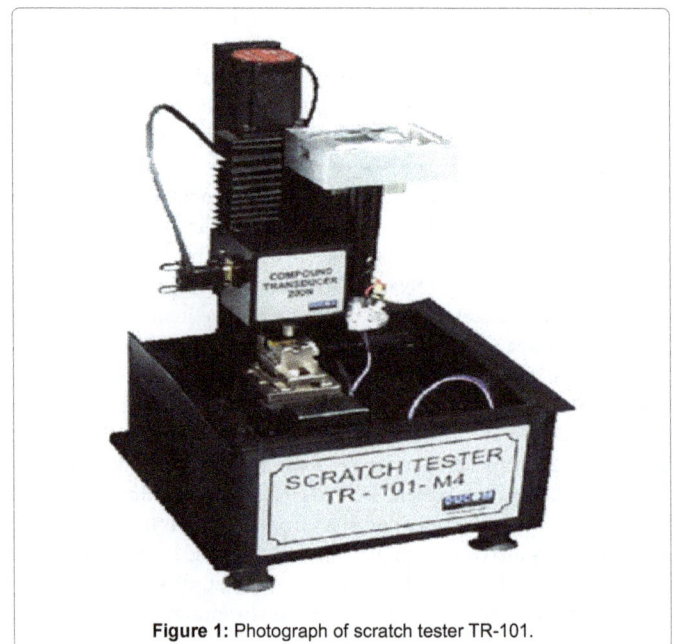

Figure 1: Photograph of scratch tester TR-101.

Elements	C	Mn	Si	P	S	Ni	Cr	Mo	Fe
Nominal	0.1-0.2	0.5-1.0	0.35	0.05	0.05	1-1.5	0.75-1.25	0.08-0.15	Bal.
Actual	0.217	0.57762	0.1895	0.0582	0.0487	0.0317	0.05306	0.00459	Bal.

Table 1: Chemical composition by wt % of EN-353 steel.

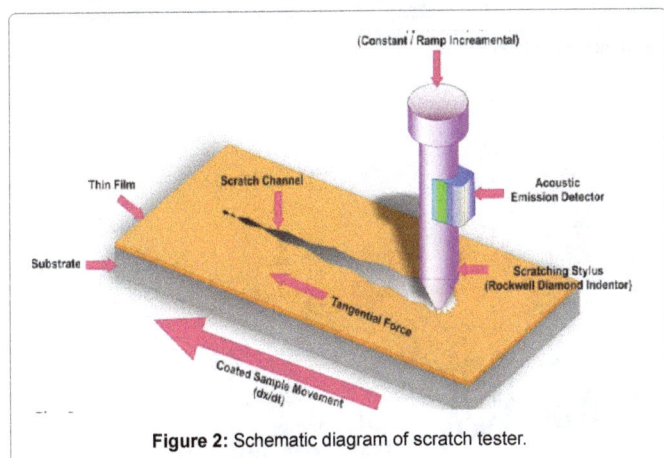

Figure 2: Schematic diagram of scratch tester.

Samples (thickness)	50 g	25 g	10 g
AlCrN (2 μm)	634	1034	1094
AlCrN (4 μm)	718	1102	1216
TiAlN (2 μm)	622	1027	1081
TiAlN (4 μm)	689	1083	1197

Table 2: Vicker's hardness of AlCrN and TiAlN coatings on MS at different loads.

Scratch Test) mode of scratch testing was used. In PLST mode the normal load was increased linearly during the test from 10 to 160 N. The indenter scratches the coated substrate linearly with increasing load progressively. Scratch adhesion testing is loaded stylus drawn a spherically tipped diamond indenter over the surface of a coated sample under a normal force increasing vertical load either stepwise or continuously until coating becomes seperates completely from the scratch channel. The value of the critical load is considered as of the coating–substrate adhesion [20,21]. The critical loads for full delamination were determined from the recorded normal force, tractional force, coefficient of friction and acoustic emission along the scracth; the respective images have also been taken.

Worn surface morphology

Scanning electron microscope (SEM-JEOL, Model-JSM 6380) is used to study the behavioral surface morphology of the coatings. SEM micrographs were taken with electron beam energy of 15 keV.

Results and Discussion

Coating microstructures

The TiAlN and AlCrN coatings have been formulated successfully by PVD (physical vapour deposition) technique on EN-353 steel. The optical micrographs of the substrate, TiAlN and AlCrN coatings indicating dense, fine grained columnar structures and are depicted in Figures 3 and 4. Figures 3a and 3b shows the microstructure of TiAlN coating is violet-grey in color and blue-grey in case of AlCrN (Figures 4a and 4b) coating. The coatings have uniform microstructure. It is evident from the microstructure that the coatings contain some pores and inclusions. The porosity for as-coated TiAlN and AlCrN coatings is 0.47% and 0.46% respectively [22].

Vicker's microhardness

Hardness values for coatings with TiAlN and AlCrN films with 2 and 4 μm thickness on EN-353 steel are summarised in Table 2. The Vicker microhardness of TiAlN and AlCrN coatings were around Hv` = 1081 to 1216, and increased with increase in coating thickness. As expected from theretical considerations, the coating hardness increased with target hardness, and film thickness, and decreased with increasing indentation load. The increase in hardness was observed with the decrease in load as listed in Table 2. With the incorporation of AlCrN film of ~2 μm to ~4 μm thick, the microhardness increases from 1094 to 1216 at 10 g load. However, the TiAlN coating Vickers microhardness was lower than that of AlCrN coating, which is about Hv = 1081 to 1197, may be because of the pores in the coatings.

The difference in the micro-hardness is in coincidence with the variation of the coating structure especially the density and porosity of the coating. Further, the influence of thickness of the AlCrN coating on the Vicker's micro-hardness was related to the coating structure (Figure 3a and 3b). Inside the thick coating (4 μm) AlCrN coating (Figure 3b) the densed AlCrN film was formed.

Crystal structure of AlCrN and TiAlN coatings

XRD spectra for AlCrN and TiAlN coating thickness of 2 and 4 μm are depicted in Figures 5 and 6 on reduced scale. XRD analysis for AlCrN coating confirmed the presence of CrN and AlN phases. Further, in case of TiAlN coating the prominent phases are a large percentage of Ti_2N along with AlN. From the XRD spectra, the grain size of the thin coatings was estimated from Scherrer formula as given in Eq. (1), and reported in Table 3. The grain size in case of TiAlN coatings (15 nm) is less than that of AlCrN coating (21 nm). Sample TiAlN coating with thickness 2 μm showed a strong preferred lattice (ε) planes have a dominant orientation at 44.92° (2θ), as shown in Figure 5. The preferred lattice plane is caused by the mobility of atoms which decreases with increasing Al/Ti ratio. In contrast, in the case of the AlCrN coating, the lattice planes (α) have a dominant orientation at 49.637° (2θ), as shown in Figure 6. The intensities of the diffraction peaks gradually decrease with increasing layer thickness, which indicates a gradual decrease in grain size of the preferred orientation. Oerlikon Balzers Ltd. India provided the data regarding hardness and the friction coefficient against steel (dry), along with the coating parameters (Table 3). The coated layer on the steel substrate has provided higher hardness as compared to the substrate. TiAlN coating showed higher hardness value than AlCrN coating as reported in Table 2.

Effect of surface roughness

Surface roughness either from the substrate or coating influences the scratch test data. Hence surface roughness of the coated surfaces prior to testing is essential. Surface roughness also influences the friction as well as wear performance of a mechanical system like sleeves used in bearings. It has also been shown from the literature on hard coatings that the rougher the surface finish, the lower will be the coating adhesion and wear resistance [1]. All morphologies exhibited domes and craters which are uniformly distributed over the entire surface. Figure 7 shows the roughness values of the TiAlN and AlCrN coated on EN-353 steel containing film thickness of 2 and 4 μm. The roughness of the TiAlN coated substrate is nearly two times higher

Coating Type	Coating thickness(μm)	Arithmetic mean roughness Ra (nm)	Maximum height Rq (nm)	Point mean roughness Rmax(nm)	Appearance of surface features
AlCrN	~ 2	235	297	2779	Irregularly spaced peaked features
TiAlN	~ 2	244	306	2340	Irregularly spaced peaked features
AlCrN	~ 4	165	206	1702	Irregularly spaced peaked features
TiAlN	~ 4	221	274	2224	Irregularly spaced peaked features

Table 3: Results demonstrated variation in the size and height of surface nano-topographs of TiAlN and AlCrN coatings.

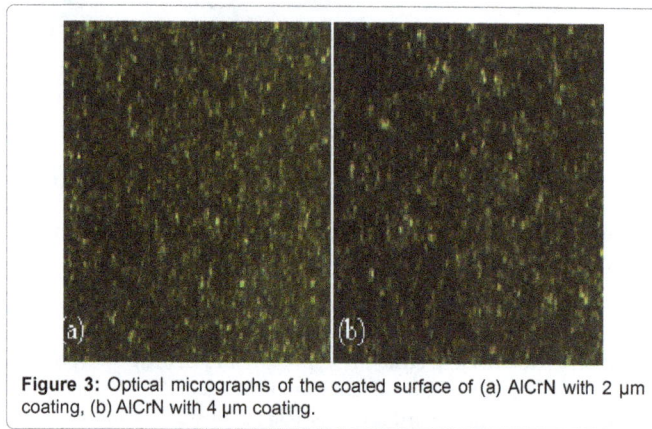

Figure 3: Optical micrographs of the coated surface of (a) AlCrN with 2 μm coating, (b) AlCrN with 4 μm coating.

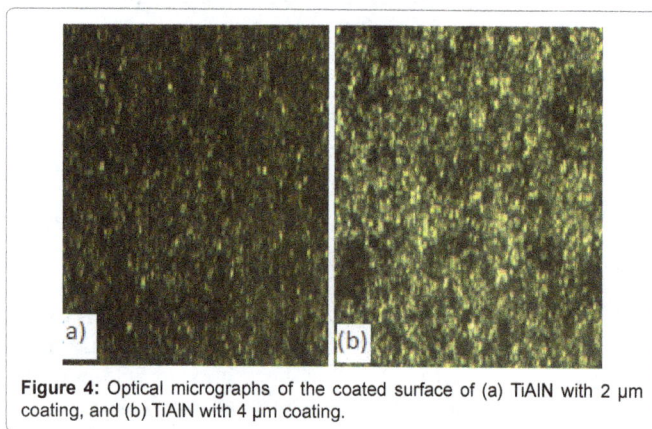

Figure 4: Optical micrographs of the coated surface of (a) TiAlN with 2 μm coating, and (b) TiAlN with 4 μm coating.

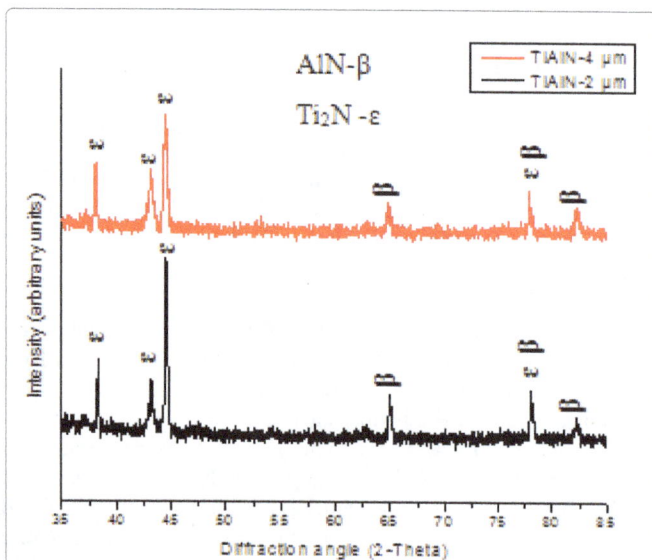

Figure 5: X-ray diffraction pattern for ~2 & ~4 μm TiAlN coating on mild steel substrate (EN-353).

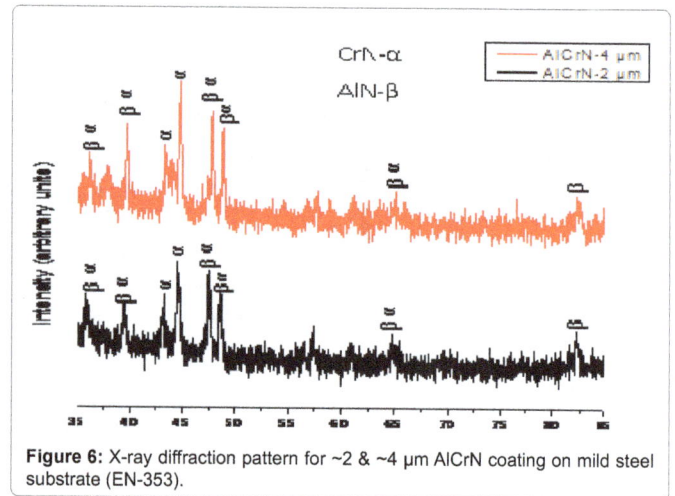

Figure 6: X-ray diffraction pattern for ~2 & ~4 μm AlCrN coating on mild steel substrate (EN-353).

than that of the AlCrN coated substrate. The Ra values are low at 2 μm coating thickness and increases with increasing coating thickness. Higher roughness values especially in 4 μm indicate the presence of little agglomeration. A higher roughness value leads to easy pull-out of coated film in wear scratch testing. Figure 7 shows the typical roughness profile of the TiAlN and AlCrN coated on EN 353 steel substrates with a Ra value of 1.2 μm and 0.93 μm respectively. The roughness profiles of TiAlN and AlCrN coated on EN 353 steel substrates are similar to the literature findings [23,24].

Figures 7a-7d shows the difference in morphology between the two films can be inferred by comparing the 2D images in Figures 7a and 7b; however, a clearer comparison of the films is afforded by viewing the 3D images in Figures 7c and 7d. Surface roughness of all the coatings increases after deposition. By increasing the coating thickness, the surface roughness turns to smoother and smoother as shown in Table 3. i.e., the area distributed with protrusions becomes larger as the coating films become thicker. The surface morphology explained that coating surface roughness results from the sharps of nucleus, from preferential nucleation at substrate inhomogeneties, from the coating thickness and preferential growth. The fluctuation of coating thickness results in the surface energy fluctuation. The 2 and 4 μm AlCrN coatings are found to give very finest asperities compared to TiAlN coatings, i.e., the morphology induced by the coating growth process and morphology of the coating thickness. The increase in coating thickness results in the fine-grained morphology of the coating surface with decreased the asperities. A surface with a higher value of Ra or Rq implies less uniformity.

Scratch resistance

Adhesion of a coating is one of the important properties for better load carrying capacity and scratch resistance of a coating system. If a coating does not adhere to the substrate, burning, cracking or peeling off the deposit would occur affecting the performance of the coating. The damage can be a cracking of the coating at the interface;

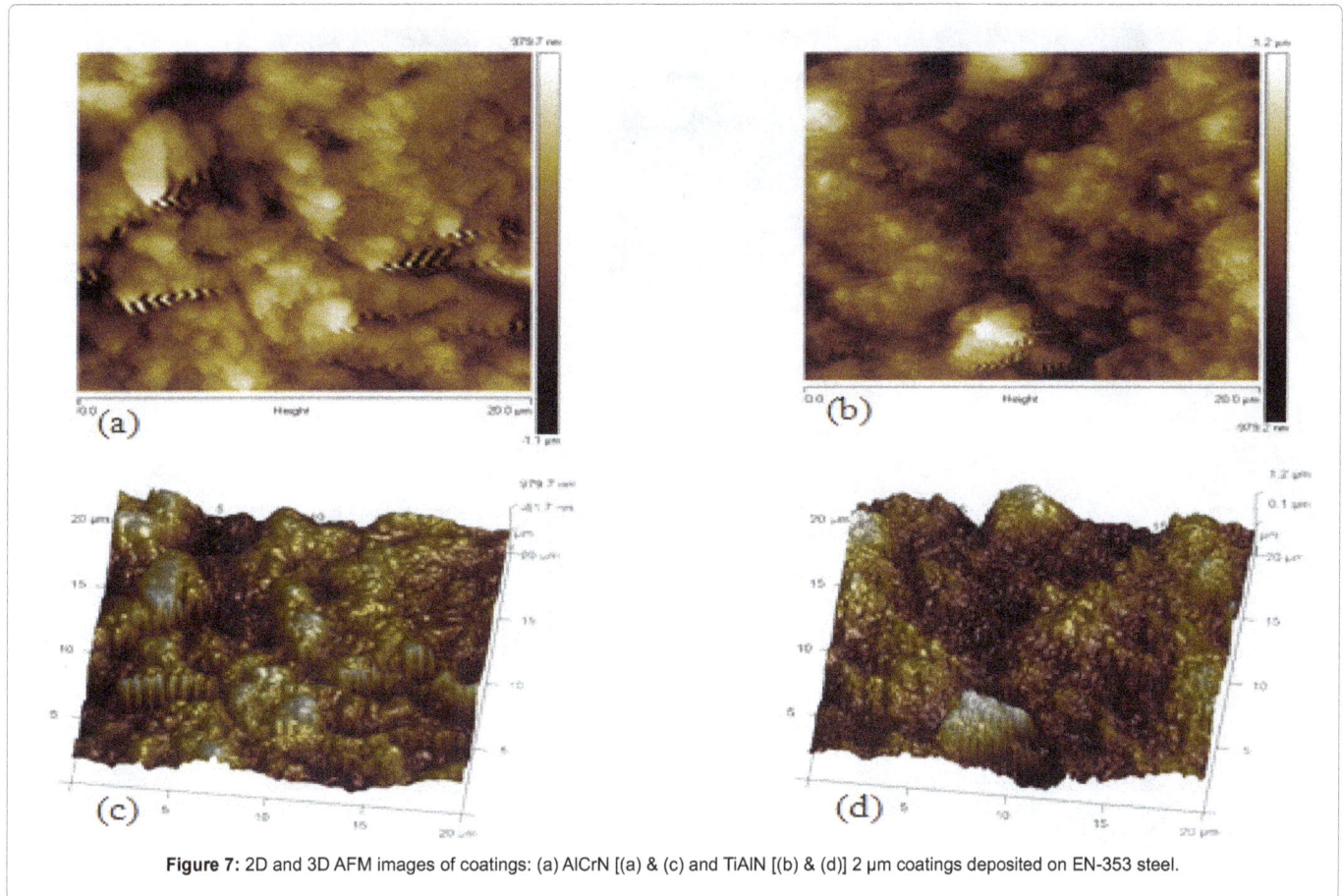

Figure 7: 2D and 3D AFM images of coatings: (a) AlCrN [(a) & (c) and TiAlN [(b) & (d)] 2 μm coatings deposited on EN-353 steel.

this indicates an adhesive type of failure. If the damage consists of chipping within the coating itself, then it is called a cohesive type of failure. The scratch test has achieved the most widespread use in assessing the adhesive strength of hard coated substrates of all the tests available for the measurement of coating adhesion. The test consists of applying an incrementally or continuously increasing load on the system at constant speed. The scratching point causes increasing elastic and plastic deformation until damage occurs in the surface region. In practice, the film is seldom removed entirely from the channel so it is convenient to define a critical load, which characterizes the mechanical adhesive strength of the coating-substrate system [25-30].

Scratch test shows variation of traction force (tangential force) and normal load (applied load) from start point to end point of the scratch. In this test, with the applied force, the traction force suddenly shows a peak and then attains a steady state value. The failure of the coating, whether cohesive/adhesive, is normally indicated by a pointed rise in the traction force. The lower value (base of the peak) of this traction force corresponds to the first sign of coating failure while the upper end (top of the peak) corresponds to the end of the failure. Scratch resistance is the normal load applied on the stylus corresponding to the above two values in traction force. It is minimum normal load required to delaminate coating from its substrate. However, due to the fact that there is a distribution of flaws at the coating substrate interface, isolated areas of coating removal do occur and the critical transitions can be somewhat subjective [31].

Number of experiments was carried out on two types of coated systems on EN-353 steel as substrate. The scratches on the coated substrates were linear progressively increasing load. Variation of the different characterizations obtained for three different sliding speeds are 0.1, 0.5 and 1.0 mm/s along with different start loads and loading rates are 10, 20 and 40 N and 2, 5 and 10 N/mm respectively are presented in Figures 8 and 9.

The critical load values of the composite TiAlN and AlCrN films, which were deposited on the EN-353 steel with 2 different coating thickness are showing relationship between normal load, coefficient of friction and scratch distance are given in Figures 8 and 9. The progressive incremental load, as shown Figure 8a was applied over the coated samples through the diamond stylus. It is observed that with increase in the scratch length the normal load is increased with respective to the sliding speed. i.e., normal load is directly proportional to the scratch distance. Initially the indenter will penetrate in the coating material. With the increase in the penetration of the indenter in the coating material the normal load increases. The scratch test is in nothing but the single point cutting tool process like shaper. As the progressive load increases, the penetration of the indenter in the substrate increases. At beginning the indenter penetrate in the coating then to the substrate ahead of the indenter goes on collecting ahead of the indenter in the form of chip. At the lower normal load of about 14 N with variation of coefficient of friction, which initiates the propagation of the microcracks and adhesive failure shows that film has begun to remove from the substrate. A strong distortion of the variation of the traction force and the coefficient of friction was observed at a higher normal load indicating that the total removal of the coating from the substrate.

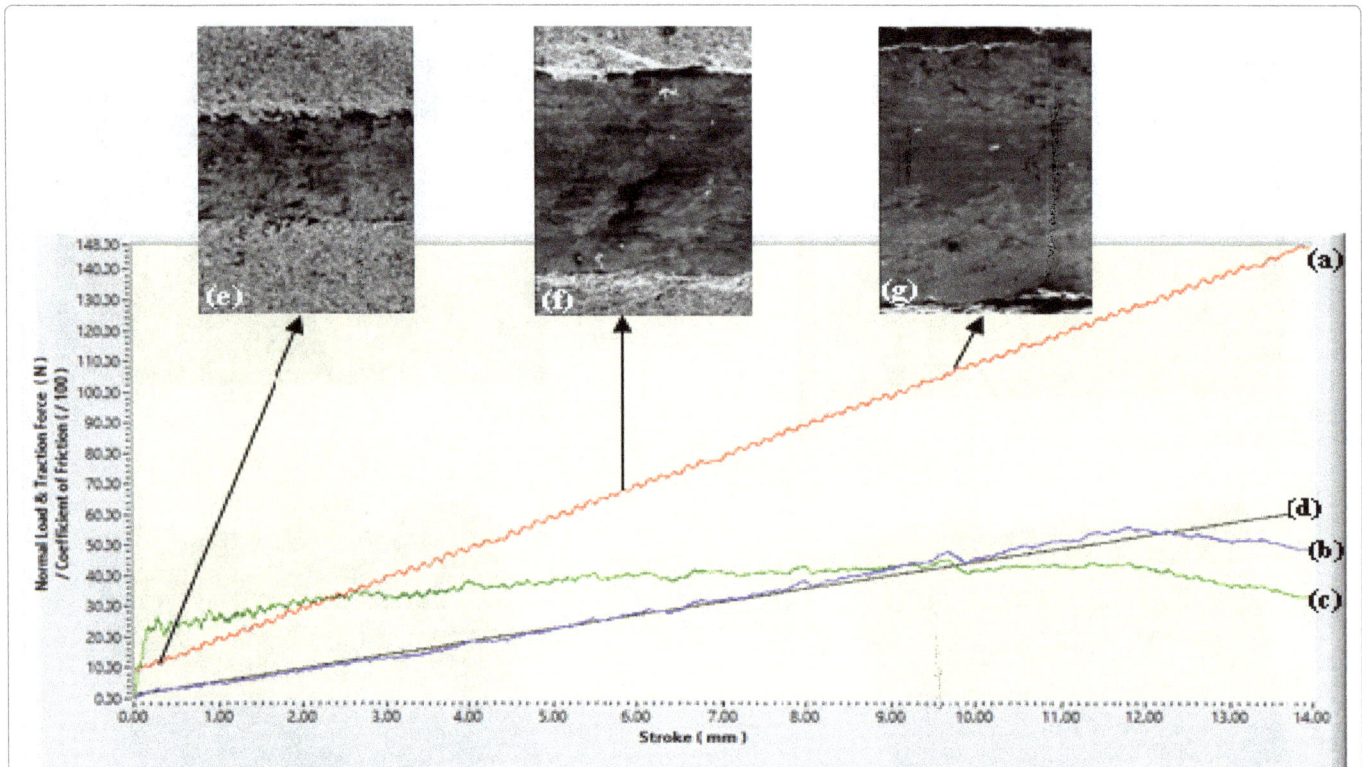

Figure 8: The typical images of scratch track along with graphs on the 2 μm TiAlN coated EN-353steel substrate at different locations. (Start Load- 10 N, Loading Rate- 10 N/mm, Scratch speed- 1.0 mm/sec and Scratch Stroke- 14 mm). a) Normal load, b) Tractional Force, c)Co-efficient of friction, d)Acoustic Emission, e)SEM, showing coating cracks at beginning, f) SEM, worn surface at middle, g) SEM, highly worn surface).

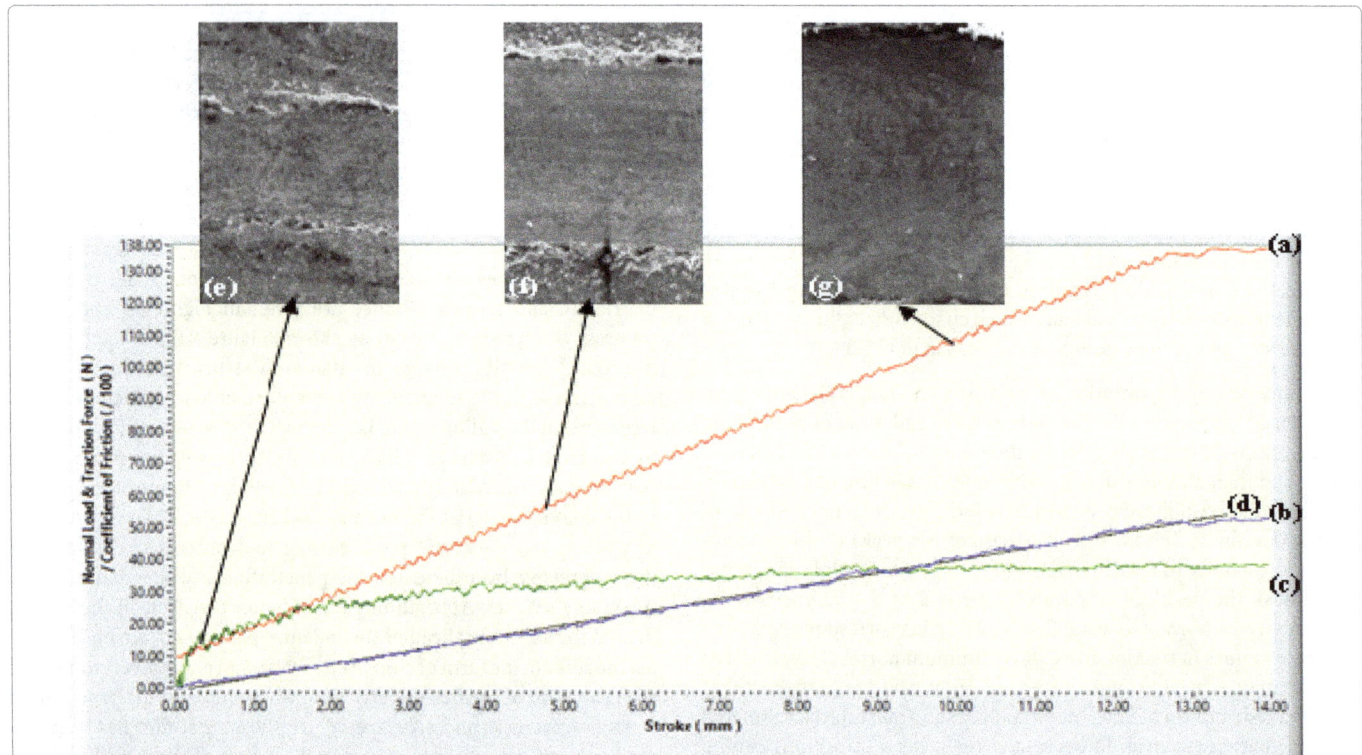

Figure 9: The typical images of scratch track along with graphs on the 2 μm TiAlN coated EN-353steel substrate at different locations. (Start Load- 10 N, Loading Rate- 10 N/mm, Scratch speed- 1.0 mm/sec and Scratch Stroke- 14 mm). a) Normal load, b) Tractional Force, c) Co-efficient of friction, d)Acoustic Emission, e) SEM, showing coating cracks at beginning, f) SEM, worn surface at middle, g) SEM, highly worn surface).

In Figure 8b indicates the variation of tractional force along scratch length on the TiAlN coating during scratch test. The traction force increased linearly with the increase in normal load and the slope changes indicates that the failure of coating material. It can see that the maximum tractional force 56 N takes place at scratch length 11.8 mm in the end part of the scratch, where the maximum depth of cut occurs [32]. Once the indenter crosses the thickness of the coating then the indenter penetrates inside the substrate material. Substrate material is softer than the coated material. Hence with further increase in the scratch length at 11.8 mm tractional force start decreases with increasing of the normal load. which may be due to the rate of collection of coating material ahead of the indenter is more when tractional force reaches a yield shear strength. This indicates that the total delamination of the coating from the substrate.

Figure 8c exhibits the coefficient of friction versus scratch distance curve, this can offer crucial clue to the determination of transition point in scratch damage mechanisms. The coefficient of friction is defined as tractional force and normal load. It is interesting to note that a changes of coefficient of friction (COF) slope that are obviously scratch damage feature, i.e., delamination, as scratch progresses. It was also observed that the fluctuation shown in COF curve, where tractional force increases gradually due to the penetration of the scratch tip into the substrate [33]. As seen from Figure 7 the COF increases initially around 0.3 mm scratch length and then remain nearly constant. The initial increase of the friction was considered to contain both adhesive and plouging friction [34]. Figure 8d shows the acoustic emission versus the scratch sliding stroke. The significance of scratch test carried out was that a direct correlation was detected about changes in the tractional force and coefficient of friction.

From Figure 9b it shows that higher tractional force were obtained for AlCrN coating than for TiAlN coating, which may be due to its higher plastic deformation resistance. Figure 9c shows lower coefficient of friction observed in AlCrN value of 0.38 as compared to the TiAlN coating which is 0.45. This confirms that the AlCrN coating has better wear resistance behaviour [5].

Graphical display of forces with almost linear increase in the traction force indicating scratch failure is of adhesive type for TiAlN and AlCrN coated EN 353 steel substrates which are shown in Figures 7 and 8, respectively. The Figure 7 also shows the traction loads corresponding to the normal loads representing the scratch failure. The AlCrN coated steel substrate shows that the failure, as indicated by the change in slope of the traction force, occurs at 110 N normal load, whereas it occurs at 90 N for the TiAlN coated steel substrate. It is also observed that the coefficient of friction of the AlCrN coated steel substrate is around 0.22 and 0.35 TiAlN coated steel substrate.

As it can be seen from the Figures 8 and 9, the frictional behaviour differs with varying sliding speed, start load and loading rate with the presence or absence of the protective layers like TiAlN and AlCrN coatings. Generally a rapid increase of the friction coefficient, reaching the local maximum value at the lowest speed, of 0.1 mm/s was detected at the primary 14 mm scratch stroke, independently of the scratch tested. Further, intermediate and higher speeds of sliding results in decreases the friction, which reaches the local minimum value promptly at a lower operational condition. Further tribological behaviour of the scratch tested depends significantly on the coating composition. AlCrN coating showed a slow increase of the coefficient of friction as compared to TiAlN coating.

For all the test coated substrates, the coefficient of friction, tangential force, normal load and acoustic emission with scratch length for all scratch tests, as shown coincide exact position with the typical topographically changes of the coatings by using and scanning electronic microscope along with the scratch track graphs.

Microscopic observations

On the basis of our experiments, we have adopted the approach whereby the adhesive wear loss is considered to be typical and significant when it occurs regularly along the scratch length (Figures 10 and 11). The coating can also be lost by cohesive failure of the substrate or by cohesive failure of the coating. These are both failures of the coating-substrate composite by a mechanism other than simple adhesive failure but both would correspond nevertheless to the failure of a part in service [35]. The scanning electron microscopy (SEM) examples shown in Figures 9e-9g was thought to be adhesive as well as cohesive failure in a TiAlN coated steel, whereas in Figures 10e-10g could be adhesive loss (flaking) in AlCrN coated steel substrate.

Scanning electron microscopy investigations of the scratched coated substrates were carried out in order to determine the mechanisms of material removal of the different coatings on substrates. SEM micrographs of the TiAlN coated EN353 steel are shown in Figures 10 and 11. During each experimental test, the normal load was increased from given start load until severe coating failure occurred (Figures 8 and 9). In most of the experiments, the critical normal load, Pc, at which coating failure initiated was detected by a sudden increase in acoustic emission and confirmed by microscopic examination.

The critical load and the corresponding coefficient of friction at the appearance of the first tensile crack can be determined around the scratch length of 114 μm, where the kinks or ripple structures occur in

Figure 10: SEM Images of scratch track for 2 μm TiAlN coated EN 353 steel substrate at different locations. (Start load- 10 N, Loading rate- 10 N/mm, Scratch speed- 1.0 mm/sec and Scratch stroke- 14 mm): (a) micrograph at 0.3 mm scratch length, (b) micrograph at 5.7 mm scratch length, and (c) micrograph at 9.1mm scratch length).

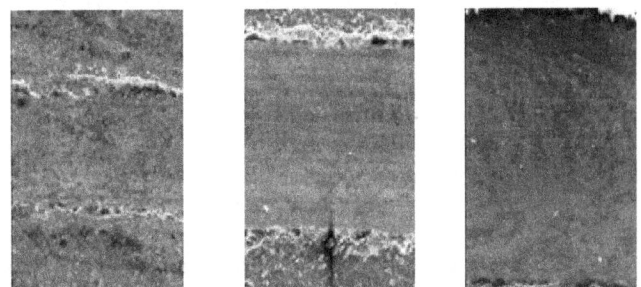

Figure 11: SEM images of scratch track for 2 μm AlCrN coated EN 353steel substrate at different locations. (Start load- 10 N, Loading rate- 10 N/mm, Scratch speed- 1.0 mm/sec and Scratch stroke- 14 mm: (a) micrograph at 0.3 mm scratch length, (b) micrograph at 4.9 mm scratch length, and (c) micrograph at 9.8 mm scratch length).

the vicinity marked by an arrow and become more remarkable as the scratch length or load increases for TiAlN coated samples. Both the critical load and the friction coefficient have been found to be 15.3N and 0.4, respectively.

It can be seen from Figure 7a that the cracks on the coating have been formed at the track without coating delamination. At higher applied loads a brittle failure mode of the coating was observed, as manifest by the formation of lateral cracks or chipping at the edges of the scratch tracks, as shown in Figure 10a. At this load, the driving force from the development of large tensile stresses at the edge of the scratch track is sufficiently high to cause the cracks to propagate across the TiAlN layer. One possible explanation for the formation of this kind of lateral crack is that the TiAlN microparticles act as stress raisers and cause the cracks to initiate. As the loading rate increases chipping at the edges of the scratch track and produces slight tensile cracks at load 60-70N, as shown in Figure 10b. This hypothesis is supported by the result of an event that is shown in Figure 10c. It is also evident that a through-thickness tensile crack travelled to the site of a macro-particle. Its propagation sheared the particle and produced the resulting circular lateral crack. The black coloration of the surface is due to the formation of oxides during wear testing.

SEM micrographs of the wear tracks of the AlCrN coated steel substrate are shown in Figures 11a-11c. It can be seen that this worn surface is characterized by scratches and plowing marks. The plowing action and particles pull-out as observed in Figure 11b are less severe for AlCrN coated steel substrate than the TiAlN coated steel substrate. The wear depth is also less for AlCrN coated steel substrate. Further, it was found that this failure mechanism was restricted to the coating, and no delamination from the substrate was observed for TiAlN coated steel substrate. Hence it can be inferred that EN-353 steel substrate has strong influence on both cohesive and adhesive strength of the TiAlN coating.

Conclusions

The TiAlN and AlCrN coatings were produced on EN-353 steel using cathodic arc evaporation method. The following conclusions can be drawn from our experimental investigations. Microstructure observations of coated steel substrates, few pores were detected optical micrographs. The thicker the deposition, the less porosity which could be due to substantial re-melting during the coating deposition. The AlCrN coating had a compound Vickers microhardness of Hv = 1094 which is higher than that of TiAlN coated samples. The compound hardness of both coatings increased with increase in coating thickness. The results of XRD data reveals that the grain size of TiAlN coating (15 nm) is less than that of AlCrN coating (21 nm). Sample TiAlN coating with thickness 2 μm showed a strong preferred lattice (ε) planes have a dominant orientation at 44.92° (2θ). The substrates coated with the AlCrN and TiAlN film, peaks protruding over the flat surface are quantitatively increased by the increase in coating thickness; the uniformity of the surface morphology can thus be improved. For the substrate with TiAlN and AlCrN coating, increase in coating thickness decreased the asperity. TiAlN coated steel substrate has higher surface roughness value and it seems that it is due to the variation in the particle size. With increase in the coating thickness, the amount of roughness also increased for TiAlN coated samples. Further, these samples showed higher wear loss and hence lower scratch resistance. The scratch resistance of AlCrN coated steel substrate was found to be superior to that of TiAlN coated samples. It could be attributed to be better adhesion of the AlCrN deposits in the steel substrate. Ploughing action and particles-pullout are the main wear mechanism of coatings.

References

1. Bautista Y, Gonzalez J, Gilabert J, Ibanez MJ, Sanz V (2011) Correlation between the wear resistance and the scratch resistance for nanocomposite coatings. Progress in Organic Coatings 70: 178-185.

2. Singer IL, Fayeulle S, Ehni PD (1996) Wear behaviour of triode-sputtered MoS₂ coatings in dry sliding contact with steel and ceramics. Wear 195: 7-20.

3. Low CTJ, Wills RGA, Walsh FC (2006) Electrodeposition of composite coatings containing nanoparticles in a metal deposit. Surface and Coating Technology 201: 371-383.

4. Gissler W, Jehn HA (1992) Advanced techniques for surface engineering. Springer Netherlands, Dordrecht, Netherlands.

5. Rickerby D, Matthews A (1991) Advanced surface coatings-A handbook for surface engineering. Springer Netherlands, Dordrecht, Netherlands.

6. Khlifi K, Cheikh Larbi AB (2014) Mechanical properties and adhesion of TiN monolayer and TiN/TiAlN nanolayer coatings. Journal of Adhesion Science and Technology 28: 85-96.

7. Totik Y (2010) Investigation of the adhesion of NbN coatings deposited by pulsed dc reactive magnetron sputtering using scratch tests. J Coat Technol Res 7: 485-492.

8. Lee YZ, Jeong KH (1998) Wear-life diagram of TiN-coated steels. Wear 217: 175-181.

9. Liu Y, Li L, Cai X, Chen Q, Xu M, et al. (2005) Effects of pretreatment by ion implantation and interlayer on adhesion between aluminum substrate and TiN film. Thin Solid Films 493: 152-159.

10. Su YL, Yao SH (1997) On the performance and application of CrN coating. Wear 205: 112-119.

11. Su YL, Yao SH, Leu ZL, Wei CS, Wu CT (1997) Comparison of tribological behaviour of three films-TiN, TiCN and CrN-grown by physical vapor deposition. Wear 213: 165-174.

12. Posti E, Nieminen I (1989) Influence of coating thickness on the life of TiN-coated high speed cutting tools. Wear 129: 273-283.

13. Guu YY, Lin JF (1996) Comparison of the tribological characteristics of titanium nitride and titanium carbonitride coating films. Surface and Coatings Technology 85: 146-155.

14. Fox-Rabinovich GS, Beake BD, Endrino JL, Veldhuis SC, Parkinson R, et al. (2006) Effect of mechanical properties measured at room and elevated temperatures on the wear resistance of cutting tools with TiAlN and AlCrN coatings. Surface & Coatings Technology 200: 5738-5742.

15. Chawla V, Chawla A, Mehta Y, Puri D, Prakash S, et al. (2011) Investigation of properties and corrosion behaviour of hard TiAlN and AlCrN PVD thin coatings in the 3 wt% NaCl solution. J Australian Ceramic Society 47: 48-55.

16. Aihua L, Jianxin D, Haibing C, Yangyang C, Jun Z (2012) Friction and wear properties of TiN, TiAlN, AlTiN and CrAlN PVD nitride coatings. Int J Refractory Metals & Hard Mat 31: 82-88.

17. Wang Y, Zhang P, Wu H, Wei D, Wei X, et al.(2014) Tribological properties of double-glow plasma surface niobizing on low-carbon steel. Tribology Transactions 57: 786-792.

18. Hedenqvist P, Olsson M, Jacobson S, Soderberg S (1990) Failure mode analysis of TiN-coated high speed steel: In situ scratch adhesion testing in the scanning electron microscope. Surface and Coatings Technology 41: 31-49.

19. Tushinsky L, IKovensky I, Plokhov A, Sindeyev V, Reshedko P (2002) Coated metal-structure and properties of metal-coating composites. Springer.

20. Xie Y, Hawthorne HM (2001) A model for compressive coating stresses in the scratch adhesion test. Surface and Coatings Technology 141: 15-25.

21. Nordin M, Larsson M (1999) Deposition and characterisation of multilayered PVD TiN/CrN coatings on cemented carbide. Surface and coatings technology 116: 108-115.

22. Nordin M, Larsson M (1999) Nano-galvanic coupling for enhanced Ag⁺ release in ZrCN-Ag films: Antibacterial application. Surface and Coatings Technology 116:108-115.

23. Chawla V (2013) structural characterization and corrosion behavior of nanostructured TiAlN and AlCrN thin coatings in 3 wt% NaCl Solution. Journal of Materials Science and Engineering 3: 22-30.

24. Huanga LY, Xua KW, Lu J (2002) Evaluation of scratch resistance of diamond-like carbon films on Ti alloy substrate by nano-scratch technique. Diamond and Related Materials 11: 1505-1510.

25. Lee LH (1991) Fundamentals of adhesion. Springer Science, Business Media, LLC, New York.

26. Gissler W, Jehn HA (1992) Scratching of materials and applications. Kluwer Academic Publishers, Italy.

27. Takadoum J (1998) Materials and surface engineering in tribology. John Wiley & Sons Inc, USA.

28. Bunshah RF (2002) Handbook of hard coatings. William Andrew, USA.

29. Lacombe R (2005) Adhesion measurement methods: theory and practice. Taylor and Frances, CRC Press, New York.

30. Tracton AA (2005) Coatings technology handbook. CRC Press, USA.

31. Rickerby D, Matthews A (1991) Advanced surface coatings: A Handbook of surface engineering. Chapman & Hall, New York.

32. Tschiptschin AP, Garzon CM, Lopez DM (2006) The effect of nitrogen on the scratch resistance of austenitic stainless steels. Trib Inter 39: 167-174.

33. Seo TW, Weon JI (2012) Influence of weathering and substrate roughness on the interfacial adhesion of acrylic coating based on an increasing load scratch test. J Mater Sci 47: 2234-2240.

34. Savas S, Danisman S (2014) Multipass sliding wear behavior of TiAlN coatings using a spherical indenter: effect of coating parameters and duplex treatment. Tribology Transactions 57: 242-255.

35. Szameitat K (1992) Hard chrome technology Modern , in a line with Okonomie and ecology. pp: 124-136.

31

Synthesis of Pyramid-Shaped NiO Nanostructures using Low-Temperature Composite- Hydroxide- Mediated Approach

Shahid T[2], Khan TM[1]*, Zakria M[1], Shakoor RI[1,4], Arfan M[2] and Khursheed S[2]

[1]*National Institute of Lasers and Optronics (NILOP), Islamabad, Pakistan*
[2]*Department of Applied Physics, Federal Urdu University of Arts, Science and Technology, Islamabad, Pakistan*
[3]*School of Physics, Trinity College Dublin (TCD), Ireland*
[4]*Department of Mechanical Engineering, Muhammad Ali Jinnah University, Islamabad, Pakistan*

Abstract

Composite-hydroxide-mediated (CHM) approach was used to synthesize NiO nanocrystals. The proposed method makes use of molten composite hydroxides; providing reaction media and lower the process temperature. Processing temperature and reaction time are the two potential parameters to control the growth of a nanomaterial. The method was used at temperatures in the range of 180-250°C and formation of the nanomaterial was monitored using XRD, SEM, EDX, FTIR, and UV-visible spectroscopy. The produced nanomaterial was purely polycrystalline with an average crystallite size in the range of 23.71-36.92 nm. Method suggested formation of pyramid shaped NiO nanocrystals in the temperature range 220-250°C. Evidence on the elemental composition, purity, and chemical bonding were obtained from EDX and FTIR analysis respectively. Estimation on direct bandgap was made from the optical analysis and found to be in the range 4.0-4.8 eV. The method is attractive and seems a cost effective route for the growth of transition metal oxides for research purpose. For further efficacy, the approach can be examined for other technologically significant nanostructures.

Keywords: CHM; Nanopyramids; NiO; SEM; Optical properties

Introduction

Nanotechnology has emerged potentially a versatile and interdisciplinary field involving many subjects of science and engineering; leading to a wide-range of new developments, innovation, and advanced significant research. Past few years are witnessed major mechanical advancements which are based on remarkable nanostructures level progress. Also, metal oxide nanomaterials including nanoparticles (NPs), nanowires, nanosheets, and nanofibers, has got an expanding attention of scientific researchers due to their potential applications in biomedical, optical, and electronic fields [1].

In the transition metal oxides, nickel oxide (NiO) is potentially important with a cubic lattice structure. It is a *p*-type semiconductor with a wide bandgap ranging from 3.6 eV to 4.0 eV. Recently, efforts have been placed to develop, characterize and describe the physio- chemical properties of NiO nanostructures. NiO has a number of intensive applications generally a pure material for specialty applications and metallurgical grade; mainly used for the manufacturing of alloys, frits, ferrites, porcelain glazes [2,3], magnetic materials [4], alkaline batteries cathode [5], anode material for Li-ion batteries [6,7], gas sensors [8], anti-ferromagnetic layers [9], solid-oxide fuel cells [10,11], drug delivery and magnetic resonance imaging (MRI) [12]. NiO use in nanoscale optoelectronics devices such as electrochromic display optical fibers, photovoltaic applications [13,14], p–n heterojunctions [15], catalysis [16,17], the electrode material for electrochemical capacitors, [18] and smart windows [19]. The requirements of these applications include many factors regarding small size and particle distribution. The factors include volume effect, the quantum size effect and the surface effect with improved properties for these various attractive applications [20]. In the literature, various methods have been reported for the synthesis of nanocrystalline NiO from different starting materials for some valuable applications. Most of these methods are; thermal decomposition [21,22], carbonyl method, sol-gel technique [23], microwave pyrolysis [24], solvothermal [25], anodic arc plasma [26], Sonochemical [27], precipitation-calcination [28] and microemulsion [29]. However, mostly these are high-temperature and

high-pressure methods; rely on surface-capping agents or demand for organic-metallic precursors and seems to be expensive for an industrial scale application. So it is always desired to seek for a simple approach to be; cost effective, workable at lower temperature, and has potential for a large-scale and controlled growth of oxide nanostructures at atmospheric pressure.

The method we have proposed here for the preparation of NiO nanomaterial is the composite- hydroxide- mediated (CHM) approach. The method of CHM is basically a technically sound, environmentally friendly methodology for creating a wide range of significant nanostructures [30]. The method is quite cheap and single step technique provides a fast nano synthesis route for nanomaterials. The synthesis is usually carried out at a lower temperature ~200°C in the ambient atmosphere [30,31]. This method has the advantages to synthesize functional nanowires, nanorods, nanobelts and several other oxide nanostructures. Recently, Khan et al. applied this method for the synthesis of CdO and ZnO nanostructures and a temperature dependent study was established [32,33]. Previously in our group, a comprehensive study was done on the feasibility of CHM approach for the synthesis of doped $Cu_{1-x}Zn_xO$ nanostructures [34]. Interesting nanostructures with flower- like features and morphological peculiarities were obtained. These structures seem to strongly depend on the content of the incorporated Zn^{+2}. Recently, C An et al. applied

*Corresponding author: Khan TM, National Institute of Lasers and Optronics (NILOP), Nilore 45650, Islamabad, Pakistan, E-mail: tajakashne@gmail.com

this method to prepare NiO nanocrystals using $NiCl_2.6H_2O$ as the source material. However, in their limited study, they showed the dependence of synthesis process on the amount of solvents used with no further understanding of optical and morphological structures [35]. The method of CHM is extensively under investigation in direction of basic understanding for research purposes. The method seems has a potential to be a viable synthesis route for the high- tech industrial scale application if it is established.

In this report we present synthesis of purely NiO nanocrystals using a cost effective CHM approach and formation of the nanostructures is established using; X-ray diffraction (XRD), Fourier-transform infrared spectroscopy (FT-IR), UV-Visible spectroscopy, scanning electron microscopy (SEM), and energy dispersive x-ray spectroscopy (EDXS). We established that well defined interesting pyramid type nanocrystals are obtained in a certain temperature range and showed a temperature-dependent mechanism of the nanostructures. This study will help to understand and control morphology of the nanomaterial with the processing temperature. The proposed approach seems to provide a fast and cheap nano-synthesis route for a verity of novel nanomaterials of technological importance for research purposes.

Experimental Methods

Synthesis of NiO nanomaterial

The synthesis of NiO was performed using CHM approach. All the chemicals were the analytical grade, used without further purification and purchased from Merk Co. The main source materials used for this experiment are; $Ni(NO_3)_2.6H_2O$, NaOH, and KOH. The synthesis procedure has been given in our previous reports [32-34]. Briefly, for the synthesis of NiO, an appropriate amount of mixed hydroxide (NaOH: KOH=51.5:48.5) were added in a Teflon beaker, and covered beaker was then placed in an electric furnace at 180°C followed by further experiments at temperatures; 200°C, 220°C, and 250°C. Until the hydroxides were totally in the molten state, the beaker was taken out from the furnace. The molten hydroxide mixture was stirred well by a Teflon bar and $Ni(NO_3)_2.6H_2O$ was added. Then Teflon beaker was put in the furnace for 24 h (processing time) at the above- described temperatures for various repeated experiments. After a reaction time of 24 h, the beaker was taken out and cooled to room temperature naturally. The product was washed with distilled water to remove the hydroxide. The black powder was obtained and collected for further characterization. A general sketch showing the procedure of the process is given in Figure 1.

The reaction mechanism to form NiO nanocrystals can be formulated as follow:

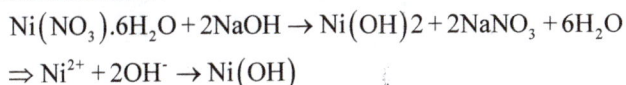

$$Ni(NO_3).6H_2O + 2NaOH \rightarrow Ni(OH)2 + 2NaNO_3 + 6H_2O$$

$$\Rightarrow Ni^{2+} + 2OH^- \rightarrow Ni(OH)$$

$Ni(OH)_2$ is chemically unstable at high temperature and splits to form NiO and water molecule. The precipitated NiO nanocrystals so formed are:

$$Ni(OH_2) \rightarrow NiO + H_2O$$

Characterization methods

X-ray diffraction (XRD) analysis of the prepared nanomaterial was carried out by PANalytical diffractometer equipped with a CuKα monochromatic radiation (λ=1.5406 Å). SEI detector in the scanning electron microscopy (SEM) and an energy- dispersive X-ray

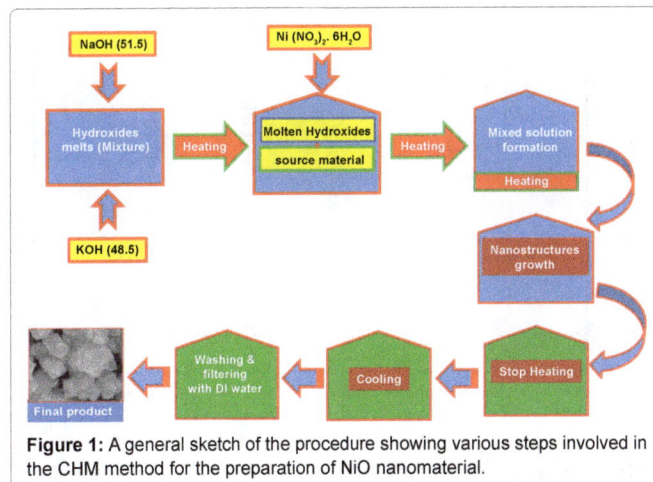

Figure 1: A general sketch of the procedure showing various steps involved in the CHM method for the preparation of NiO nanomaterial.

spectroscopy (EDX) was used to analyze the morphology and chemical composition of the as- prepared NiO.

FTIR spectrum was studied on a Perkin Elmer FTIR spectrometer in the spectral range of 400-4000 cm^{-1}. The UV-visible absorption spectrum analysis was carried out in the spectral range 250 - 800 nm. The optical bandgap was estimated by using UV-visible absorption spectroscopy.

Results and Discussion

XRD analysis

The X-rays diffraction patterns of as- prepared samples of NiO nanostructures prepared at different temperatures; 180°C, 200°C, 220°C and 250°C are presented in Figure 2a. The sharp reflection peaks in the XRD patterns indicate the growth of nanosized crystallites of NiO (Bunsenite) nanostructures according to JCPDS# 00-071-1179, which crystallizes in the cubic structure. For the 180°C sample, the peak positions appearing at 2θ=37.25°, 43.27°, 62.85°, 74.96°, and 79.4° are corresponding to (111), (200), (220), (311) and (222) crystal planes respectively. All the reflections are indexed to face- centered cubic (FCC) of NiO phase with the lattice parameters a=b=c=0.41841 nm. The cubic structure NiO prepared at 200°C shows reflections at 2θ=37.16°, 43.18°, 62.68°, 75.16° and 79.15° for the (111), (200), (220), (311) and (222) orientations with lattice parameters a=b=c=0.41885 nm. Similar reflections are also observed for the samples prepared at 220°C and 250°C. The intensities of (111), (200) and (220) reflections are much stronger compared to other reflections in the XRD pattern. This gives strongly oriented crystallites in these planes directions. However, the growth along (200) is more prominent. The sharpness and strong intensity of these very certain peaks indicate a good crystalline nature of the product. The peak broadening in XRD pattern is a characteristic of the produced nanostructures. The average crystallite size of NiO is estimated by the well-known Debye-Sherrer formula using peak broadening [36],

$$D = K\lambda / \beta cos\theta \qquad (1)$$

Here β denotes "full width at half maximum" of the diffraction peak in radian units, K is the correction factor and its value is assumed to be 0.94 for FWHM of the crystals, λ is the wavelength in unit of nm of Cu-Kα radiations and θ is the angular position in radian units (the Bragg's angle). The calculated values of crystallite sizes, corresponding to each plane along with FWHM are given in Table 1. The lattice parameter a, is calculated by the formula:

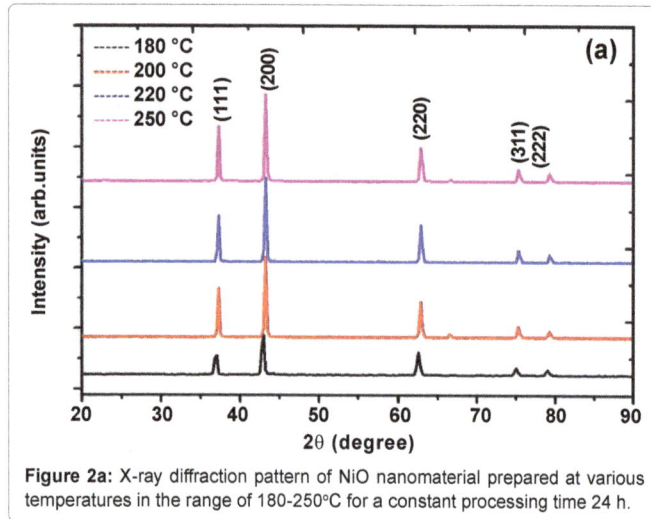

Figure 2a: X-ray diffraction pattern of NiO nanomaterial prepared at various temperatures in the range of 180-250°C for a constant processing time 24 h.

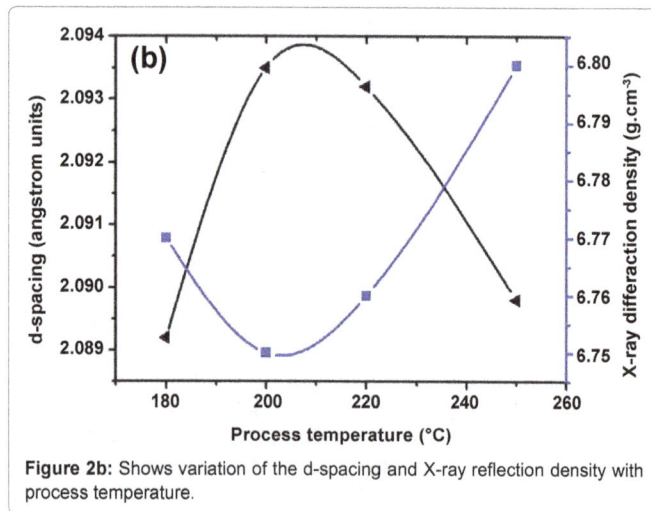

Figure 2b: Shows variation of the d-spacing and X-ray reflection density with process temperature.

$$\alpha = \frac{n\lambda}{2sin\theta\left(h^2 + k^2 + l^2\right)^{1/2}} \qquad (2)$$

Then the cell volume of prepared cubic NiO nanostructures was estimated by the formula;

$$V_{cell} = a^3 \qquad (3)$$

The X-ray density was calculated using molecular weight and volume of unit cell of the samples by the following formula [37];

$$\rho_{X-ray} = \frac{ZM}{V_{cell} N_A} \qquad (4)$$

Where "Z" is the number of molecules per formula unit, "M" is the molar mass, "N_A" have its usual meanings and the V_{cell} is volume of a unit cell. All the calculated structural parameters are listed in the Table 1. It can be clearly seen that the lattice parameters have good agreement with the standard values. A small decrease is seen for the samples from 180°C to 220°C, where at 250°C the lattice parameter is very close to standard value. The cell volume also has the same tendency but values of the X-ray density has increased with processing temperature. The average size of crystallites obtained are; 23.71 nm (at 180°C), 30.81 nm (at 200°C), 31.01 nm (at 220°C), and 36.92 nm (at 250°C). The increase in crystallite size with rising reaction temperature

is due to the viscosity of the hydroxide melts. As the viscosity of the hydroxide melts is greater at temperatures higher than 180°C, this causes slow nucleation and crystallization which result in formation of bigger nanosized crystallites. The variation in d-spacing and diffraction density with processing temperature are displayed in Figure 2b. The value of d-spacing shows an increasing trend in the temperature range 180-220°C followed by a reverse dropping trend at 250°C. Reflection density shows an opposite behavior to that observed for d-spacing in the same temperature range.

SEM characterization

Figures 3a and 3b shows SEM images and EDX of NiO nanomaterial prepared at different temperatures. These images give information on the morphology and composition of nanosized NiO. It seems that at temperatures 180°C and 200°C the prepared samples have no well- defined morphological structure. This predicts that a minimum temperature is required in the CHM process for different materials to starts to grow and nucleate in well- defined morphological peculiarities and structure. This also shows a temperature- dependent mechanism of nanomaterial synthesis by the CHM method. This temperature dependent nucleation can also be associated with the viscosity of the melts which is strongly influenced by the temperature factor and suggests viscosity is playing a significant role in the formation and growth of a nanomaterial in the CHM approach. As the temperature is raised to 220 and 250°C, a significant change is observed in the nanostructure of as- prepared samples. Interesting nanopyramid type structures are obtained with a high density. The color line circle in the image for 250°C shows two enclosed nanopyramids with its magnified marked image alongside EDS to manifests a clear visual picture of the

Processing Temperature	a (Å)	V (Å³)	Density (g.cm⁻³)	Crystallite size (nm)	Bandgap (eV)
180°C	4.1841	73.24	6.77	23.71	4.8
200°C	4.1885	73.48	6.75	30.81	4.5
220°C	4.1858	73.34	6.76	31.01	4.4
250°C	4.1791	72.99	6.8	36.92	4.0
NiO*	4.1780	72.93	6.803	-	4

Table 1: Structural and optical parameters of CHM prepared NiO calculated by XRD and absorption spectroscopy.

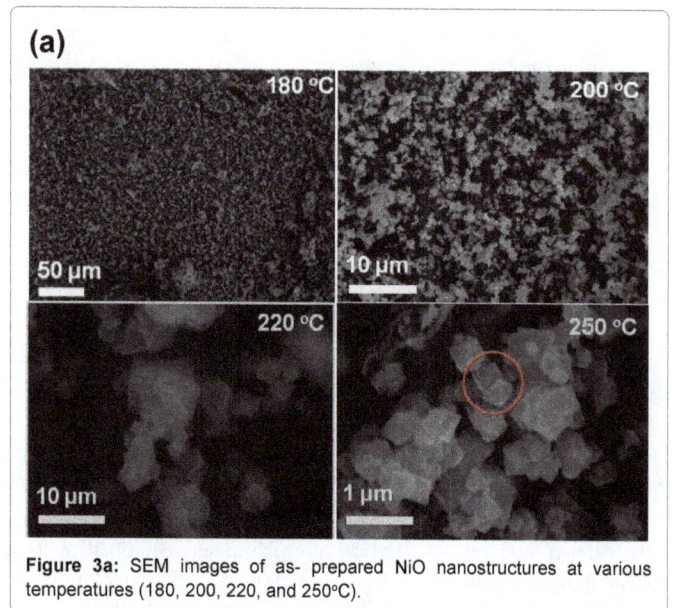

Figure 3a: SEM images of as- prepared NiO nanostructures at various temperatures (180, 200, 220, and 250°C).

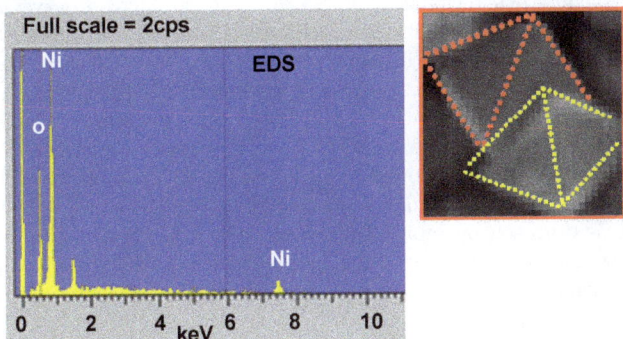

Figure 3b: EDS spectra taken on the sample showing Ni and O elements in the samples.

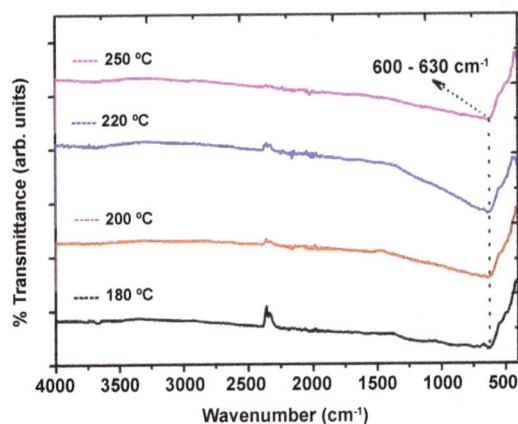

Figure 4: FT-IR spectra of NiO prepared at temperatures 180, 200, 220, and 250°C and for a constant reaction time (24 h).

prepared nanopyramids. The length of sides of these nanopyramids is in the range of 0.3–0.5 μm with random orientation. These NiO crystallites self-organized nanopyramids are high- density assemblies with a clear appearance. For the chemical composition, energy-dispersive X-ray (EDX) was performed on each sample and it was confirmed each sample showed the existence of Ni and O elements. This gives a strong signature on the chemically pure nature of the prepared product by the CHM method.

FTIR studies

The FTIR spectra were studied in the spectral range of 400-4000 cm^{-1} shown in Figure 4. FTIR is carried out in order to understand the chemical and structural roots of the prepared samples of NiO and the effects of reaction materials used. The absorption band in the region of 600–630 cm^{-1} is the characteristic stretching vibration band of NiO and is an obvious signature on the formation of NiO. The broadness of this band indicates the presence of NiO nanocrystals in the prepared samples. The observed vibration band for all the samples is well-consisted with the literature reported vibration band of NiO [38]. The small double peak at 2362 and 2346 cm^{-1} is caused by the presence of CO_2 molecules. The intensity of these bands decreases with increasing temperature [39,40]. FTIR study strongly supports the XRD results in respect of purity of the samples showing the formation of a single phase NiO.

UV- visible spectroscopy

This technique is based on the fact that an electron is moved from valence band to conduction band due to the absorbed photon having energy larger than the band gap energy of semiconductor. The electronic transition happened due to absorbed photon provides rich information about the type of transition. When the electron momentum is conserved the transition is direct otherwise indirect and energy of the transitions is measured using Tauc's relation [41,42].

UV-visible spectroscopy was performed to investigate the optical behavior of the nanostructures and to make an estimation of the bandgap. The optical absorption spectra of NiO samples at 180°C, 200°C, 220°C and 250°C were taken at room temperature by dispersing them in absolute ethanol which is a low absorption medium [43]. UV-vis absorption spectra of the samples are shown below in Figure 5a. A slight shift towards higher wavelength for the absorption edges was observed as the process temperature increased. This shift indicates a decrease in bandgap value, which can be attributed to an increase in the particle size. The value of absorption edges of NiO prepared at 180°C, 200°C, 220°C, 250°C were 295, 317, 321, 322 nanometer respectively. Also, it can be seen from the Figure 5 that there is an exponential decrease in the intensity of absorption with the increase in wavelength. This decreasing behavior in intensity is attributed for many semiconductors and may be due to many reasons similarly to internal electric fields within the crystal, deformation of lattice due to strain caused by imperfection and inelastic scattering of charge carriers by phonons [44]. Absorption coefficient (α) associated with the strong absorption region of the sample was calculated from the following relation [45]:

$$\alpha = 2.303 A / t \qquad (5)$$

Here t' stands for the path length in cm and is taken to the cuvette length of 1 cm. According to the data of the absorption spectra, the optical bandgap (Eg) of NiO can be estimated by using the following Tauc equation [46]:

$$(\alpha \upsilon)^n = B(h\upsilon - E_g) \qquad (6)$$

Where hv is photo energy, α is absorption coefficient, B is a constant relative to the material and n is either 2 for direct band gap material or 1/2 for an indirect band gap material. According to the equation, the optical bandgap for the absorption peak can be obtained by extrapolating the linear portion of the $(\alpha h\upsilon)^n$ – hν curve to zero of the energy axis shown in Figure 5b. The corresponding bandgap

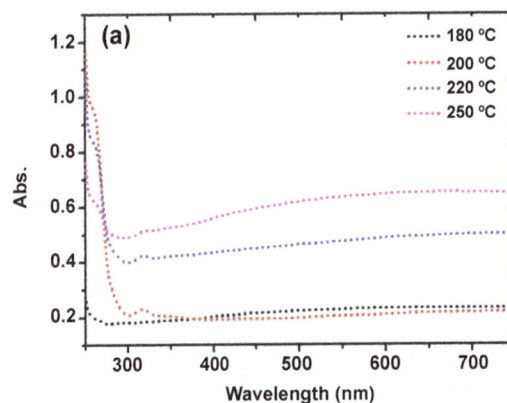

Figure 5a: UV- vis. absorption spectra of the CHM prepared NiO nanomaterial at various temperatures in the range of 180-250°C.

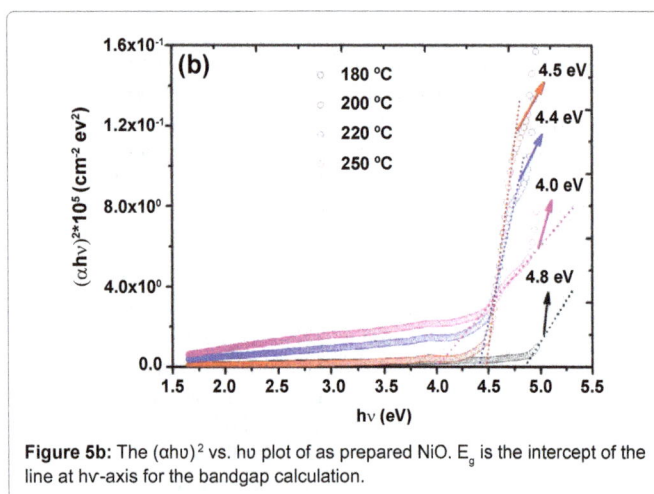

Figure 5b: The $(\alpha h\upsilon)^2$ vs. $h\upsilon$ plot of as prepared NiO. E_g is the intercept of the line at $h\upsilon$-axis for the bandgap calculation.

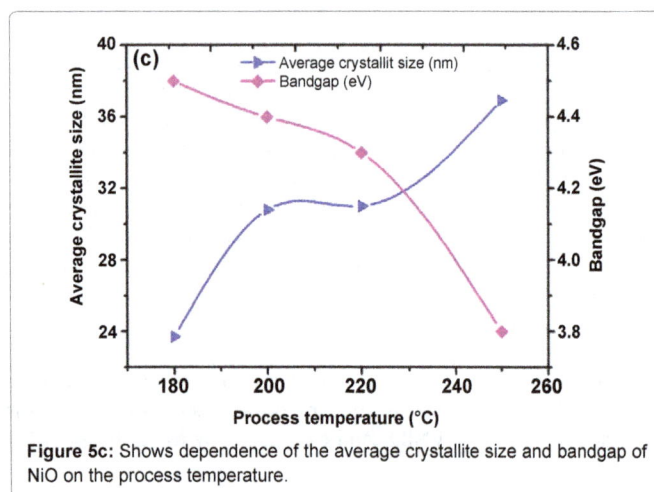

Figure 5c: Shows dependence of the average crystallite size and bandgap of NiO on the process temperature.

energies of produced nano-NiO at different temperature are given in Table 1. No linear relation was found for $n = 1/2$, suggesting that the as-synthesized NiO is semiconductor material with a direct transition at this energy [47]. The bandgap energies calculated are; 4.8 eV (at 180°C), 4.5 eV (at 200°C), 4.4 eV (220°C), and 4.0 eV (at 250°C). These values well consist reported by Khashan et al. [48] and Rifaya et al. [49]. This bandgap is the p-d character of NiO has been investigated by a range of spectroscopic techniques including optical absorption. The increasing trends of the bandgap energy upon the decreasing particles size can be associated nanosized structures. This effect is likely due to the chemical defects or vacancies present in the intergranular regions generating new energy level to reduce the band gap energy [50]. The NiO bandgap showed an increase with processing temperature with an overall maximum increased value up to 4.5 eV. The variation in E_g with processing temperature along with the crystallite size is depicted in Figure 5c.

Conclusion

In conclusion, pyramid- shaped NiO nanostructures were synthesized by using a low temperature and feasible CHM approach. A minimum limit on the processing temperature was established to nucleate NiO with a well- defined morphological structure. At temperatures below 220°C, NiO was presenting no definite morphology; while at temperatures in the range 220-250°C well-

defined pyramid shaped morphological structures were obtained. The nanostructures were polycrystalline with an estimated average crystallite size in the range of 23.71 - 36.92 nm. The increase in average crystallite size assumed was caused by viscosity effect of the melts; playing a significant role in the nucleation process. Morphology showed a temperature dependent nature and was influenced by the processing temperature. The product was a pure crystalline material; composed of Ni and O elements with Ni-O stretching vibration bonding. The process temperature showed influence on the optical properties. The direct bandgap estimated was tuned in the range of 4.8-4.0 eV with the rise in process temperature. The present study would be beneficial as a new cost-effective route for the development of novel nanomaterials and temperature dependent morphological structures for prospective application in optoelectronics.

Acknowledgements

The research work carried-out in FUUAST. All the authors greatly acknowledge their parent institutes for this research work.

References

1. Thierry B, Majewski P, Ngothai Y, Shi Y (2007) Preparation of monodisperse functionalized nanoparticles. Internat J Nanotechnology 4: 5.

2. Jana NR, Chen YF, Peng XG (2004) Size- and shape-controlled magnetic (cr, mn, fe, co, ni) oxide nanocrystals via a simple and general approach. Chem Mater 16: 3931-3935.

3. Sasi B, Gopchandran KG, Manoj PK, Koshy P, Prabhakara Rao P (2002) Preparation of transparent and semiconducting NiO films. Vacuum 68: 149-154.

4. Thota S, Kumar J (2007) Sol-gel synthesis and anomalous magnetic behaviour of NiO nanoparticles. J Phys Chem Solids 68: 1951-1964.

5. Needham SA, Wang GX, Liu HK (2006) Synthesis of NiO nanotubes for use as negative electrodes in lithium ion batteries. J Power Sources 159: 254-257.

6. Chiu KF, Chang CY, Lin CM (2005) The electrochemical performance of bias-sputter deposited nanocrystalline nickel oxide thin films toward lithium. J Electrochem Soc 152: A1188-A1192.

7. Idirs NH, Wang JZ, Chou S, Zhong C, Rahman MM, et al. (2011) Effects of polypyrrole on the performance of nickel oxide anode materials for rechargeable lithium-ion batteries. Journal of Materials Research 26: 860-866.

8. Hotovy I, Huran J, Spiess L, Hascik S, Rehacek V (1999) Preparation of nickel oxide thin films for gas sensors applications. Sens Actuators B Chem 57: 147-152.

9. Bi H, Li S, Zhang, Du Y (2004) Ferromagnetic-like behavior of ultrafine NiO nanocrystallites. J Magn Magn Mater 277: 363-367.

10. Yoshito WK, Ussui V, Lazar DR, Paschoal JOA (2010) Synthesis of nickel oxide - zirconia composites by coprecipitation route followed by hydrothermal treatment. Material Science Forum 661: 977-982.

11. Marrero-Lopez DD, Ruiz-Morales JC, Pena-Martinez J, Canales-Vazques J, Nunez P (2008) Preparation of thin layer materials with macroporous microstructure for SOFC applications. J Solid State Chem 181: 685-692.

12. Richardson JR, Yiagas DI, Turk B, Forster K, Twigg MV (1991) Origin of superparamagnetism in nickel oxide. J Appl Phys 70: 6977-6982.

13. Lin SH, Chen FR, Kai JJ (2008) Electrochromic properties of nano-structured nickel oxide thin film prepared by spray pyrolysis method. Applied Surface Science 254: 2017-2022.

14. Borgstrom M, Blart E, Boschloo G, Mukhtar E, Hagfeldt A, et al. (2005) Sensitized hole injection of phosphorus porphyrin into NiO: toward new photovoltaic devices. J Phys Chem B 109: 22928-22934.

15. Chrissanthopoulos A, Baskoutas S, Bouropoulos N, Dracopoulos V, Yannopoulos SN (2011) Synthesis and characterization of ZnO/NiO p–n heterojunctions: ZnO nanorods grown on NiO thin film by thermal evaporation. Photonics and Nanostructures-Fundamentals and Applications 9: 132-139.

16. Alejandre A, Medina F, Salagre P, Fabregat A, Sueiras JE (1998) Characterization and activity of copper and nickel catalysts for the oxidation of phenol aqueous solutions. Applied catalysis B 18: 307-315.

17. Dooley KM, Chen SY, Ross JRH (1994) Stable nickel-containing catalysts for the oxidative coupling of methane. J Catal 145: 402-408.

18. Zhang F, Zhou Y, Li H (2004) Nanocrystalline NiO as an electrode material for electrochemical capacitor. Mater Chem Phys 83: 260-264.

19. Bodurov G, Stefchev P, Ivanova T, Gesheva K (2014) Investigation of electrodeposited NiO films as electrochromic material for counter electrodes in Smart Windows. Mater Lett 117: 270-272.

20. Granqvist CG (1995) Handbook of inorganic electrochromic materials. Elsevier, Amsterdam.

21. Wang W, Liu Y, Xu C, Zheng C, Wang G (2002) Synthesis of NiO nanorods by a novel simple precursor thermal decomposition approach. Chem Phys Lett 362: 119-122.

22. Hosny NM (2011) Synthesis, characterization and optical band gap of NiO nanoparticles derived from anthranilic acid precursors via a thermal decomposition route. Polyhedron 30: 470-476.

23. Li X, Zhang X, Li Z, Qian Y (2006) Synthesis and characteristics of NiO nanoparticles by thermal decomposition of nickel dimethylglyoximate rods. Solid State Commun 137: 581-584.

24. Xiang L, Deng XY, Jin Y (2002) Experimental study on synthesis of NiO nano-particles. Scr Mater 47: 219-224.

25. Wang Y, Ke JJ (1996) Preparation of nickel oxide powder by decomposition of basic nickel carbonate in microwave field with nickel oxide seed as a microwave absorbing additive. Mater Res Bull 31: 55-61.

26. Anandan K, Rajendran V (2011) Morphological and size effects of NiO nanoparticles via solvothermal process and their optical properties. Solid State Electron 14: 43-47.

27. Wei Z, Qiao H, Yang H, Zhang C, Yan X (2009) Characterization of NiO nanoparticles by anodic arc plasma method. J Alloys Compd 479: 855-858.

28. Mohseni Meybodi S, Hosseini SA, Rezaee M, Sadrnezhaad SK, Mohammadyani D (2012) Synthesis of Wide Bandgap Nanocrystalline NiO Powder via a Sonochemical Method. Ultrason Sonochem 19: 841-845.

29. Deng XY, Chen Z (2004) Preparation of nano-NiO by ammonia precipitation and reaction in solution and competitive balance. Mater Lett 58: 276-280.

30. Hu C, Xi Y, Liu H, Wang ZL (2009) Composite-hydroxide-mediated approach as a general methodology for synthesizing nanostructures. Journal of Material Chemistry 19: 858-868.

31. Peng WQ, Cong GW, Qu SC, Wang ZG (2005) Synthesis of shuttle-like ZnO nanostructures from precursor ZnS nanoparticles. Nanotechnology 16: 1469.

32. Khan TM, Shahid T, Zakria M, Shakoor RI (2015) Optoelectronic properties and temperature dependent mechanisms of CHM approach for the synthesis of CdO nanomaterials. Electron Mater Lett 11: 366-373.

33. Khan TM, Zakria M, Shakoor RI, Ahmad M, Raffi M (2015) Mechanisms of composite-hydroxide-mediated approach for the synthesis of functional ZnO nanostructures and morphological dependent optical emissions. Advanced Materials Letters 6: 592-599.

34. Shahid T, Arfan M, Ahmad W, BiBi T, Khan TM (2016) Synthesis and doping feasibility of composite-hydroxide-mediated approach for the $Cu_{1-x}Zn_xO$ nanomaterials. Advanced Materials Letters 7: 561-566.

35. An C, Wang R, Wang S, Liu Y (2008) A low temperature composite-hydroxide approach to NiO nanocrystals. Mater Res Bull 43: 2563-2568.

36. Abdeen AM, Hemeda OM, Assem EE, Elsehly MM (2002) Structural, electrical and transport phenomena of Co ferrite substituted by Cd. J Magn Magn Mater 238: 75-83.

37. Anandan K, Rajendran V (2011) Morphological and size effects of NiO nanoparticles via solvothermal process and their optical properties. Mater Sci Semicond Process 14: 43-47.

38. El-Kemary M, Nagy N, El-Mehasseb I (2013) Nickel oxide nanoparticles: Synthesis and spectral studies of interactions with glucose. Materials Science in Semiconductor Processing 16: 1747-1752.

39. Snopatin GE, Yu M, Matveeva M, Butsyn GG (2006) Effect of SO_2 impurity on the optical transmission of As_2S_3 glass. Inorg Mater 42: 1388-1392.

40. Willardson R, Beer A (1967) Optical Properties of III-V Compounds. Academic Press, New York. pp: 318-400.

41. Dressel M, Gruner G (2002) Electrodynamics of solids optical properties of electron in matter. Cambridge University Press, USA.

42. Dharmaraj N, Prabu P, Nagarajan S, Kim CH, Park JH, et al. (2006) Synthesis of nickel oxide nanoparticles using nickel acetate and poly(vinyl acetate) precursor. Mater Sci Eng B 128: 111-114.

43. Mallick P, Sahoo CS, Mishra NC (2012) Structural and optical characterization of NiO nanoparticles synthesized by sol-gel route. AIP Conf Proc 1461: 229-232.

44. Kumar H, Rani R (2013) Structural and optical characterization of ZnO nanoparticles synthesized by microemulsion route. Int Lett Chem Phys Astron 14: 26-36.

45. Tauc J (1966) Optical properties of solids. Academic Press Inc., New York.

46. Hwang SS, Vasiliev AL, Padture NP (2007) Improved processing and oxidation-resistance of ZrB2 ultra-high temperature ceramics containing SiC nanodispersoids. Mater Sci Eng A 464: 216-224.

47. Khashan KS, Sulaiman GM, Ameer FAKA, Napolitano G (2016) Synthesis, characterization and antibacterial activity of colloidal NiO Nanoparticles. Pak J Pharm Sci 29: 541-546.

48. Tasker PW (1979) The stability of ionic crystal surfaces. J Phys C: Solid State Phys 12: 4977-4984.

49. Nowsath Rifaya M, Theivasanthi T, Alagar M (2012) Chemical capping synthesis of nickel oxide nanoparticles and their characterizations studies. Nanoscience and Nanotechnology 2: 134-138.

50. Song X, Gao L (2008) Facile synthesis of polycrystalline NiO nanorods assisted by microwave heating. J Am Ceram Soc 91: 3465-3468.

32

Self-Recovering Section of RPV Steel Radiation Embrittlement Kinetics as Indication of Material Smart Behavior

Evgenii K*

Kurchatov Sq 1, Moscow 123182, Russia

Abstract

Influence of neutron irradiation on reactor pressure vessel (RPV) steel degradation are examined with reference to the possible reasons of the substantial experimental data scatter and furthermore -nonstandard (non-monotonous) and oscillatory embrittlement behavior. In our glance this phenomenon may be explained by presence of the wavelike recovering component in the embrittlement kinetics.

We suppose that the main factor affecting steel anomalous embrittlement is fast neutron intensity (dose rate or flux), flux effect manifestation depends on state-of-the-art fluence level. At low fluencies radiation degradation has to exceed normative value, then approaches to normative meaning and finally became sub normative. Data on radiation damage change including through the ex-service RPVs taking into account chemical factor, fast neutron fluence and neutron flux were obtained and analyzed.

In our opinion controversy in the estimation on neutron flux on radiation degradation impact may be explained by presence of the wavelike component in the embrittlement kinetics. Therefore flux effect manifestation depends on fluence level. At low fluencies radiation degradation has to exceed normative value, then approaches to normative meaning and finally became sub normative. As a result of dose rate effect manifestation peripheral RPV's zones in some range of fluencies have to be damaged to a large extent than situated closely to core.

Moreover as a hypothesis we suppose that at some stages of irradiation damaged metal have to be partially restored by irradiation i.e. neutron bombardment. Nascent during irradiation structure undergo occurring once or periodically transformation in a direction both degradation and recovery of the initial properties. According to our hypothesis at some stagem(s) of metal structure degradation neutron bombardment became recovering factor. Therefore self-recovering section of RPV steel radiation embrittlement kinetics is an indication of material smart behavior. As a result oscillation arise that intern lead to enhanced data scatter.

Disclosure of the steel degradation oscillating is a sign of the steel structure cyclic self-recovery transformation as it take place in self-organization processes. This assumption has received support through the discovery of the similar "anomalous" data in scientific publications and by means of own additional experiments.

Data obtained stimulate looking-for ways to management of the structural steel radiation stability (for example, by means of nano-structure modification for radiation defects annihilation intensification) for creation of the intelligent self-recovering material.

Expected results:

- radiation degradation theory and mechanisms development,

- more adequate models of the radiation embrittlement elaboration,

- methods and facility development for usage data of the accelerated materials irradiation for forecasting of their capacity for work in realistic (practical) circumstances of operation,

- search of the ways for creating of the stable under neutron irradiation self-recovery smart materials.

Keywords: Reactor pressure vessel; Steel; Self-recovering section; Embrittlement kinetics; Material smart behavior

Introduction

Operating PWR's reactor pressure vessels are subject to multi-factor influence. It is practically impossible to reproduce some of this factors (long-time bias, e.g.) in the framework of experimental investigations including RPV surveillance specimens tests. Detailed information that can be obtained by means of taking through RPV wall samples immediate from the decommissioned RPVs is more representative than received by any another ways and therefore has a highest value.

Testing of the specimens presented a unique opportunity for qualifying consequences of long-term irradiation, multi-factor influence and dose rate (flux) effect on actual RPV properties. Data on radiation damage change through the ex-service RPV walls taking into account chemical factor and neutron flux were obtained.

Controversy in the opinions on neutron flux on radiation degradation impact may be explained by presence of the wavelike component in the embrittlement kinetics that in turn is an indication of material intelligent behavior at some sections of RPV steel radiation embrittlement kinetics [1,2].

---

*Corresponding author: Evgenii K, Kurchatov Sq., 1, Moscow 123182, Russia
E-mail: ekrasikov@mail.ru

Experimental Procedures

Along with routine investigations in Russia systematic research on actual radiation embrittlement of the decommissioned PWR pressure vessel via through samples (trepans) has been carried out.

The earliest commercial PWR prototype unit Novovoronezh-1 (NV1) RPV after 20 years of operation was trepanned in 1987. Then Novovoronezh-2 (NV2), the oldest PWR type experimental reactor-prototype ERP and, finely, nuclear icebreaker NIB) «Lenin» RPVs also were trepanned. The most interesting and unexpected data were discovered during trepans of the first nuclear ship-icebreaker «Lenin» investigation.

Chemical analyses of the icebreaker RPV material were carried out with FSQ «Baird» optical emission spectrometer. Results can be seen in Table 1. Fast (E>0.5MeV) neutron fluence evaluation was based on the specific Mn-54 and Nb-93m activities of the vessels steel and on the Nb-93m activity of the RPV cladding.

The RPV materials radiation degradation (embrittlement) was determined by finding the ductile-to-brittle transition temperature shift (TTS). RKP-300 impact pendulum test machine for standard Charpy specimens testing was used.

The general dependence of the TTS on radiation embrittlement coefficient (REC) A_F is as follows:

$$TTS = A_F \times F^n,$$

where A_F is the radiation embrittlement coefficient and F is the fast (E≥0,5MeV) neutron fluence in units of $10^{18} sm^{-2}$ (fluence factor), n-coefficient (~1/3).

For Russian RPV Cr-Mo-V base steel $A_F = A \times (P + 0,07Cu)$, where A_F- coefficient (800 at 270°C), P and Cu-are the weight concentrations of these elements.

Test Results

The unexpected results of the icebreaker RPV weld and base metal studying are given in Figure 1, where one can recognize that the actual (measured) radiation embrittlement coefficients of the trepan materials for the periphery (remote) zone of the vessel are significantly higher that for the inner part. One can see also that hardness measurement and Charpy impact testing results agree.

Data Analysis and Discussion

First impression from foregoing decommissioned PWR pressure vessel material properties study- enhanced degradation rate at low neutron fluxes. Examples for icebreaker base metal and weld metal are demonstrated in Figures 2 and 3. Registered facts denote that known as «flux effect» factor was in action. Unexpected circumstance however is reduced embrittlement zone appearance that follows after of previous part of enhanced embrittlement. It is seen by means of TTS on neutron flux and fluence dependencies observation.

For the sake of correctness, it is necessary to underline that the first mention concerning distinction between test reactors and low-lead-factor (surveillance) data had appeared as early as 1980 (Figure 4) [3].

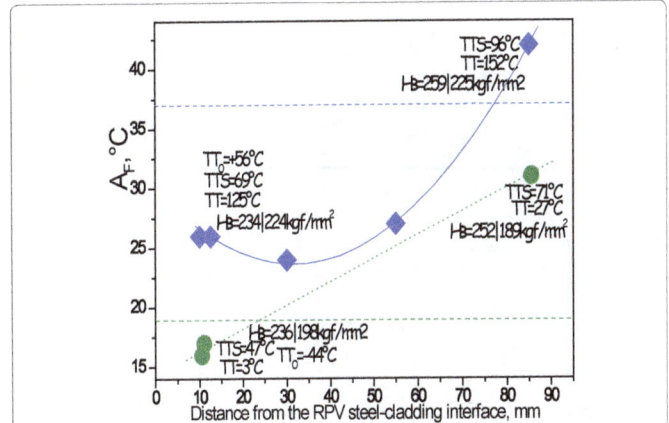

Note: H_B = 252|189 kgf/mm² - hardness values at as-received condition|after annealing 650°C/2h; TT- Transition temperature; TT0 - transition temperature in initial state.

Figure 1: Radiation embrittlement coefficient AF and hardness value HB through the icebreaker RPV wall distribution.

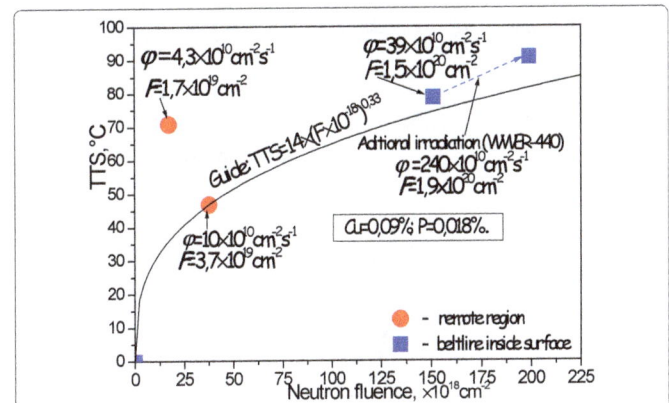

Figure 2: Comparison of the TTSs between «remote» and «inside» icebreaker RPV base metal specimens.

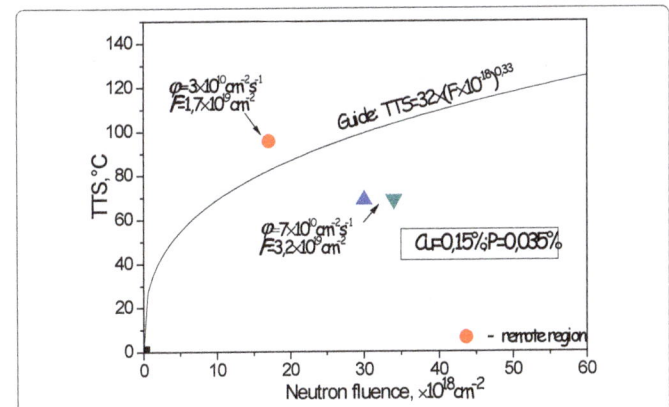

Figure 3: Comparison of the TTSs between «remote» and «inside» icebreaker RPV weld metal specimens.

Summing the previously mentioned on the subject discussed one might conclude, that depending of the fluence level reached manifestation of the «flux effect» in reference to Guide pattern may be quite different, namely: negative, positive and «two zeros» as represented in Figure 5. Situation looks like famous parable «blind men and an elephant», where a group of blind men (or men in the dark)

Material	C	Mn	Si	P	Cu	Mo	Ni	Cr	V
Weld	0,05	1,03	0,41	0,035	0,15	0,49	0,17	1,39	0,15
Base	0,17	0,45	0,28	0,018	0,09	0,67	0,35	2,75	0,09

Table 1: Chemical composition of the icebreaker RPV materials under study [%%wt.].

touch an elephant to learn what it is like. The story is used to indicate that reality may be viewed differently depending upon one's perspective, suggesting that what seems an absolute truth may be relative due to the deceptive nature of half-truths.

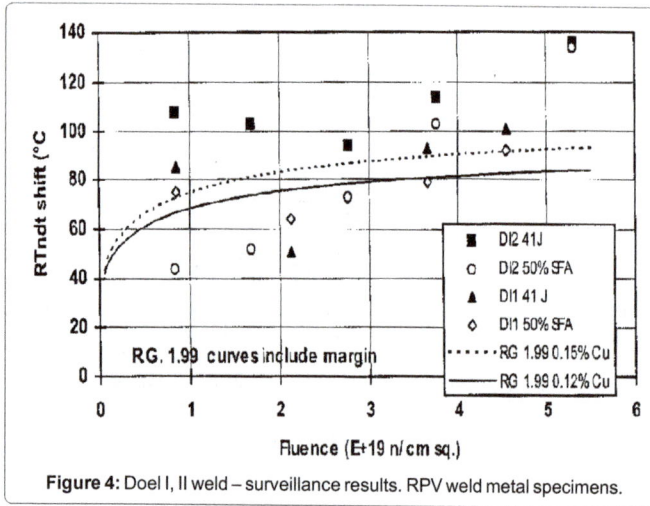

Figure 4: Doel I, II weld – surveillance results. RPV weld metal specimens.

Regulatory Guide dependence (formula) has appeared as a result of forced irradiation in test reactors. One can recognize however those specific properties of metal and actual reactor environment may deform this ideal trajectory to the extent that curve monotony character damage. The very wonderful fact is the enhanced degradation *after effect* of the temporary weakening of the embrittlement appearance. Possible comprehensible explanation is as follows: the radiation-induced copper-rich precipitates nature (dimensions and concentration) alteration. Evidently, we have fixed phenomenon similar to observed in [4] where neutron irradiation in some range of doses improves the mechanical properties of the unirradiated mild steel (Figure 6).

It is seen that irradiation of unirradiated (initial condition) steel up to dose of $\sim 2,0 \times 10^{18} cm^{-2}$ along with strengthening lead to more than 2-fold ductility increase.

In accordance with sketch of Figure 5 flux effects manifestation depends on fluence level. At low fluencies radiation degradation has to exceed normative value, then approaches to normative meaning and finally became sub normative. As a result of dose rate effect manifestation peripheral RPV's zones in some range of fluencies have to be damaged to a large extent than situated closely to core.

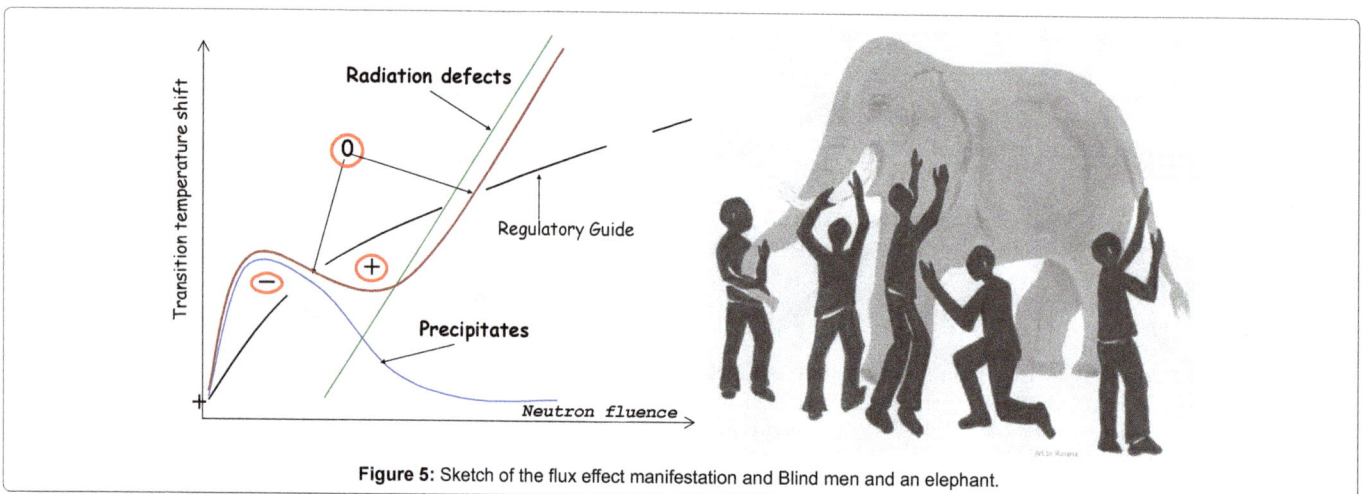

Figure 5: Sketch of the flux effect manifestation and Blind men and an elephant.

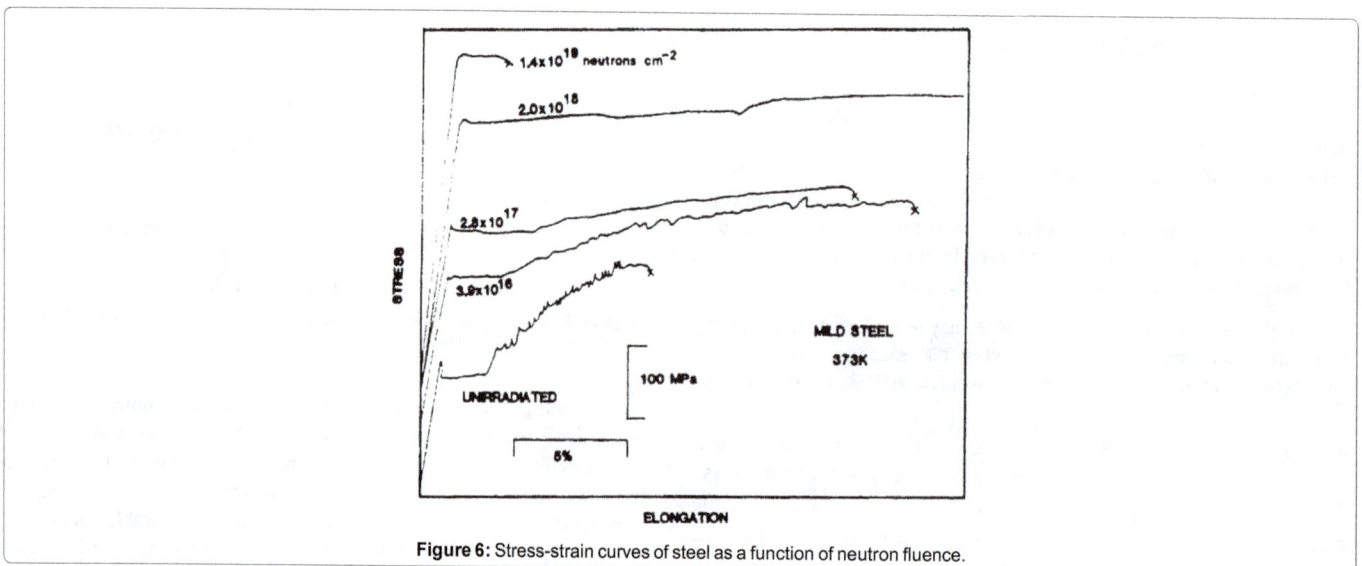

Figure 6: Stress-strain curves of steel as a function of neutron fluence.

Conclusion

Influence of neutron irradiation on reactor pressure vessel (RPV) steel degradation are examined with reference to the possible reasons of the substantial experimental data scatter and furthermore – nonstandard (non-monotonous) and oscillatory embrittlement behavior. In our glance this phenomenon may be explained by presence of the wavelike recovering component in the embrittlement kinetics.

We suppose that the main factor affecting steel anomalous embrittlement is fast neutron intensity (dose rate or flux), flux effect manifestation depends on state-of-the-art fluence level. At low fluencies radiation degradation has to exceed normative value, then approaches to normative meaning and finally became sub normative. Data on radiation damage change including through the ex-service RPVs taking into account chemical factor, fast neutron fluence and neutron flux were obtained and analyzed.

In our opinion controversy in the estimation on neutron flux on radiation degradation impact may be explained by presence of the wavelike component in the embrittlement kinetics. Therefore flux effect manifestation depends on fluence level. At low fluencies radiation degradation has to exceed normative value, then approaches to normative meaning and finally became sub normative.

As a result of dose rate effect manifestation peripheral RPV's zones in some range of fluencies have to be damaged to a large extent than situated closely to core. This finding recently was confirmed by German scientists [5].

Moreover as a hypothesis we suppose that at some stages of irradiation damaged metal have to be partially restored by irradiation i.e. neutron bombardment. Nascent during irradiation structure undergo occurring once or periodically transformation in a direction both degradation and recovery of the initial properties. According to our hypothesis at some stage(s) of metal structure degradation neutron bombardment became recovering factor. Therefore self-recovering section of RPV steel radiation embrittlement kinetics is an indication of material smart behavior. As a result oscillation arise that in tern lead to enhanced data scatter. In this case we have to consider irradiation as a recovery factor.

Disclosure of the steel degradation oscillating is a sign of the steel structure cyclic self-recovery transformation as it take place in self-organization processes. This assumption has received support through the discovery of the similar "anomalous" data in scientific publications and by means of own additional experiments.

Data obtained stimulate looking-for ways to management of the structural steel radiation stability (for example, by means of nano - structure modification for radiation defects annihilation intensification) for creation of the intelligent self-recovering material.

Expected results:

- radiation degradation theory and mechanisms development,

- more adequate models of the radiation embrittlement elaboration,

- methods and facility development for usage data of the accelerated materials irradiation for forecasting of their capacity for work in realistic (practical) circumstances of operation,

- search of the ways for creating of the stable under neutron irradiation self-recovery smart materials.

Foregoing hypothetical assumptions on "low-dose effects" in terms "radiation embrittlement contains oscillatory component" and "radiation annealing of the radiation embrittlement" is questionable and needs additional experimental verification and profound scientific study.

The information gained would be relevant to the RPV degradation mechanisms consideration and understanding, also to possible current PWR generation lifetime extension evaluation.

References

1. Krasikov E, Nikolaenko V (2004) Dose rate effect on WWER-440/213 RPV materials embrittlement. Atomic Energy 97: 177-182.

2. Fabry A, Gerard R (1996) In-service embrittlement of the pressure vessel welds at the Doel I and II NPPs. In Gelles D, Nanstad R (Eds.) Effects of radiation on materials ASTM STP 1270, West Conshohocken, PA, Ann Arbor.

3. Marston TU (1980) Radiation embrittlement significance of its effects on integrity and operation of PWR pressure vessels. Nuclear Safety 21: 724-730.

4. Murty KL (1984) Is neutron irradiation exposure always detrimental to metals (steels)? Nature 308: 51-52.

5. Viehrig H, Valo M (2012) Fracture Mechanics Characterization of the Beltline Welding Seam of the Decommissioned WWER-440 RPV of NPP Greifswald Unit 4. International Journal of Pressure Vessels and Piping 89: 129-136.

33

Translational Development of Biocompatible X-Ray Visible Microspheres for Use in Transcatheter Embolization Procedures

Benzina A[1], Aldenhoff YB[2], Heijboer R[3] and Koole LH[2,4*]

[1]Faculty of Health, Medicine and Life Sciences, Maastricht University, UniSingel 50, 6229 ER Maastricht, The Netherlands
[2]Interface Biomaterials BV, B. Lemmensstraat 364, 6163 JT Geleen, The Netherlands
[3]Zuyderland Medisch Centrum, Henri Dunantstraat 5, 6419 PC Heerlen, The Netherlands
[4]Department of Biomedical Engineering, University of Malaya, Jalan Universiti, 50603 Kuala Lumpur, Malaysia

Abstract

Embolization is a minimally invasive treatment that specifically blocks the arterial blood flow into a target blood vessel bed, which is usually a benign or malignant tumor. The aim of the procedure is to shrink the tumor and/or to retard its growth. Embolization the injection of embolic particles via a catheter tube, of which the tip has been navigated carefully (under X-ray guidance) into an arterial branch that exclusively feeds the tumor, and no surrounding healthy tissues. Most of the clinical experience with embolization relates to treatment of leiomyomata (benign tumors growing in the wall of the uterus). There is solid evidence that catheter-based embolization of leiomyomata provides a fully acceptable therapeutic alternative for much more demanding surgical procedures (i.e., hysterectomy and myomectomy). Embolization offers much faster recovery, possible options to become pregnant, and considerable cost saving. There are several commercial brands of embolization agents, suitable to treat leiomyomata. We hypothesized, some years ago, that these products are suboptimal, and that embolization of leiomyomata may be improved further through better engineering of the embolic particles. We developed injectable radiopaque polymer microspheres, which can be monitored during and after the embolization procedure. The embolic microbeads are X-ray traceable, and this has been achieved without compromising other essential properties, such as structural stability and excellent biocompatibility. Herein, we describe new the features of the new embolic microspheres, as observed in preclinical experiments and in the first clinical cases. It is mentioned briefly that this work became an example of successful translation: it has led to a new medical device (Class-IIB) that is now CE-certified and commercially available throughout Europe.

Keywords: Embolization; Radiopacity; X-ray visibility; Polymer microspheres

Introduction

The hallmark of minimally invasive therapy is the endovascular coronary stent, which secures that a coronary atherosclerotic lesion remains open after percutaneous translumenal angioplasty [1]. Stenting has become the preferred revascularization modality in patients with coronary single-vessel or low-risk multivessel disease. The technique is minimally invasive, fast, relatively cheap, and associated with faster patient recovery. Over the years, coronary stenting has seen many technical improvements, in part due to the exploitation of improved biomaterials [2,3]. Recently this has culminated in the development of polymer bio-eroding and drug-eluting vascular scaffolds [4,5].

During the last years, comparable minimally invasive techniques have gained importance in other fields as well. An important example is found in gynaecology, particularly in the treatment of benign tumors that grow in the uterus wall (leiomyomata) [6]. This disease, too, can be treated effectively in a minimally invasive manner, i.e. through controlled targeted injection of embolic particles (diameter around 500 μm) into the arterial vessel tree of each fibroid, via a catheter tube [7-10]. The particles are usually spherical (microspheres), but they can also be irregular [11]. The procedure is known as TACE (TransArterial ChemoEmbolization), and is performed by an interventional radiologist. Embolization is carried out under real-time X-ray fluoroscopic guidance in a dedicated angiosuite, which is comparable to the facility that is used for coronary stenting. There is good evidence that embolization of leiomyomata provides a genuine alternative for the two surgical techniques which are used classically: myomectomy (which is not always possible, depending on shape and location of the myomas), and hysterectomy (which involves radical excision of the

uterus and all benign tumors growing therein) [12-15]. Embolization offers significant advantages in terms of patient comfort, and recovery is fast. Psychological burden, which is inevitably associated with hysterectomy, can largely be avoided. Several cases of pregnancy after embolization of leiomyomata have been reported [16,17], but the actual fertility rate after this treatment is still uncertain [18].

It is important to underline that embolization is also rapidly gaining importance in the treatment of malignant tumors, particularly those in the liver or in the kidney. However, embolization of malignant tumors usually stimulates angiogenesis, leading to the formation of new arteries guiding the arterial blood around the embolic obstacles. Hence, embolization of malignant tumors must be accompanied by local or systemic chemotherapy.

Our interest in embolization started when we realized that chemical synthesis in the context of (bio)materials science offers possibilities to add functionalities to the injectable embolizing particles. We (and others) hypothesized that efficacy and safety of TACE for the treatment of leiomyomata can be enhanced when the embolic microspheres would

*Corresponding author: Koole LH, Interface Biomaterials BV, B. Lemmensstraat 364, 6163 JT Geleen, The Netherlands, E-mail: koole.leo@gmail.com

be radiopaque (i.e. detectable via X-ray fluoroscopy) [19-26]. Note that X-ray fluoroscopy is used in every procedure anyway, for roadmapping during navigation of the catheter's tip, and also to determine the procedure's end point. Note, furthermore, that all existing commercial products for embolization use embolic particles that consist of classical polymers (such as polyvinyl alcohol), which are radiolucent [24]. We reasoned that use of radiopaque microspheres will enable interventional radiologists to actually monitor the synthetic emboli *in situ*. We discussed this idea extensively with >20 interventionalists/TACE experts, and found broad consensus that this feature would help to enhance accuracy and safety of targeted embolization. According to recent literature, the idea is rapidly gaining acceptance [23-26]. Previously, we have described the preparation of our radiopaque microspheres, of which the key elements can be summarized as follows [20]:

- Only reactive monomers belonging to the methacrylate family are used. Linear and crosslinked poly (methacrylate)s are widely used in permanent implants (bone cements, intraocular lenses), and these materials are well-known for their long-term biocompatibility and stability.

- Microspheres are manufactured through suspension polymerization. Subsequently, the particles are size-sorted through automated sieving.

- Radiopacity is introduced through the use of a methacrylate monomer that contains covalently bound iodine (Figure 1a), and hydrophilicity is introduced through use of the monomer 2-hydroxy-ethylmethacrylate (HEMA).

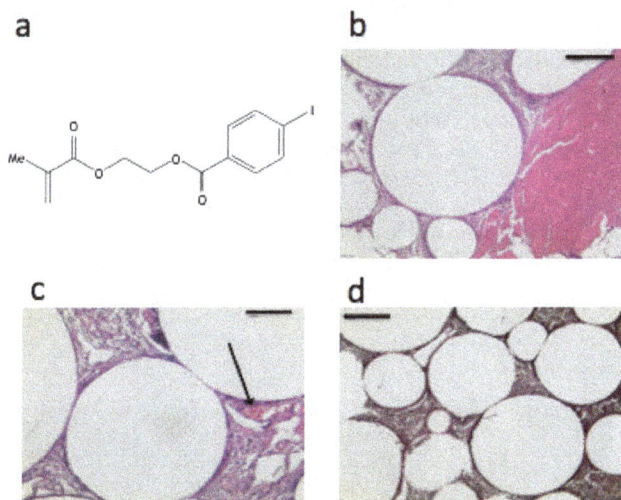

Figure 1: (a) Structural formula of the reactive monomer that is used in the manufacture of the radiopaque emblic microspheres. Note the methacryl moiety on the left side of the formula, and the aromatic ring with the covalently bound iodine. (b) Light microscopy of implanted microspheres + surrounding tissue (H/E staining), after 28 days of implantation. Around the microspheres, a mild inflammatory reaction is observed. Histiocytes have accumulated at the surface of the microspheres, especially in regions where there is no direct contact between the particles and the surrounding muscular tissue (bar = 200 μm). (c) As (b), but now at slightly larger magnification. Invaded histiocytes are clearly visible. The arrow points at a capillary blood vessel (filled with erythrocytes), which has formed to perfuse the newly formed tissue (bar = 100 μm). (d) As (b) and (c), while Elastin-van Gieson's stain was used. Note the formation of a thin collagenous capsule around most of the microspheres. This is a minimal fibrotic response, showing that the particles are well accepted in the host tissue (bar = 200 μm).

Here, we describe the essential features of our radiopaque microspheres: (i), *in vivo* biocompatibility, and (ii), X-ray imaging, both under realistic preclinical experimental conditions, and in a particular clinical situation. We also report briefly that this work provides an example of successful translation of research: the new radiopaque microspheres provided the basis for a new CE-certified medical device (class IIB) for embolization, which is now commercially available throughout Europe.

Materials and Methods

Injectability and X-ray imaging

Both kidneys were explanted from a cadaver of a rabbit that was sacrificed in a completely different experiment. The explantation was done within 20 min after sacrifice. The renal artery was prepared free, and the tip of a 20-G needle was inserted carefully into the arterial lumen. The kidneys were first flushed with saline. Then, a suspension of the microspheres (diameter range 400-600 micrometer, 20 mg microspheres in 1.6 ml) was carefully injected. The microspheres were carried along with the injection fluid, into the vascular tree of both kidneys. The kidneys were immediately frozen (-20°C) and stored until X-ray imaging. Images were recorded on a Phoenix Nanomex Imaging System (manufactured by General Electric).

In vivo biocompatibility study

This was performed by BSL BIOSERVICE Scientific Laboratories GmbH, Planegg, Germany. This company is certified according to the Principles of Good Laboratory Practice and accredited according to 90/385/EWG 93/42/EWG, and DIN EN ISO/IEC 17025:2005. The study complies with internationally accepted guidelines and recommendations regarding biological testing of medical devices:

- ISO 10993-1:2009 "Evaluation and testing within a risk management process"

- ISO 10993-6:2007 "Tests for local effects after implantation"

- ISO 10993-12:2007 "Sample preparation and reference materials":

- USP Biological Reactivity Tests, *In Vivo*, Implantation Test, current version

- OECD Series on principles of Good Laboratory Practice and compliance monitoring. Document No 13 ENV/JM/MONO (2002) [9].

Nine animals (healthy female New Zealand White Rabbits) were used (three animals per time point, i.e. 7 days, 14 days and 28 days). The rabbits were purchased from Charles River Deutschland (97633 Sulzfeld, Germany). The animals were derived from a controlled full-barrier maintained breeding system (SPF). The animals were bred for experimental purposes, according to Art. 9.2(no. 7) of the German Act on Animal welfare [26].

USP reference standard high-density poly(ethylene) (Promochem GmbH, lot no. 046) was used as the negative control material. The control samples were prepared according to the guideline ISO 10993-6, i.e. the material (film with thickness 1.0 mm) was processed by heating in a validated autoclave (121°C, 20 min). Then, the samples were cut out of the film (circular, 10 mm diameter, volume approximately 80 μL). The test samples were radiopaque iodine-containing microspheres in the diameter range 200-800 micrometer. These particles were also processed by heating in the autoclave (vide supra). In all cases, the trocar was filled up with a quantity of microspheres corresponding to approximately 80 µl. Pre- and post-surgery; the animals were housed in an air-conditioned room. An adequate acclimatization time of at least 5 days was maintained. The animals were housed in ABS-plastic rabbit cages with a floor surface of 4200 cm². The temperature was 18 ± 3°C, and the relative humidity was 55 ± 10 %. The artificial light was automatically switched on and off; 12 h light and 12 h dark. The air exchange was 10 x per hour at least. The animals had free access to autoclaved hay and to Altromin 2123 (maintenance diet for rabbits, which is rich in crude fibre). The animals also had free access to tap water (drinking water, municipal residue control, microbiological controls at regular intervals). The animals were anaesthetized with ketamine (Pharmanovo, lot no. 23116, exp. Date 07/2012), and xylazine (Riemser, lot no. 000660/1, exp. Date 12/2011). The fur on the back of the test animals was shaved on both sides of the spinal column. Care was taken to avoid mechanical irritation and trauma. Then, the implantation area was washed with antiseptic solution. The test items and control material were implanted into the muscular tissue, approximately 2.5 cm. away from the midline, and approximately 2.5 cm. apart from each other. The test items were implanted on the left side of the spinal columns, the control items in the right side. A sufficient number of implant sites was used to yield 10 test specimens and 10 control specimens for assessment. The implantation period was either 7 days, 14 days, or 28 days. Post-implantation, the animals were observed at least once daily. At the end of each experimental period, the animals were euthanized with an overdose of anaesthetic. After examination and macroscopic evaluation, the test and control material implant sites were excised together with sufficient unaffected tissue, to enable the evaluation of the biological response. The tissues were fixed in a 10 % formalin-buffered solution.

Tissue samples were received by BMP Laboratory for Medical Material Testing GmbH (Aachen, Germany). This company is accredited by the Zentralstelle der Lander fur Gesundheitsschutz bei Arzneimitteln und Medizinprodukten ZLG-P-585.00.08). The samples were first cut in three equal parts, and each part was placed in a HistoTec box. The samples were dehydrated in alcohol, and then embedded in paraffin. In total, 90 samples were processed in this way: (15 control samples + 15 test samples) * 3 parts per samples. Sections of each specimen (thickness 4 µm) were cut (microtome), and stained with either hematoxylin and eosin (H&E) or Elastica von Gieson (EvG).

Results and Discussion

In vivo biocompatibility

Photomicrographs of microspheres and surrounding muscular tissue are shown in Figure 1b-1d (follow-up 28 days); the embedded microspheres appear as circular regions. Figure 1b shows the mild foreign-body reaction that is observed at the interface of the microspheres and the host tissue. There is some accumulation of histiocytes, T-lymphocytes and some foreign-body giant cells. Connective tissue formation was minimal with some collagen fibers and fibrocytes around each microsphere. Almost no fibrotic reactions were encountered, and no granulocytes or plasma cells were found. The tissue reactions are similar to those observed after 7 days or 14 days of implantation, with one exception: the density of capillaries was larger after 28 days, compared to 7-days and 14-days follow-up. This reflects the flexible and slightly compressible nature of the particles, in vivo. Figure 1c is a photomicrograph of the same slide, now at larger magnification. Cells and fibrotic tissue surrounding the microspheres are clearly seen. In addition, a capillary blood vessel, filled with

erythrocytes (arrow) is noted. The formation of capillaries reveals that the microspheres became integrated in the host tissue. Microspheres that are in contact with each other may deform slightly, as is seen in Figure 1b and 1c. Figure 1d shows a representative photomicrograph of a similar tissue sample; this slide was treated with a mixture of picric acid and fuchsin (van Gieson's stain), which stains elastic fibers black, and collagen fibers dark-red. The very thin collagen layer around some of the microspheres shows that only a minimal encapsulation response has occurred [27].

Ex vivo embolization

"Artificial embolization" was achieved with two freshly explanted rabbit kidneys. These were perfused, via the renal artery, with a 1.8 ml of a suspension of the radiopaque microspheres (20 mg microspheres, diameter range 400-600 µm). Immediately thereafter, the kidneys were frozen and stored until X-ray imaging (Figure 2a). This revealed how the microspheres were distributed throughout the vascular beds (Figure 2b). Most of the microspheres are aligned in one of the major arteries, while other microspheres entered side branches. The data provide an example of non-specific embolization.

Preclinical in vivo embolization

Catheter-based embolization of the left kidneys of two living sheep was performed by an experienced interventional radiologist. Both procedures proceeded smoothly. Figure 3a shows a kidney upon perfusion with contrast, but prior to the injection of microspheres. Note that Figure 3a is a digital subtraction image, i.e. it is the difference between the X-ray images before and shortly after (several seconds) injection of contrast [25]. Hence, the image only provides information about the distribution of the injected contrast fluid; all other X-ray absorbing parts of the body are, in fact, eliminated. Numerous arterial branches within the kidney are seen clearly. The organ's contour is clearly visible as well, and this reveals that the contrast nicely flows throughout the entire organ. Figure 3b is technically the same, although this image was recorded near the endpoint of the embolization procedure. Note that Figure 3a and 3b are markedly different. In Figure 3b, the contrast is seen to accumulate in the larger arteries. The contrast hardly reaches the cortical regions of the kidney, and the organ's contour is almost invisible now. Furthermore, many small arteries are not discernable in Figure 3b. Apparently, these are no longer perfused with contrast fluid, indicating that embolization was successful.

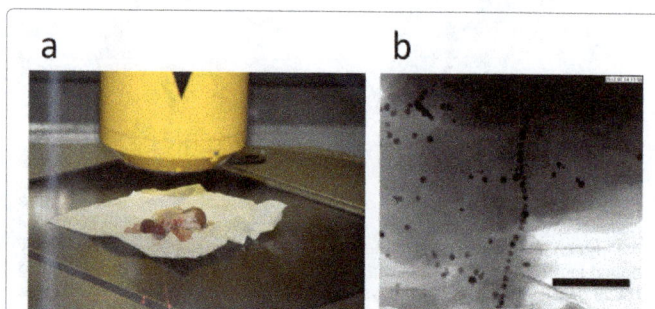

Figure 2: (a) Frozen explanted rabbit kidney, photographed during X-ray imaging in the Nanomex X-ray imaging instrument. The kidneys were explanted immediately after sacrifice of the animal, and immediately perfused, via the renal artery, with a suspension of the embolic microspheres (diameter range 400-600 µm; 1.8 ml, 20 mg microspheres). The organs containing the microspheres were frozen and stored until imaging was performed. (b) Representative X-ray image (shadow), showing the distribution of the radiopaque microspheres throughout the kidney's arterial vessel bed. Bar=5000 µm.

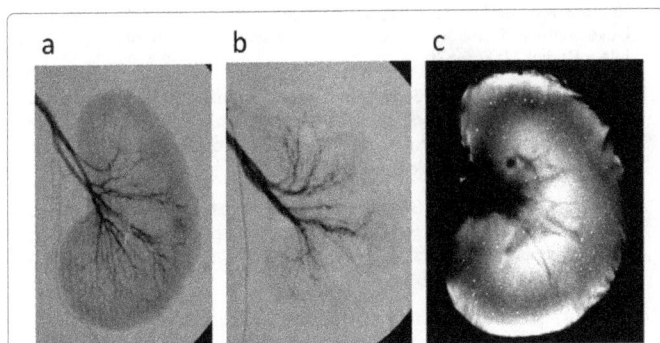

Figure 3: X-ray images, obtained during and after the embolization of the left kidney in a living sheep model. (a) Digital subtraction X-ray image of the sheep's left kidney, recorded prior to embolization. The vasculature is visible in detail; even small arteries near the organ's cortex can be discerned. (b) Same image as (a), but now recorded near the end-point of embolization. Perfusion of contrast is diminished clearly, and many small arteries are no longer seen. Dotted X-ray absorbing patterns are noted, which can possibly be ascribed to accumulated radiopaque embolic microspheres. (c) Slice abstracted from the 3D computed tomography image of the explanted embolized sheep kidney (thickness 2 mm). The radiopaque particles are seen clearly, as white dots. Note that most of the particles are found close to the cortex, where they became arrested due to gradual narrowing of the arterial branches in the kidney's arterial vessel tree.

Another striking feature seen in Figure 3b is the dotted appearance of contrast, particular in pre-cortical regions. We assume that these dots represent the radiopaque embolizing particles. We could not prove this unambiguously, given the limited sensitivity and spatial resolution that could be obtained with the imaging equipment in the animal clinic. The embolized kidneys were explanted from the animals, immediately after sacrifice. The organs were frozen, with the aim to examine them later via X-ray 3D computed tomography at higher spatial resolution. These measurements showed the embolic particles *in situ*; Figure 3c shows a horizontal slice (thickness 2 mm), which was abstracted from the data set. This image, which is representative for the entire data set, represents the center of one of the left kidneys. Note that there is no X-ray contrast fluid in the arteries any more. In Figure 3c, the microspheres are clearly visible as white (X-ray absorbing) small dots. Most of the microspheres became arrested near the cortex of the kidney [28,29]. This nicely confirms the common view, that embolic particles are carried along with the blood stream after leaving the catheter's mouth, to become arrested downstream in the arterial tree, once the diameter of the arterial branch becomes equal to their diameter (which is, in this case, in the range 400-600 µm, vide supra).

Conclusions

The new iodine-containing polymethacrylate crosslinked radiopaque microspheres are suitable for use in clinical embolization procedures. There are no concerns regarding injectability, biocompatibility or stability *in situ*. The new microspheres provide a level of X-ray visibility that-in principle-allows the interventional radiologist to localize the embolic particles during and after the embolization procedures. The extent of "visibility" depends on several factors, such as: spatial distribution of the microspheres, diameter size of the microspheres (for small- sized microspheres only ensembles of the particles will be detectable), proximity of bone tissue, quality of the X-ray imaging equipment, etc.). The idea that the use of X-ray visible embolic particles may translate into advantages for patients, clinicians and heath care systems is clearly growing stronger [24-26]. It is, however, still unclear for which type of embolizations the feature will be

most advantageous. We anticipate that the feature will be particularly helpful in pre-operative embolization of cerebral tumors, which can be done to prevent bleeding complications during surgery. Monitoring of the embolic particles may help to prevent non-target embolization, and can help to judge whether or not total embolization has been achieved. The latter is important to maximize the chances for success. The new radiopaque microspheres are used in a new CE-certified medical device for embolization, called X-Spheres. Therefore, this work provides a quite unusual example of translational development of a new medical device, classified within the highest-but-one risk group (IIB), on the basis of a new synthetic polymer biomaterial that was originally developed in an academic laboratory.

Competing Interests

The authors Y.B.A. and L.H.K. declare to have stock in the company Interface BIOmaterials BV (Geleen, The Netherlands). This company is the manufacturer of the new embolization product X-Spheres; this product utilizes radiopaque embolic microspheres.

Acknowledgements

We thank Dr. U. Mueller, Professor B. Klosterhalfen and mr. C. Musters for advice and technical assistance. We thanks prof. P. Lohle (Elisabeth Hospital Tilburg, the Netherlands) for performing the kidney embolizations in the sheep models. This study was financed, in part, by the Interreg IV-A project 'BioMiMedics' (www.biomimedics.org) (Interreg IV-A 1.2-2010-03/063). The Universities of Maastricht, Liege (Belgium), Hasselt (Belgium) and Aachen (Germany; RWTH and Fachhochschule), as well as several local regional biotechnological enterprises cooperated in 'BioMiMedics'. This particular study was financed through generous contributions of the EU (through INterreg IV-A), the Government of the Province Dutch Limburg, the Dutch National Ministry of Economic Affairs, Agriculture and Innovation, Maastricht University, The Limburg Bank for Industry Innovation (LIOF), and the Company INterface BIOmaterials BV (Geleen, The Netherlands).

References

1. Iqbal J, Gunn J, Serruys PW (2013) Coronary stents: Historical development, current status and future directions. Br Med Bull 106: 193-211.

2. Sternberg K, Grabow N, Petersen S, Weitschies W, Harder C, et al. (2013) Advances in coronary stent technology--active drug-loaded stent surfaces for prevention of restenosis and improvement of biocompatibility. Curr Pharm Biotechnol 14: 76-90.

3. Nikam N, Steinberg TB, Steinberg DH (2014) Advances in stent technologies and their effect on clinical efficacy and safety. Med Devices (Auckl) 7: 165-178.

4. Garg S, Bourantas C, Serruys PW (2013) New concepts in the design of drug-eluting coronary stents. Nat Rev Cardiol 10: 248-260.

5. Brugaletta S, Garcia-Garcia HM, Onuma Y, Serruys PW (2012) Everolimus-eluting ABSORB bioresorbable vascular scaffold: present and future perspectives. Expert Rev Med Devices 9: 327-338.

6. Khan AT, Shehmar M, Gupta JK (2014) Uterine fibroids: current perspectives. Int J Womens Health 6: 95-114.

7. Spies JB (2013) Current evidence on uterine embolization for fibroids. Semin Intervent Radiol 30: 340-346.

8. Lopera J, Suri R, Kroma GM, Garza-Berlanga A, Thomas J (2013) Role of interventional procedures in obstetrics/gynecology. Radiol Clin North Am 51: 1049-1066.

9. Bulman JC, Ascher SM, Spies JB (2012) Current concepts in uterine fibroid embolization. Radiographics 32: 1735-1750.

10. Van der Kooij SM, Ankum WM, Hehenkamp WJ (2012) Review of nonsurgical/minimally invasive treatments for uterine fibroids. Curr Opin Obstet Gynecol 24: 368-375.

11. Salazar GM, Petrozza JC, Walker TG (2009) Transcatheter endovascular techniques for management of obstetrical and gynecologic emergencies. Tech Vasc Interv Radiol 12: 139-147.

12. Van der Kooij SM, Bipat S, Hehenkamp WJ, Ankum WM, Reekers JA (2010) Uterine artery embolization vs hysterectomy in the treatment of symptomatic

uterine fibroids: 5-year outcome from the randomized EMMY trial. Am J Obstet Gynecol 203: 105 e1-e13.

13. Van der Kooij SM, Hehenkamp WJ, Birnie E, Ankum WM, Mol BW, et al. (2013) The effect of treatment preference and treatment allocation on patients' health-related quality of life in the randomized EMMY trial. Eur J Obstet Gynecol Reprod Biol 169: 69-74.

14. Hehenkamp WJ, Volkers NA, Birnie E, Reekers JA, Ankum WM (2006) Pain and return to daily activities after uterine artery embolization and hysterectomy in the treatment of symptomatic uterine fibroids: results from the randomized EMMY trial. Cardiovasc Intervent Radiol 29: 179-187.

15. Hehenkamp WJ, Volkers NA, Donderwinkel PF, De Blok S, Birnie E, et al. (2005) Uterine artery embolization versus hysterectomy in the treatment of symptomatic uterine fibroids (EMMY trial): peri- and postprocedural results from a randomized controlled trial. Am J Obstet Gynecol 193: 1618-1629.

16. Pisco JM, Duarte M, Bilhim T, Cirurgião F, Oliveira AG (2011) Pregnancy after uterine fibroid embolization. Fertil Steril 95: 1121-1126.

17. Pinto Pabón I, Magret JP, Unzurrunzaga EA, García IM, Catalán IB, et al. (2008) Pregnancy after uterine fibroid embolization: follow-up of 100 patients embolized using tris-acryl gelatin microspheres. Fertil Steril 90: 2356-2360.

18. Torre A, Paillusson B, Fain V, Labauge P, Pelage JP, et al. (2014) Uterine artery embolization for severe symptomatic fibroids: effects on fertility and symptoms. Hum Reprod 29: 490-501.

19. Saralidze K, Van Hooy-Corstjens CS, Koole LH, Knetsch ML (2007) New acrylic microspheres for arterial embolization: combining radiopacity for precise localization with immobilized thrombin to trigger local blood coagulation. Biomaterials 28: 2457-2464.

20. Saralidze K, Knetsch ML, Van der Marel C, Koole LH (2010) Versatile polymer microspheres for injection therapy: aspects of fluoroscopic traceability and biofunctionalization. Biomacromolecules 11: 3556-3562.

21. Sharma KV, Dreher MR, Tang Y, Pritchard W, Chiesa OA, et al. (2010) Development of "imageable" beads for transcatheter embolotherapy. J Vasc Interv Radiol 21: 865-876.

22. Galperin A, Margel S (2006) Synthesis and characterization of new micrometer-sized radiopaque polymeric particles of narrow size distribution by a single-step swelling of uniform polystyrene template microspheres for X-ray imaging applications. Biomacromolecules 7: 2650-2660.

23. Stampfl U, Sommer CM, Bellemann N, Holzschuh M, Kueller A, et al. (2012) Multimodal visibility of a modified polyzene-F-coated spherical embolic agent for liver embolization: Feasibility study in a porcine model. J Vasc Interv Radiol 23: 1225-1231.

24. Duran R, Sharma K, Dreher MR, Ashrafi K, Mirpour S, et al. (2016) A novel inherently radiopaque bead for transarterial embolization to treat liver cancer -a preclinical study. Theranostics 6: 28-39.

25. Tacher V, Duran R, Lin M, Sohn JH, Sharma KV, et al. (2015) Multimodality imaging of ethiodized oil-loaded radiopaque microspheres during transarterial emboliozation of rabbits with VX2 liver tumors. Radiology 16: 141624.

26. Johnson CG, Tang Y, Beck A, Dreher MR, Woods DL, et al. (2016) Preparation of radiopaque drug-eluting beads for transcatheter chemoembolization. J Vasc Interv Radiol 27: 117-126.

27. Shlansky-Goldberg RD, Rosen MA, Mondschein JI, Stavropoulos SW, Trerotola SO, et al. (2014) Comparison of polyvinyl alcohol microspheres and tris-acryl gelatin microspheres for uterine fibroid embolization: results of a single-center randomized study. J Vasc Interv Radiol 25: 823-832.

28. Salazar GM, Petrozza JC, Walker TG (2009) Evaluation and management of acute vascular trauma. Tech Vasc Interv Radiol 12: 139-147.

29. German Animal Welfare Act (2009).

34

Using Flixweed Seed as a Pore-former to Prepare Porous Ceramics

Hedayat N and Du Y*

College of Applied Engineering, Sustainability and Technology (CAEST), Aeronautics & Technology Building (ATB), Kent State University, Lefton Esplanade, Kent, Ohio, USA

Abstract

Flixweed (*Descurainia Sophia* L.) seeds, known as an herbal medicine, for the first time are used as a promising pore-forming agent (pore-former) in ceramic technology. Flixweed seeds were selected because of their unique constant shape (oblong, 1.2 mm long with the aspect ratio of about 2) and narrow size distribution as well as their low-cost. Porous zirconia ceramics have been fabricated using flixweed seeds by tape casting technique. The dried tape-cast cut into disk-shaped pieces and were fired at 1400°C for 2h, resulting in porous zirconia disks with a bulk density of 3.96 g/cm^3, total porosity of 34.6 ± 0.9% (open porosity 25.5 ± 0.7%, closed porosity 9.1 ± 0.3%) and a linear shrinkage of 21.5 ± 0.3%. The pore shape and size were similar in shape and size to the original pore-former.

Keywords: Pore-former; Porous ceramics; Microstructure; Tape casting; Flixweed seeds

Introduction

Porous ceramics are used for an ever-expanding range of applications in bone tissue engineering, membrane separation, and catalytic reactors. Different applications of porous ceramics have been described in a book edited by Scheffler and Colombo [1]. A number of processing routes are used to prepare porous ceramics including partial sintering [2], gel casting [3], sol-gel technique [4], dry foaming method [5], and the incorporation of pyrolizable pore-forming agents (pore-formers) that burn out during firing. Various types of pyrolizabale pore-formers have been examined as sacrificial templates or fugitive materials to obtain the different shape and size of pores [6-11]. The pyrolizable pore-formers have behaved as template in forming the pores that correspond closely to the shape and size of original pore-former [12]. The number and nature of the pores (open pores and closed pores) can be determined by the combination of characterization techniques including Archimedes method, scanning electron microscopy (SEM) imaging, optical microscopy, focused ion beam (FIB)-SEM studies, mercury porosimetry, and gas permeability tests [13].

The pore diameter size that is closely connected to the size of the pore-former is determinable by microscopic image analysis or three dimensional (3D) tomography, but the pore size determined by mercury intrusion is the size of interconnections between open pores and refers to pore throat size, which is usually in the range of about 1–10 μm [14]. Pore-formers with the diameter of about 3 μm or smaller have significant impact on the sintering kinetics and shrinkage, and pore-formers with the diameter of about 20 μm are used to tailor the porosity and to improve the gas diffusion in porous electrodes for solid oxide fuel cells [7]. The pore size can be controlled by varying the following variables: pore-former size [12]; thermal decomposition profile of the pore-former [15]; and volume ratio of pore-former/ceramic particles [12,16]. Hu et al. [17] reported that the porosity and also the number of large pores are increased with the pore-former loading. A composite pore-former containing two or more pore-formers in the different size range can be used to improve the porous structure and adjust the shrinkage [17,18]. Small pores (<100 μm) have been generated using common pore-formers such as different graphite types [9], different starch types [6], polymethyl methacrylate (PMMA) [7], carbon microspheres [8], cellulose [10], and paper-fibers [19]. Larger pore-formers that can generate larger pores are desirable for many purposes such as acoustic and thermal insulation materials, lightweight structured ceramics, and bone tissue ingrowth into bio-

ceramic implants [14]. The pores incorporated into the zirconia ceramics would decrease the thermal conductivity owing to the air entrapped in the pores. The higher porosity results in the lower thermal conductivity [20].

Flixweed (*Descurainia sophia* L.) seeds, a known medical herb commonly used in traditional medicine [21], is one of the most abundant weeds in North America and China [22]. Felixweed seed is very small, dark yellow or brown, possess an uneven surface in a stretched oval form, one end of which is cut and maintains a transparent yellowish ring [23]. Flixweed seeds might be a potentially interesting and unique pore-former for ceramics, and there is no report so far on the use of flixweed seeds as a pore-former in ceramic fabrication. In the present study, we report the first results on the use of flixweed seeds to fabricate porous ceramics. A composite pore-former containing microcrystalline cellulose and flixweed seeds together used to generate zirconia ceramic with a hierarchical porosity including small and large pores derived from the burnout of microcrystalline cellulose and flixweed seeds. Cellulose as a natural pore-former has a number of advantages including easy availability and processing (narrow decomposition temperature range of 300-350°C), and low-cost [24]. Alumina ceramics prepared using poppy seeds [14,25], and commercially milled coffee [25] are compared to the samples fabricated using flixweed seeds by tape casting.

Materials and Methods

Reagents and apparatus

Flixweed seeds were purchased from Sadaf (Soofer Co., Inc., USA), and microcrystalline cellulose (PH-301) was provided by FMC BioPolymer, USA. TZ-3Y powder was obtained from Tosoh, Japan. Methyl ethyl Ketone (MEK, ≥99.0%) and polyethylene glycol (PEG 200) were purchased from Sigma-Aldrich. Hypermer KD-1, Butvar

---

*Corresponding author: Du Y, College of Applied Engineering, Sustainability and Technology (CAEST), Aeronautics & Technology Building (ATB), Kent State University, 1400 Lefton Esplanade, Kent, Ohio 44242, USA
E-mail: ydu5@kent.edu

(polyvinyl butyral (PVB), B-98) and Ethanol (94-96%) were prepared form Tape Casting Warehouse, Inc., USA, Solutia, Inc., and Alfa Aesar, respectively. A lab roll ball mill (Tencan, Model No.: QM-5, China) using yttria stabilized zirconia (YSZ) cylinders as milling media (5×5 mm, Inframat Advanced Materials) was used for the ball milling of slurry. Tape casting was performed using a Lab scale tape caster (Richard E. Mistler, Inc.), in which slurry was cast onto a polyethylene carrier film (Mylar sheet). A muffle furnace (Across International, NJ) was used for the firing of samples. The S-2600N scanning electron microscope that was used to take the SEM images is from Hitachi, Japan.

Procedure

In the present study, tape casting (TC) or doctor blade process that is a low-cost and simple method to fabricate thin flat sheets of ceramics was used. Figure 1 depicts a schematic representation of fabricating porous zirconia ceramic using flixweed seeds by tape casting. Zirconia tape was produced by tape casting of the slurry that was prepared by (i) weighing the TZ-3Y powder in the required amount to prepare a 60 wt% zirconia suspension, (ii) dispersing the TZ-3Y and dispersant powder (KD-1) in a binary solvent system of ethanol and methyl ethyl ketone with the ratio of EtOH/MEK : 34/66 wt% for 3h (dispersion step), (iii) introducing tape casting additives such as binder (B-98), plasticizer I (S-160), and plasticizer II (PEG 200), and (iv) ball milling the resulting mixture for 24h (thickening step). Finally, flixweed seeds were mixed, and dispersed using a mechanical stirrer to make the dispersion uniform for 15 min. The flixweed seeds were mixed and dispersed right after ball milling and before de-airing to prevent the breakage of the flixweed seeds during ball milling, and to take advantage of the presence of solvents that could help to dispersion of the seeds before de-airing. De-airing under a mechanical vacuum is necessary to remove the extra solvent and also to prevent the formation of air bubbles during the casting. A mechanical vacuum pump and a vacuum chamber were used to perform the de-airing during magnetic stirring. Upon de-airing, the mechanical stirring continued for further 5 min. to make the dispersion more uniform. The slurry cast at the thickness of 3000 μm, and the casting speed was 2 mm/s. The tape-cast dried for 48h, and the thickness of dried tape was 1000 μm. The tape-cast was cut into 25 mm diameter disks, and fired at 1400°C for 2h, resulting in disks without visible cracks.

Results

Figure 2a shows a SEM image of flixweed seeds with typical oblong shape. Figure 2b shows a SEM image of flixweed seeds and the network texture on the surface. Table 1 presents the maximum Feret diameter (length), minimum Feret diameter (width) and aspect ratio of flixweed seeds obtained by image analysis software (Image J) that revealed the particle size distribution of flixweed seed is narrow.

Table 2 lists the density of ingredients used. The density of flixweed seed is required to calculate the volume percentage of pore-former in the system and it was determined by floating flixweed seeds in a sugar solution (saccharose). The addition of sugar to the water in which flixweed seeds had settled down continued until the flixweed seeds exhibited buoyancy. The measured density of flixweed seeds in this study is 1.14 ± 0.04 g/cm^3.

Figure 3 shows the SEM micrograph of the pores generated by burning out of flixweed seeds. As expected, the shape of the pores due to flixweed seed is oblong. The maximum Feret diameter, the minimum Feret diameter, and the aspect ratio of the pores derived from the burnout of flixweed seeds are 740 μm, 380 μm, and 1.9, respectively. Figure 4 shows the SEM micrograph of the pores generated by burning out of microcrystalline cellulose. The pore diameter of the pores derived from the burnout of microcrystalline cellulose is 0.3 μm.

Archimedes relations are used to calculate the bulk density (D_b), and open porosity (P_o) of composites. W_D, W_S and W_I are the weight of dry, weight of saturation and weight of immersed in distilled water (D_{water} = 1 g/cm^3), respectively.

$$D_b = \frac{W_D}{W_S - W_I} \times 1g / cm^3 \tag{1}$$

$$P_o = 100 \left(\frac{W_S - W_D}{W_S - W_I} \right) \tag{2}$$

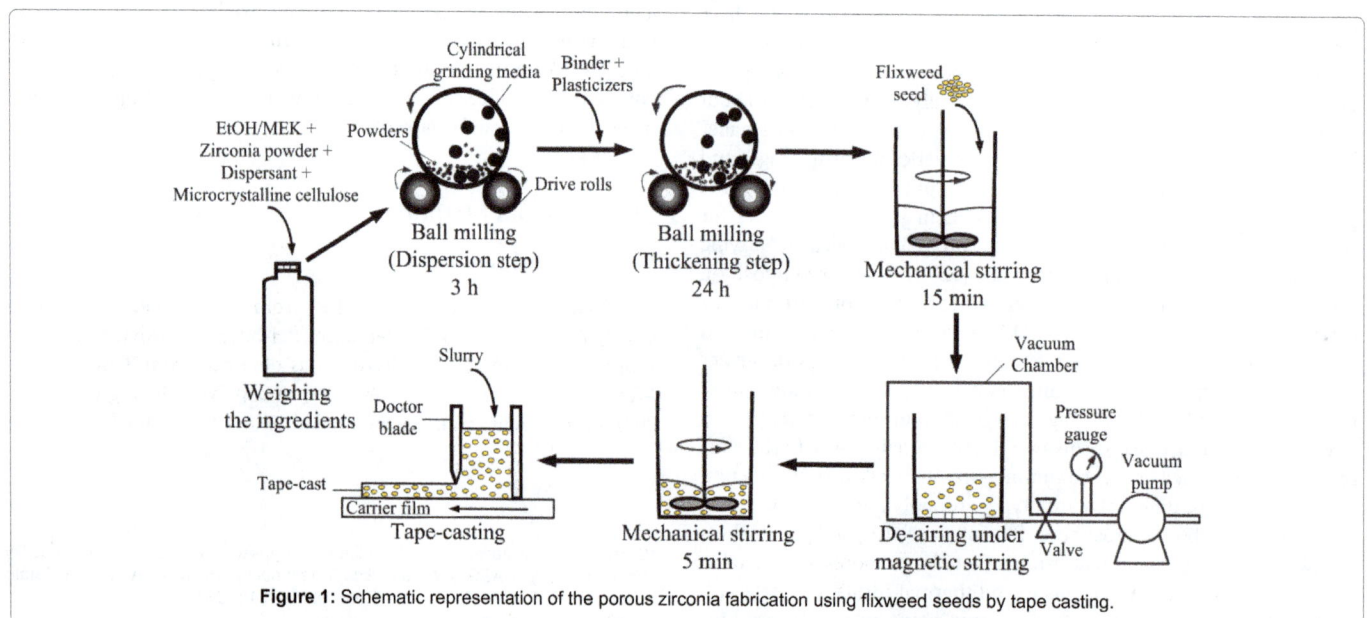

Figure 1: Schematic representation of the porous zirconia fabrication using flixweed seeds by tape casting.

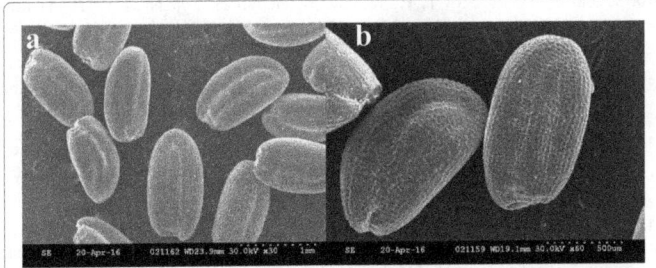

Figure 2: (a) SEM micrograph of flixweed seeds with typical oblong shape; (b) SEM micrograph of flixweed seeds demonstrating the network texture on the surface, which is the reticulum of wider-than-long pits, like corn on the cob.

Feature	Arithmetic mean value and standard deviation
Maximum Feret diameter (μm)	1277 ± 23
Minimum Feret diameter (μm)	680 ± 21
Aspect ratio	1.88 ± 0.08

Table 1: Size and shape characteristics of flixweed seeds (Feret diameters and aspect ratios determined by image analysis).

Ingredient	Density (g/cm³)
TZ-3Y	6.05
EtOH	0.79
MEK	0.81
PVB	1.08
KD-1	0.88
PH-301	0.37
Flixweed seeds	1.14
S-160	1.12
PEG 200	1.12

Table 2: Densities of the ingredient used.

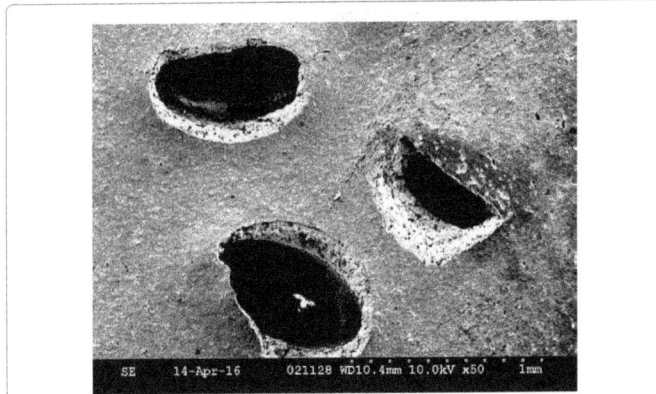

Figure 3: SEM micrograph of large pores derived from the burnout of flixweed seeds after firing at 1400°C.

The total porosity (P_T) is obtained from the following equation:

$$P_T = 1 - \frac{D_b}{D_{th}} \qquad (3)$$

Where,

D_{th} is the theoretical density of zirconia (TZ-3Y), i.e., 6.05 g/cm³.

Table 3 summarizes the microstructural characteristics of fired porous zirconia ceramics with pores derived from microcrystalline cellulose and flixweed seeds. Microcrystalline cellulose and flixweed seeds were added to 165 g of the as-prepared zirconia suspension in

the amounts of 5.2 and 1.2 g that according to the densities presented in Table 2 corresponds to a total pore-former volume of 28.1% on dry green tape basis. The total porosity of fabricated porous zirconia is 34.6% that is 6.5% higher than the pore-former content of dry green tape. This extra porosity must be a result of dispersant burnout (4.2 vol% on dry green tape basis) and also from the binder burnout (18.8 vol. % on dry green tape basis). Živcová et al. [25] used traditional slip casting (TS) into plaster molds (cylindrical rods, diameter 5 mm) and starch consolidation casting (SC) using metal molds (cylindrical rods, diameter 7 mm) for the fabrication of porous alumina ceramics. In the starch consolidation casting (SC), starch can be used as pore-former and at the same time as binder [14,25]. Commercially milled coffee, and poppy seeds are of potential interest due to their specific size. Table 4 compares the alumina ceramics prepared using commercially milled coffee, and poppy seeds, and the zirconia ceramics prepared

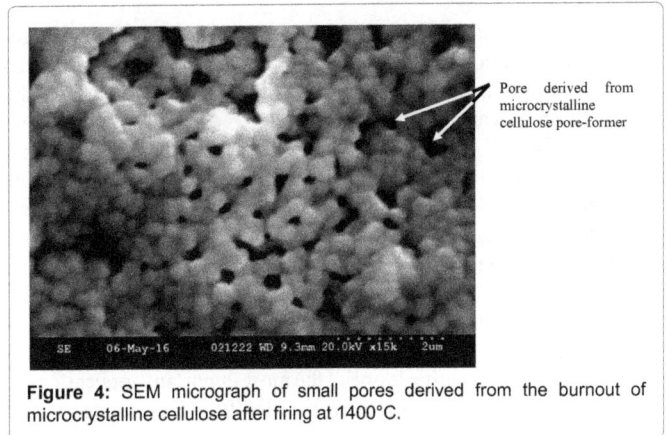

Figure 4: SEM micrograph of small pores derived from the burnout of microcrystalline cellulose after firing at 1400°C.

Feature	Arithmetic mean value and standard deviation
Bulk density (g/cm³)	3.96 ± 0.05
Apparent density (g/cm³)	5.31 ± 0.03
Open porosity (%)	25.5 ± 0.7
Total porosity (%)	34.6 ± 0.9
Closed porosity (%)	9.1 ± 0.3
Linear shrinkage (%)	21.5 ± 0.3
Volumetric shrinkage (%)	44.5 ± 0.5
Weight loss (%)	8.1 ± 0.2

Table 3: Microstructural characteristics of porous zirconia with large and small pores due to flixweed seeds and microcrystalline cellulose fired in 1400°C.

Pore-former	Fabrication method	Pore-former [vol.%]	Open porosity [%]	Total porosity [%]	Characteristic diameter of pore-former [μm]	Ref.
Potato starch	SC	10	22.0 ± 0.3	25.8 ± 0.3	46.3	[25]
		30	32.9 ± 1.3	37.0 ± 0.5		
Coffee	TS	19.7	25.7 ± 1.5	32.7 ± 0.9	405	[25]
	SC	36.7 (1)	44.6 ± 0.7	46.6 ± 0.4		
Poppy seeds	SC	32.3 (2)	29.6 ± 1.6	38.7 ± 1.3	max Feret: 1038	[14]
					min Feret: 1265	
Flixweed seeds	TC	28.1 (3)	25.5 ± 0.7	34.6 ± 0.9	max Feret: 1277	This work
					min Feret: 680	

Note:
(1) 15.6 vol.% coffee + 21.1% Potato starch
(2) 15.8 vol.% Poppy seed + 16.5 vol.% Potato starch
(3) 2.1 vol.% Flixweed seed + 26.0 vol.% microcrystalline cellulose

Table 4: Alumina ceramics prepared using different pore-formers compared to the samples fabricated in this study.

using flixweed seeds as pore-former. The data for alumina ceramics prepared using potato starch is also presented because in alumina ceramics prepared using coffee and poppy seed by starch consolidation method potato starch performs as binder. Fabriation of the zirconia with hierarchical porosity using flixweed seed by tape casting has generated ceramics with the total porosity of about 35% that is in the same range compared to other methods and pore-formers. About 26% of the formed pore volumes in the prepared porous zirconia are closed, but they are functional for uses such as acoustic and thermal insulation. The lamination of ceramic sheets prepared by tape casting and multi-layer tape casting provides the advantage of manufacturing ceramic sheets with gradual porosity. The controlled pore size and porosity enables the technique to fabricate ceramic multi-layers with different thermal conductivity in each layer.

Conclusion

In this study, it is shown that flixweed seeds are a potential pore-former for ceramic preparation. The defect-free burnout of flixweed seeds was feasible using a conventional firing profile. The density measurement revealed that the density of flixweed seed is a slightly higher than that of water. Therefore, not only for organic slurry formulations, but also for aqueous slurry formulations flixweed seeds are appropriate pore-former compared to many synthetic polymeric pore-formers that are exhibiting strong buoyancy effects at the same size. Among the main advantages for the use of flixweed seeds as a pore-former in ceramic technology are easy handling, defect-free burnout, low-cost, ready availability in the constant oblong shape with narrow size distribution and well-defined aspect ratio (about 2). Considering that the thermal barrier coating is one of the most important applications of zirconia-based ceramics [26], the compatibility of flixweed seeds with zirconia and its large size highlights the use of this natural pore-former.

Acknowledgements

The SEM data were obtained at the SEM lab of the Characterization Facility of the Liquid Crystal Institute, Kent State University. The authors acknowledge the FMC BioPolymer, USA for providing microcrystalline cellulose (PH-301).

References

1. Scheffler M, Colombo P (2006) Cellular ceramics: Structure, manufacturing, properties and applications. John Wiley & Sons, USA.

2. Deng ZY, Yang JF, Beppu Y, Ando M, Ohji T (2002) Effect of agglomeration on mechanical properties of porous zirconia fabricated by partial sintering. Journal of the American Ceramic Society 85: 1961-1965.

3. Chen X, Hong L (2010) An in situ approach to create porous ceramic membrane: Polymerization of acrylamide in a confined environment. Journal of the American Ceramic Society 93: 96-103.

4. Larbot A, Fabre JP, Guizard C, Cot L (1988) Inorganic membranes obtained by sol-gel techniques. Journal of Membrane Science 39: 203-212.

5. Gonzenbach UT, Studart AR, Tervoort E, Gauckler LJ (2007) Macroporous ceramics from particle-stabilized wet foams. Journal of the American Ceramic Society 90: 16-22.

6. Gregorová E, Pabst W, Bohačenko I (2006) Characterization of different starch types for their application in ceramic processing. Journal of the European Ceramic Society 26: 1301-1309.

7. Horri BA, Selomulya C, Wang H (2012) Characteristics of Ni/YSZ ceramic anode prepared using carbon microspheres as a pore former. International Journal of Hydrogen Energy 37: 15311-15319.

8. Panthi D, Choi B, Tsutsumi A (2012) Performance enhancement of strontium-doped lanthanum manganite cathode by developing a highly porous microstructure. Journal of Applied Electrochemistry 42: 953-959.

9. Sarikaya A, Dogan F (2013) Effect of various pore formers on the microstructural development of tape-cast porous ceramics. Ceramics International 39: 403-413.

10. Suzuki T, Yamaguchi T, Hamamoto K, Sumia H, Fujishiroa Y (2011) Low temperature densification process of solid-oxide fuel cell electrolyte controlled by anode support shrinkage. Rsc Advances 1: 911-916.

11. Bose S, Das C (2015) Sawdust: from wood waste to pore-former in the fabrication of ceramic membrane. Ceramics International 41: 4070-4079.

12. Boaro M, Vohs JM, Gorte RJ (2003) Synthesis of highly porous yttria-stabilized zirconia by tape-casting methods. Journal of the American Ceramic Society 86: 395-400.

13. Simwonis D, Naoumidis A, Dias FJ, Linke J, Moropoulou A (1997) Material characterization in support of the development of an anode substrate for solid oxide fuel cells. Journal of Materials Research 12: 1508-1518.

14. Gregorová E, Pabst W (2007) Porous ceramics prepared using poppy seed as a pore-forming agent. Ceramics international 33: 1385-1388.

15. Sarikaya A, Petrovsky V, Dogan F (2012) Effect of the anode microstructure on the enhanced performance of solid oxide fuel cells. International Journal of Hydrogen Energy 37: 11370-11377.

16. Tang F, Fudouzi H, Uchikoshi T, Sakka Y (2004) Preparation of porous materials with controlled pore size and porosity. Journal of the European Ceramic Society 24: 341-344.

17. Hu J, Lü Z, Chen K, Huang X, Ai N, et al. (2008) Effect of composite pore-former on the fabrication and performance of anode-supported membranes for SOFCs. Journal of Membrane Science 318: 445-451.

18. Kim M, Lee J, Han JH (2011) Fabrication of anode support for solid oxide fuel cell using zirconium hydroxide as a pore former. Journal of Power Sources 196: 2475-2482.

19. Pan WP, Lü Z, Chen KF, Zhu XB, Huang XQ, et al. (2011) Paper-fibres used as a pore-former for anode substrate of solid oxide fuel cell. Fuel Cells 11: 172-177.

20. Liu R, Wang C (2013) Effects of mono-dispersed PMMA micro-balls as pore-forming agent on the properties of porous YSZ ceramics. Journal of the European Ceramic Society 33: 1859-1865.

21. Nimrouzi M, Sadeghpour O, Imanieh MH, Ardekani MS, Salehi A, et al. (2015) Flixweed vs. polyethylene glycol in the treatment of childhood functional constipation: a randomized clinical trial. Iranian Journal of paediatrics 25: e425.

22. Li W, Liu X, Khan MA, Kamiya Y, Yamaguchi S (2005) Hormonal and environmental regulation of seed germination in flixweed (Descurainia sohia). Plant growth regulation 45: 199-207.

23. Amin G (1990) Traditional herbal drugs in Iran. Iranian ministry of health publications, Tehran.

24. Calahorra M, Cortázar M, Eguiazábal JI, Guzmán GM (1989) Thermogravimetric analysis of cellulose: effect of the molecular weight on thermal decomposition. Journal of Applied Polymer Science 37: 3305-3314.

25. Živcová Z, Gregorová E, Pabst W (2008) Alumina ceramics prepared with new pore-forming agents. Processing and Application of Ceramics 2: 1-8.

26. Loganathan A, Gandhi AS (2012) Effect of phase transformations on the fracture toughness of t'yttria stabilized zirconia. Materials Science and Engineering: A 556: 927-935.

35

Total Excess Lifetime Cancer Risk Estimation from Enhanced Heavy Metals Concentrations Resulting from Tailings in Katsina Steel Rolling Mill, Nigeria

Bello S[1]*, Muhammad BG[1] and Bature B[2]

[1]Department of Physics, Faculty of Natural and Applied Sciences, Umaru Musa Yar'adua University Katsina, Nigeria
[2]Department of Mathematics, Faculty of Natural and Applied Sciences, Umaru Musa Yar'adua University Katsina, Nigeria

Abstract

Soil samples were randomly collected from the dump-yard of Katsina steel rolling mill and were analyzed for the presence and concentrations of the carcinogenic heavy metals namely: Chromium (Cr), Arsenic (As), Cadmium (Cd), Cobalt (Co) and Lead (Pb) using flame atomic absorption spectrophotometry instrumental method. The obtained concentrations were used to estimate the excess lifetime cancer risk due to exposure from these metals using models provided by the United State Environmental protection Agency for the population ages. The total estimated excess lifetime cancer risk due to exposure from these heavy metals via ingestion, inhalation and dermal pathways was found to be in the range of 2.73E-04 to 9.23E-07 for children, 6.07E-07E-07 to 5.64E-02 for adults and were majorly contributed by Chromium (Cr). These range clearly indicated the existence of values far above the USEPA recommended threshold of 1.00E-06 and consequently indicating that there is high risk of lifetime cancer development in the inhabitants around the study area.

Keywords: Excess lifetime cancer; Heavy metals; Annual daily intake; Cancer risk; Exposure pathways

Introduction

It has been established that heavy metals originate from natural sources at concentrations mostly within the safe limit [1,2]. In urban systems, human activities contribute to the enhancement of the natural concentrations of this heavy metals through activities such as traffic, industrial and weathering of buildings and pavements [3,4]. Rigorous monitoring of heavy metals concentrations is necessary to avert their potential of continuous exposure in order to prevent damage to health. Steel production by iron extraction from metal scraps generates waste that are of serious environmental concern when deposited on soil [5-7] have established the existence of high levels of some heavy metals in the tailings from steel scraps. The relatively large surface area of soil fine particles facilitates heavy metals absorption and binding to iron and organic matter [8,9]. Polluted soils when blown by wind can cause aerial dispersion of these heavy metals [10]. These heavy metals may get in to the human system through various exposure pathways such as direct ingestion of soils and dust inhalation. It is important to study the bioavailability of these heavy metals in order to understand the possible effect on biota and particularly on human health [7,8,10-15]. Most metals are very toxic when they exist in excess and might be capable of causing major health effects such as developmental retardation, kidney damage, neurological and immunological effects as well as several types of cancer [16]. The importance of risk quantification through identifying, defining and characterizing adverse exposure consequences cannot be overemphasized [17]. Arsenic (As), Chromium (Cr), Lead (Pb) is known to be toxic to humans and was classified as carcinogens. Human exposure to heavy metals has unsurprisingly increased over the last few decades worldwide [18]. The use of synthetic products such as batteries, pesticides, paints and industrial/domestic wastes can result in heavy metal contamination of agricultural and urban soils [18]. The rapid urbanization and industrialization of the world have increased heavy metal emissions and consequent human exposures to them. Arsenic and its compounds are used in herbicides, pesticides and insecticides which may form part of the exposure sources to humans in addition to air, water, cigarette smoking and contaminated food

[19,20]. Lead contamination may result from industrial sources such as manufacturing activities and lead smelting [21]. Lead and some other heavy metals remains a major hazard for human health because of their inherent nature of accumulation and non-bio-degradability especially when they accumulate in the body tissues faster than the body's detoxification pathway can dispose of them [22,23]. Acute poisoning from heavy metals occurs through ingestion and dermal contact. Exposure to heavy metals is normally chronic due to food chain transfer and repeated long term contact with them can cause cancer [24]. It has been revealed that waste disposal as an integral part of industrial activities may be directly linked to the increase in the metal load of the ambient environment by virtue of metal bearing wastes introduction [25,26]. This work therefore investigates the carcinogenic risk values due to exposure to chromium, arsenic, cadmium, cobalt and lead concentrations in soils from Katsina steel rolling dump-yard using united states environmental protection agency guidelines.

Materials and Methods

Study area and sample analysis

Figure 1 shows the study area from where all the soil samples were collected. The collected soil samples were prepared and analyzed using standard flame atomic absorption spectrophotometry method as described in [27,28].

*Corresponding author: Suleiman Bello, Department of Physics, Faculty of Natural and Applied Sciences, Umaru Musa Yar'adua University Katsina, Nigeria
E-mail: suleiman.bello@umyu.edu.ng

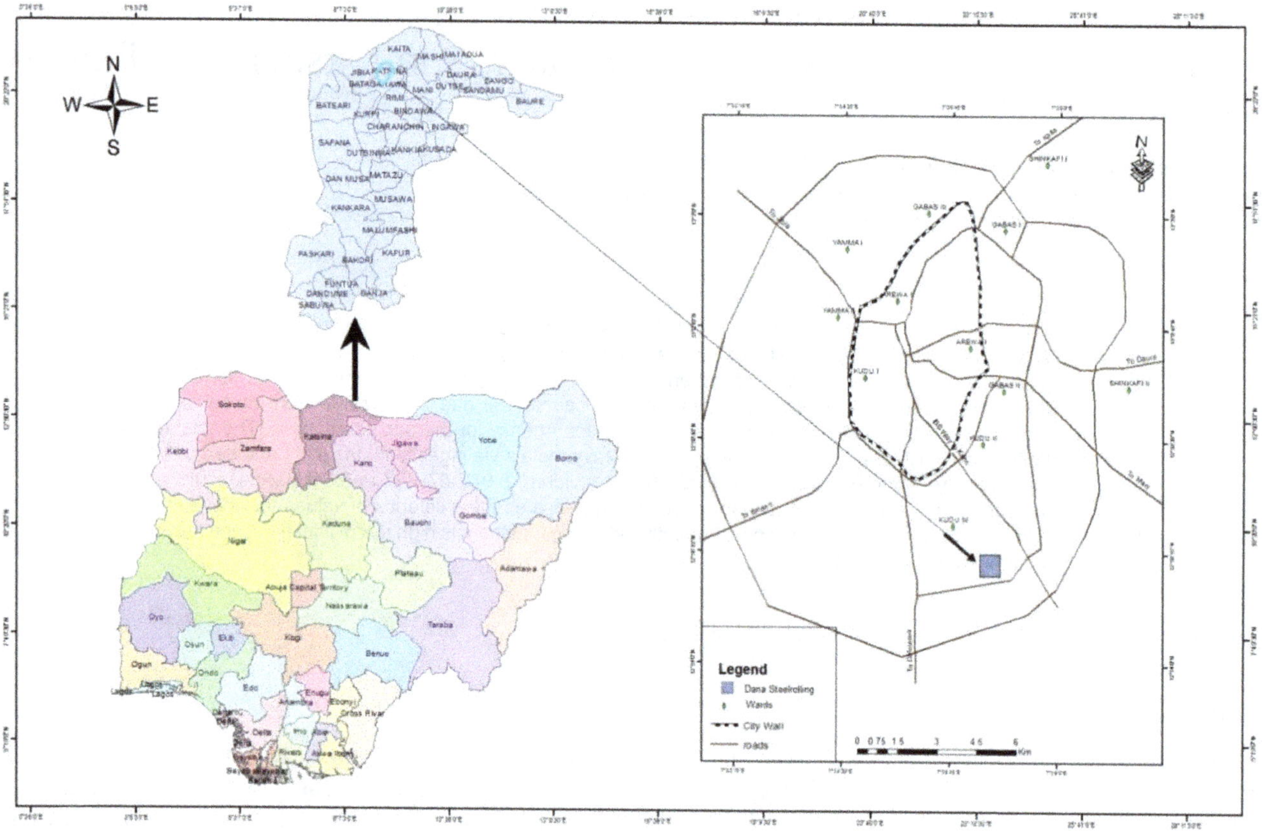

Figure 1: Map of the study area.

Total excess lifetime cancer risk assessment

Carcinogenic risk assessment: Carcinogenic risk assessment was carried out in the following chronological order: Identification of the hazard, assessment of exposure, dose-response (toxicity) assessment and risk characterization as suggested by Namgung and Xia [20,29]. Identification of hazard was done by taking the carcinogenic heavy metals as hazards for the population and obtaining their concentrations using flame atomic absorption spectrophotometry method. Assessment of exposure was done by estimating the frequency, intensity and duration of human exposures to the studied heavy metals separately for adults and children because of their physiological and behavioral differences [30]. Cancer slope factors were the toxicity index used in this assessment. Risk characterization was carried out by integrating all the gathered information in order to quantitatively estimate the excess lifetime cancer risk of children and adults [31]. Annual daily intake values were calculated for the various exposure pathways using eqns. (1)-(3) as recommended in ref. [32].

Ingestion of heavy metals through soil

$$ADIing = C * IR * EF * ED * \frac{CF}{BW * AT} \quad (1)$$

Where,

ADI_{ing}: Average daily intake of heavy metals ingested from soil in mg/kg-day;

C: Concentration of heavy metal in mg/kg for soil;

IR: Ingestion rate in mg/day;

EF: Exposure frequency in days/year; ED: exposure duration in years;

BW: Body weight of the exposed individual in kg;

AT: Time period over which the dose is averaged in days;

CF: Conversion factor in kg/mg.

Inhalation of heavy metals via soil particulates

$$ADIinh = Cs * IRair * EF * ED * \frac{CF}{BW * AT * PEF} \quad (2)$$

Where,

ADI inh: Average daily intake of heavy metals inhaled from soil in mg/kg-day;

CS: Concentration of heavy metal in soil in mg/kg;

IRair: Inhalation rate in m^3/day;

PEF: Is the particulate emission factor in m^3/kg;

EF, ED, BW and AT are as defined earlier in eqn. (1) above.

Dermal contact with soil

$$ADIderm = Cs * SA * FE * AF * ABS * EF * ED * \frac{CF}{BW * AT} \quad (3)$$

Where,

ADI derm: exposure dose via dermal contact in mg/kg/day;

CS: concentration of heavy metal in soil in mg/kg; SA=exposed skin area in cm^2;

FE: fraction of the dermal exposure ratio to soil;

AF is the soil adherence factor in mg/cm^2; ABS=fraction of the applied dose absorbed across the skin;

EF, ED, BW, CF and AT are as defined earlier in eqn. (1) before. Table 1 provides the exposure parameters that are used for health risk assessment for a typical residential exposure scenario [31-33].

Total excess lifetime cancer risk assessment: Carcinogenic risk assessment was carried out by estimating the incremental probability of an individual developing cancer over his lifetime as a result of exposure to the identified carcinogens. The excess lifetime cancer risk was calculated from the following equation:

$$Risk_{pathway} = \sum_{k=1}^{n} ADI_k CSF_k \qquad (4)$$

Where, Risk is a unit less probability of an individual developing cancer over a lifetime. ADI_k (mg/kg/day) and CSF_k are the average daily intake and the cancer slope factor respectively for the Kth heavy metal, for n number of heavy metals. The slope factor converts the estimated daily intake of the heavy metal averaged over a lifetime of exposure directly to incremental risk of an individual developing cancer [31]. The total excess lifetime cancer risk for an individual was finally calculated by summing the average contribution of the individual heavy metals for all the pathways (ingestion, inhalation and dermal) using the following equation:

$$Risk_{total} = Risk_{ing} + Risk_{inh} + Risk_{dermal} \qquad (5)$$

Where Risk (ing), Risk (inh) and Risk (derm) are the risks contributions through ingestion, inhalation and dermal pathways. The carcinogenic risk assessment was calculated using cancer slope factors provided by Table 2 below [31-33].

Results and Discussion

Carcinogenic health risk of heavy metals for adults and children

The concentrations of heavy metals (mg/kg) in the analyzed soil samples from Katsina steel rolling mill dumpsite were used for the computations of annual daily intake values (mg/kg/day) using the models provided by eqns. (1), (2) and (3) for ingestion, inhalation and dermal pathways respectively. The exposure parameters provided by Environmental protection agency were used for the computation [34-37]. The obtained annual daily intake values were subjected to

Heavy metal	Ingestion CSF	Dermal CSF	Inhalation CSF
As	1.50E+00	1.50E+00	1.50E+01
Pb	8.50E-03	-	4.20E-02
Cd	-	-	6.30E+00
Cr	5.00E-01	-	4.10E+01
Co	-	-	9.80E+00

Table 2: Cancer slope factors (CSF) in (mg/kg/day)-1 for the different heavy metals.

descriptive statistics using MS Excel 2010 and the Mean, minimum and maximum values corresponding to each heavy metal for a particular receptor (adult and children) via a particular pathway were presented in Table 3. The obtained annual daily intake values were further used for the computations of cancer risk using eqns. (4) and (5) and the cancer slope factors provided by ref. [29] in Table 2. The total excess lifetime cancer risk in adults and children for each pathway due to exposure from all the studied heavy metals was also calculated and the results were also subjected to descriptive statistics with the mean, minimum and maximum presented in Table 4.

The calculated risk indices were compared with the United States environmental protection guidelines for maximum cancer risk of 1E-06. Based on this guideline, it was found that the values of cancer risks for Cr were seriously above the limits for all the exposure pathways (ingestion, inhalation, dermal) in both adults and children implying that both population ages are at serious risk of developing cancer in their lifetime due to Cr exposure. The mean cancer risk values of Cr were found to be 9.654E-03 and 3.045E-06 in adults via ingestion and inhalation pathways respectively with maximum values of 5.63E-02 and 1.778E-05 respectively. For children the mean cancer risk values were estimated to be 4.51E-05 and 1.421E-06 for ingestion and inhalation pathways respectively with maximum values of 2.63E-04 and 8.295E-06. For Pb some cancer risk values were too high for both adults and children in ingestion pathway with maximum values of 4.08E-06 and 7.62E-06 for adults and children respectively. For As the cancer risk values were found to be too high in some samples for ingestion in children with maximum values of 1.22E-06. The cancer risk due to Cd and Co was found to be within the requirement for all the samples in all the exposure pathways. The total cancer risk values due to ingestion pathway in adults and children were found to be above the requirement and were majorly contributed by Cr, Pb and As in both adults and children. For the inhalation pathway, the total cancer risk values were found to be above the requirement with major contribution mainly from Cr. For dermal, the cancer risk values due to As were all within the requirement indicating no risk to members of population. The total excess lifetime cancer risk was found to have maximum and minimum values of 2.73E-04 and 9.23E-07 for children, 5.64E-02 and 6.07E-07 for adult (Table 3).

Conclusions

Soil samples were collected from Katsina steel rolling mill and analyzed using flame atomic absorption spectrophotometry instrumental method for the presence and concentrations of the carcinogenic heavy metals Arsenic (As), Chromium (Cr), Cadmium (Cd), Cobalt (Co) and Lead (Pb). The obtained concentrations were used to obtain the corresponding annual daily intake values through the exposure pathways of ingestion, inhalation and dermal contact. The obtained annual daily intake values were further used for the carcinogenic risk values. It is evident from the obtained results that there is very high probability that the inhabitants around the steel rolling mill will develop one type of cancer or another in their lifetime. This alarming situation should be regularly monitored for cancer health

Parameter	Symbol	Unit	Child	Adult
Body weight(BW)	BW	Kg	15	70
Exposure frequency(EF)	EF	days/year	350	350
Exposure duration(ED)	ED	Years	6	30
Ingestion rate(IR)	IR	mg/day	200	100
Inhalation rate(IRair)	IRair	m^3/day	10	20
Skin surface area(SA)	SA	cm^2	2100	5800
Soil adherence factor(AF)	AF	mg/cm^2	0.2	0.07
Dermal absorption factor(ABS)	ABS	None	0.1	0.1
Dermal exposure ratio(FE)	FE	None	0.61	0.61
Particulate emission factor (PEF)	PEF	m^3/kg	1.3E+09	1.3E+09
Conversion factor (CF)	CF	kg/mg	E-06	E-06
Averaging time (AT)	AT	Days	365 × 70	365 × 70

Table 1: Exposure parameters used for the assessment of carcinogenic health risk.

Parameter	Receptor	Statistical parameter	Average daily intake values (ADI) for heavy metals in (mg/kg/day).				
			Pb	As	Cr	Cd	Co
ADI_{ing} (mg/kg/day)	Adult	Mean	5.93E-05	3.31E-07	4.83E-04	8.82E-06	3.66E-05
		Minimum	N/D	2.52E-07	N/D	2.35E-07	3.99E-06
		Maximum	4.80E-04	4.34E-07	2.82E-03	1.84E-05	4.83E-05
	Children	Mean	1.11E-04	6.18E-07	9.01E-04	1.65E-05	6.83E-05
		Minimum	N/D	4.71E-07	N/D	4.38E-07	7.45E-06
		Maximum	8.97E-04	8.11E-07	5.26E-03	3.44E-05	9.01E-05
ADI_{inh} (mg/kg/day)	Adult	Mean	9.13E-09	5.09E-11	7.43E-08	1.35E-09	5.63E-09
		Minimum	N/D	3.88E-11	N/D	3.61E-11	6.14E-10
		Maximum	7.39E-08	6.68E-11	4.34E-07	2.83E-09	7.42E-09
	Children	Mean	4.26E-09	2.38E-04	3.47E-08	6.33E-10	2.63E-09
		Minimum	N/D	1.81E-11	N/D	1.69E-11	2.87E-10
		Maximum	3.45E-08	3.12E-11	2.02E-07	1.32E-09	3.46E-09
ADI_{derm} (mg/kg/day)	Adult	Mean	1.40E-05	8.20E-08	1.20E-04	2.18E-06	9.07E-06
		Minimum	N/D	6.25E-08	N/D	5.82E-08	9.89E-07
		Maximum	1.19E-04	1.08E-07	6.98E-04	4.57E-06	1.20E-05
	Children	Mean	1.42E-05	7.92E-08	1.15E-04	2.11E-06	8.75E-06
		Minimum	N/D	6.04E-08	N/D	5.62E-08	9.55E-07
		Maximum	1.15E-04	1.04E-07	6.74E-04	4.41E-06	1.15E-05

N/D means not detected.

Table 3: Descriptive statistics of Average daily intake (ADI) values in mg/kg/day for adults and children in soils from Katsina steel rolling mill dumpsite for carcinogenic risk calculations.

Parameter	Receptor	Statistical parameter	Cancer risk values					Total excess lifetime cancer risk.
			Pb	As	Cd	Cr	Co	
Risk (Ingestion)	Adult	Mean	5.04E-07	4.97E-07	-	9.65E-03	-	9.66E-03
		Minimum	N/D	3.79E-07	-	N/D	-	4.23E-07
		Maximum	4.08E-06	6.52E-07	-	5.63E-02	-	5.63E-02
	Children	Mean	9.41E-07	9.27E-07	-	4.51E-05	-	4.69E-05
		Minimum	N/D	7.07E-07	-	N/D	-	7.89E-07
		Maximum	7.62E-06	1.22E-06	-	2.63E-04	-	2.650E-04
Risk (Inhalation)	Adult	Mean	3.83E-10	7.64E-10	8.55E-09	3.05E-06	5.52E-08	3.11E-06
		Minimum	N/D	7.83E-10	2.28E-10	N/D	6.02E-09	6.571E-08
		Maximum	3.10E-09	1.00E-09	1.79E-08	1.78E-05	7.28E-08	1.78E-05
	Children	Mean	1.79E-10	3.57E-10	2.66E-11	1.42E-06	2.58E-08	1.45E-06
		Minimum	N/D	2.72E-10	7.08E-13	N/D	2.81E-09	2.54E-08
		Maximum	1.45E-09	4.68E-10	5.56E-11	8.30E-06	3.40E-08	8.32E-06
Risk (Dermal)	Adult	Mean	-	1.23E-07	-	-	-	1.23E-07
		Minimum	-	9.39E-08	-	-	-	9.39E-08
		Maximum	-	1.61E-07	-	-	-	1.61E-07
	Children	Mean	-	1.19E-07	-	-	-	1.19E-07
		Minimum	-	9.05E-08	-	-	-	9.05E-08
		Maximum	-	1.56E-07	-	-	-	1.56E-07

Table 4: Descriptive statistics of calculated cancer risk values for adults and children in soils from Katsina steel rolling mill dumpsite.

related problems in the inhabitants around the area. It is therefore recommended that immediate remediation action should be started on the site to bring down the concentrations to the bearable limits and that future steel rolling mill tailings should be properly disposed-off far away from the residential and commercial areas.

References

1. United States Department of Agriculture (USDA) (2000) Heavy Metal Soil Contamination. Soil Quality Institute, Natural Resources Conservation Service, Urban Technical Note No. 3: 1-7.

2. Strömberg U, Lundh T, Skerfving S (2008) Yearly measurements of blood lead in Swedish children since 1978: the declining trend continues in the petrol-lead-free period 1995-2007. Environmental Research 107: 332-335.

3. Chen TB, Zheng YM, Lei M, Huang ZC, Wu HT, et al. (2005) Assessment of heavy metal pollution in surface soils of urban parks in Beijing, China. Chemosphere 60: 542-551.

4. Wang X, Sato T, Xing B, Tao S (2005) Health risks of heavy metals to the general public in Tianjin, China via consumption of vegetables and fish. Science of the Total Environment 350: 28-37.

5. Candeias C, Silva EFD, Salgueiro AR, Ávila PF, Coelho P, et al. (2011) Modelling the impact of Panasqueira mine on the ecosystems and human health: a multidisciplinary approach. In International Conference on Occupational and Environmental Health: ICOEH 2011.

6. Abidemi OO (2011) An assessment of soil heavy metal pollution by various allied artisans in automobile workshop in Osun state, Nigeria. Electronic Journal of Environmental, Agricultural and Food Chemistry 10.

7. Radojevic M, Bashkin VN (1999) Practical environmental analysis. Royal Society of Chemistry.

8. Rasmussen PE (1998) Long-range atmospheric transport of trace metals: the need for geoscience perspectives. Environmental geology 33: 96-108.

9. Wei Y, Han IK, Shao M, Hu M, Zhang J, et al. (2009) PM2. 5 constituents and

oxidative DNA damage in humans. Environmental Science and Technology 43: 4757-4762.

10. Chen X, Wright JV, Conca JL, Peurrung LM (1997) Evaluation of heavy metal remediation using mineral apatite. Water, Air, and Soil Pollution 98: 57-78.

11. Banza CLN, Nawrot TS, Haufroid V, Decrée S, De Putter T, et al. (2009) High human exposure to cobalt and other metals in Katanga, a mining area of the Democratic Republic of Congo. Environmental research 109: 745-752.

12. Bosso ST, Enzweiler J (2008) Bioaccessible lead in soils, slag, and mine wastes from an abandoned mining district in Brazil. Environmental Geochemistry and Health 30: 219-229.

13. Douay F, Pruvot C, Roussel H, Ciesielski H, Fourrier H, et al. (2008) Contamination of urban soils in an area of Northern France polluted by dust emissions of two smelters. Water, Air, and Soil Pollution, 188: 247-260.

14. Juhasz AL, Weber J, Smith E (2011) Impact of soil particle size and bioaccessibility on children and adult lead exposure in peri-urban contaminated soils. Journal of Hazardous Materials 186: 1870-1879.

15. Ettler V, Kříbek B, Majer V, Knésl I, Mihaljevič M (2012) Differences in the bioaccessibility of metals/metalloids in soils from mining and smelting areas (Copperbelt, Zambia). Journal of Geochemical Exploration 113: 68-75.

16. Mudgal V, Madaan N, Mudgal A, Singh R, Mishra S (2010) Effect of toxic metals on human health. Open Nutraceut J 3: 94-99.

17. Sun Y, Zhou Q, Xie X, Liu R (2010) Spatial, sources and risk assessment of heavy metal contamination of urban soils in typical regions of Shenyang, China. Journal of Hazardous Materials 174: 455-462.

18. U.S. Environmental Protection Agency. Risk Assessment Guidance for Superfund Volume I (2004) Human Health Evaluation Manual (Part E, Supplemental Guidance for Dermal Risk Assessment); USEPA: Washington, DC, USA.

19. American Conference of Government Industrial Hygienists (2003) Documentation of the arsenic, elemental and inorganic Compounds except arsine TLV. In threshold limit values for chemical substances and Physical Agents and Biological Exposure Indices. Cincinnati, Ohio, ACGIH Worldwide.

20. Namgung UK, Xia Z (2001) Arsenic induces apoptosis in rat cerebellar neurons via activation of JNK3 and p38 MAP kinases. Toxicology and Applied Pharmacology 174: 130-138.

21. Roussel H, Waterlot C, Pelfrêne A, Pruvot C, Mazzuca M, et al. (2010) Cd, Pb and Zn oral bioaccessibility of urban soils contaminated in the past by atmospheric emissions from two lead and zinc smelters. Archives of Environmental Contamination and Toxicology 58: 945-954.

22. Berlin M, Uberg S (1963) Accumulation and retention of mercury in mouse 111: An auto radiographic compensation of methyl mercury dicyanidiamide with organic mercury Arch. Environ Health 6: 610-616.

23. Garrett NE, Garrett RB, Archdeacon JW (1972) Placental transmission of mercury to the fatal rate. Toxicology and Applied Pharmacology 22: 649-654.

24. International Occupational Safety and Health Information Centre (1999). Metals. In: Basics of Chemical Safety, Chapter 7. Geneva: International Labour Organization.

25. Agency for Toxic Substance and Disease Registry (ATSDR) (2003) Case studies in Environmental Medicine, Lead toxicity.

26. Occupational Safety and Health Administration (2003) Substance data sheet for occupational exposure to lead. Standard 1910.1025. OSHA, Washington, DC.

27. Bello S, Zakari YI, Ibeanu IGE, Muhammad BG (2015) Evaluation of heavy metal pollution in soils of Dana Steel limited dumpsite, Katsina State, Nigeria using Pollution load and degree of contamination indices.

28. Bello S, Zakari YI, Ibeanu IGE, Muhammad BG (2016) Characterization and assessment of heavy metal pollution levels in soils of Dana steel limited dumpsite, Katsina state, Nigeria using geo-accumulation, Ecological Risk and Hazard Indices 5: 49-61.

29. https://www.atsdr.cdc.gov/.

30. Vassilakos C, Veros D, Michopoulos J, Maggos T, O'Connor CM (2007) Estimation of selected heavy metals and arsenic in PM 10 aerosols in the ambient air of the Greater Athens Area, Greece. Journal of Hazardous Materials 140: 389-398.

31. U.S. Environmental Protection Agency. Risk Assessment Guidance for Superfund Volume 1(1989) Human Health Evaluation Manual (Part A); Office of Emergency and Remedial Response: Washington, DC, USA.

32. U.S. Environmental Protection Agency (1991) Human Health Evaluation Manual, Supplemental Guidance: Standard Default Exposure Factors; USEPA: Washington, DC, USA.

33. Department of Environmental affairs: The Framework for the Management of Contaminated Land, South Africa.

34. Marfo BT (2014) Heavy metals contaminations of soil and water at Agbogbloshie Scrap Market, Accra (Doctoral dissertation).

35. Goyer RA (1996) Results of lead research: prenatal exposure and neurological consequences. Environmental Health Perspectives 104: 1050.

36. Kong S, Lu B, Ji Y, Zhao X, Bai Z, et al. (2012) Risk assessment of heavy metals in road and soil dusts within PM 2.5, PM 10 and PM 100 fractions in Dongying city, Shandong Province, China. Journal of Environmental Monitoring 14: 791-803.

37. Yukselen MA, Alpaslan B (2001) Leaching of metals from soil contaminated by mining activities. Journal of Hazardous Materials 87: 289-300.

36

Two Binary Liquid Critical Mixtures Belong to Class of Universality

Ata BN and Abderaziq IR*

Physics Department, An-Najah Ntional University, Nablus Weast Bank, Israel

Abstract

The dynamic shear viscosity of a binary liquid mixture phenol–water has been measured at different temperatures (50.0°C ≤ T ≤ 75.0°C) and different concentrations (0.00% up to 100.00% by weight of phenol). The critical temperature T_c and critical concentration x_c are found to be 67.0°C and 33.90% by weight of phenol respectively, the critical mass density ρ_c is measured to be 0.8952 g/cm³. The critical and background amplitudes of specific heat at constant pressure are calculated to be 78.12 J/kg.K and 85.29 J/kg.K respectively. The pressure derivative of the critical temperature along the critical line $T_c'T$ is calculated to be 9.722 ×10⁻⁶ K/Pa.

In addition, dynamic shear viscosity of binary liquid mixture phenol–cyclohexane has been measured at different temperatures (14.0°C ≤ T ≤ 30.0°C) and different concentrations (2.00% up to 39.70% by weight of phenol). The critical temperature T_c and critical concentration x_c are found to be 17.0°C and 2.70% by weight of phenol respectively; the critical mass density ρ_c is measured to be 0.7627 g/cm³. The critical and background amplitudes of isobaric thermal expansion coefficient α_{pc} and α_{pb} are calculated to be 8×10⁻⁶ K⁻¹, 6×10⁻⁴ K⁻¹ respectively. The pressure derivative of the critical temperature T_c' for the binary is calculated to be 2.8572 × 10⁻⁸ K/Pa. The universal quantity R^+_ξ for the binary liquid critical mixture phenol–water is calculated to be 0.2716 ± 0.0005. In addition, the universal quantity R^+_ξ for the binary liquid critical mixture phenol–cyclohexane is calculated to be 0.2699 ± 0.0001. The calculated values of the universal quantity R^+_ξ are in a good agreement with the theoretical value of R^+_ζ which is equal 0.2710. The two binary liquid critical mixture belong to the class of universality "Two–Scale–Factor Universality".

Keywords: Binary liquid; Critical mixture; Homogeneous

Introduction

Binary liquid mixtures and critical point

Mixtures are the product of a mechanical blending or mixing of chemical substances like elements and compounds, without chemical bonding or other chemical changes, so that each ingredient substance retains its own chemical properties [1]. Mixtures can be either homogeneous or heterogeneous. A homogeneous mixture is a type of mixture in which the composition is uniform and every part of the solution has the same properties. A heterogeneous mixture is a type of mixture in which the components can be seen, as there are two or more phases present.

Binary liquid mixtures are combination of two pure liquid substances, which have a limited solubility of each one in the other [2]. Critical point is the point at which phase transition occurs at certain temperature called critical temperature and concentration [3]. The critical point represents the boundary between regions of homogeneous and heterogeneous behavior in phase diagrams for mixtures [4]. Hypothesis of universality greatly reduces the variety of different types of critical behavior by classifying all systems into a small number of equivalence classes [5].

The phenomenological theory of scaling has been extremely useful in understanding critical phenomena in model systems and in real materials [5]. The first characteristic of a universality class is that all the systems have the same critical exponents. In addition, the equation of state, the correlation functions and other quantities become identical near criticality, provided one matches the scales of the order parameter, the ordering field, the correlation length and the correlation time [5].

A property of hyper scaling or hyper universality (Two–Scale–Factor Universality) applies to systems in the universality classes of fluctuation-dominated (i.e., non-mean-field) critical behavior. These ideas were first developed phenomenologically and later confirmed by explicit renormalization group (RG) calculations [5]. The RG theory of critical phenomena has elucidated the mathematical mechanism for scaling and universality, and has provided a number of calculational tools for estimating universal properties [5].

Theory

Viscosity

The viscosity of a fluid is a measure of its resistance to gradual deformation by shear stress or tensile stress [6]. Viscosity is affected by different factors such as temperature, shear rate, catalyst, pressure, molecular weight concentration and storage age [7].

Mode coupling theory and shear viscosity of binary mixtures

The mode coupling theory explains the behavior of the binary mixtures at the critical temperature and concentration. The mode coupling approach of Kawasaki and Perl and Ferrell predicts a critical anomaly of the dynamic shear viscosity coefficient according to the law [8,9].

$$\frac{\eta - \eta 0}{\eta} = \frac{\Delta \eta}{\eta} = A \ln \zeta + A \ln q_D \qquad (2.1)$$

Where η_0 the noncritical part of the measured shear viscosity.

A is a constant which was calculated by D'Arrigo and given by A ~ 0.054 [10], and q_D is Debye momentum cutoff. The dynamic shear viscosity is temperature dependent at the critical concentration which

***Corresponding author:** Abdelraziq IR, Physics Department, An-Najah Ntional University, Nablus Weast Bank, Israel, E-mail: ashqer@najah.edu

is given by the power law [11,12].

$$\eta = \eta_0 t^{-x_\eta \nu} \tag{2.2}$$

Where t is the reduced temperature $t = \dfrac{T - T_c}{T_c}$, x_η and ν are critical exponents where $x_\eta \nu = 0.04$ [10,11].

Two–scale–factor universality

The Two–Scale–Factor Universality has been used in modern theories to explain the critical phenomena of binary liquid mixtures by predicting R_ζ [13]. The Two–Scale–Factor Universality hypothesis states that the critical amplitudes do not depend on three different scales of parameter (length, temperature, external field) but only on two scales of parameter [14]. Most of the observed quantities depend only on the dimensionalities of the space (d) and of the order parameter (n) [15].

The fluid and binary mixtures transitions belong to the same class of universality d=3, n=1[16]. All binary liquid mixtures with critical mixing points belong to the same universality class. The universality concept offers the possibility to relate critical amplitudes of these systems. The exponents are universal and related by the so-called scaling laws [16].

The amplitudes of the correlation length, thermal expansion and specific heat can be deduced using the universal amplitude combinations [15,17-19]. Correlation length is a measure of the distances over which the spin–spin (or density–density) correlations in the system extend [20]. The correlation length of a binary mixture at critical composition exhibits an anomalous behavior conforming to the following exponential law:

$$\xi = \xi^+_0 t^{-\nu}, T > T_c \tag{2.3}$$

Where ν is a critical exponent which accepted to be 0.630 ± 0.001,¶ ξ^+_0 is the critical amplitude and t is the reduced temperature $t \dfrac{T - T_c}{T_c}$, = where T_c is the critical temperature [21].

The specific heat at constant pressure c_p in zero field is singular and is given by:

$$c_p = c^+_{pc} t^{-\alpha} + c^+_{pb}, T > T_c \tag{2.4}$$

Where c_{pc} and c_{pb} are the critical and background amplitudes of the specific heat and $\alpha = 0.11$ is the critical exponent [6, 22, 23, 24].

The asymptotic behavior of the thermal expansion α_p can be represented by power law of the form,

$$\alpha_p = \alpha^+_{pc} t^{-\alpha} + \alpha^+_{pb}, T > T_c \tag{2.5}$$

Where α_{pc} and α_{pb} are the critical and the background amplitudes of the thermal expansion [16]. With these three amplitudes ξ^+_0, c^+_{pc}, and a^+_{pc} it is possible to construct a quantity, denoted R^+_c, which is universal in the same sense as critical indices are universal. This quantity is defined as:

$$R^+_\xi = \xi_0 \left(\frac{\acute{a} c_{pc} \rho_c}{K_B} \right)^{1/d} = \xi_0 \left(\frac{\acute{a} T_c \alpha_{pc}}{T'_c K_B} \right)^{1/d} \tag{2.6}$$

Where d=3 is the dimension of the space, K_B is Boltzman's constant, ρ_c is the density of the critical mixture at critical temperature T_c and concentration $T'_c = \dfrac{dT_c}{dp}$ is the pressure derivative of the critical temperature along the critical line [16]. The theoretical value of the

universal constant $R^+_\xi = \nu \left(\dfrac{n}{4\pi} \right)^{1/d}$ in three dimensions for n = 1, d = 3 and $\nu = 0.64$ which equal 0.2710 [16].

Experimental

Methodology

In this work two binary mixtures were used the phenol–water binary mixture and phenol–cyclohexane binary mixture. The viscosities were measured for both at different temperatures and concentrations. The critical temperature, concentration, heat capacity at constant pressure and density were measured of each mixture.

Experimental Apparatus

Viscosity Apparatus:

•   **Capillary Viscometer:** is a device used to measure the viscosity of the liquid with a known density by measuring the time for a known volume of the liquid to flow through the capillary under the influence of gravity [25].

•   **Brookfield Digital Viscometer Model DV-I+:** It measures the viscosity of a liquid in centipoises with accuracy ± 1%. It is used to measure the dynamic viscosity from 1 up to $1.33 \times 10_7$ cP.

Density Apparatus: A pycnometer is used to measure the density of the mixtures and pure substances. The Analytical Balance HR-200: It is used to measure the mass in gm unit with accuracy of ± 0.00005%.

Temperature Apparatus: Digital Prima long Thermometer it is used to measure the temperature of the samples with accuracy ± 1%, the range of the temperature from -20 up to 100°C. Julabo F25-MV Refrigerated and Heating Circulator is used to control the temperature of the samples with accuracy 1%.

The calorimeter: It is an instrument used to measure the heat of chemical reactions or physical changes as well as heat capacity. A calorimeter has been constructed with glass pyrex beaker instead of the aluminum, and a nichrome resistance wire covered by a U- tube glass. This calorimeter has been constructed to avoid the reaction of phenol with metals.

Results and Discussion

Viscosity measurements

Phenol–Water Binary Mixture: The results of the dynamic viscosity η as a function of temperature for different concentrations of phenol–water binary mixture are evaluated. The dynamic shear viscosities of phenol–water binary mixture are plotted as a function of temperature at different concentrations of phenol in appendix B. The critical temperature occurs when the two phases of the binary mixture become one phase which appears as anomaly at 67.0°C for the concentration 33.90% by weight of phenol, as shown in Figure 1. In addition, the mixture was visually observed as one phase at the critical temperature and concentration.

Phenol–cyclohexane binary mixture: The results of the dynamic viscosity η as a function of temperature for different concentrations of phenol–cyclohexane binary mixture are evaluated. The dynamic shear viscosities of phenol–cyclohexane binary mixture are plotted as a function of temperature at different concentrations of phenol in appendix B.

The critical temperature occurs when the two phases of the binary

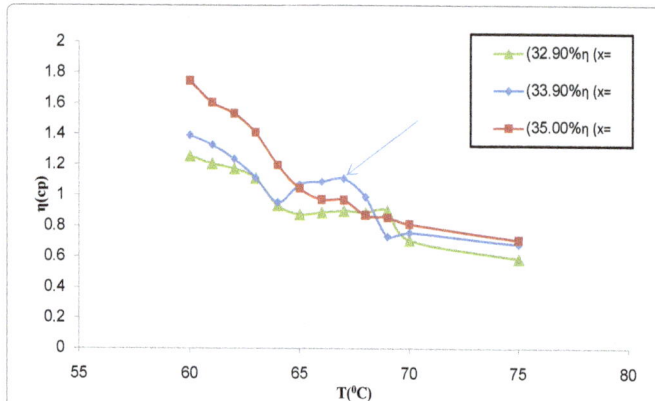

Figure 1: The dynamic shear viscosity of phenol–water mixture as a function of temperature at concentrations 32.90%, 33.90% and 35.00% by weight of phenol.

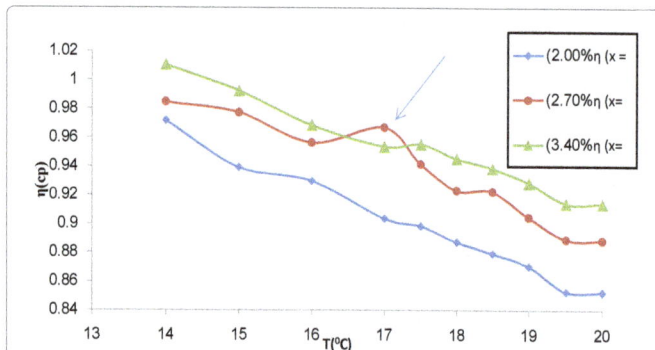

Figure 2: The dynamic shear viscosity of phenol–cyclohexane mixture as a function of temperature at concentrations 2.00%, 2.70% and 3.40% by weight of phenol.

mixture become one phase which appears as anomaly at 17.0°C for the concentration 2.70 % by weight of phenol, as shown in Figure 2. In addition, the mixture was visually observed as one phase at the critical temperature and concentration.

Mass density measurements

Phenol–Water Binary Mixture: The critical mass density ρ_c = 0.8952 g/cm³ ± 0.0001 at the critical concentration x_c = 33.90% by weight of phenol and the critical temperature T_c = 67.0°C for the binary mixture phenol–water.

Phenol–Cyclohexane Binary Mixture: The critical mass density ρ_c = 0.7627 g/cm³ ± 0.0001 at the critical concentration x_c = 2.70% by weight of phenol and the critical temperature T_c = 17.0°C for the binary mixture phenol–cyclohexane.

Specific heat measurements

Phenol–water binary mixture: The specific heat at constant pressure c_p for the binary mixture phenol–water is calculated using the relation:

$$W = IVt = (m_r c_r + m_b c_b + m_g c_g + m_p c_p)\Delta T$$

The specific heat at constant pressure is given by relation (2.4) which is

$$c_p = c^+_{pc}\tau^{-\alpha} + c^+_{pb} \qquad (2.4)$$

Where t is the reduced temperature and α is the critical exponent which equals 0.11. The specific heat at constant pressure and $\tau^{-\alpha}$ at different temperatures are presented in Table 1.

The specific heat at constant pressure is plotted with $\tau^{-\alpha}$ in Figure 3, using the relation

$$c_p = c^+_{pc}\tau^{-\alpha} + c^+_{pb}$$

The slope of the line is c_{pc} = 78.12 ± 0.04 J/kg.K,, which is the critical amplitude of specific heat at constant pressure of the binary mixture phenol–water, and the intercept is c_{pb} = 85.29 ± 0.02 J/kg.K which is the background amplitude of specific heat at constant pressure.

Calculation of the universal quantity R^+_ξ

The Two–Scale–Factor Universality relation (2.6),

$$R^+_\xi = \xi_0 \left(\frac{\alpha c_{pc}\rho_c}{K_B}\right)^{1/d} = \xi_0 \left(\frac{\alpha\, T_c\, \alpha_{pc}}{T'_c\, K_B}\right)^{1/d} \qquad (2.6)$$

The universal quantity R^+_ξ can be calculated using the first term

$$R^+_\xi = \xi_0 \left(\frac{\alpha c_{pc}\rho_c}{K_B}\right)^{1/d}$$

Where the critical exponent α= 0.11, Boltzmann's constant K_B = 1.38×10⁻²³ J/K, and the dimensionality d = 3.

Phenol–water binary mixture: The universal quantity R^+_ξ can be calculated for the binary mixture phenol–water to be 0.2716 ± 0.0125 by substituting the values of c_{pc} = 78.12 J/kg.K, ρ_c = 0.8952 gm/cm³,

Temperature (°C)	$\frac{T - T_c}{T_c}$	$\tau^{-\alpha}$	c_p (J/kg.K)
68.0	0.014925	1.588065	30.4
68.5	0.022388	1.518792	29.9
69.0	0.029851	1.471482	29.4
69.5	0.037313	1.435803	28.6
70.0	0.044776	1.407294	27.2
71.0	0.059701	1.363458	23.5
72.0	0.074627	1.330398	19.5
73.0	0.089552	1.303982	16.4
74.0	0.104478	1.282057	13.7
75.0	0.119403	1.263364	10.9

Table 1: Specific heat data of binary mixture phenol - water.

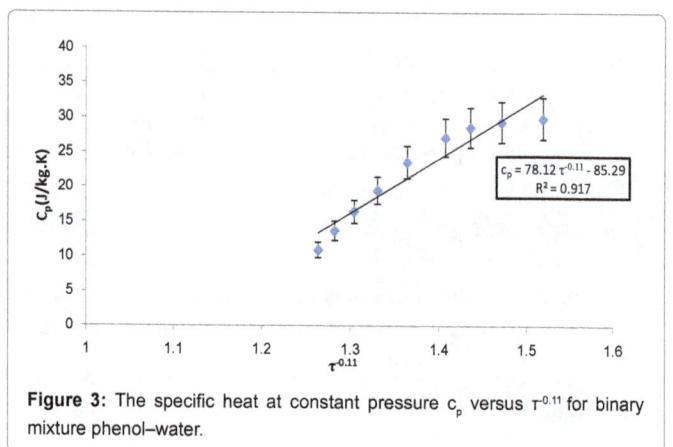

Figure 3: The specific heat at constant pressure c_p versus $\tau^{-0.11}$ for binary mixture phenol–water.

and $\xi_0 = 3.3$Å [26]. The theoretical value of $R^+_\zeta = v(\frac{n}{4\pi})^{\frac{1}{d}}$ in three dimensions for n = 1, d = 3 and v = 0.64 , equal 0.2710.

Phenol–cyclohexane binary mixture: The universal quantity R^+_ξ can be calculated for the binary mixture phenol–cyclohexane to be 0.2699 ± 0.0125 by substituting the values of c_{pc} = 106.60 J/kg.K [27], ρ_c = 0.7627 gm/cm³, and ξ_0 = 3.12Å. The theoretical value of R^+_ζ = 0.2710. This indicates that the two binary liquid critical mixtures phenol–water and phenol–cyclohexane belong to the same class of universality "Two–Scale–Factor Universality".

Calculation of T'_c

The Two–Scale–Factor Universality relation (2.6),

$$R^+_\xi = \xi_0 \left(\frac{\alpha c_{pc} \rho_c}{K_B}\right)^{1/d} = \xi_0 \left(\frac{\alpha T_c \alpha_{pc}}{T'_c K_B}\right)^{1/d} \tag{2.6}$$

The pressure derivative of the critical temperature along the critical line T'_c can be calculated using the second term

$$R^+_\xi = \xi_0 \left(\frac{\alpha T_c \alpha_{pc}}{T'_c K_B}\right)^{\frac{1}{d}}$$

Phenol–water binary mixture: The pressure derivative of the critical temperature along the critical line T'_c for the binary mixture phenol–water can be calculated to be 9.722×10^{-6} K/Pa. The values of T_c = 340 K, α_{pc} = 0.002 K⁻¹ [28], ξ_0 = 3.3Å [26] and R^+_ξ = 0.2716.

Phenol–cyclohexne binary mixture

The critical amplitude of isobaric thermal expansion coefficient α_{pc} is need to calculate T'_c. The isobaric thermal expansion coefficient α_p can be calculated using the relation:

$$\alpha_p = \frac{1}{V}\left(\frac{dV}{dT}\right)_p \tag{4.1}$$

Where V: is the volume, T: is the temperature and P: is the pressure. Equation 4.1 could be expressed in another form by applying $V = \frac{m}{\tilde{n}}$.

$$\alpha_p = -\frac{1}{\rho}\left(\frac{d\rho}{dT}\right)_p = \alpha_p = \rho\left(\frac{d\rho^{-1}}{dT}\right) \tag{4.2}$$

Where, m : is the mass and ρ: is the density of the binary mixture. The values of the density and its reciprocal at different temperatures above the critical point are presented in Table 2. The reciprocal of the density is fitted with the corresponding temperatures and the slope $\left(\frac{d\rho^{-1}}{dT}\right)$ is determined from Figure 4.

T(K)	Mass density ρ(gm/cm³)	ρ^{-1}(cm³/gm)
290.5	0.7376	1.3561
291.0	0.7374	1.3565
291.5	0.7372	1.3568
292.0	0.7370	1.3572
292.5	0.7368	1.3574
293.0	0.7367	1.3579
293.5	0.7364	1.3589
294.0	0.7359	1.3592
295.0	0.7357	1.3602
298.0	0.7352	1.3616

Table 2: The mass density and its reciprocal values at different temperatures for the binary mixture phenol - cyclohexane.

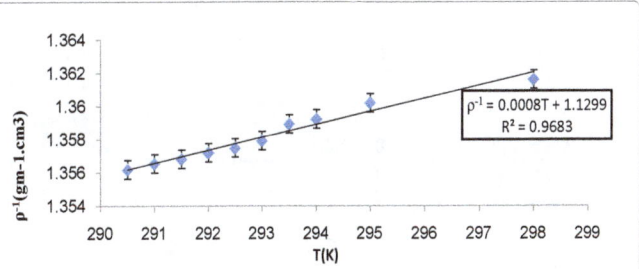

Figure 4: The reciprocal of density for the binary mixture phenol–cyclohexane as a function of temperature.

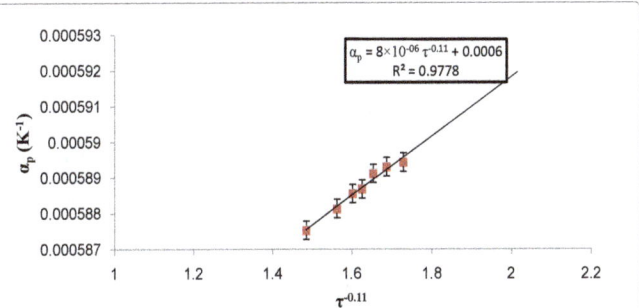

Figure 5: The isobaric thermal expansion coefficient for the binary mixture phenol–cyclohexane as a function of $\tau^{-0.11}$.

The slope from Figure 5 is $\frac{d\rho^{-1}}{dT} = 0.0008 \frac{cm^3}{g.K}$

The critical and the background isobaric thermal expansion coefficients can be determined by linear fitting of the isobaric thermal expansion coefficient α_p versus $\tau^{-\alpha}$, where $\tau = \frac{T - T_c}{T_c}$ and α =0.11 depending on the relation:

$$\alpha_p = \alpha_{pc}\tau^{-\alpha} + \alpha_{pb} \tag{2.4}$$

The values of α_p are calculated using equation (4.2) and the data are presented in Table 3. The data of the isobaric thermal expansion coefficient α_p are plotted versus $\tau^{-0.11}$ as shown in Figure 5. The slope of the line represents the critical isobaric thermal expansion coefficient α_{pc} = 8×10^{-6} K⁻¹, and the intercept of the line represents the background isobaric thermal expansion coefficient α_{pb} = 6×10^{-4} K⁻¹.

The pressure derivative of the critical temperature along the critical line T'_c for the binary mixture phenol–cyclohexane is calculated to be 2.8572×10^{-8} K/Pa. The values of T_c = 290 K, α_{pc} = 6×10^{-6} K⁻¹, ξ_0 = 3.12Å [26] and R^+_ξ = 0.2699.

Conclusion

The dynamic shear viscosity of two binary liquid mixtures phenol–water and phenol–cyclohexane has been measured at different temperatures and concentrations. The critical temperature and critical concentration for the binary liquid mixture phenol–water were T_c=67.0°C and x_c=33.90 by weight of phenol. The critical density ρc for the binary liquid mixture phenol–water was found to be 0.8952 g/cm3 at the critical temperature and concentration.

The specific heat at constant pressure cp of the binary liquid mixture phenol–water has been measured; the critical cpc and back ground cpb amplitudes of the specific heat at constant pressure have been calculated to be 78.11 J/kg.K and 85.29 J/kg.K respectively.

T(K)	$a_p(K^{-1})$	$T^{-0.11}$
292.0	0.000589	1.728823
292.5	0.000589	1.686904
293.0	0.000589	1.65341
293.5	0.000589	1.62561
294.0	0.000589	1.601907
295.0	0.000588	1.563065
298.0	0.000588	1.484307

Table 3: The isobaric thermal expansion coefficient at different temperatures for the critical mixture of phenol - cyclohexane.

Parameter	This work	Previous works
T_c	67.0°C	66.4°C [a] 69.0°C [b]
x_c	33.90%	34.6% [a] 34.0% [b]
ρ_c	0.8952g/cm³	-
cp_c	$78.117 \frac{J}{k_g \cdot K}$	-
cp_b	$85.292 \frac{J}{k_g \cdot K}$	-
T'_c	9.722 ×10⁻⁶ K/Pascal	-
R^+_ξ	0.2716	-

Table 4: Summary of the measured and calculated results in this work and previous works for phenol - water binary.

Parameter	This work	Previous work
T_c	17.0°C	-
x_c	2.70%	-
ρ_c	0.7627g/cm³	-
a_{pc}	8×10⁻⁶ K⁻¹	
a_{pb}	6×10⁻⁴ K⁻¹	
T'_c	2.8572×10⁻⁸ K/Pa	-
R^+_ξ	0.2699	-

Table 5: Summary of the measured and calculated results in this work for phenol - cyclohexane binary mixture.

The pressure derivative of the critical temperature along the critical line $T\_c^{\wedge\prime}$ is calculated for the binary mixture phenol–water to be 9.722×10-6 K/Pa.

The critical temperature and critical concentration for the binary liquid mixture phenol–cyclohexane have been measured, the results were $T\_c=17.0$°C and $x\_c=2.70\%$ by weight of phenol. The critical density ρc for the binary liquid mixture phenol–cyclohexane is found to be 0.7627 g/cm³ at the critical temperature and concentration. The pressure derivative of the critical temperature along the critical line $T\_c^{\wedge\prime}$ is calculated for the binary mixture phenol–cyclohexane to be 2.8572 × 10-8K/Pa. The isobaric thermal expansion coefficient αp for phenol–cyclohexane binary mixture are calculated at different temperatures, the critical αpc and back ground αpb amplitudes is determined to be 8×10^{-6} K-1, 6×10^{-4} K-1 respectively. The measured and calculated parameters for phenol–water binary mixture are summarized in Table 4. The measured and calculated parameters for phenol–cyclohexane binary mixture are summarized in Table 5. The universal quantity R+ξ for the binary liquid critical mixture phenol–water is calculated to be 0.2716 ± 0.0005. In addition, the universal quantity R+ξ for the binary liquid critical mixture phenol–cyclohexane is calculated to be 0.2699 ± 0.0001. The calculated values of the universal quantity R+ξ are in a

good agreement with the theoretical value of R+ζ which is equal 0.2710. As a result the two binary liquid critical mixtures belong to the class of universality "Two–Scale–Factor Universality".

References

1. Atkins P, Julio P (2010) Atkins' physical chemistry. (8th Eds) Oxford university press.

2. Popiel WJ (1964) Laboratory manual of physical chemistry.

3. Cheung A (2011) Phase transitions and collective phenomena. Cavendish Laboratory-University of Cambridge, Cambridge.

4. Gil L, Otin SF, Embid JM, Gallardo MA, Blanco S, et al. (2008) Experimental setup to measure critical properties of pure and binary mixtures and their densities at different pressures and temperatures Determination of the precision and uncertainty in the results. J of Supercritical Fluids 44: 123-138.

5. Domb C, Lebowitz JL (2000) Phase transitions and critical phenomena. Academic Press.

6. Symon KR (1971) Mechanics. (3rd Ed) Addison-Wesley publishing company Inc.

7. Lide DR (2005) Handbook of chemistry and physics. (86th Eds) BocaRaton (FL).

8. Senger JV (1972) Critical phenomena, proceedings of the international school of physics enrico Fermi. Course LI. MS Green Academic, New York.

9. Perl R, Ferrell RA (1972) Some topics in non-equilibrium critical phenomena. Physical Review A6: 23-58.

10. D'Arrigo G, Mistura L, Tartagila P (1977) Concentration and temperature dependence of viscosity in the critical mixing region of aniline–cyclohexane. Chemical Physics 66: 74-80.

11. Klein H, Woermann D(1978) Analysis of light-scattering and specific heat data of binary liquid mixtures in terms of the two-scale-factor universality. Physical Chemistry 82: 1084-1086.

12. Abdelraziq IR (2002) Concentration and temperature dependence of shear viscosity of the critical mixture of nitroethane and 3-methylpentane. An-Najah Univ J Res 16: 117-124.

13. Hohenberg PC, Aharony A, Halperin BI, Siggia ED (1976) Two-scale-factor universality and the renormalization group. Physical Review B13 7: 2986-2996.

14. Bervillier C, Godrèche C (1980) Universal combination of critical amplitudes from field theory. Physical Review B 21: 5427.

15. Zalczer G, Bourgou A, Beysens D (1983) Amplitude combinations in the critical binary fluid nitrobenzene and n-hexane. Physical Review A 28: 440.

16. Abdelraziq IR (2003) Two-scale-factor universality of binary liquid critical mixtures. Pakistan Journal of Applied Sciences 3: 142-144.

17. Clerke EA, Sengers JV, Ferrell RA, Bhattacharjee JK (1983) Pressure effects and ultrasonic attenuation in the binary liquid mixture 3-methylpentane and nitroethane near the critical point. Phys Rev A 27: 2140.

18. Jacobs DT (1986) Turbidity in the binary fluid mixture methanol–cyclohexane. Physical Review A 33: 2605-2611.

19. Bloemen E, Thoen J , Van Dael W (1980) The specific heat anomaly in triethylamine–heavy water near the critical solution point. J Chem Phys 73: 4628-4635.

20. Pathria RK, Beale PD (2011) Statistical mechanics. (3rd Eds) Elsevier.

21. Souto-Caride M, Troncoso J, Peleteiro J, Carballo E, Romani L (2006) Estimation of critical amplitudes of the correlation length by means of calorimetric and viscosimetric measurements. Chemical Physics 324: 483–488.

22. Bhatacharjee JK, Ferrell RA (1981) Dynamic scaling theory for the critical ultrasonic attenuation in binary liquids. Phys Rev A 24: 1643-1646.

23. Iwanowski I (2007) Critical behavior and crossover effects in the properties of binary and ternary mixtures and verification of the dynamic scaling conception. Dissertation, Georgia Augusta University.

24. Bhattacharjee JK, Kaatze U, Mirzaev SZ (2010) Sound attenuation near the demixing point of binary liquids: interplay of critical dynamics and noncritical kinetics. Reports on Progress in Physics 73.

25. Generalic E (2014) Glass capillary viscometer. Croatin-English Chemistry Dictionary and Glossary.

26. Abdelraziq IR (2015) Unpublished work.

27. Hussein GF, Ashqer I, Saadeddin I (2015) Critical behavior of the density of binary liquid mixture cyclohexane–phenol. An-Najah National University.

28. Reehan M, Ashqer I, Abu-Jafar M (2015) Critical behavior of the ultrasonic attenuation for the binary mixture of water and phenol. An-Najah Nationl University.

29. Howell OR (1932) A study of the system water - phenol: I densities. Proceedings of the Royal Society of London A 137: 418-433.

30. Krishnan RS (1935) Molecular clustering in binary liquid mixtures. Proceedings of the Indian Academy of Science 1: 915-927.

37

Use of the Additive Based on Amorphous Silica-Alumina in the Adhesive Dry Mixes

Loganina VI* and Zhegera CV

Department of "Quality management and construction technologies" Penza State University of Architecture and Construction, Russia

Abstract

This article proves the possibility of using amorphous aluminosilicate as a modifying additive for the adhesive dry mixes. The data is given on the microstructure and chemical composition of the amorphous aluminosilicates. This article described the character changes in the rheological properties of cement-sand mortar, depending on the percentage of additives. The model of cement stone strength using synthetic additives in the formulation is illustrated. The results of physical and mechanical properties of tile adhesive made on the basis of the developed adhesive dry mix formulations are described.

Keywords: Dry mixes; Amorphous silica-alumina; Plastic strength; Tile adhesive; Cement

Introduction

One of the priorities of modern building materials science is the development of effective building materials. To regulate the technical and operating characteristics of dry mortar formulation is administered in their structure various modifying agents [1-8].

Most of the modifiers used in the formulation of domestic dry construction mixtures, are coming from abroad, which significantly increases the cost of dry mixes and makes production dependent on imported supplies. In this regard need to the development of domestic of production the modifiers. As the modifying agent of domestic production is proposed to use synthetic zeolites as structure-forming and water-retaining additive for dry construction mixtures.

Previous studies have confirmed the efficacy of synthetic zeolites as a modifying agent for cement and lime dry mixes [9-19].

Materials and Methods

We received amorphous silica-alumina their precipitation from the solution of aluminum sulfate of technical $Al_2(SO_4)_3$ with the addition of sodium silicate followed by washing the precipitate with water. Then, the resulting precipitate was dried.

Adhesive strength was determined by testing the samples fon stretching by tearing instrument ИР 50-57 with traverse moving speed 35 mk/c.

Plastic strength or yield stress of the mixture was determined by plastometer KP-3. Plastic strength determined by the formula:

$$\eta = \tau = \tau_0 = k * \frac{P}{h^2},\qquad(1)$$

Where η: Plastic strength;

τ: Shear stress;

τo: Yield stress;

k: coefficient depending on the value of the vertex angle of the cone; for the metal cone with an apex angle of 30° - k=1,116;

P: The weight of the movable part of the device (load);

h: Depth of immersion of the cone in the mortar mixture.

Research Results

Additive based on amorphous silica-alumina is a powder of white color with a high specific surface component S_{sp}=68.6 m²/g. Microstructure and chemical composition of the amorphous aluminosilicate examined via analytical scanning electron microscopy (Figure 1 and Table 1).

It was found that the microstructure of the synthetic additives is characterized by particles of round shape, dimensions 5,208-5,704 μm, but the particles are present also oblong form, size 7.13-8.56 μm.

Analyzing the data in Table 1 revealed that predominate chemical elements O, Si, Na, S, and Al in chemical composition amorphous aluminosilicates-containing 60.69%, 31.26%, 24.23%, 18.69% and 8.29% respectively. The preponderance of this element has a positive

Figure 1: Microstructure of amorphous aluminosilicate.

*Corresponding author: Loganina VI, Department of "Quality management and construction technologies" Penza State University of Architecture and Construction, Russia, E-mail: loganin@mail.ru

Content	Chemical elements additives weight, (%)				
	O	Na	Al	Si	S
Maximum	60.69	24.23	8.29	31.26	18.69
Minimum	36.73	8.61	1.10	7.92	0.68

Table 1: Chemical composition of the admixture.

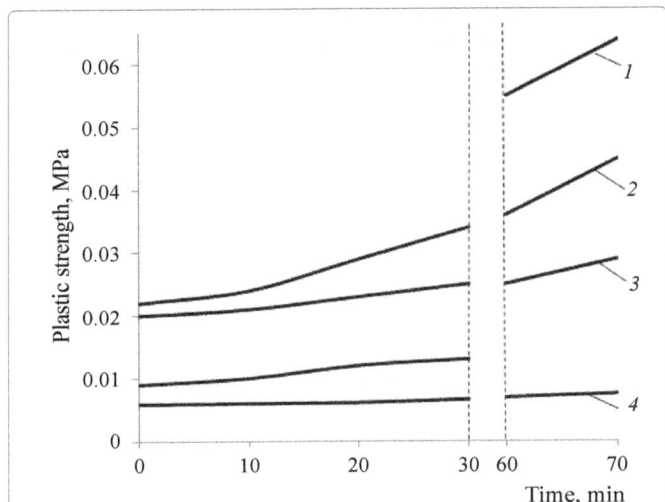

Figure 2: Kinetics set plastic strength cement-sand mortar: 1 - with a synthetic additive (30% by weight of cement) 2 - with a synthetic additive (20% by weight of cement), 3 - with the use of synthetic additives (10% by weight cement), 4 - a control sample (without the use of synthetic additives).

Name of indicator	Value of the indicator	
	Designed composition	The prototype (no additives)
Density of the mix, (kg/m³)	1800	1670
The correction time, (min)	20	30
Water retention, (%)	97.8-99.3	95.0-97.0
Slipping tile no more than, (mm)	0.3	0.5
Frost resistance of tile adhesive	F50	F50
Frost resistance of contact zone	$F_{кз}50$	$F_{кз}50$
Adhesion strength, R_{adg}, (MPa)	более 1.4	1.1
Cohesive strength, R_{kog}, (MPa)	2.2	1.6
Adhesion strength in shear, (MPa)	0.92	0.6

Table 2: Physical and mechanical properties of tile adhesive.

M400 Portland cement, sand fractions (mm) 0.63-0.315: 0.315-0.16 in a ratio of (%) 80: 20, amorphous aluminosilicates, plasticizer Kratasol PFM and redispersible powder Neolith P 4400) are given in the Table 2.

Conclusion

Determined, that the use of a binder with additive amorphous aluminosilicate leads to higher values the plastic strength in early and late periods hardening. The given model of cement stone strength in the presence of synthetic additives and plasticizer Kratasol PFM proved the efficiency of the use of amorphous aluminosilicate as a modifying additive. This modifying additives a structure formation and improves the physical and mechanical properties of tile adhesive.

effect on the formation of cement stone structures with used synthetic additives.

The effect of amorphous aluminosilicate was investigated to modify the rheological properties of cement-sand mortar. For research we used Wolski portlandcement M400, and sand deposits of Ukhta in the ratio 1:2. In Figure 2 given results of these studies are presented.

Analysis Figure 2 showed that the introduction in the cement-sand mortar the additive based on amorphous aluminosilicate leads to higher values of plastic strength aged 20 min after curing compared to the control sample is 1.9-4.7 times (depending on the content the additives). Thus, the sample aged for 20 minutes from the beginning of solidification has strength was 0.0061 MPa, while the sample with using of amorphous aluminosilicate (20% by weight of cement) – 0.023 MPa.

It is obvious that, when introduced into the formulation of cement-sand mortar additives based on amorphous aluminosilicate period of hardening cement-sand mortar reduced, that is, the admixture has a water-holding capacity.

On the basis of mathematical research and experimental design [20] constitute the model of cement stone strength. The factors affecting the change in strength of cement paste were investigated size of the specific surface additives (x_1), the percentage of synthetic additives (x_2) and the percentage of plasticizer Kratasol PFM (x_3). After the analysis of the experimental data and the exclusion of insignificant coefficients of the regression equation, the model of cement stone strength expressed by the formula:

$$y = 24,477 + 2,6433 \cdot x_1 - 1,0383 \cdot x_2 - 1,3303 \cdot x_3 - 4,5958 \cdot$$
$$(x_1)^2 - 3,3495 \cdot (x_2)^2 - 2,0564 \cdot (x_3)^2 \qquad (2)$$
$$+0,69375 \cdot x_1 x_2 + 0,36875 \cdot x_1 x_3 + 0,31875 \cdot x_2 x_3$$

Physical and mechanical properties of tile adhesive (includes the

References

1. Loganina VI, Zhernovsky VI, Sadovnikova MA, Zhegera KV (2013) The addition of aluminosilicate-based cement systems. Eastern European Journal of advanced technologies 5: 11-68.

2. Loganina VI, Petukhova NA, Gorbunov VN, Dmitrieva TN (2009) Prospects for the manufacture of organo-mineral supplements on the basis of domestic raw materials. Proceedings of the higher educational institutions Building, pp: 936-939.

3. Jenni A, Holzer M, Zurbriggen M, Herwegh M (2005) Influence of polymers on microstructure and adhesive strength of cementitious tile adhesive mortars. Cement and Concrete Research 35: 35-50.

4. Stark U, Reinold M, Muller A (2003) Neue Methoden zur Messung der Korngrobe und Kornform von Mikro bis Marko. Weimar pp: 11369-1380.

5. Taramasso M, Perego G, Notari B (1980) Proceedings of the 5th International Conference on Zeolites (Ed.L.V.C.Rees) 2: 40.

6. Mirsky YV, Mitrofanov MG, Dorogochinskiy AZ (1964) New adsorbents-molecular sieves. Grozny: Chechen-Ingush, p: 385.

7. Doroshenko YM, Shanah JI (1989) Processes of structure and properties of cement stone with polymeric modifiers. 15 Szilikatip esszilikattund konf SILICONF R.I Budapest, pp: 273-276.

8. Yiqing F, Lianzhang C (1991) Effect of aqueous boric acid (H3B03) treatment on catalytic performance of HZSM-5 zeolite catalysts. Shiyou Jiagong 7: 44.

9. Loganina VI, Zhegera KV (2014) Evaluating the effectiveness of the use of synthetic aluminum silicates in cement systems. Academic Gazette UralNIIproekt RAASN, pp: 384-387.

10. Loganina VI, Ariskin MV, Karpova OV, Zhegera KV (2015) Evaluation of crack resistance of the finishing layer based on adhesive dry mixture with synthetic aluminum silicates. Building Materials, pp: 1086.

11. Loganina VI, Makarova LV, Tarasov RV, Sadovnikova MA (2014) Composition limy binder with the use of the synthesized alumiosilicates for dty construction blends. Advanced Materials Research 977: 34-37.

12. Loganina VI, Makarova LV, Tarasov RV, Zhegera KV (2014) The composition cement binder with the use of the synthesized alumosilicates. Advanced Materials Research 10223: 6.

13. Loganina VI, Makarova LV, Tarasov RV, Ryzhov AD (2014) The limy composite

binder with the use of the synthesized alumosilicates. Applied Mechanics and Materials 662: 11-14.

14. Loganina VI, Ryzhov AD (2015) Structure and properties of synthesized additive based onamorphous aluminosilicates. Case Studies in Construction Materials 3: 132-136.

15. Mumpton FA (1999) Larocamagica: Uses of natural zeolites in agriculture and industry. PNAS 96: 7346-7347.

16. Andrejkovicova S, Ferraz L, Velosa AL, Silva AS, Rocha F (2012) Air Lime mortars with incorporation of sepiolite and synthetic zeolite pellets. ActaGeodyn Geomater 9: 79-91.

17. Aiello R, Collela C, Sersale R (1971) Zeolite Formation from Synthetic and Natural Glasses. Advances Chem Ser 101: 51-62.

18. Barrer MR, Cole JF, Sticher H (1968) Chem Soc Inorg Phys Theoret, p: 102475.

19. Broussard L, Shoemker DP (1960) The Structures of Synthetic Molecular Sieves. Journal of The A merican Chemical Society 82: 1041-1051.

20. Loganina VI, Tarasov RV, Zhegera KV (2014) Optimization of tile adhesive using synthesized aluminosilicate. Regional architecture and engineering Penza, pp: 44-48.

38

Wetting of Olivine Sand against Steel Alloys

Rastgoo Oskoui P[1]* and Payam RO[2]

[1]*Department of Materials Engineering, Faculty of Mechanical Engineering, University of Tabriz, Tabriz, Iran*
[2]*National Iranian Copper Industries Company, Tabriz, Iran*

Abstract

Olivine sand is used in steel plants operating with Electric Arc Furnaces as Eccentric Bottom Tap Hole. The free opening rate is mainly determined by the performance of the tap hole filler sand. A free opening occurs when steel flows freely from the tap hole to ladle once the tap hole is opened. Various parameters such as sintering behavior, particle size distribution, wettability, etc. affect the performance of the sand. One of the important factors affecting the performance of sand is wetting the sand by molten steel. Olivine sand is a type of sand that used in electric arc furnaces. Wetting of Olivine sand against FeO alloys were presented. The wetting characteristics of liquid FeO alloys in a matrix of the Olivine sand at air pressure and temperature of 1650°C were studied by determining the liquid metal-Olivine contact angles. The median wetting angle values from textually equilibrated samples were found 100°. These results suggest that the steel melt forms in isolated pockets at grain corners or on grain boundaries. This will Increase the permeability of sand And the performance of the tap hole filler sand.

Keywords: Electric arc furnace; EBT sand; Olivine; Wetting

Introduction

In an electric steelmaking, there are great concerns about the operational variables that limit production in the continuous casting process. One of these concerns is free opening the tap hole with very good flowability and permeability of sand.

In order to avoid the contact between molten steel and the clouser tap hole used filler sand. Different types of filler sands can be used into tap hole. Among the important factors in the selection of sand are refractoriness, particle-size distribution, flowability and wettability [1].

Olivine sand is one of the most widely used materials. Its Lower free-silica content and strong resistance to metal attack along with the refractory properties caused that Olivine sand to be used in eccentric bottom tap filler [2].

Pore morphology in EBT filling sand equilibrated with steel molten and significantly controls many physical properties such as elasticity, Thermal conductivity and permeability of the EBT sand [3,4].

Immiscible metal liquids play an important role in controlling the dispersal of steel liquid penetration in the Tap Hole. Knowledge of the wetting properties of such melts allows accurate prediction of their topology on the grain scale, which in turn provides insight into the mechanisms for migration of the steel liquid, and whether the final distribution of steel is likely to be interconnected melt network or whether the steel is isolated pockets [5].

In textually equilibrated liquid-solid systems flow ability of liquid is mostly controlled by the melt fraction and the solid-solid (γ_{ss}) and solid-liquid (γ_{sl}) interfacial energies of the phases involved [6]. The geometry of a melt pocket in the solid matrix is determines by the ratio of the interfacial energies. Interfacial energies equilibrated the contact angle between the melt and the confining grains, known as the wetting angle θ [7,8].

$$\frac{\gamma_{ss}}{\gamma_{sl}} = 2\cos\frac{\theta}{2} \qquad (1)$$

If the wetting angle θ is less than 60°, called wetting boundary and an interconnected melt network formed and melt can migrate through the solid matrix. If, on the other hand, θ>60°, melt will be formed to in isolated pockets [9,10].

The wetting angle of Olivine in the Presence of iron melts at 1370-1410°C is very greater than the maximum allowable wetting angle for melt connection have large wetting angles [8]. Pure liquid Fe has a high wetting angle (>100°) in silicate matrix [11].

Experimental Section

The run products consist of Olivine sand and Iron sulfide. Olivine sand from the RHI Company was used as a material matrix. Micro particles of Olivine sand produced in an agate mortar and pestle and sieved through screens to achieve near uniform grain size distribution of particles. The size fractions used in the experiments were between 70 and 60 mesh (210 to 250 μm). Reagent grade powders of FeS as a source of FeO was used. Raw materials were mixed together at a weight ratio of 50-50%. And then in order to make the initial adhesion between particles were added 5% polyvinyl alcohol by weight.

A piston cylinder apparatus was used for experiments conducted at pressures of up to 30 tons. Samples are placed in the dryer for one day and then the tube furnace maintained at a temperature of 1650°C. Heating and cooling rates of the furnace were selected 4 and 7°C/min. In order to simulate experimental conditions and working conditions eccentric bottom tap sand was used air atmosphere.

Microstructural characterization was carried out by a scanning electron microscope (Mira3 Tescan, Czech Republic). Chemical analysis was performed simultaneously with SEM, using energy dispersive spectroscopy (EDS). Wetting angle iron liquid against matrix was measured by using SEM images.

In BSE images with clearly visible liquid FeO pocket-Olivine triple junctions the wetting angles were measured with the angle tool

*Corresponding author: Rastgoo Oskoui P, Independent Researcher, Department of Materials Engineering, Faculty of Mechanical Engineering, University of Tabriz, Tabriz, Iran, E-mail: parisa.oskoui@gmail.com

of AutoCAD software. In this study the number of angles measured in each sample containing Olivine matrix ranges from 150 to 260. Since the two-dimensional section images taken through melt-grain junctions in various orientations a single measurement will give an apparent angle value, which may be an over or underestimation of the true angle measured perpendicular to the Olivine-iron liquid contact. To account for this effect, the average value of the measured angle population in a sample was taken as this has been shown to be a good approximation of the true wetting angle [12].

X-ray diffraction (XRD) analysis (Cu lamp, λ=1.54 Å, 40 KV, 30 mA, Siemens D5000 model) was carried out on the samples. In order to determine the wettability Olivine sand, grains were kept in the tube furnace with temperature 165°C for 1 h.

Result and Discussion

XRD analyse

Figure 1 presents the XRD patterns of as received Olivine sand that crushed an agate mortar and pestle and sieved. As it seems in Figure 1 the solely detected crystalline phases are Olivine and Forsterite.

Anisotropy of Olivine free surface energy

In the isotropic equilibrium theory of partial melts it is supposed that the surface energies of crystal phases are isotropic. For isotropic solids, Surface tension is independent of surface direction. When isolated from other crystalline grains, their equilibrium shapes, derived from minimizing surface free energy, are spheres. For solids with anisotropic crystal structures, surface tensions are functions of the orientations of their respective planes, which are generally expressed by Miller indices [13].

Surface energy on the surface of the crystal grains can be expressed using the theory periodic bond chain (PBC) developed by Hartman and Perdok. Here's surface energy is defined as energy per unit area. An area that is required to divide an infinite crystal into two Half along a specified plane [14-16]. In PBC theory the crystal structure of systematic searches for chains with a period of strong bonds of network. For Forsterite Olivine, where $[SiO_4]$ tetrahedra are treated as crystallizing units, chains are constructed such that only MgO bonds are broken. If at least two connected chains are found in the same plane, the crystal face corresponding to this plane is called an F (flat) face The higher the bond energy within the plane, the lower the surface energy of the individual crystal face, and the more important the crystal face becomes for the crystal habit [17]. t'Hart applied PBC theory to Olivine and calculated surface energies for different bond models ranging from purely covalent to purely ionic. The predicted crystal habit is elongated along the c axis and slightly tabular on (010) [15].

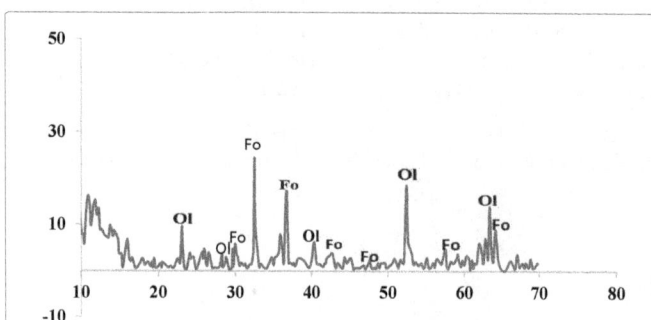

According to the surface energy values, the stability order of the Olivine faces was found to be (010) < (120) < (001) < (101) < (111) < (021) < (110) [18].

SEM image of Olivine grain Maintained in the tube furnace at 165°C for 1 h is shown in Figure 2. Because of the anisotropy of crystal structure of Olivine, The molten manufactured in all directions will not be able to wetting grains where melt separates the grains labelled I and II, but the same grain boundary of grain III with grain VI is dry.

Measurement of the wetting angle

In the heating process of the sample, oxidation of FeS has occurred. In Figure 3 EDS spectra of isolated region of Fe and O is shown. When FeS is heated in the oxidizing atmosphere, two reactions are done. In the first stage reaction to take place is the transformation of FeS to magnetite or hematite

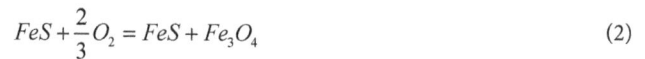

$$FeS + \frac{2}{3}O_2 = FeS + Fe_3O_4 \qquad (2)$$

Or

$$2FeS + \frac{3}{4}O_2 = Fe_2O_3 + S_2 \qquad (3)$$

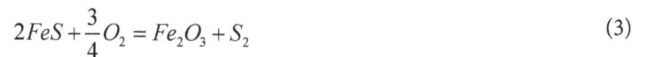

Sponge iron is seen due to gas emission from regions containing iron. Melt pockets in the spongy rim FeO is shown in Figure 3a. The size of the melt pockets is in the range of a few hundred nanometres up to several microns.

By doing these two reactions and produce molten iron working conditions will be similar to work conditions of eccentric bottom tap in electric arc furnace.

The particles were analysed for elemental variations by energy dispersive spectroscopy (EDS) through the use of scanning electron microscopy (SEM).

In SEM image areas included FeO are heavier than Olivine and as a result heavier than FeO region which backscatter more efficiently appear brighter than lighter Olivine region in a backscattered electron image.

The SEM micrographs in Figure 4 illustrate that the distribution of molten. Onions such as S and O have significant effects on the liquid Fe surface tension. Iron sulfide melted at 1109°C and at 1350°C Decomposition reaction is completed and converted to Fe_3O_4. In this case, compared to the previous condition anion to cation ratio of the melt increased. Oxygen-poor melts in which the anion to cation less,

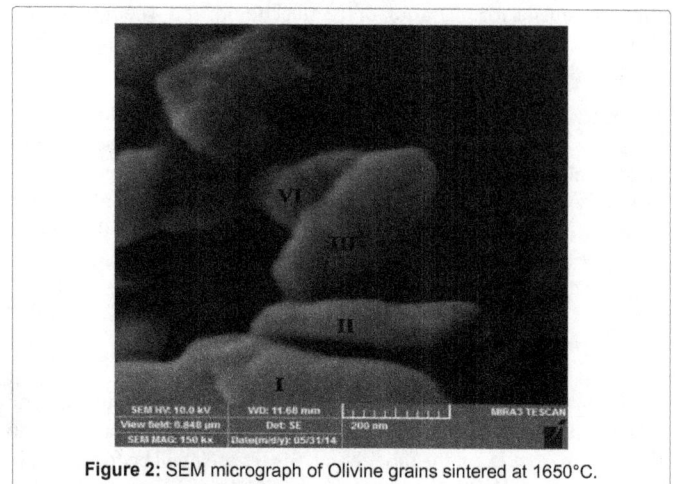

Figure 1: X-ray diffraction patterns for EBT sand.

Figure 2: SEM micrograph of Olivine grains sintered at 1650°C.

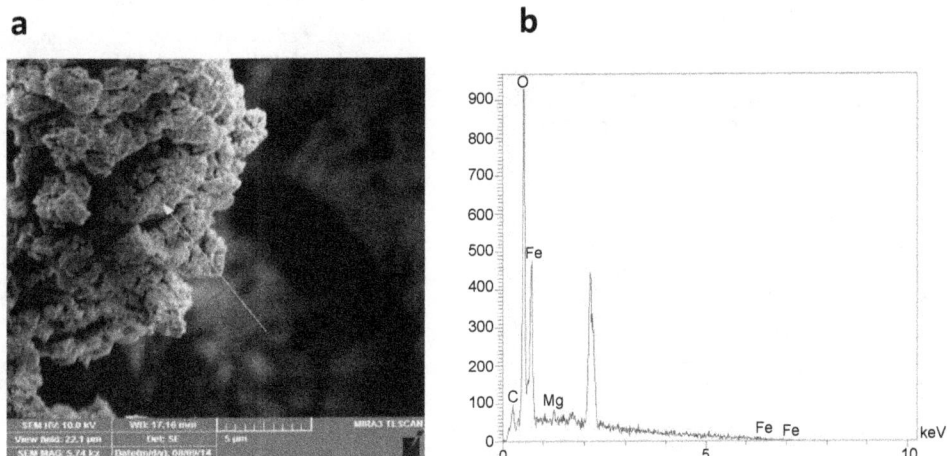

Figure 3: (a) SEM micrograph of isolated iron pockets sintered at 1650°C, (b) EDS spectra of a.

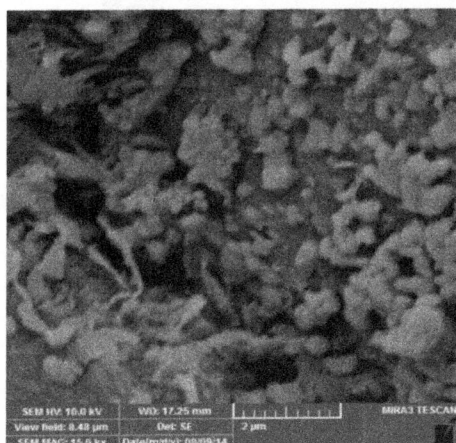

Figure 4: SEM micrograph of Olivine grains with 50% by weight of iron sulfide sintered at 1650°C (brighter phase is iron melt forms in isolated pockets and darker phase is Olivine).

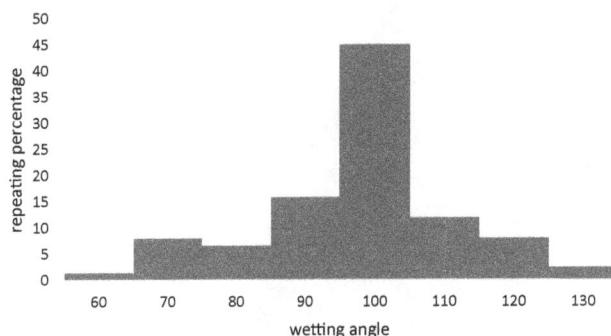

Figure 5: Diagram of repeating percentage distributions for various interfacial angles.

because oxygen-cation bonding is not complete, It has high surface energy. As a result, so do not be perfect wetting of the particles. The increased anion to cation in this way that The surface free energy is dependent on the presence of surface atoms that completely surrounded by arrangements of the periodic structure, therefore the unsaturated bonds are formed.

Unsaturated bonds lead to a reduction in surface energy of the crystal at the interface, As a result of increased wettability. Despite the increase in the ratio of anion to cation, the melt does not have the ability to wetting Olivine matrix. Repeating percentage distribution of FeS-Olivine dihedral angle is shown in Figure 5 and the wetting angle by using SEM micrograph obtained 100°. Thus the irons liquid had not the ability to the wetting Olivine sand and remain isolated.

Conclusion

In this study the wettability of sand by Steel molten was investigated. The wettability angle of sand was obtained at 100°. Leads the molten iron does not have the ability to wetting used Olivine sand in eccentric bottom tap furnace and the liquid permeability in the tap hole reduced. As a result EBT Tap hole Filler in EAF operated steel plants can achieve up to 98% Free Opening without requirement of poking or oxy lancing.

Due to non-wetting of Olivine by Steel liquid has been appropriate selection in steel plants operating with Electric Arc Furnaces as Eccentric Bottom Tap Hole (EBT) Filler during melting operations.

References

1. Cruz RTD, Pelisser GF, Bielefeldt WV, Bragança SR (2016) Free Opening Performance of Steel Ladle as a Function of Filler Sand Properties. Materials Research 19: 408-412.

2. Kho TS, Swinbourne DR, Blanpain B, Arnout S, Langberg D (2010) Understanding stainless steelmaking through computational thermodynamics Part 1: Electric arc furnace melting. Mineral Processing and Extractive Metallurgy 119: 1-8.

3. Takei Y (2002) Effect of pore geometry on Vp/Vs: From equilibrium geometry to crack. Journal of Geophysical Research: Solid Earth 107(B2).

4. Schmeling H (1985) Numerical models on the influence of partial melt on elastic, an elastic and electric properties of rocks. Part I: Elasticity and anelasticity. Physics of the Earth and Planetary Interiors 41: 34-57.

5. Watson EB, Brenan JM (1987) Fluids in the lithosphere, 1. Experimentally-determined wetting characteristics of CO_2H_2O fluids and their implications for fluid transport, host-rock physical properties, and fluid inclusion formation. Earth and Planetary Science Letters 85: 497-515.

6. Bargen N, Waff HS (1986) Permeabilities, interfacial areas and curvatures of partially molten systems: results of numerical computations of equilibrium microstructures. Journal of Geophysical Research: Solid Earth 91: 9261-9276.

7. Bulau JR, Waff HS, Tyburczy JA (1979) Mechanical and thermodynamic constraints on fluid distribution in partial melts. Journal of Geophysical Research: Solid Earth 84(B11): 6102-6108.

8. Jurewicz SR, Jurewicz AJ (1986) Distribution of apparent angles on random sections with emphasis on dihedral angle measurements. Journal of Geophysical Research: Solid Earth 91(B9): 9277-9282.

9. Terasaki H, Frost DJ, Rubie DC, Langenhorst F (2005) The effect of oxygen and sulphur on the dihedral angle between Fe-O-S melt and silicate minerals at high pressure: Implications for Martian core formation. Earth and Planetary Science Letters 232: 379-392.

10. Yoshino T, Walter MJ, Katsura T (2003) Core formation in planetesimals triggered by permeable flow. Nature 422: 154-157.

11. Jurewicz SR, Jones JH (1995, March) Preliminary results of olivine/metal wetting experiments and the direct measurement of metal phase interconnectivity. In: Lunar and Planetary Science Conference (Vol. 26).

12. Jurewicz SR, Jurewicz AJ (1986) Distribution of apparent angles on random sections with emphasis on dihedral angle measurements. Journal of Geophysical Research: Solid Earth 91(B9): 9277-9282.

13. Landau LD, Lifshitz EM (1980) Statistical Physics, part 1, Pergamon, New York.

14. Hartman P, Perdok WG (1955) On the relation between structure and morphology, I, Acta Crystallogr 8: 49-52.

15. Hartman P, Perdok WG (1955) On the relation between structure and morphology II, Acta Crystallogr 8: 521-524.

16. Hartman P, Perdok WG (1955) On the relation between structure and morphology III. Acta Crystallogr 8: 525-529.

17. Bennema P (1983) Crystal graphs, connected nets, roughening transition and the morphology of crystals. In abstracts of papers of the american chemical society (vol. 186, no. Aug, pp. 157-coll). 1155 16TH ST, NW, WASHINGTON, DC 20036: AMER CHEMICAL SOC.

18. Bruno M, Massaro FR, Prencipe M, Demichelis R, De La Pierre M, et al. (2014) Ab Initio calculations of the main crystal surfaces of forsterite (Mg2SiO4): A preliminary study to understand the nature of geochemical processes at the olivine interface. The Journal of Physical Chemistry C 118: 2498-2506.

39

Visible Light assisted Photocatalytic Acitivity of Zinc Titanate in Presence of Metallic Sodium

Sirajudheen P*

Department of Chemistry, WMO Imam Gazzali Arts and Science College, Koolivayal, Panamaram, Wayanad, Kerala, India

Abstract

Zinc titanate powder was synthesizd by an organic free co-precipitation peroxide technique. Zinc chloride and titanium (IV) isopropoxide were used as the primary material with the ratio of Zn:Ti was 1:1 and the later compound was prepared by mixing titanium chloride with isopropanol. The stoichiometric ratio of the synthesized zinc titanate was measured by using AAS. The synthesized powder was calcined at 800°C for 3 hours using the X-ray diffraction and the calcined zinc titanate powder crystalline phase formation was found to be cubic. FTIR was used for studying the bonding characteristics of ZnO and TiO_2. Disintegration temperature was analyzed by means of Thermogravimetric analysis. The photocatalytic activity was measured based on the degradation of methyl orange in aqueous solution in the presence of metallic sodium. The results showed that $ZnTiO_3$ particle exhibited good photocatalytic activity under visible range radiation.

Keywords: Zinc titanate; Cubic phase; Methyl orange; Photocatalytic activity

Introduction

Zinc titanate ($ZnTiO_3$), has a promising material as a gas sensor [1] (for ethanol, NO, CO, etc.), paint pigment [2], catalyst [3] and etc. Also $ZnTiO_3$ has been reported as a material with an exceptional electrical properties which could be a suitable contender as microwave resonator [4]. This material has a dielectric constant of 19, quality values of 30.0 GHz and the temperature coefficients of the resonant frequency of -55 ppm/qC [5]. The structures of titanium dioxide (TiO_2), $Zn_2Ti_3O_8$, $ZnTiO_3$ and Zn_2TiO_4 comprised of TiO_6 octahedra. In rutile and in $ZnTiO_3$, the connection of the TiO_6 octahedra results chains and/or layers. As a result of this resemblance, $ZnTiO_3$ is formed only in the presence of rutile [6].

Zinc titanates are generally synthesized by characteristic solid state reactions at high temperatures [7]. Due to some confines of solid-state synthesis, such as large grain size and uncontrolled and irregular morphologies, different kinds of alternative techniques have been depicted including mechano-chemical activation [8,9], molten salt synthesis [10], a semi chemical route combined with vigorous micro beads milling [11,12] and sol-gel method. But the sol-gel methods are generally complicated and the reagents used are expensive. In wet chemical processes for preparing highly quality of powders, better homogeneity, control morphology and smaller particle size are preferred.

The co- precipitation reaction technique has exciting characteristics such as its simplicity, its relatively low cost and the fact that it usually results in products with the preferred structure and composition [13]. The effort aimed to synthesize $ZnTiO_3$ powder by the co- precipitation peroxide method. The phase change was studied by X-ray diffraction (XRD). The chemical characterization of zinc titanate encompasses analysis of elements using AAS. The photocatalytic activity was measured based on the degradation of methyl orange in aqueous solution in the presence of metallic sodium.

Materials and Methods

Materials

Anhydrous Zinc chloride and Titanium chloride in isopropanol were used as starting materials for zinc titanate synthesis. Hydrogen peroxide and ammonia solutions from Merck are also used. All the chemicals were used without further purification. The crystalline structure of the samples was determined by powder XRD using BRUKER D8 Advance X-ray diffractometer using CuKα radiation. FT-IR spectra were recorded using KBr pellet method on Shimadzu IR affinity spectrophotometer in the range from 4000 cm^{-1} to 400 cm^{-1}. TGA experiments were carried out using SDT Q600 V20.9 instrument. The decolorization of methyl orange was measured by using UV-Visible Spectrophotometer of Shimadzu at 465 nm by incorporating the dye solution with zinc titanate in presence of metallic sodium, exposed to sunlight for different irradiation time, using UV-Visible Spectrophotometer. During the irradiation time, the solution was bubbled with air and the initial concentration of dye was 10.0 ppm.

Preparation of the sample

0.2 mol (0.272 g in 100 ml) zinc chloride was dissolved in 100 ml of water and 0.2 mol of titanium chloride in isopropanol is also prepared up to 100 ml and the two solutions are mixed together to get a homogenous solution. Then 15 ml of hydrogen peroxide and 20 ml of ammonia solutions are mixed and 165 ml of water is added to it. The homogeneous solution having zinc chloride and titanium tetra chloride are added to the second solution drop by drop by means of a burette. A precipitate was developed in the beaker it was allowed to settle for some time and filtered. The residue obtained was washed with water dried by using owen and calcined in Muffle furnace at 800°C about half an hour in a silica crucible [14]. The zinc titanate formed was powdered and used for further experiments.

*Corresponding author: Sirajudheen P, Department of Chemistry, WMO Imam Gazzali Arts and Science College, Koolivayal, Panamaram, Wayanad, Kerala, India
E-mail: sirajpalliyalil@gmail.com

Photocatalytic activity

The photocatalytic efficiency of the synthesized powders was calculated by deprivation of a model aqueous solution of methyl orange in presence of metallic sodium. The initial concentration of methyl orange aqueous solution was 10 ppm exposed under visble light irradiation. During analysis varying amount of zinc titanate sample was added to methyl orange solution and it was exposed under visible light irradiation with constant weight (0.02 g) of sodium. During the irradiation time, the solution was bubbled with air. The photo-bleaching of methyl orange was strongly affected by the presence of zinc titanate in the present of metallic sodium. As the contact time increases, the degradation of methyl orange was also increases.

Result and Discussion

Results of structural and thermal analyses

The X-ray diffraction patterns of prepared $ZnTiO_3$ calcinated at different temperatures are shown in Figure 1. Neither ZnO nor TiO_2 phases were observed in the spectra. At 600°C, the samples undergo solid state reaction to form cubic structure. On increasing the calcination temperature from 600°C to 700°C, the crystallainity of cubic phase $ZnTiO_3$ also increases It was evidenced from the increased intensity peaks of $ZnTiO_3$ at 30.45° and peak at 35.37° [15]. When using the same precursors [16] at 800°C, several peaks related to the hexagonal $ZnTiO_3$ appears, but the cubic crystals are still dominant.

The Differential Thermal Analysis (DTA) curve of the formerly heat treated at 400°C sample elucidates two endothermic (120 and 380°C) and two exothermic peaks at 450 and 540°C. The exothermic lines might related to the combustion of the organic residues and crystallization of the sample with development of $ZnTiO_3$ phase (540°C) and the endothermic effects can be ascribed to the dehydratation of the sample. Both exothermic peaks (at 450°C and 540°C) are not observed in the DTA curve of the formerly heat treated sample at 500°C. This shows that the last exothermic effect is related to the crystallization process which is in good agreement with the X-ray diffraction results (Figure 2). The results are well-matched to those obtained by Hosono et al. and Wang et al. [15,16].

The Fourier Transform Infrared Spectrum (FTIR) of the prepared zinc titanate ($ZnTiO_3$) samples was shown in Figure 3. Dominant bands at 730 cm⁻¹, 490 cm⁻¹ and 450 cm⁻¹ along with a sharp band at 420 cm⁻¹ are observed in the IR spectra of $ZnTiO_3$. It is identified that,

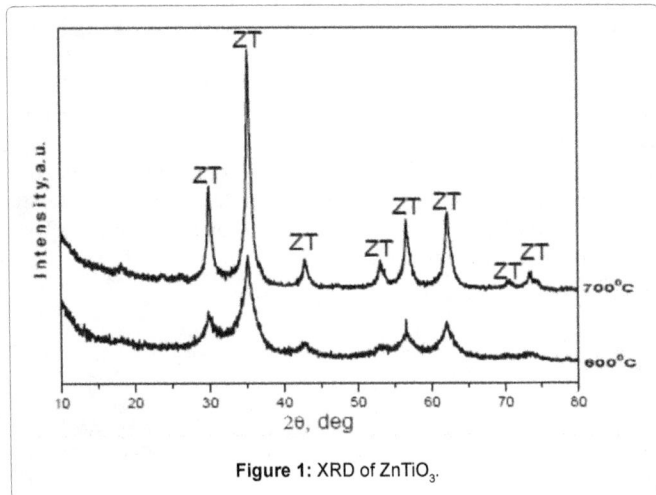

Figure 2: DTA curve of $ZnTiO_3$.

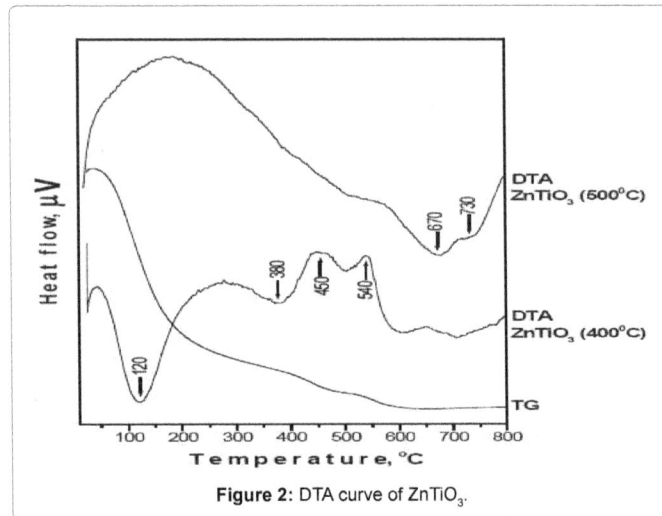

Figure 3: FTIR spectra of $ZnTiO_3$.

the bands in the absorption range 700-400 cm⁻¹ might be associated to the vibrations of TiO_6 units in $ZnTiO_3$ [16,17]. According to the X-ray diffraction data for this sample, only one phase ($ZnTiO_3$) was noticed. It is also known that, bands corresponding to ZnO_n polyhedra are in the same absorption range [18,19]. The IR spectrum of $ZnTiO_3$ is connected to the increased intensity of the band centered at 420 cm⁻¹. As it is known this band is typical for the vibrations of TiO_6 units [20,21] and the weak band near 450 cm⁻¹ could be related to the Ti-O stretching vibrations in $ZnTiO_3$ [17,22]. A characteristic band at 730 cm⁻¹ appeared at 800°C. This can be assigned to the Zn-O-Ti bond structure in cubic $ZnTiO_3$ [23].

Photocatalytic activity

The photodegradation of methyl orage was performed by adding varying amount of zinc titanate particle in 10.0 ppm methyl orange solution exposed under visible light irradiation [16] by taking 0.02 g of sodium. Figure 4 shows the photobleaching of the dye with different amount of zinc titanate exposed to visible light irradiation for different duration [24]. If the dye concentration is increased, the number of dye molecule in the solution is also increased which eventually affect the degradation rate [16]. The photo-bleaching of methyl orange was

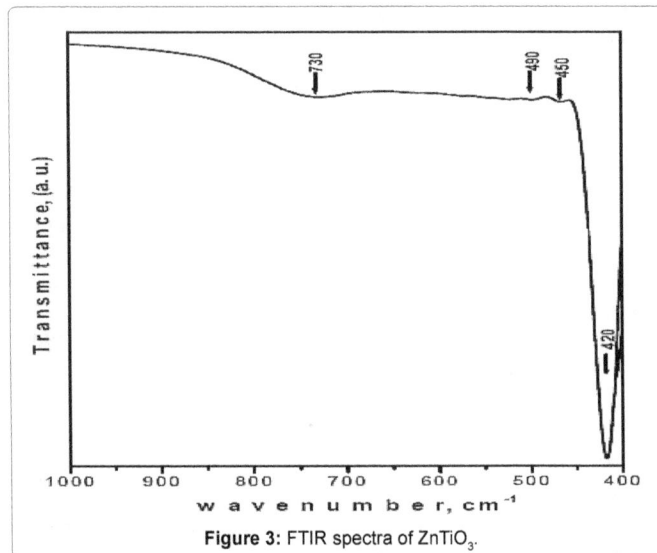

Figure 1: XRD of $ZnTiO_3$.

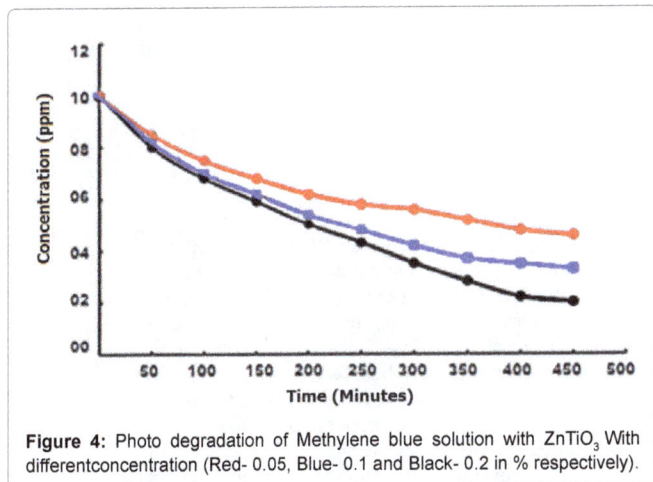

Figure 4: Photo degradation of Methylene blue solution with ZnTiO₃ With differentconcentration (Red- 0.05, Blue- 0.1 and Black- 0.2 in % respectively).

strongly affected by the presence of zinc titanate present in the system [25]. As the exposure time increases, the degradation of dye also increases [26]. This is due to increased photon absorption by zinc titanate which is catalysed by metallic sodium, which leads more transfer of photoelectrons and photo holes between valance band and conduction band and there by generating hydroxyl radical and super radical oxygen in the photo catalysis cell. These radical ions are responsible for the degradation of the methyl orange [24,27]. As the concentration of zinc titanate increased, more absorption of photon from the visible light leads to more degradation of the dye. The concentration of methyl orange was decreased to 2.0 ppm when solution containing 0.2% of zinc titanate under the visible light exposure of 4.15 hours.

Conclusion

The zinc titanate powders were prepared by co- precipitation peroxide method and its physical properties are studied by using XRD, FTIR and TGA. The stoichiometric ratio of the synthesized zinc titanate was measured by using AAS, it showed that the prepared zinc titanate contains about 40% of zinc and 23.68% of titanium and the compound displays dominant cubic structure. The concentration of methyl orange was diminished from 10.0 ppm to 0.2 ppm by introducing 0.2% zinc titanate in presence of 0.02 g of sodium exposed to sun light for 90 minutes duration. The photo bleaching of the dye under visible light with ZnTiO₃ confirms the photocatalytic activity of zinc titanate under visible light condition.

References

1. Obayashi H, Sakurai Y, Gejo T (1976) Perovskite-type oxides as ethanol sensors. J Solid State Chem 17: 299-303.

2. McCord AT, Saunder HF (1945) U.S. Patent 2739019. Ceram Abstr 24: 155.

3. Bartram SF, Slepetys RA (1961) Compound Formation and Crystal Structure in the System ZnO-TiO₂. J Am Ceram Soc 44: 493-499.

4. Kim HT, Nahm S, Byun JD, Kim Y (1999) Low-Fired (Zn,Mg)TiO₃ Microwave Dielectrics. J Am Ceram Soc 82: 3476-3480.

5. Kim HT, Kim SH, Nahm S, Byun JD, Kim YH, et al. (1999) Low-Temperature Sintering and Microwave Dielectric Properties of Zinc Metatitanate-Rutile Mixtures Using Boron. J Am Ceram Soc 82: 3043-3048.

6. Liu Z, Zhou D, Gong S, Li H (2009) Studies on a basic question of zinc titanates. J Alloy Compd 475: 840-845.

7. Chang YS, Chang YH, Chen IG, Chen GJ, Chai YL, et al. (2004) Synthesis, formation and characterization of ZnTiO₃ ceramics. Ceramic International 30: 2183-2189.

8. Botta PM, Aglietti EF, Port López JM (2004) Mechanochemical effects on the kinetics of zinc titanate formation. J of Mat Science 39: 5195- 5199.

9. Labus N, Obradović N, Srećković T, Mitić V, Ristić MM (2005) Evolution of Mullite from a Solgel Precursor. Science of Sintering 37: 115-122.

10. Xing X, Zhang C, Qiao L, Liu G, Meng J (2006) Facile Preparation of ZnTiO₃ Ceramic Powders in Sodium/Potassium Chloride Melts. Journal of the American Ceramican Society 89: 1150-1152.

11. Kim HT, Kim SH, Nahm S, Byun JD, Kim YH (1999) Correlation Between Temperature Coefficient of Resonant Frequency and Tetragonality Ratio. Journal of the American Ceramic Society 82: 3043-3048.

12. Liu X, Gao F, Zhao L, Zhao M, Tian C (2007) Efects of V₂O₅ addition on the phase- structure and dielectric properties of zinc titanate ceramics. Journal of Electroceramics 18: 103-109.

13. Patil KC, Aruna ST, Ekambaram S (1997) Combustion synthesis. Cur Op in Solid State & Mat Sci 2: 158-I65.

14. Sirajudheen P, Gireesh V, Sanoop KB (2015) Data reconciliation and gross error analysis of self powered neutron detectors: comparison of PCA and IPCA based models. Int J of Adv in Mat Sc and Eng 4: 33-39.

15. Hosono E, Fujihara S, Onuki M, Kimura T (2004) Low-Temperature Synthesis of Nanocrystalline Zinc Titanate Materials with High Specific Surface Area. J Amer Cer Soc 87: 1785-1788.

16. Wang SF, Gu F, Lu MK, Song CF, Liu SW, et al. (2003) Preparation and characterization of sol–gel derived ZnTiO₃ nanocrystals. Mater Res 38: 1283-1288.

17. Shabalin BG (1982) Polyol method for the preparation of nanosized Gd₂O₃, boehmite and other oxides. Mineral 4: 54-61.

18. Mancheva M, Iordanova R, Dimitriev Y (2011) Mechanochemical synthesis of nanocrystalline ZnWO4 at room temperature. J Alloys Compd 509: 15-20.

19. Andres-Verges M, Martinez-Gailego M (1992) Reactivity and molecular structure of silicon carbide fibres derived from polycarbosilanes. J Mater Sci 27: 3756-3762.

20. Murashkevich A, Lavitkaya A, Barannikova T (2008) Signs of phase transitions and a thermooptic memory effect in absorption spectra of (N(CH₃)⁴)2Zn0.8Ni0.2Cl₄ solid solutions. J Appl Spectr 75: 730-734.

21. Yurchenko E, Kustovar G, Bacanov S (1981) Vib Spec of inorganic comp Nauka.

22. Yamaguchi O, Morimi M, Kawabata H, Shimizu K(1987) Formation and Transformation of ZnTiO₃. J Amer Ceram Soc 70: c97-c98.

23. Budigi L, Nasina MR, Shaik K, Amaravadi S (2015) Structural and optical properties of zinc titanates synthesized by precipitation method. J Chem Sci 127: 509-518.

24. Li Y, Sun S, Ma M, Ouyang Y, Yan W (2008) Kinetic study and model of the photocatalytic degradation of rhodamine B (RhB) by a TiO₂-coated activated carbon catalyst: Effects of initial RhB content, light intensity and TiO2 content in the catalyst. Chem Eng J 142: 147-155.

25. Rauf MA, Meetani MA, Hisaindee S (2011) An overview on the photocatalytic degradation of azo dyes in the presence of TiO₂ doped with selective transition metals. Desalination 276: 13-27.

26. Daneshvar N, Salari D, Khataee AR (2003) Photocatalytic degradation of azo dye acid red 14 in water: investigation of the effect of operational parameters. J of Photochem and Photobio A: Chem 157: 111-116.

27. Fujishima A, Zhang X, Tryk DA (2008) TiO₂ photocatalysis and related surface phenomena. Surf Sci Rep 63: 515-582.

40

Universality of Graphene as 2-D Material

Solanki Bhaumikkumar Ketansinh*

Mechanical Engineering, Chhotubhai Gopalbhai Patel Institute of Technology, India

Abstract

World always asks for compact and effective equipment. For that we need suitable materials. Now think about the thinnest material which has extraordinary properties. Graphene is real two dimensional material. It has lots of useful properties like conductivity, high strength and good flexibility. This Material has lots of potential, but still there is no more researches occupied on Graphene. This research paper is about to introduce the Graphene to the present world and to shows the impotency of Graphene in future. But question arise that which type of equipment or machines need a Graphene, where we can use Graphene to increase the effectiveness of equipment. It has carbon as raw material so it is cheap, eco-friendly and sustainable. It is much stronger, transparent and good conductive material which can use in Conductors, Transistor, Heat spreader and Interconnect wires of Integrated circuits, super capacitors and also useful as micro sensor and actuator. In future with use of Graphene composite, we will able to make space elevator of 36000 km altitude. But still it has some challenges that we have to solve. Challenges like, High prize at this moment (1 cm² = 3700 rupees), Sensitive to environment with no effective passivation, High quality thin films still lacking of reproducibility, Large scale transfer of films still irreproducible, Sheet resistance still too high, No band gap which a transistor requires (turn off problems). This paper give you total overview of the all basic information require using Graphene in the future. This paper has all basic information regarding to every properties of the Graphene. Here all properties are equally justified on the bases of application too.

Keywords: 2-D Material; Graphene; Universal material; Super capacitors; Space elevator; Heat spreader; Interconnect wires of Integrated circuits; The thinnest material

Introduction

Graphene has carbon as core material. It is a one kind of atomic structure of carbon. Diamond and graphite both made of carbon but because of them atomic structure they are different. In diamonds 1 carbon atom connected with 4 other atoms and in Graphite 1 carbon is connected with other 3 atoms. In graphite it has layer structure, in each layer Carbons is connected with extremely strong bond. But between layers there are Vander Val Bond which easy to fracture. Graphene are discovered from graphite, one layer of graphite is known as Graphene which is harder than diamonds. Graphene can make by polishing process of Graphite but only 10micrometer thickness can be achieved. In 2004 Scotch tape is use to produce few layers of Graphene. This experiment was not for fabrication of Graphene it was just to verify its unique properties.

Graphene production with help of scotch tape

Figure 1 represents scotch tape experiment [1].

High Potential Application of Graphene

It has unique features; it integrates good electrical, optical, mechanical, thermal and chemical properties in one material. Like other Nano materials Graphene is more important.

This 2-D metal is use for,

- Transparent Conductor in Opt-electronics

- Biological applications

- Sensor and Actuators

- Integrated Circuits

- Composite materials

- Energy applications and many more.

Properties of Graphene

Electrical properties

Graphene is Semi-metal (conductor) so electron can easily accelerate in Graphene.

$$\mu = \frac{V}{E} \quad (1)$$

Where, μ= Mobility

V= Drift velocity of electrons

E= External electric field

Graphene has mobility at room temperature is $2*10^8$ unit which is higher than below material (Table 1). Conductivity is determined by Mobility and Density of material.

Materials	$\mu(Cm^2/[Vs])$
Si	1000-2000
GaAs	9000
InP	5000-7000
InAs	33000
InSb	78000

Table 1: Mobility of different materials [22].

***Corresponding author:** Solanki Bhaumikkumar Ketansinh, Mechanical Engineering, Chhotubhai Gopalbhai Patel Institute of Technology, India
E-mail: bhaumiksolanki20495@gmail.com

$$\sigma = ne\mu \quad (2)$$

Where; σ = Conductivity

n = Carrier Density

e = Elementary charge

μ = Mobility

Graphene is 65% more conductive than copper

Optical properties

Most of metal conductor are optically opaque because free electron in conductor have a screening effect, preventing the photons to pass through but Graphene is one of the transparent conductors known to us. It is just one atom thick and it absorbs only 2-3% of incident line so it is theoretically transparent (Figures 2-4).

Thermal properties

Graphene is excellent thermal conductor. It has highest thermal conductivity in all material, even higher than diamonds. It has thermal conductivity >3000 W/m/K. It is also isotropic ballistic thermal conductance, means same effect in all direction [2].

Mechanical properties

It is the thinnest material ever (0.34 nm). It has breaking strength is 130 GPa, 100 time greater than steel (It is pressure so convert it into normal force per cross sectional area so we found that it is strongest material ever). Also this material is stretchable up to 20% of its initial length; $1m^2$ can sustain weight of 4 kg. It is completely flexible due to its atomic thickness.

Chemical and biological properties

Graphene are chemically stable, it can't attack by acids or basis although very strong acids or basis. It is strongest in nature. Graphene pump energy is 607 KJ/module where Diamonds has 347 KJ/module. At room temperature it is stable with oxygen but at high temperature near to 700°C Graphene convert into Carbon Dioxide. Surface of it can easily modify by oxygen or nitrogen containing fundamental groups. It use as a substance to be interface with various Biomolecules and Cells. It is largely Biocompatible (this property is under investigation) [3].

Applications of Graphene

Transparent conductor in optoelectronics

It has extraordinary conductivity, high mobility, high transitivity so one of the best conductor. It use as light emetic diode (LED). It will use as a key element in solid state lighting in future society and that LED will more efficient than traditional. It will also useful for touch screen so in future there are flexible touch screens and flexible mobile or laptop. It also uses as transparent light absorbers. Like, transparent solar cell. Recently Indium teen oxide use in solar cell but indium is rarely available, brittle and it is also not transparent for UV. Graphene are low cost, eco-friendly and large area of production. Technical threshold relatively low compare with transistor in integrated circuits [4].

Transistor of integrated circuits

IC are common for our modern information society, mobile phone laptop all base on IC. Transistor is key element in IC. Transistor is tunable resistor which can amplify and switch electronic signals and electric power. Most up transistors are made of silicon, from the definition of mobility we can see that given electric field can drift electron in higher velocity. So

Graphene is use for Ultra-Fast transistor. So in future silicon will replace by Graphene, but Graphene is not a semi-conductor. Ideal transistors are made of semi conductive materials so that current can completely switch off. If current will not switch off completely than it carries lots of energy so it become over heated. Scientists are trying to open band gape of Graphene so ultra-fast Graphene transistor can turn off.

Heat spreader and interconnect wires of IC

Today's electronic and photonic systems are generating big amount of heat during process because more functionality. Like, Data servers of USA are 50% use to cool the system, 20% for computing and another percentage for the driving cable and Communication. Graphene is

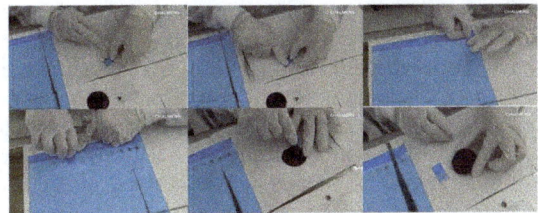

Figure 1: Skotch tape experiment.

Figure 2: Transparency of graphene.

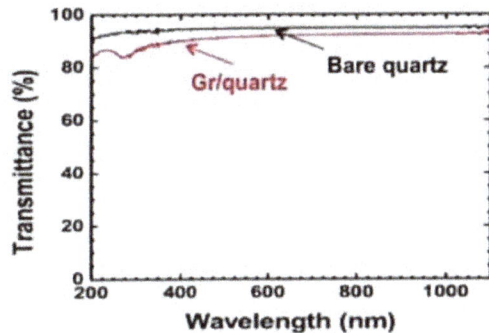

Figure 3: Graphene is transparent for UV and UR rays [22].

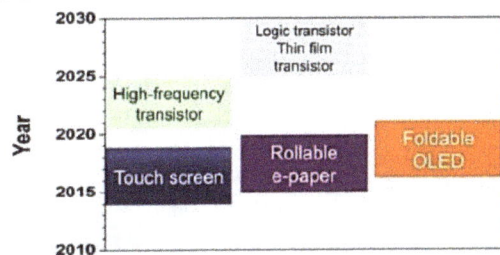

Figure 4: Future of Graphene in electronics and optoelectronics.

integrated between chip and thermal interface. Heat transfer from Heat sink to thermal interface material than dissipated in air. Graphene as interconnect wires of IC, Best conductor, lightweight & thinner also very stable and without electron migration. Electron migration means material transport caused by gradual moment of irons in the conductor due to momentum transport between conducting electron and diffusing metal atoms [5]. It has drawback when many metal inter connect such as copper but not for Graphene. It use as wires either in layer is not as could as emplaned conductivity. Other materials like CNTs (carbon Nano tube) is use to connect Graphene wires between laboring levels (Figures 5 and 6).

Energy applications

Graphene will use as super conductive plates of Capacitors, which are double layer capacitors. In super capacitor two conductive plates separated by dialectics layers. After the charging it will be more stable because of electrostatic charges. It also uses as electrode material for lithium ion batteries. It can store more energy for long time and charge discharge process is very quick. Graphene is electrically 10-100 times more conductive than activated carbon. Graphene base electrodes have 60% more density from recent capacitors. It also use for hydrogen storage, Mono layer Graphene can give hydrogen storage ratio up to 7.7% which useful for hydrogen vehicles (Figures 6 and 7).

Sensor and actuators

Graphene is mechanically strong so it is use full for Nano scale sensing and also useful as actuator in Nanoelecromechanical System (NEMS). It can able to sense 1 atomic change in element. It will use for Mass sensor, Gas sensor, DNA sensor, PH sensor (Figure 8).

Graphene composites

By adding other materials in Graphene atom we can able to make multi-tasking materials. Now Graphene and Epoxy composites are

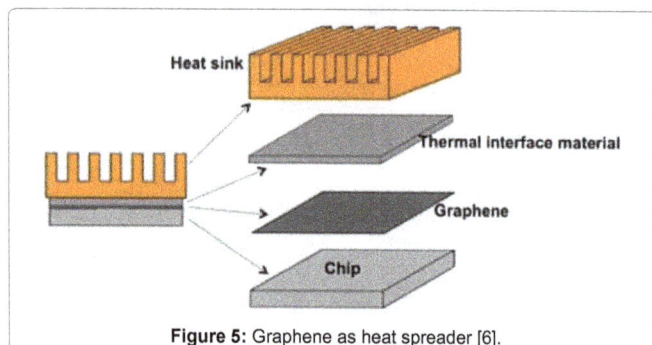

Figure 5: Graphene as heat spreader [6].

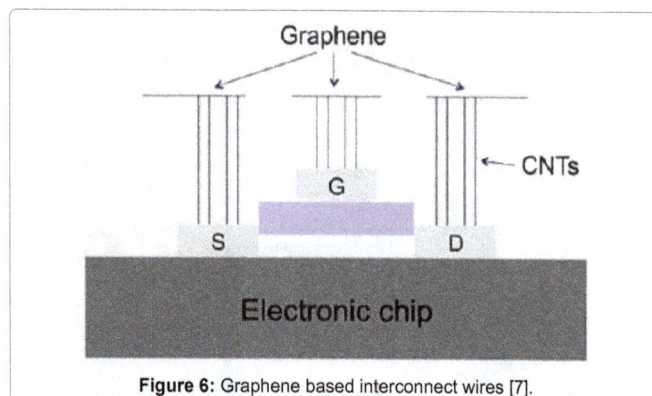

Figure 6: Graphene based interconnect wires [7].

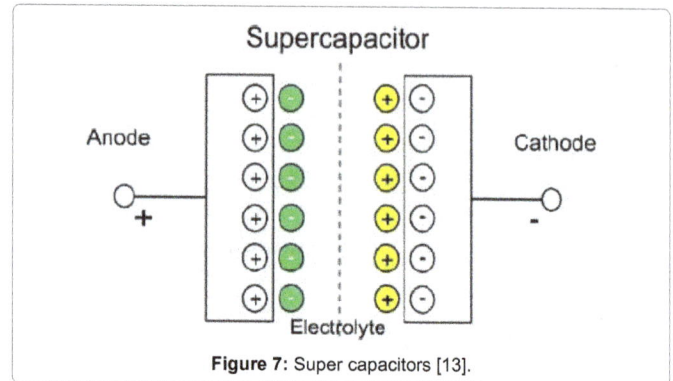

Figure 7: Super capacitors [13].

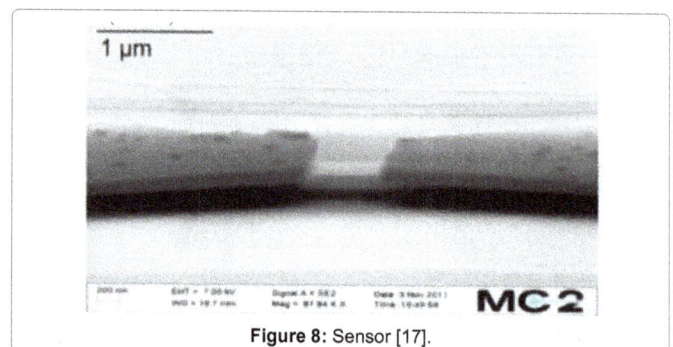

Figure 8: Sensor [17].

together extremely stronger, stiffer component. This material will use in future to make space elevator. The elevator which is connected to earth to space. It has 36000 km altitude. Means space travel without using rocket. In future it will be used to deliver a drug in body. It will use to impact drug directly inside the living cells. It will be a breakthrough for Medical filed (Figure 9).

Bio-applications

In future it will be used to deliver a drug in body [6-8]. It will use to impact drug directly inside the living cells. It will be a breakthrough for Medical filed (Figure 10).

Lattice of Monolayer Graphene

Different arrangement of atoms gives different play ground to electrons (Figure 11). Graphene's category is hexagon lattice which connect 1 atom with other 3 atoms (Figure 12). In Graphene there is no atom at center of hexagon. In Graphene atom A is connected with 2 B atoms at left side one is upper side and one at lower side and one atom at right side. B atom is connected with two A atoms at right side one at upper side and another at lower side and third atom at left side. This type of lattice is 'complex lattice' (Figure 13). If we observe only A atoms, it makes a hexagon lattice, Which can visible in above figure. One set of sub lattice has hexagon structure with one center. It called as Probe lattice. In above figure we can see that there are only two atoms. One is atom A and another is atom B (Figure 14).

Why only 2 A. carbon atoms are there?

The carbon has 6 electrons. Electron nearer to nucleus has low energy and high energy electron in outer orbits. Arrow shows the spine of electrons [9]. There is space for 4 electrons in second orbit. Electrons are connected with covalent bond. There is main two type of covalent bond, one is π bond and another is σ bond. π bond is intersecting for Graphene (Figure 15 and Table 2).

Graphene Composite

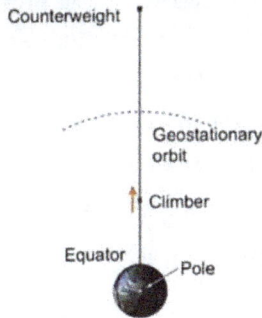

Figure 9: Space elevator [19].

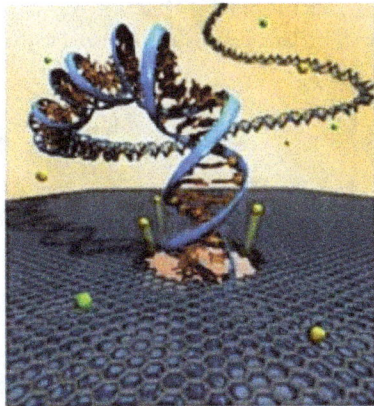

Figure 10: Graphene impact drug on live cells [18].

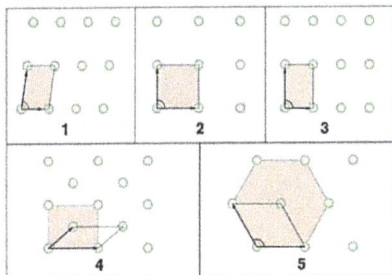

Figure 11: Lattices for 2-D material [20].

σ bond	∏ bond
Heavily overlapped	Slightly overlapped
Stable, not very reactive	Unstable and reactive
Head to head connection	Shoulder to shoulder connection

Table 2: Covalent bonds.

Hybridization of Carbon

Hybridization is denoted as SP^1, SP^2, or SP^3 (Figure 16).

Why atoms perform hybridization process?

Because of hybridization process energy of electron reduces so atom become more stable. After hybridization Graphene become SP^2 $2P_z^1$ and Diamond become SP^3 (Figure 17). Graphene make 3 σ bonds (which are strongest bond) with angle of 120° other $2P_s^1$ are localized bond. It is to share electron with other atoms. It is the reason for more

conductivity. If we make more layers than it make Vanderwall bond with each other and it became graphite [10-12].

Lattice staking graphene to graphite

1. A-B-A stacking = 1.42°A

2. A-B-C stacking = 1.42°A at right and 1.42°A at left too (Figure 18).

K-Space

To understand different properties of solid materials, it difficult with Spherical space (space of x, y, z dimensions). For that we have to take different space, we have to assume one space which known as K-space. In K-space K = 2π/λ, is the (circular) wavelength, K is the vector (Figure 19).

K = Unit 1 over length and K space = moment space.

Graphene's periodic lattice

K space is use to do study of different properties of Graphene. There is two lattice one is direct lattice another is Reciprocal lattice. When we rotate direct lattice at 90° angle it become reciprocal lattice. Atom, at the center is known as super lattice. The corners which we can see in the above figure are not part of the Reciprocal lattice (Figure 20).

Why Graphene is conducting the current?

Electron always tries to stay at low energy state. On the basic of this property of electron some materials become conductor or semiconductor, some of them are insulator. It is also depends upon the band gap. Delivery of electrons depends on the rule, first come first served. Electrons with higher energy level are at the outer orbits. It is known as Fermi level (highest occupied electrons energy level) (Figure 21).

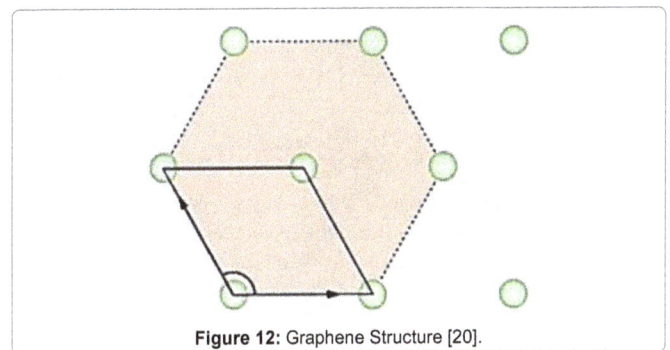

Figure 12: Graphene Structure [20].

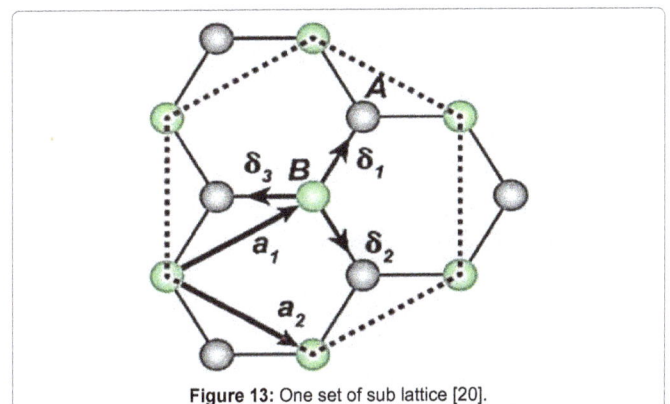

Figure 13: One set of sub lattice [20].

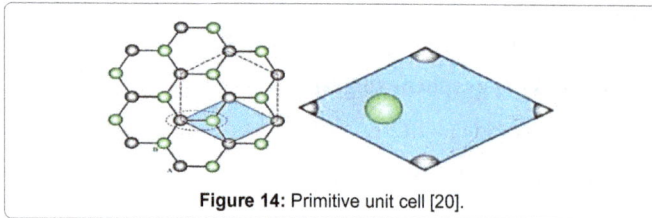

Figure 14: Primitive unit cell [20].

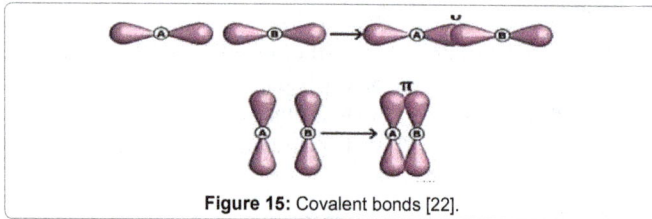

Figure 15: Covalent bonds [22].

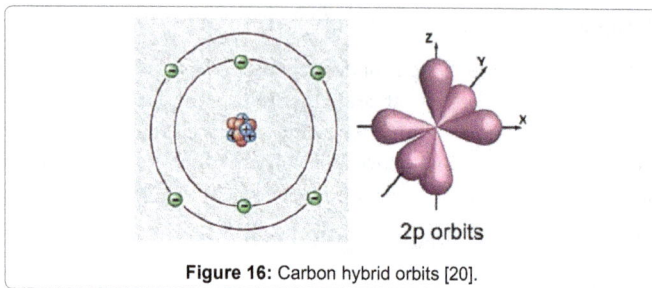

2p orbits

Figure 16: Carbon hybrid orbits [20].

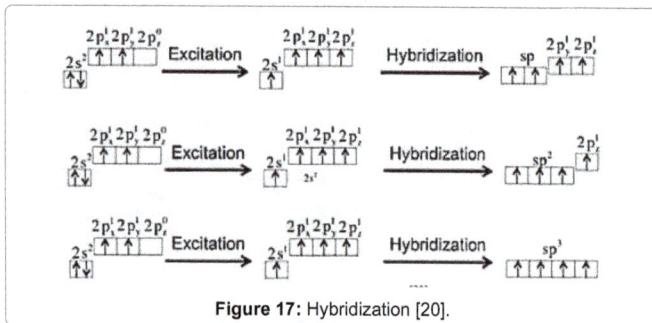

Figure 17: Hybridization [20].

There is to type of band. One is energy band and another is forbidden band. Energy band are allowed band and forbidden band are not allowed band [11-15]. Forbidden band is sandwiched between two energy bands. Highly full occupied allow band is valance band and first allowed band about the valance band is called as conduction band. On the basis of these bands we can differentiate conductor, semi-conductor and isolator in different category.

Metal: Valance band and conduction band partly occupied.

Semi-metal: Valance band and conduction band are touching or having overlaps.

Insulator: Band gap occupied at all Insulator.

Semi-conductor: It has small band gap, Electron may exit from valance bond to fill the gap of conduction bond.

For net current we need to fill band partially. When, m atoms come closer to form a solid, one energy level split into m levels (Figure 22) [16]. Upper side there are conduction band and lower side it is a valence band. SP^2 electrons are connected with σ bond 2 other bond for electric conductivity. Both bands are touches at Fermi level. Graphene

is conductor with zero band gaps. If we add electrons than it go up to Fermi level, it create N type Graphene. If we remove electron than graph goes beloved the Fermi level, it create P type Graphene. Graphene has linear Dispersion (Figures 23 and 24).

Unique linear energy dispersion

- Liner dispersion with approximately ± 0.6 eV.

- Massless Dirac particle with (Effective) rest mass = 0.

- Special theory of relativity plays a role.

- Unusual properties of quantum electrodynamics can be observed at much smaller speed.

Photon has linear dispersion. Graphene atoms are act same like a photon atoms but at lower speed [17-19]. It is properties of quantum electrodynamics where relativity effect can be observe without need to accelerate the particle to close speed of light. Graphene is good for a study of relative effect. Graphene is a semi-metal with no overlapping.

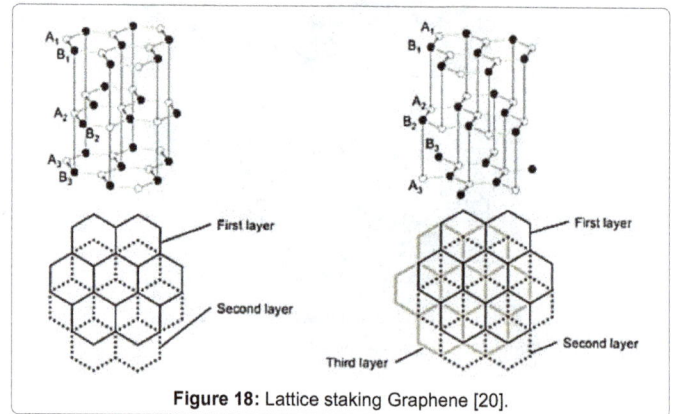

Figure 18: Lattice staking Graphene [20].

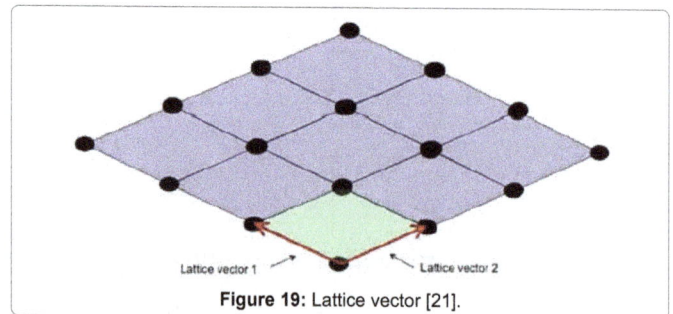

Figure 19: Lattice vector [21].

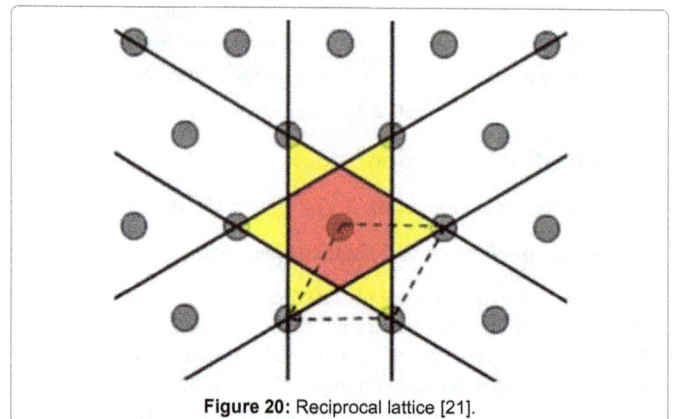

Figure 20: Reciprocal lattice [21].

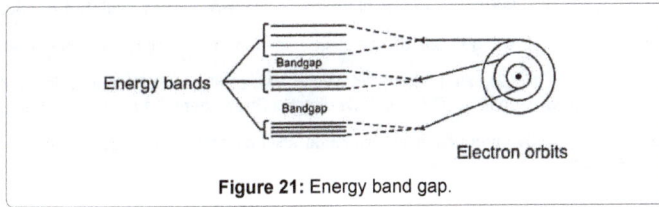

Figure 21: Energy band gap.

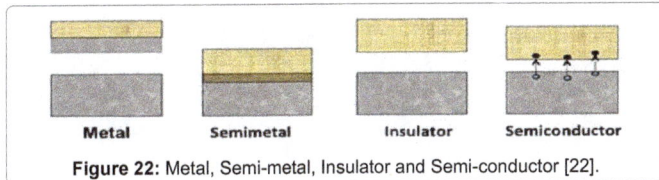

Figure 22: Metal, Semi-metal, Insulator and Semi-conductor [22].

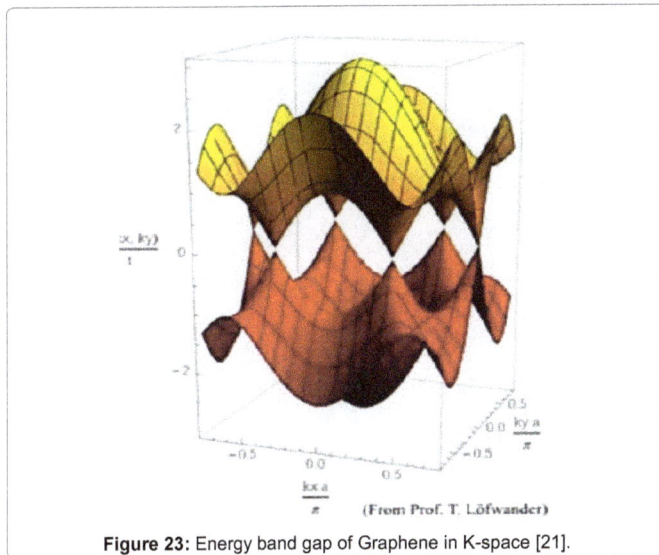

(From Prof. T. Löfwander)

Figure 23: Energy band gap of Graphene in K-space [21].

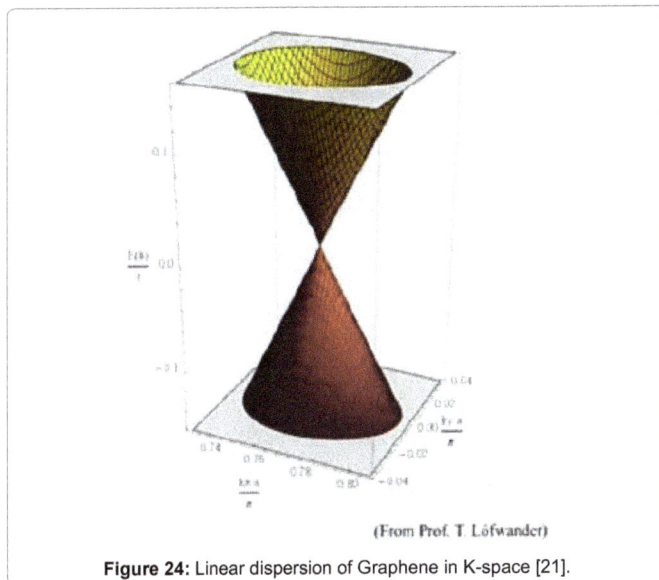

(From Prof. T. Löfwander)

Figure 24: Linear dispersion of Graphene in K-space [21].

Advantages of Graphene

- Extraordinary properties in one material

- High potential in both, fundamental studies and application

- Abundant raw material carbon: Cheap, eco-friendly and sustainable.

- Planer structure that is compatible with traditional semi-conductor takes place of silicon.

- Much stronger material than steel [20,21].

Challenges of Graphene

- High prize at this moment (1 cm^2 = 3700 rupees).

- Sensitive to environment with no effective passivation but has only 10 year history.

- High quality thin films still lacking of reproducibility.

- Large scale transfer of films still irreproducible.

- Sheet resistance still too high.

- No band gap which a transistor requires, (so turn off problems) [22].

Conclusion

Nearest in future, Graphene will take over most up places of the other materials. It will be the most useful material in every filed. Above all examples are just begging of the list where Graphene will be used in future. After the solution of the challenges it will be an undefeatable material. So here I conclude that the most of problems which are accurse because of Size, Strength, Conductivity, Transitivity. That all problems can solved by the use of Graphene. Medical filed, Space elevator, Transistor of Integrated Circuits, Heat Spreader & Interconnect Wires of IC, Transparent Conductor in Optoelectronics, Energy Applications, Sensors, Actuators, Graphene composites all become more effective after the use of Graphene.

References:

1. Geim AK (2010) Random walk to Graphene. Int J Mod Phys B 25: 4055.

2. Bunch JS (2008) Mechanical and electrical properties of Graphene sheets.

3. Brun SJ, Thomsen MR, Pedersen TG (2014) Electronic and optical properties of Graphene and Graphene Antidot structures. Journal of Physics: Condensed Matter 26: 265301.

4. Pop E, Varshney V, Roy AK (2012) Thermal properties of Graphene: Fundamentals and applications. MRS Bull 37: 1273-1279.

5. Bonaccorso F, Sun Z, Hasan T, Ferrari AC (2010) Graphene photonics and optoelectronics. Nature Photonics 4: 611-622.

6. Schiwerz F (2010) Graphene transistors. Nature Nanotechnology 5: 487-496.

7. Vaziri S, Lupina G,Henkel C, Smith AD, Östling M, et al. (2013)A Graphene-based hot electron transistor. Nano Lett 13: 1435-1439.

8. Sorokin PB, Chernozatonskii LA (2013) Semiconductor nanostructures based on Graphene. Physics Uspekhi 56: 105-122.

9. Kim S, Nah J, Jo I, Shahrjerdi D, Colombo L, et al. (2009) Realization of a high mobility dual gated Graphene field-effect transistor with Al$_2$O$_3$ dielectric. Appl Phys Lett 94: 062107-062103.

10. Mehr W, Lemme MC, Vaziri S, Lupina G, Henkel C, et al. (2012) Vertical transistor with a Graphene base IC and electrical device 33: 691-693.

11. Geim AK, Novoselov KS (2007) The rise of Graphene. Nature Materials 6: 183-191.

12. Das TK, Prusty S (2013) Recent advances in applications of Graphene 41-45.

13. Kannappan S, Kaliyappan K, Manian RK, Pandian AS, Yang H, et al. (2013) Graphene based supercapacitors with improved specific capacitance and fast charging time at high current density.

14. Bruke A (2000) Ultracapacitors: why, how, and where is the technology. J Power Sources 91: 37-50.

15. Kotz R, Carlen M (2000) Principles and applications of electrochemical capacitors. Electrochimica Acta 45: 2483-2498.

16. Amin KR, Bid A (2014) Graphene as a sensor. 430-434.

17. Kochmann S (2013) Graphene as a sensor material.

18. Kumar A, Lee CH (2013) Synthesis and biomedical applications of Graphene: Present and future trends. Intech open science 56-67.

19. Aravind PK (2006) The physics of the space elevator. Am J Phys 75: 125-130.

20. Mccann E, Koshino M (2013) The electronic properties of bilayer Graphene.

21. Wang YY, Ni ZH, Yu T, Shen ZX, Wang HM, et al.(2008) Raman studies of monolayer Graphene: The substrate effect. J Phys Chem C 112: 10637-10640.

22. Sun J (2015) Introduction to Graphene science and technology. Chalrmers, Sweden.

Permissions

List of Contributors

Frihat MH, Al Quran FMF and Al-Odat MQ
Department of Mechanical Engineering, Al-Huson University College, Al-Balqa Applied University, Al-Huson-Irbid, Jordan

Radic D, Gargallo L and Leiva A
Pontificia Universidad Catolica de Chile Santiago, Chile

Zeng Tao and Chen Yanqiang
College of Mechanical Engineering, Sichuan University of Science and Engineering, Zigong, 643000, China

Luo Yun-Rong and Lin Haibo
College of Mechanical Engineering, Sichuan University of Science and Engineering, Zigong, 643000, China
Key Lab in Sichuan Colleges on Industry Process Equipments and Control Engineering, Zigong, China

Fu Lei
College of Mechanical Engineering, Sichuan University of Science and Engineering, Zigong, 643000, China
School of Aeronautics and Astronautics, Sichuan University, Chengdu 610065, PR China

Pradhan SK
Department of Physics, National Institute of Technology, Kurukshetra, Haryana, India

Khan NT and Jameel J
Department of Biotechnology, Faculty of Life Sciences and Informatics, Balochistan University of Information Technology, Engineering and Management Sciences, Balochistan, Pakistan

Zavodinsky V
Institute for Materials Science of the Russian Academy of Sciences 153 Tikhookeanskaya str., Khabarovsk, 680042, Russia

Kabaldin Y
Nizhny Novgorod State Technical University, 24 Minin str. Nizhny Novgorod, 603013, Russia

Moutcine A, Maallah R and Chtaini A
Equipe Electrochemistry and Molecular Inorganic Materials, University Sultan Moulay Slimane, Faculty of Science and Technology of Beni Mellal, Morocco

Akhramez S and Hafid A
Laboratoirre of Organic Chemistry and Analytical, Sultan Moulay Slimane University, Faculty of Science and Technology of Beni Mellal, Morocco

Kedelbaev BS, Korazbekova KU and Kudasova DE
M. Auezov South Kazakhstan State University, Kazakhstan

Wahab ZA and Talib ZA
Department Of Physics, Universiti Putra Malaysia, 43400 UPM Serdang, Selangor, Malaysia

Raouf RM
Department Of Physics, Universiti Putra Malaysia, 43400 UPM Serdang, Selangor, Malaysia
Materials Engineering Department, College of Engineering, Al-Mustansiriyah University, Baghdad, Iraq

Ibrahim NA
Department Of Chemistry, Faculty of Science, Universiti Putra Malaysia, 43400 UPM Serdang, Selangor, Malaysia
Materials Processing and Technology Laboratory, Institute of Advanced Technology, Universiti Putra Malaysia, 43400 UPM Serdang, Selangor, Malaysia

Wang Q, Ning H and Pillay S
Department of Materials Science and Engineering, University of Alabama at Birmingham, Birmingham, USA

Vaidya U
Department of Mechanical, Aerospace and Biomedical Engineering, University of Tennessee, Knoxville, USA

Nouha K and Tyagi RD
Université du Québec, Institut national de la recherche scientifique, Centre Eau, Terre and Environnement, 490 de la Couronne, Québec, G1K 9A9, Canada

Hoang NV
Institute of Environmental Technology, Vietnam Academy of Science and Technology, 18 Hoang Quoc Viet, Hanoi, Vietnam

Zargari S and Rahimi R
Department of Chemistry, Iran University of Science and Technology, Iran

Rahimi A
Department of Civil Engineering, Iran University of Science and Technology, Iran

Hafs A and Aitbara A
University of El Tarf, B.P. 73, El Tarf, 36000, Algeria

Benaldijia A
University of Annaba, B.P. 12, Annaba, 23000 Algeria

Mbarki R
Department of Mechanical Engineering, The Australian College of Kuwait, Meshraf, Kuwait

Borvayeh L
Department of Mathematics, The Australian College of Kuwait, Meshraf, Kuwait

Sabati M
Department of Electrical Engineering, The Australian College of Kuwait, Meshraf, Kuwait
The Dr. John T. Macdonald Foundation Biomedical Nanotechnology Institute, Department of Radiology, University of Miami, USA

Talwar DN
Department of Physics, Indiana University of Pennsylvania, Pennsylvania, USA

Becla P
Department of Materials Science and Engineering, Massachusetts Institute of Technology, Cambridge, Massachusetts, USA

Sánchez LM
Environmental engineer and Research teacher, University of Meta, Villavicencio, 500001, Colombia

Quiñonez MF
Architect and Research teacher, University of Meta, Villavicencio, 500001, Colombia

Bergmann CP
Ceramic Materials Laboratory, Federal University of Rio Grande do Sul, Av. Osvaldo Aranha 99, sl. 705C, CEP 90035-190, Porto Alegre, RS, Brazil

Panta PC
Ceramic Materials Laboratory, Federal University of Rio Grande do Sul, Av. Osvaldo Aranha 99, sl. 705C, CEP 90035-190, Porto Alegre, RS, Brazil

Department of Chemistry and Physics, University of Santa Cruz do Sul, Santa Cruz do Sul, Brazil, Av. Independência 2293, CEP 96815-900, Santa Cruz do Sul, Brazil

Sánchez LG, Portal AJC, Salazar Y Caso De Los Cobos JMG and Martínez García JA
Departamento de Ciencia de Materiales e Ingeniería Metalúrgica, Facultad de Ciencias Químicas, Universidad Complutense de Madrid (U.C.M.), Madrid, España

Valenzuela FP
Museo del Cobre de Cerro Muriano, Córdoba. 14350 Obejo, Córdoba, España

Sharma R
Department of Applied Sciences, Model Institute of Engineering and Technology, Jammu (J&K), India

Bolandhemat N and Rahman M
Department of Physics, Faculty of Science, University Putra Malaysia, 43400 UPM Serdang, Selangor, Malaysia

Shuaibu A
Department of Physics, Faculty of Science, University Putra Malaysia, 43400 UPM Serdang, Selangor, Malaysia
Department of Physics, Faculty of Science, Nigerian Defence Academy, P.M.B 2109, Kaduna, Nigeria

Kozhabayevich KM, Dossanuly SR and Zhanabayevich OA
M. Auezov South Kazakhstan State University, Shymkent City, Kazakhstan

Abdelraziq IR and Nierat TH
Physics Department, An-Najah National University, Nablus, Palestine

Mongelli GF
Department of Chemical Engineering, Case Western Reserve University, USA

Ltifi I, Ayari F, Hassen Chehimi DB and Ayadi MT
Laboratory of Applications of Chemistry to Resources and Natural Substances and the Environment (LACReSNE), Department of Chemistry, Faculty of Science of Bizerte, University of Carthage, Tunisia

Singare S, Shenggui C and Nan Li
School of Mechanical Engineering, Dongguan University of Technology, Dongguan, Dongguan 523808, China

Netopilík M and Trhlíková O
Institute of Macromolecular Chemistry, Academy of Sciences of the Czech Republic, Heyrovský Sq. 2, 162 06 Prague 6, Czech Republic

Dessalew Berihun
Department of Urban Environmental Management, Kotebe Metropolitan University, Addis Ababa, Ethiopia

Mohammed IK and Kasim Uthman ISAH
Department of Physics, Federal University of Technology, Minna, Nigeria

Yabagi JA
Department of Physics, Ibrahim Badamasi Babangida University Lapai, Niger State, Nigeria

Taufiq S
Department of Preliminary Studies, Umaru Waziri Federal Polytechnics, Birnin Kebbi, Kebbi State, Nigeria

Heredia AS
Research Center on Engineering and Applied Sciences - (IICBA)

Aguilar PAM and Ocampo AM
Department of Engineering, Autonomous Univiersity of Morelos State, Mexico

Chandrashekhar A
Department of Mechanical Engineering, Basaveshwar Engineering College, Bagalkot, India

Kabadi VR and Bhide R
Department of Mechanical Engineering, Nitte Meenakshi Institute of Technology, Bangaluru, India
Application Support Centre, Oerlikon Balzers Coating India Limited, Bhosari, Pune, Maharashtra, India

Khan TM and Zakria M
National Institute of Lasers and Optronics (NILOP), Islamabad, Pakistan

Shahid T, Arfan M and Khursheed S
Department of Applied Physics, Federal Urdu University of Arts, Science and Technology, Islamabad, Pakistan

Shakoor RI
National Institute of Lasers and Optronics (NILOP), Islamabad, Pakistan
Department of Mechanical Engineering, Muhammad Ali Jinnah University, Islamabad, Pakistan

Evgenii K
Kurchatov Sq 1, Moscow 123182, Russia

Benzina A
Faculty of Health, Medicine and Life Sciences, Maastricht University, UniSingel 50, 6229 ER Maastricht, The Netherlands

Aldenhoff YB
Interface Biomaterials BV, B. Lemmensstraat 364, 6163 JT Geleen, The Netherlands

Heijboer R
Zuyderland Medisch Centrum, Henri Dunantstraat 5, 6419 PC Heerlen, The Netherlands

Koole LH
Interface Biomaterials BV, B. Lemmensstraat 364, 6163 JT Geleen, The Netherlands
Department of Biomedical Engineering, University of Malaya, Jalan Universiti, 50603 Kuala Lumpur, Malaysia

Hedayat N and Du Y
College of Applied Engineering, Sustainability and Technology (CAEST), Aeronautics & Technology Building (ATB), Kent State University, Lefton Esplanade, Kent, Ohio, USA

Bello S and Muhammad BG
Department of Physics, Faculty of Natural and Applied Sciences, Umaru Musa Yar'adua University Katsina, Nigeria

Bature B
Department of Mathematics, Faculty of Natural and Applied Sciences, Umaru Musa Yar'adua University Katsina, Nigeria

Ata BN and Abderaziq IR
Physics Department, An-Najah Ntional University, Nablus Weast Bank, Israel

Loganina VI and Zhegera CV
Department of "Quality management and construction technologies" Penza State University of Architecture and Construction, Russia

Loganina VI and Zhegera CV
Department of "Quality management and construction technologies" Penza State University of Architecture and Construction, Russia

Rastgoo Oskoui P
Department of Materials Engineering, Faculty of Mechanical Engineering, University of Tabriz, Tabriz, Iran

Payam RO
National Iranian Copper Industries Company, Tabriz, Iran

Sirajudheen P
Department of Chemistry, WMO Imam Gazzali Arts and Science College, Koolivayal, Panamaram, Wayanad, Kerala, India

Solanki Bhaumikkumar Ketansinh
Mechanical Engineering, Chhotubhai Gopalbhai Patel Institute of Technology, India

Index

A

Ab Initio Simulation, 25
Activated Carbon, 115, 120-121, 130, 135, 196
Adsorption, 28, 30-33, 61, 89, 115-121, 125, 130-135
Adsorption Isotherms, 33, 116, 118, 121, 130, 134
Ammonium Hydroxide, 87
Amorphous Silica-alumina, 184
Anphiphilicity Isotherms, 5
Archaeological, 90-91, 94
Armament, 90
Artificial Magnetite, 90-94

B

Binary Liquid, 178, 181-183
Bioconstruction Biomass, 84
Bioflocculation, 48, 52-53, 55-57
Burnishing, 1-4, 88

C

Carrier Concentration, 75, 77, 95, 98
Castor Oil, 106-110
Cationic Dye, 115, 121
Cellulose Acetate Butyrate (CAB), 35
Chemical Contamination, 28
Cloacibacterium Normanense, 48, 50-52, 55
Combustion Processes, 63, 65
Coprecipitation, 87, 89, 158
Critical Mixture, 178, 182
Cubic Phase, 191-192
Cyclic Voltammetry, 28, 30

D

Density Functional Theory, 25, 99-100
Designed Prosthesis, 122
Dielectric, 5, 11, 16-17, 19-20, 69, 75, 77-80, 82-83, 102, 191, 193, 199
Digital Image Correlation, 139, 141, 144
Dopants Influence, 25-26
Dye-sensitized Solar Cell, 136, 138

E

Ebt Sand, 187-188
Electric Arc Furnace, 187-189
Ellipsometry, 16-20, 75, 79, 82-83
Embolization, 164-168
Embrittlement Kinetics, 160, 163

Entomological Analysis, 84
Epilayers, 75, 77
Excess Lifetime Cancer, 173-175

F

Ferroalloys, 31-33
Flexoelectric, 69, 73
Fusarium Oxysporum, 21-23

G

Geometric Frustration, 99
Gold Thin Films, 16, 18-19

H

Hardness, 1-4, 63, 86, 145-147, 152
Heat Spreader, 194-196, 199
Heavy Metals, 48, 130, 173-177
Hydrogenation, 31-33
Hydrophobic Effect, 112
Hyphaene Thebaica, 136, 138

I

Infectiousness, 103-105
Iron Oxide, 87-89

L

Laser Speckle, 139, 144
Leishmania, 103-104
Long Carbon Fibers, 42
Low Cycle Fatigue, 12, 15

M

Magnetron Sputtering, 145, 152
Mechanical Alloying (MA), 63
Metal-semiconductor (MS), 95
Mycosynthesis, 21, 23

N

Nanopyramids, 154, 156-157
Nanostructure, 58, 60-61, 67, 136, 156
Nickel Oxide (NIO), 154
Noncrystalline, 63

O

Orinoquia, 84

P

Phase Transitions, 9, 11, 99, 182

Photocatalytic, 30, 58-62, 191-193
Piezoelectric, 69, 73-74
Plastic Strength, 184-185
Poly(benzylitaconate)s, 5
Polyarylamide, 42, 46
Polymer Blend, 35
Polymeric, 11, 35-38, 41, 48, 56-58, 112, 168, 172, 185
Polysulfone (PSF), 35, 40
Pore-former, 169-170, 172
Porous Ceramics, 169, 172
Predominant Failure Mode, 12
Pseudo Dielectric Functions, 16

R
Radiopacity, 164-165, 168
Reactor Pressure Vessel, 160, 163
Reflectivity, 75, 77-82, 140
Rheology, 48-49, 51, 106, 110

S
Scanning Electron Microscope (SEM), 12-13, 35, 45
Schottky Barrier Diode, 95, 98

Scratch Testing, 145-148
Semiconductors, 58, 75, 77-78, 83, 157
Separation Mechanism, 125, 128
Shift Modulus, 25-26
Silver Nanoparticles, 21-24
Size-exclusion Chromatography, 125, 128
Sorbitol, 31-32
Space Elevator, 194, 196-197, 199-200
Spectroscopic Ellipsometry, 16, 20, 75, 79, 82-83
Staphylococcus Aurous, 28
Static Elasticity, 5
Surface Roughness, 1-4, 146-148

T
Thermoplastic, 35, 41-42, 46-47
Tomography, 122, 167, 169

V
Viscosity, 48-49, 55, 106-111, 125, 136, 156, 158, 178, 180-182

Z
Zinc Titanate, 191-193